Boron Separation Processes

Boron Separation Processes

NALAN KABAY

Ege University, Chemical Engineering Department, Faculty of Engineering, Izmir, Turkey

MAREK BRYJAK

Wrocław University of Technology, Faculty of Chemistry, Department of Polymer and Carbon Materials, Wrocław, Poland

NIDAL HILAL

Center for Water Advanced Tehnologies and Environmental Research (CWATER), College of Engineering, Swansea University, Swansea, United Kingdom

ELSEVIER

Amsterdam • Boston • Heidelberg • London • New York
Oxford • Paris • San Diego • San Francisco • Sydney • Tokyo

Elsevier
Radarweg 29, PO Box 211, 1000 AE Amsterdam, Netherlands
The Boulevard, Langford Lane, Kidlington, Oxford OX5 1GB, UK
225 Wyman Street, Waltham, MA 02451, USA

ISBN: 978-0-444-63454-2

British Library Cataloguing in Publication Data
A catalogue record for this book is available from the British Library

Library of Congress Cataloging-in-Publication Data
A catalog record for this book is available from the Library of Congress

For information on all **Elsevier** publications visit
our web site at http://store.eslevier.com

Printed and bound in Poland

CONTENTS

EDITORS' PREFACE

Due to the increasing demand for delivery of safe potable or irrigation water and limited available water resources, many suppliers have to face a problem to find some alternatives urgently. Hence, seawater, brackish water, and contaminated surface waters have become a target as new water resources for these activities which need fresh water. However, these alternatives may contain some trace contaminants that have not been noted so far and their removal has been not considered at the technological level. Boron is one of the target species in the list of such unwanted contaminants. This element is found in seawater at the level of 4–7 mg/L, depending on the region, and in underground water at higher levels.

The World Health Organization (WHO) set the limit for boron as 0.3 mg/L in drinking water in 1993. However, in 2011, the Drinking-Water Quality Committee of WHO revised the Boron Guideline Value for potable water as 2.4 mg/L. A major limiting factor for the presence of boron in water is related to the possibility of plant damage rather than human health-related concerns. Although boron is a vital element for plant growth in trace quantities as a micronutrient and it is delivered as fertilizer, it can be detrimental to some plants at higher concentrations. According to the published literature, excess boron reduces fruit yield, induces premature ripening, and causes massive leaf damage. Therefore, boron limits for agricultural water is still kept between 0.3 and 1 mg/L, depending on the country.

The main goals of this book are to focus the attention on boron-related problems and to present some challenges for safe water production in order to invoke appropriate actions in efficient innovative directions. For these reasons, this book is divided into four sections that show the impact of boron for our life, adsorption methods for water deboronation, the use of membrane processes for boron removal from water, and recent studies on process optimization.

SECTION 1—BORON IN THE ENVIRONMENT

Chapter 1 introduces the history of boron discovery, the environmental chemistry of the element and its biogeochemistry. The reader gets information on the impact of nature to bind and transform boron, as well as the toxicology of this element.

Chemistry of boron in aqueous solution is the subject of Chapter 2. The authors discussed the distribution of boron on the Earth and the paths that boron enters into the aquatic environment. Contents of boron in surface waters, underground waters, and seawater are presented to give the general layout of the boron problem. Special attention is directed to the chemistry of boron containing compounds and the legal regulations of the boron concentration in potable water.

Boron has been considered to have a negative effect on animal reproduction and development. However, boron-mediated unfavorable effects in males have not been proven for humans. It is the subject of discussion presented in Chapter 3, where the authors, based on the recently published epidemiological studies, have provided valuable data for highly boron-exposed workers in China and Turkey. The results indicated that human boron exposure, even at the highest rate, are too low to reach the blood concentrations that would cause adverse effects on the reproduction system.

SECTION 2—REMOVAL OF BORON BY ION EXCHANGE AND ADSORPTION PROCESSES

This section combines chapters dealing with one of the oldest methods, which is adsorption for mitigation of boron level in aqueous solutions. Chapter 4 presents mechanisms of boron sorption on ion exchangers and gives the fundamental information on sorption equilibrium and kinetics. The reader can find there some data on the formation of polyborates and their impact on ion exchange processes.

Authors of Chapter 5 focused on the use of chelating adsorbents for boron recovery. They describe properties of chelating resins and fibers bearing N-methyl-D-glucamine ligands. Using the mechanism of boron chelation, the authors elucidate such sorbent properties as adsorption rate, pH-related uptake, and adsorption capacity. The case studies presented in the chapter show application of the described materials for boron recovery from geothermal water and salt lake brines.

The use of natural inorganic materials for boron removal is the subject of Chapter 6. The authors provided plenty of data on the adsorption of boron on different minerals. This chapter also addresses some studies of organic natural matter. Some data on soil or humic acids sorption are delivered at the end of the chapter.

Chapter 7 is dedicated to the description of the kind of chelating materials containing iminobis-propylenediol ligands for boron binding. Three forms of tailored adsorbent are presented: linear polymer, crosslinked beads, and hairy function resins. These materials can be used at different applications such as polymer-enhanced ultrafiltration systems, batch adsorption, and fix bed columns.

SECTION 3—REMOVAL OF BORON BY MEMBRANE PROCESSES

The chapters in this section describe the use of membrane processes for boron removal. Chapter 8 can be considered as an introduction to the other contributions. It gives an insight to the direct use of RO membranes for desalination, application of UF and MF membranes for systems when boron is complexed/adsorbed on coupling agents or the use of other membrane processes.

The supplementary data on the use of various membrane systems for desalination of seawater are presented in Chapter 9. The authors discussed the integrated systems, ion-exchange systems, hybrid systems, ED systems, and others. The final SWOT analysis allows us to understand the way to make the final selection of the best system.

The next chapter, Chapter 10, deals with details on various hybrid systems. It presents some fundamentals on membrane-enhanced hybrid processes when coupling agents form complexes with boron and, as large substances, are removed by membrane filtration. Such hybrid systems as molecule-enhanced membrane separation (MEMS), polymer-enhanced ultrafiltration (PEUF), micellar-enhanced ultrafiltration (MEUF), colloid-enhanced ultrafiltration (CEUF), and suspension-enhanced microfiltration (SEMF) are discussed. Some hints for regeneration of coupling agents are also presented.

Chapter 11 is dedicated to the application of ion exchange membranes for water deboranation. It discusses the effects of membranes, boron bearing species and process parameters. Three methods are considered by the authors: electrodialysis (ED), Donnan dialysis (DD) and electrodeionization (EDI). Finally, the chapter provides cost evaluation for processes employing ion-exchange membranes.

Chapter 12 deals with boron removal from geothermal water. In this chapter, application of membrane separation methods such as reverse osmosis, sorption-membrane filtration hybrid, and electromembrane (ED, EDI) methods for desalination of geothermal water and boron removal are discussed.

SECTION 4—SIMULATION AND OPTIMIZATION STUDIES

More complex studies on process simulation and optimization are gathered in the last section. Authors of Chapter 13 present the problem for full-scale RO units. In this chapter, fundamentals of mechanistic predictive models are discussed for predicting boron rejection in pilot and full-scale plants.

Chapter 14 shows the method for the reduction of boron in the permeate for a single RO pass at high pH values. The authors model the process with respect to its high boron removal taking into account the acid—base equilibria and speciation of boron.

Comparison of the existing technologies for boron removal is presented in Chapter 15. The investigation was performed by means of model of analytical hierarchical process using the Hasse diagram technique and DART software. The Hasse diagram serves as a tool for environment quality assessment.

Optimization of the suspension-enhanced microfiltration system is the topic of Chapter 16. In the designed hybrid system, the suspension of fine particles of boron selective resins was concentrated on a cross-flow ceramic module and on a polymer submerged module. The comparison to a fix bed system points at utilization of much lower amounts of sorbent and chemicals in the hybrid system.

The last chapter, Chapter 17, presents application of response surface methodology for optimization of process parameters for boron adsorption. Using that methodology, one is able to select the best conditions to run a particular process for any types of feed water.

We would like to express our sincere thanks to the authors who contributed to this book. The editors of this book would like to thank Elsevier, especially Christine McElvenny, Editorial Project Manager and Mohana Priyan Rajendran, Project Manager of Book Production for the kind help in production of this book. We thank Dr Daniel Johnson, Swansea University, UK for his patience and skill as the proof reader of this book.

Nalan KABAY, Marek BRYJAK, Nidal HILAL
nalan.kabay@ege.edu.tr
marek.bryjak@pwr.wroc.pl
N.Hilal@swansea.ac.uk
Izmir (Turkey), Wroclaw (Poland), Swansea (UK)

CONTRIBUTORS

Nurşen Başaran
Hacettepe University, Faculty of Pharmacy, Department of Toxicology, Sıhhiye, Ankara, Turkey

Ulker Beker
Chemical Engineering Department, Yildiz Technical University, Istanbul, Turkey

Niyazi Bicak
Department of Chemistry, Istanbul Technical University, Istanbul, Turkey

Amos Bick
7 Harey-Jerusalem St. Ganey-Tikva, Israel

Marek Blahušiak
Slovak University of Technology, Institute of Chemical and Environmental Engineering, Radlinského, Bratislava, Slovakia

Hermann M. Bolt
Leibniz Research Centre for Working Environment and Human Factors (IfADo), Dortmund, Germany

Marek Bryjak
Wrocław University of Technology, Faculty of Chemistry, Department of Polymer and Carbon Materials, Wrocław, Poland

Nawaf Bin Darwish
Centre for Water Advanced Technologies and Environmental Research (CWATER), College of Engineering, Swansea University, Swansea, UK

Yalçın Duydu
Ankara University, Faculty of Pharmacy, Department of Toxicology, Tandoğan, Ankara, Turkey

Piotr Dydo
Silesian University of Technology, Gliwice, Poland

Gary L. Foutch
Oklahoma State University Chemical Engineering Department, Engineering North Stillwater, OK, USA

Viatcheslav (Slava) Freger
Chemical Engineering Department, Technion—Israel Institute of Technology, Haifa, Israel

Nidal Hilal
Centre for Water Advanced Technologies and Environmental Research (CWATER), College of Engineering, Swansea University, Swansea, UK

Nalan Kabay
Ege University, Chemical Engineering Department, Faculty of Engineering, Izmir, Turkey

Jae-Hong Kim
Department of Chemical and Environmental Engineering, Yale University, New Haven, CT, USA

Victor Kochkodan
Centre for Water Advanced Technologies and Environmental Research (CWATER), College of Engineering, Swansea University, Swansea, UK

Fyodor S. Kot
Faculty of Civil and Environmental Engineering, Technion—Israel Institute of Technology, Haifa, Israel

Tomasz Koźlecki
Faculty of Chemistry, Wrocław University of Technology, Wrocław, Poland

Ori Lahav
Faculty of Civil and Environmental Engineering, Technion—Israel Institute of Technology, Haifa, Israel

Jidong Lou
Oklahoma State University Chemical Engineering Department, Engineering North Stillwater, OK, USA

Oded Nir
Faculty of Civil and Environmental Engineering, Technion—Israel Institute of Technology, Haifa, Israel

Syouhei Nishihama
Department of Chemical Engineering, The University of Kitakyushu, Kitakyushu, Japan

Gideon Oron
J. Blaustein Institutes for Desert Research, Ben-Gurion University of the Negev, Kiryat Sde-Boker, Israel; Faculty of Industrial Engineering and Management, Ben-Gurion University of the Negev, Beer Sheva, Israel; The Environmental Engineering Program, Ben-Gurion University of the Negev, Beer Sheva, Israel

H. Onder Ozbelge
Chemical Engineering Department, Middle East Technical University, Ankara, Turkey

Pyung-Kyu Park
Department of Environmental Engineering, Yonsei University, Wonju-Shi, Gangwon-Do, South Korea

Izabela Polowczyk
Faculty of Chemistry, Wrocław University of Technology, Wrocław, Poland

Abraham (Avi) Sagiv
Chemical Engineering Department, Technion—Israel Institute of Technology, Haifa, Israel

Štefan Schlosser
Slovak University of Technology, Institute of Chemical and Environmental Engineering, Radlinského, Bratislava, Slovakia

Raphael (Rafi) Semiat
Chemical Engineering Department, Technion—Israel Institute of Technology, Haifa, Israel

Bahire Filiz Senkal
Department of Chemistry, Istanbul Technical University, Istanbul, Turkey

Hilla Shemer
Chemical Engineering Department, Technion—Israel Institute of Technology, Haifa, Israel

Esra Bilgin Simsek
Chemical & Process Engineering Department, Yalova University, Yalova, Turkey; Chemical Engineering Department, Yildiz Technical University, Istanbul, Turkey

Marian Turek
Silesian University of Technology, Gliwice, Poland

Levent Yilmaz
Chemical Engineering Department, Middle East Technical University, Ankara, Turkey

Kazuharu Yoshizuka
Department of Chemical Engineering, The University of Kitakyushu, Kitakyushu, Japan

Hasan Zerze
Chemical Engineering Department, Middle East Technical University, Ankara, Turkey

CHAPTER 1

Boron in the Environment

Fyodor S. Kot
Faculty of Civil and Environmental Engineering, Technion—Israel Institute of Technology, Haifa, Israel

1.1 BORON HISTORY, SOURCES, CHEMISTRY, AND APPLICATIONS

Boron (B) as an individual chemical element was first isolated in 1808 by Joseph-Louis Gay-Lussac and Louis-Jacques Thénard in France and, independently, by Sir Humphry Davy in England. In fact, neither had produced the pure element, which is almost impossible to obtain owing to its high melting point (about 3400 K). Eventually, Weintraub in the USA produced totally pure B by sparking a mixture of B chloride and hydrogen.[1] The material of B obtained in this way was found to have very different properties to those previously reported, described originally by Laubengayer et al.[2]

In spite of its small atomic weight B is much scarcer in space than H, He, and C. In chondrites, content of B was found to vary from 0.5 to 1.4 mg/kg.[3] Deficiency of B in space caused its relative deficiency in the earth less than 1 mg/kg in the upper mantle. However, the element is enriched in the lithosphere—about 10 mg/kg in the continental crust and in seawater there is 4.5 mg/kg. In the earth's crust, B accumulates mostly in granitoides and pegmatites. Due to volatility of its compounds, B is a noticeable element in volcanic activity; B compounds are emitted to the atmosphere, they accumulate in the thermal waters and enter groundwaters. Boron endogenic ores relate to postmagmatic processes—skarn, forming borosilicates—datolite ($CaBSiO_4OH$) and borate-ludvigite ($(Mg,Fe)Fe(BO_3)O_2$).[4]

Boron is the only nonmetal in Group 13 of the Mendeleev Periodic Table and it has many similarities to its close neighbor carbon and its diagonal relative silicon. Thus, like C and Si, B shows a marked propensity to form covalent molecular compounds, but it sharply differs from them as it has one less valence electron than the number of valence orbitals. This is referred to as an "electron deficiency," and has a dominant effect on the behavior of B in chemical processes. Elements of this type usually adopt metallic bonding, but the small size and high ionization energies of B result in covalent rather than metallic bonding. Boron normally has a coordination number of either three or four in naturally occurring compounds.

Free elemental B does not exist in nature. The most important oxidation state is B^{3+}. The small highly polarizing B^{3+} cation does not exist under chemically significant conditions. When it comes about B in rocks, it is almost always about B complexes with oxygen. Ordinary exceptions to this generalization are ferrucite ($NaBF_4$), avogadrite

Boron Separation Processes
ISBN 978-0-444-63454-2, http://dx.doi.org/10.1016/B978-0-444-63454-2.00001-0

((K,Cs)BF_4), and barberiite (NH_4BF_4), which have been reported from Mount Vesuvius, Italy.[5] Borates, such as boric acid, boric oxide, and sodium borates are stable, except for under dehydration at high temperatures.[6]

Boron is unique among elements in structural complexity of its allotropic modifications. It is second only to carbon in its ability to form element bonded networks. Vast numbers of organic compounds containing B—O are known.[7] The B atom can be surrounded by innumerable combinations of groups, including acytoxy (RCOO—), peroxo (ROO—), halogeno (X—), and hydrido, in either open or cyclic arrays.[8,9] Simple alcohols react with boric acid to give esters $B(OR)_3$. The partially esterified species $(RO)_2BOH$ and $ROB(OH)_2$ are probably also involved. Polyhydric alcohols form cyclic esters with boric acid.[10] Organoboron compounds include B—N compounds, because B—N is isoelectronic with C—C.[11,12] Organoboron complexes occur in plants and are most likely present in animal and human tissues. Experimental evidence suggests these organoboron complexes are the result of interaction with either —OH or $-NH_2$ groups.[13] The stability of B—N complexes of biological relevance remains to be shown. Thioborates of the type $B(SR)_3$, $R'B(SR)_2$, and $R'_2(SR)$ are well documented.[8,14] There are also a growing number of binary B sulfides and B—sulfur anions, which may form chains, rings, and networks.[8] Comprehensive reviews of known and probable natural B-containing compounds may be found elsewhere.[8,13,15–17]

Boron compounds have been utilized since the early times.[15] The Babylonians have been credited with importing borax (tinkar) over 4000 years ago for use as a flux for working gold. Mummifying, medicinal, and metallurgic applications of B are sometimes attributed to the ancient Egyptians. None of this very old borax history has been verified, but solid evidence exists that borax was first used in the eighth century in Hejaz, western Arabia having been brought there by Arab traders. The use of borax flux by European goldsmiths dates to about the twelfth century. The earliest source of borax was from lakes in Tibet. The borax was transported in bags tied to sheep, which were driven over the Himalayas to India.

In modern times, B compounds are widely utilized in industry and agriculture. Glass production and detergent production are the main users of B. Other uses include in metal alloys, fire retardants, and chemical fertilizers. For example, the major U.S. industry consumption of B in 1990[18] is presented in Figure 1.1.

Between 2001 and 2005, B (as B_2O_3) consumption rose by 4.7% per year, when it reached 1.8×10^9 kg. The global economic crisis of late 2008 and recession of 2009 negatively affected sectors vital for B consumption, such as the construction and automotive industries. The moderate economic recovery in 2010 created steady growth in B production and consumption. The consumption is expected to increase in 2011 and the coming years, spurred by strong demand in the Asian and South American agricultural, ceramic, and glass markets. In particular, B consumption in the global fiberglass industry was projected to increase by 7% annually through 2013, spurred by a projected 19%

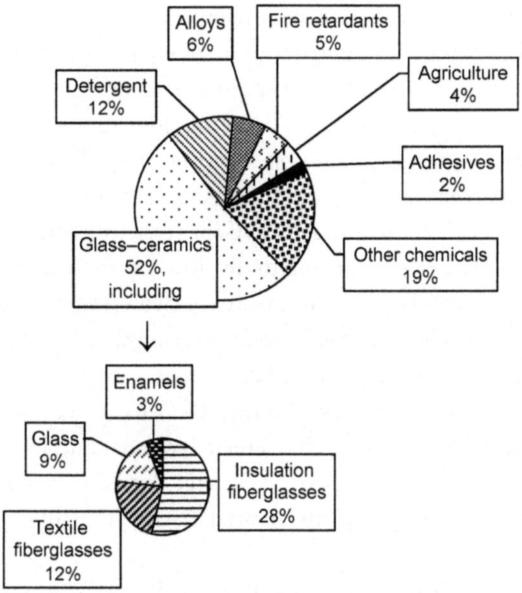

Figure 1.1 Boron consumption by the U.S. industry. *(According to Ref. 18).*

increase in Chinese consumption. World consumption of borates was projected to reach 2.0×10^9 kg of B_2O_3 by 2014. Demand for borates was expected to shift slightly away from detergents and soaps toward glass and ceramics.[19]

1.2 BORON SOURCES AND CYCLES IN THE ENVIRONMENT

1.2.1 A Requisite Note on Boron Analysis

Boron is a difficult element to analyze. Several techniques are routinely employed to analyze B in geochemical and biological matrices, but only few are sensitive enough to study B at trace level (<1 ppm).[20,21] Sah and Brown[22] have made a comprehensive review of published methods of sample preparation, determinant purification, and determination of B in samples. The most common methods for determining B concentrations are spectrophotometric and plasma-source spectrometric methods. Although most spectrophotometric methods are based on colorimetric reactions of oxy-B complexes with azomethine-H, curcumin, or carmine, other colorimetric and fluorometric methods have also been used to some extent. Most of these methods in general suffer from interferences and have low sensitivity and precision, but the method with azomethine-H is probably an exception—it was found to be both sensitive (the detection limit is 0.02 mg/L) and selective.[23,24] The application of nuclear reaction and atomic emission spectrometric methods has remained limited because these methods have poor sensitivity and suffer from serious memory effects and interference. Out of a large number of published nuclear reaction methods, only thermal neutron-prompt γ-spectrometry has been of practical use. This technique provides a nondestructive analysis for >0.05 μg of B in intact samples,[25] which makes it especially useful for some medical applications. However, this method is time consuming and is not suitable for detecting low and trace B concentrations. Inductively coupled plasma optical emission spectrometry (ICP-OES) has created a new dimension in B determination because of its simplicity and sensitivity. However, it suffers from interference and is not adequately sensitive for some nutritional and medical applications involving animal tissues that are naturally low in B. Among nuclear techniques, ^{11}B-NMR seems to be a powerful tool, not only for detecting borate cross-linked biomolecules, but also for analyzing the type of borate complex.[26] Size-exclusion chromatography with parallel inductively coupled plasma mass spectrometry (ICP-MS) and refractometric detection was the primary technique used to investigate metal complexes with polysaccharides, and was used to characterize water-soluble B compounds in radish roots.[27a] Gaspar et al.[27b] developed a Fourier transform ion cyclotron resonance mass spectrometry technique to identify natural B-organic (fulvic) complexes in samples from peat.

Additionally, when examining references on the trace quantities of B in the environment, one should take into consideration the clean technique in sampling, storage and treatment of the samples to avoid questionable data.[20] That is why data on trace elements, including B, published before the introduction of clean procedures since approximately the mid-1970s should be considered carefully.

1.2.2 Sources, Sinks, and Environmental Cycles

Volcanic and geothermal activity has been long supposed to be the main B input source to the atmosphere and ocean, and that rock weathering and anthropogenic activity (due to fossil fuel burning, mining, and agricultural fires) represent only a minor contribution (Tables 1.1 and 1.2). Two sources have been suggested as the significant B inputs to the ocean: (1) volatilization associated with arc volcanogenesis and (2) mobilization associated with compaction-induced dewatering of accretionary prisms in subduction zones.[29,41] Kopf and Deyhle[42] identified the significance of B fluid flux through active mud volcanoes.

However, published data vary dramatically as our knowledge on the B sources and sinks suffers from significant uncertainties and critical assumptions (Ref. 40 for review). As a result, the obtained evaluations differ, often by orders of magnitude. Little is known about B fluxes from submarine volcanic and geothermal activity, and information on volatile B compounds is scarce. This is probably the reason for estimated imbalances within the major B environmental cycles: (1) into and out of the atmosphere and (2) into and out of the ocean. At this stage it seems to be impossible to put the data on B environmental fluxes together and to give a realistic numerical picture of B cycles in the biosphere.[29]

Table 1.1 Boron Sources and Sinks via Natural and Anthropogenic Activity, 10^9 kg B/year

Source/Flux	Estimation	Reference
Natural Input		
Volcanic activity*	0.01–2.1	
Hydrothermal activity	0.13	28
	0.004–0.042	29
	0.08	30
Chemical weathering	0.043	30
	0.026	31
Physical weathering	0.15	30
Fluids from subduction zones	0.02	32
Desorbable boron in marine sediments	0.1	33
Anthropogenic Input		
Boron mining	0.4	34
	0.31	30
	1.36	31
Biomass burning	0.26–0.43	30
	0.24	31
Fossil fuels combustion	0.24	31
Coal combustion	0.20	37,38
Output		
Sedimentation	0.47	30
Biogenic carbonate sink	0.064	33
Organic matter burial	0.014	30
Biogenic silica sink	0.013	36,37
Altered oceanic crust sink	0.14	33
Low-temperature hydrothermal sink	0.08–0.27	30

*See Table 1.2 for more details.
From Ref. 40.

Table 1.2 Boron—Gaseous and Particulate forms in Terrestrial Volcanic Emanations to Atmosphere from Different Sources, 10^9 kg B/year

Gaseous Emission	Particulate	Reference
0.01	–	37
2.1	0.0003	36
0.24	0.21	39
0.017–0.022	0.00022–0.00047	38

From Ref. 40.

Argust[35] allotted the major stores and reservoirs of B in the biosphere (Table 1.3). When it enters the biosphere, B is involved in a series of major and minor environmental cycles and transformations. Primarily, the global B biogeochemical cycle is driven by a large flux through the atmosphere derived from sea aerosols.[35,38] Boron fluxes in the environment are governed by three major cycles: (1) the atmosphere–ocean/land cycle, (2) marine ecosystems, and (3) soil–plant terrestrial ecosystems (Figure 1.2). The two latter are controlled by biological processes of phytoplankton and plant primary

Table 1.3 Boron Major Stores and Reservoirs, kg B, in Order of Magnitude

Stores and Reservoirs	Content
Continental and oceanic crust	10^{18}
Oceans	10^{15}
Groundwater	10^{11}
Ice	10^{11}
Coal deposits	10^{10}
Commercial borate deposits	10^{10}
Biomass	10^{10}
Surface waters	10^{8}

After Ref. 35.

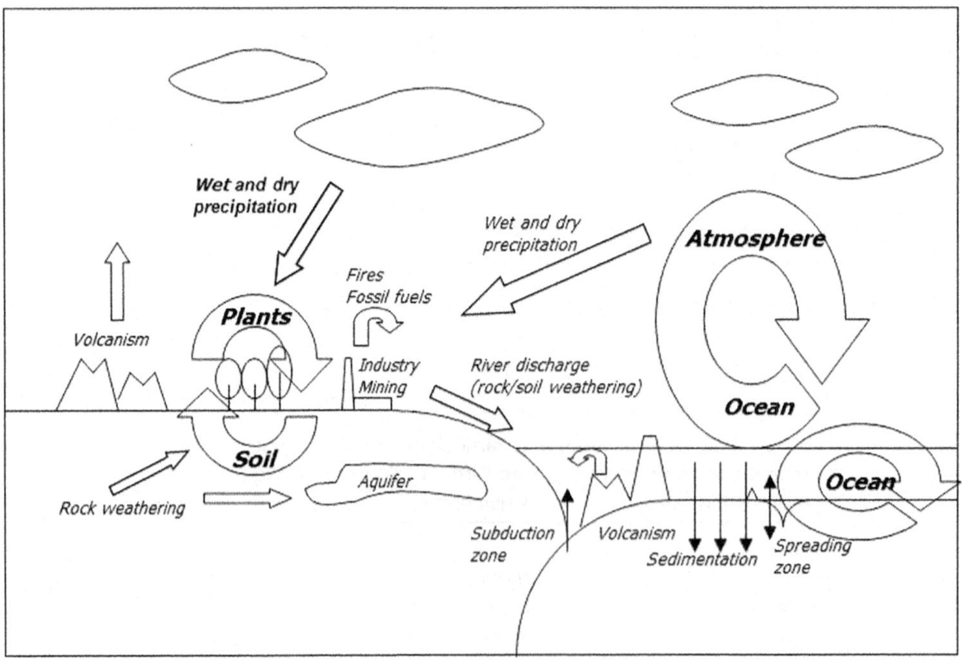

Figure 1.2 Schematic diagram of boron turnover in the environment. *(From Ref. 40).*

production and utilization. Park and Schlesinger[38] evaluated B turnover in these cycles as 4.8×10^9 and 4.4×10^9 kg B/year, respectively, while Klee and Graedel[30] estimated B mobilization by terrestrial plant primary production as 12.9×10^9 kg B/year. These two cycles determined lesser B environmental fluxes. The terrestrial soil–plant system determines: (1) B flux with drainage from the soil to aquifers and surface waters—0.43–1.3×10^9 kg B/year[35] and, consequently, (2) B outflow with river discharge to the ocean—0.53–0.63×10^9 kg B/year, and (3) output with soil-bearing aerosols—0.017–0.033×10^9 kg B/year.[38] It is also the primary source of B fluxes from combustion of biomass and, in geologic time, combustion of fossil fuels. The marine ecosystems are the source of B coprecipitated with biogenic carbonates and silica. Organic matter burial in the ocean contributes to B sediment fluxes and transformations (Table 1.1). The third major B flux is governed by B turnover between the ocean and the atmosphere; it includes the lesser atmosphere–land–ocean B turnover. The total input of B to the atmosphere from sea salt aerosols was calculated by Argust[35] as from 1.3 to 4.5×10^9 kg B/year; Park and Schlesinger[38] estimated this value as from 1.0 to 2.3×10^9 kg B/year. The counter flux from the atmosphere to the ocean and land occurs by wet and dry precipitation.

An ancient and still unresolved problem of B geochemistry is the enrichment of B in shales (up to 100 mg/kg) compared to the granitic continental crust (9–10 mg/kg).[44] It has been suggested that this enrichment originates in the sorption of seawater B on clay and organic sediments through several weathering and deposition cycles.[45] Other rocks and formations may also be enriched with B: clays (100 mg/kg), carbonate rocks (up to 350 mg/kg), and deep ocean carbonaceous (550 mg/kg) and clayey (2300 mg/kg) sediments.[44]

Terrestrial humid landscapes are poor in B, as a rule. In semiarid and arid landscapes, content of B is higher due to higher evapotranspiration rate; here, B may accumulate as polyborate anions forming poorly soluble Ca and Mg salts. In dry steppes and deserts, landscapes enriched in B have been described.[4,46,47] These are areas of salt-bearing strata, modern and ancient volcanic activity. In some areas, soils and lakes are highly enriched in B—"borate ecosystems" in California and Tibet; B ores of exogenic types containing borax ($Na_2B_4O_7 \cdot 10H_2O$), asharite ($MgHBO_3$), etc. are formed there.[4] During geological history, a progressive concentration and differentiation of B in the biosphere occurred: B accumulated in the sedimentary rocks in the form of B ores and B bound in clays[48] and organogenic formations—coals (as much as 500 mg/kg),[49] oils (up to 70 mg/kg),[50] oil shales (61–175 mg/kg),[51,52] etc. High B content may be found in waters related to rock metamorphism.[6]

Boron mobilized by human activity may have a noticeable contribution to the environment. The major technologic-derived B sources are the following: (1) mining, (2) biomass burning, including deforestation, charcoal, combustion of agricultural residues, man-made fires, incineration of wastes, and (3) fossil fuel combustion (Table 1.1).

However, B is considered to be an unperturbed element, with less then 15% of the total element's natural mobilization.

1.3 BORON IN ATMOSPHERE, NATURAL WATERS, AND SOIL

1.3.1 Atmosphere

Boron is a variable constituent in the atmosphere with concentrations usually between 0.2 and 300 μg/L.[36,53,54a] It is commonly suggested that much of the gaseous fraction of B may exist as H_3BO_3. Rose-Koga et al.[54a] data on lichen indicate that the atmospheric residence time of gaseous B is about 16 times of that for particulate B, which agrees with previous estimates obtained from rain data of 19–36 days and 2–6 days for gaseous and particulate B, respectively.[36] Rain and snow are considered to be the major agents removing B from the atmosphere, although gas exchange with the oceans has also been proposed.[36,53,54a] Gaseous and particulate anthropogenic contributions may be locally or regionally important.

A quantitative understanding of B in the atmosphere, however, does not yet exist; in particular, the atmospheric evolution of seawater-derived B remains enigmatic.[54a] Boyd and Walley[54b] for the Southeastern U.S. found that concentration of B in rainwater generally increased during the autumn to a peak in February and then declined during the spring, reaching a minimum in the summer. Boron concentrations in small showers were generally higher than values for heavier rains. Demuth and Heumann[55] found a significant dependence of the B concentration in rainwater on meteorological conditions, but not on the season of sampling, or not on the anthropogenic influences.

In situ measurements of the chemical compositions of individual aerosol particles in the stratosphere above eastern-central and southern North America reveal B in about 4% of spectra, at all altitudes and latitudes. The presence of −O and −OH functional groups as well as a small but nonzero concentration of organic matter as potential B binders in aerosols has been noted.[56] Investigations of Gaillardet et al.[57] indicate that the formation of volatile organic complexes of B may also be expected.

The anthropogenic release of B to the atmosphere is considered to occur mainly in the vapor form of boric acid. Some B halogens, such as BF_3 and BCl_3, may be produced during biomass burning, fossil fuel combustion, incineration, and manufacturing.[58]

1.3.2 Natural Waters

In the hydrosphere, the main B reservoir is seawaters, containing the element on the level of 10^{15} kg that is by far dominates other reservoirs—groundwater (10^{11} kg), ice (10^{11} kg), and surface waters (10^{8} kg).[35] Boron is one of the major constituent in seawaters, referred as about 4.5 mg/L, and in some groundwaters of active volcanic and geothermal activities, where B concentration may reach extremely high level, while in freshwaters B contents does not usually exceed 0.5 mg/L (Table 1.4). In seawater, equilibrium reactions of B contribute to alkalinity and to the buffering of pH.[72] Elevated concentrations of B are common in oil field waters[73] suggestive of organically derived B.

According to Perelman,[4] B is a highly mobile element in the natural waters ("active water migrant")—according to the relatively high ratio of B concentration in the natural waters to that in the earth's crust ("factor of water migration").

Table 1.4 Content of Boron in Natural and Contaminated Waters, mg/L

Waters	Concentration	Reference
Continental geothermal waters	Up to 1080	59
Waters of active volcanic and geothermal activities	0.2–72	60
Atmospheric precipitation		
Rains, Southeastern U.S.	<0.001–0.095	54b
Summer, May–August	<0.001–0.0035	
September–April	0.0015–0.095	
Rains, Germany, Switzerland	0.0003–0.007	56
Rains, Paris, France	0.002	61
Rains, Southern Asia	0.0003–0.009	62
Snowpack	0.0001–0.002	
Surface fresh		
Average	0.01–0.1	34
River water, average	0.018	63
River suspended matter	70 mg/kg	
River water, average	0.009–0.30	62
Rivers, Great Britain		
Rural regions	0.015–0.096	64
Agricultural regions	0.193–0.387	
Industrial/urban regions	0.120–0.459	
Rhine and Meuse rivers, The Netherlands	0.04–0.20	65
Rivers, northern France	0.10 (<0.01–0.93)	
Rivers, Seine River basin, northern France	0.050 (0.002–0.091)	61
Rivers, streams, Liaoning, China		66
Pristine	0.002–0.51	
Contaminated	0.039–25.1	
Groundwater		
Mean	0.017–1.90	34
Average	0.04	4
Temperate climate	0.045	
Mountain regions	0.02	
Permafrost	0.01	
Mediterranean basin (Cyprus, Greece, Tuscany, SE Spain, S Israel, W Turkey)	3–13	67–70
Seawater, average	4.5	71

From Ref. 40.

1.3.3 Seawaters

As early as 1932, Goldschmidt et al.[74] observed that the B concentration of seawater is proportional to the total salt concentration. Since, it was commonly assumed that the B concentration can be evaluated by measuring only the salt concentration. The ratio of B (mg/kg) to Cl^- showed an average value of 0.232 at 35.0‰ salinity, while for the upper sea layers, where contributions from land run-off, atmospheric precipitation may take place, this value was in the range of 0.222–0.255.[75] That is, B is considered to be a conservative element in the seawater, depending on the total salinity.

Brunskill et al.[76] stressed that improvements in the oceanic budget of B is needed because this element holds good potential for estimation of several interesting geochemical processes on land and sea. In the global ocean budget, B has dominant inputs from rivers, it has oceanic residence times $>10^6$ year, and it has poorly known or variable removal mechanisms, and in addition B displays environmental stable isotopic fractionation. These factors should improve our knowledge of the history of major cycles of climate change, continental weathering, and oceanic geochemical budgets.[77]

Boron is commonly agreed to occur in seawater as free boric acid and the borate ion $B(OH)_4^-$.[78] According to Bassett,[79] B in natural waters can occur as free borates (undissociated boric acid and partly dissociated borate anion), polyborates complexed with transitional metals, and as fluoroborate complexes. The aqueous chemistry of borates depends on their concentration and pH. Investigation of Gast and Thompson[80] strongly suggested the existence of soluble B-organic matter complexes in seawater. These complexes were believed to form with *cis*-type polyhydric organic compounds[81]; such compounds reacted with boric acid and gave low results with the mannitol method of B analysis. Complete degradation of the organic complexes to free the combined borate ions appears feasible only through vigorous oxidation.[82] Though, Williams and Strack[83] have concluded from a consideration of the stability complexes that borate-organic compounds will be present only in negligible proportions at the very low levels of carbohydrates prevailing in the sea. On the other hand, B, along with molybdenum and vanadium, in the form of oxyanions is fixed by humic matter (but the process evidently involves transformation of the anions to cations).[84] Humic matter, being resistant to microbial decomposition, can probably remain in the water indefinitely without being mineralized. It is well known that humic matter forms complexes with a variety of nonhumus organic compounds, including amino acids, carbohydrates, etc. North,[85] investigating seawaters and sediments, detected a complex mixture of primary amines such as proteins, peptides, and amino acids, including glycine. Some of these compounds form strong and specific complexes with borates. Simple biochemical compounds "fixed" to humic colloids may thereby be protected from microbial decomposition.[86,87] The organic matter of deep ocean waters is extremely stable.[88,89] To this, the uptake experiments with *Platymonas* showed that about half the natural primary amines in seawater can be taken up by certain phytoplankton cells. Either these molecules are truly in a "free" dissolved state, or the complexes of amino acids with other substances does not prevent uptake by algal cells.

Bottom sediments are an important part of aquatic ecosystems; they are notable as an active chemical interface and also for their highly intensive microbiotic activity.[4,90] Most published works on B in bottom sediments emphasized on the element interaction and binding with clay minerals, with most B incorporated firmly into their crystalline lattice.[41,91] From the analyses of pore waters and sediment size fractions, Spivack et al.[41] concluded that it is unlikely that there is any significant, irreversible fixation of B into aluminosilicate detritus at low temperatures in the marine environment. It is most probable that B is fixed in clays only during burial metamorphism.[92] However, the high B content of metalliferous sediments from the Bauer Basin suggests that B is incorporated into authigenic silicates where they are forming on the sea floor. Organic matter of bottom sediments may be an important underestimated carrier for B. For example, a study by Williams et al.[93] has demonstrated that kerogen (refractory organic matter fraction) in the Gulf of Mexico coast basin contains a significant amount of organically bound B (about 140 mg/kg). Although kerogen makes up only a small percentage of the sedimentary material, its significance as a source of B may be high.

1.3.4 Freshwaters

Away from the active volcanic and geothermal zones, the major B sources to the surface waters are precipitation and soil and underground water drainage. The B input may originate from different natural and anthropogenic sources dependent on the character of regional rocks, aquifers, soils, the distance from the sea and the local climate conditions, and, significantly, on the local industrial and agricultural pressure. Boron concentration in unpolluted freshwaters does not usually exceed 0.5 mg/L (Table 1.4) and maximum acceptable concentration for B in drinking water for different countries (Table 1.5). The evaluated contribution of the B sources varies much in the published works. Jahiruddin et al.[99] found that in unpolluted areas of northeastern Scotland, B inputs in precipitation originated from ocean spray, along with dry deposition, approximately balance B outputs with river waters, but not in agricultural lands. For the

Table 1.5 Maximum Acceptable Concentrations for Boron in Drinking Water, mg/L

Country/Organization	Concentration	Reference
Canada	5.0	94
The European Union	1.0	95
Russia	0.5	96
The USA	0.3	97
WHO	0.3	98

Himalayan Rivers, Rose et al.[62] found that the isotopic composition of dissolved B results from isotopic fractionation induced by reactions in soils. Chetelat et al.[61] applied the isotopic technique to coastal rivers of French Guiana and discovered that most dissolved B originated from seawater and a biomass-derived component, presumably, biomass burning. Investigations with similar techniques in the basin of the Seine River, France, which is highly affected by human activity, found that the "total dissolved" B originated from three distinct sources: urban effluents (about 65%), agricultural effluents (<10%), while rainwater contributed about 25%, reaching up to 30% during high water floods.[100] Lemarchand and Gaillardet[101] examined waters of the Mackenzie River basin, northwestern Canada and they found that the B content appears to be regulated, in the largest part, by reactions involving silicate rocks, namely clays. The other contributions of rainwaters and the dissolution of carbonate and evaporite rocks can locally control the B budget but they remain of second importance at the regional to continental scales. It also appears that dissolved B in rivers is regulated by input of groundwater.

Brunskill et al.[76] have made an extensive integrated mass balance for dissolved and particulate B in the river-sea mixing zone, including a continental shelf and slope—using the example of the Gulf of Papua. The authors concluded that about 66% of river input of B was in the particulate phase, and low "dissolved" B (<0.45-μm fraction) concentrations in freshwater were conservatively mixed with higher concentrations of B in seawater across the salinity gradient. Removal of B to inner-shelf sediments was about 83% of the total river input, indicating a small export of B to the sea. About half of the dissolved B input from rivers is sorbed to particles and trapped in inner-shelf sediments.

Investigations of B forms in freshwaters are traditionally focused on inorganic "dissolved" B species like boric acid and borates. Though, some authors believe that such approach is not adequate toward natural waters—marine, river, lake, underground, meteoric (and soil). Natural waters are complex (and highly dynamic) polycomponent bodies impacted with immanent living activity and life-bearing compounds; they could hardly be described adequately in terms of thermodynamics that are applicable for inorganic laboratory solutions.[4] Moreover, B concentrations measured in conditionally dissolved fractions, i.e., in the filtrates obtained with paper and membrane filters, cannot be regarded as real dissolved B species (*more detailed on this pressing problem*[4,102–105]). Reactivity of natural waters is defined by immanent association of balancing colloidal (including "living colloids"), suspended organic and mineral, and dissolved components, complicated by biota activity, which is an integral part of natural waters ("quasi-living" systems—according to Perelman[102b]).

Natural waters are rich in potential carrying and binding species for B, such as organic and humic compounds, clays and clay–organic complexes, iron/manganese amorphous minerals, etc. Freshwaters always contain a generous amount of water–mobile organic matter, of which components are bacterial peptidoglycans, aromatic amines, peptides and amino acids, hydrocarbons/tannins, polyuronic acids.[106] One could expect a noticeable quantity of humus/organic-bound B, inherited from draining soils and underlying bottom sediments, from aquatic organism excretion and remnant decomposition, and swamp discharge. Chauveheid and Denis[107] showed a linear correlation between B and dissolved organic carbon for ground and surface waters. The authors found that such a correlation could be used to define natural water pristine conditions.

Perelman[4,102b] believed that the prevalence of the "pure chemical" approach over the biogeochemical approach in hydrochemistry is the result of domination of specialists with a chemical educational base.

The information on B in freshwater bottom sediments is scarce, in spite of their important role in the river and lake chemistry. Bottom sediments are an integral product of the weathering processes and as such they reflect a (bio)geochemical situation of the draining area. Jaquet et al.[108] found that B occurs more in coarser lake sediment fractions (micas, feldspars, tourmaline) influenced by river discharge, but also in clay-size fraction (<2 μm). Data obtained by Stewart et al.[109] showed that the content of B in lake sediments are not very different from soils of the feeding basin, but the extractable level is higher. Mun and Zhaimina[110] found that B concentrations of interstitial waters of the sediments are higher in fresh lake and marine bottom sediments enriched in organic matter.

The exciting question of B forms and mobility in natural waters, including soil and sediment pore waters still awaits further investigations.

1.3.5 Groundwater

In most groundwaters, B is present naturally at concentrations of <1 mg/L (Table 1.4). Boron enrichment can result from hydrothermal influence on infiltrating waters[111], human pollution (e.g., Refs 79,112,113), evaporative concentration, especially with respect to irrigation water (Refs 114,115 and references therein), dissolution of evaporites[111], the presence of residual seawater, and mineral weathering. High B content in groundwaters may be of natural origin, such as in some Mediterranean areas: data collected from regional databases for more than 6000 selected groundwater points (the BOREMED project[67]) reveal that about 10% of the investigated water sources have B levels exceeding 1 mg/L, i.e., the new EU Drinking Water Directive. As such, some groundwaters in Cyprus, the Chalkidiki peninsula, Greece, and in Tuscany, Italy are currently unusable. High B concentrations have also been reported for groundwaters in Southeastern Spain,[69] the coastal aquifer shared between Israel and the Gaza Strip,[68] western Turkey,[70] etc.

Overexploitation of the groundwater basins, particularly during the tourist season, has resulted in the lowering of groundwater tables and increasing seawater intrusion into the aquifers. Countries, such as Cyprus and Israel have shut down hundreds of wells along the coastline that were used primarily for drinking water. Facing a shortage of suitable drinking water, many countries have had either to look for alternative sources, such as imported water, or to implement costly technological solutions, such as desalination. But as these well-publicized projects move forward to address salinity and groundwater contamination, another problem looms: B contamination.

The application of B compounds, especially sodium perborate ($NaBO_3 \cdot nH_2O$) as a bleaching agent in detergents, leads to an enrichment of B in wastewaters. Vengosh et al.[116] were the first to utilize the isotopic composition of B for tracing sewage effluent and contaminated groundwater: anthropogenic B in wastewater is isotopically distinct from natural B in groundwater and thus can be utilized to identify the source of contamination.

One of the most important results of the BOREMED project is that it is challenging the conventional perception that the B contamination in the Mediterranean is due to human pollution.[67] However, for years, companies added B to detergents because it is an excellent bleaching agent—thus resulting in the formation of B-rich sewage. Moreover, B is not removed during standard sewage treatment processes and even treated wastewater typically has high B concentrations. That is why, because B contamination in the investigated cases comes from natural geochemical background pollution and hence cannot be prevented, the only way to address the B problem is through treatment of the drinking water.

Boron contamination is a very sensitive issue for irrigation in arid and semiarid areas. The U.S. Salinity Laboratory developed a rating table that indicates the permissible B concentrations in irrigation water for three classes of plants according to their sensitivity for B (Table 1.6).

1.3.6 Soil

Boron is involved in important soil processes[117]: the intrasoil biological cycle; humification; isomorphic substitution of clay minerals and the formation of colloids; illuviation; and hydrogenic accumulation (meadow soils, solonchaks). Goldschmidt[118] found that B is concentrated in the uppermost humus layer of forest soils. Viets[119] supposed that

Table 1.6 Rating of Irrigation Water for Various Crops, on the Bases of B Concentration in Water, mg/L[210].

Class of Water		Crops		
Rating	**Grade**	**Sensitive**	**Semitolerant**	**Tolerant**
1	Excellent	<0.33	<0.67	<1.00
2	Good	0.33–0.67	0.67–1.33	1.00–2.00
3	Permissible	0.67–1.00	1.33–2.00	2.00–3.00
4	Doubtful	1.00–1.25	2.00–2.50	3.00–3.75
5	Unsuitable	>1.25	>2.50	>3.75

much of the B in soils is complexed with humus, probably as adsorbed, chelated, or complexed ions, where its release to plants is presumably dependent on soil moisture. Berger and Pratt[120] stated that a large part of the total B in soils is held in the organic matter in tightly bound compounds that have been formed in the growing plants themselves—within the major soil–plant circulation; B in organic matter is primarily released in an available form through the action of microbes.

The movement of mobile B compounds in soils follows the water flux. In cool humid climates, B is leached downward in soil profiles, whereas in soils of warm humid, or arid and semiarid regions, B tends to concentrate in surface horizons. Boron may also be concentrated in soil horizons enriched in illitic clays or sesquioxides.

Of all micronutrient deficiencies in plants, B deficiency is the most widespread. On the other hand, B excess and toxicity is an important disorder that can limit plant growth in soils of arid and semiarid environments. Although of considerable agronomic and environmental importance and despite exponential growth of special publications on the topic, our understanding of B behavior and its fate in soil ecosystems remains limited.

Boron concentration in regular soils varied approximately in the range of 10–100 mg/kg (Figure 1.3). Boron deficiency is most likely in coarse textured soils of humid regions. Temporary deficiency of available B can be triggered by liming of acid soils due to increased B adsorption at higher soil pH.[122] In arid and semiarid areas, B toxicity is a result of high levels of B in soils due to high evaporation rate together with additions of B via irrigation water.[123] The highest naturally occurring concentrations of soil B are in soils derived from marine evaporates and marine argillaceous sediments. In addition, human pollution may increase soil B to levels that are toxic for plants. The most important source is irrigation water, but others include waste from surface mining, fly

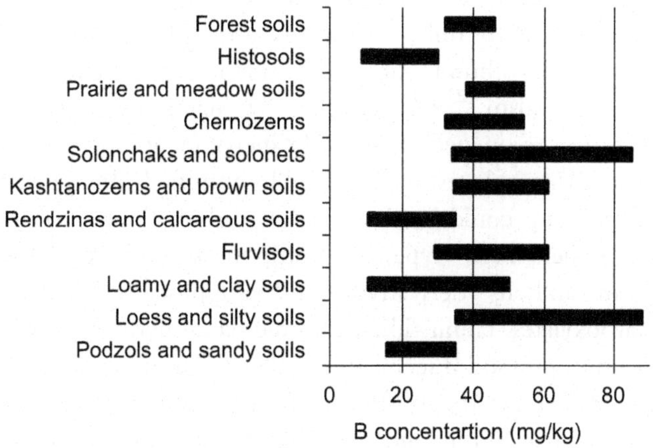

Figure 1.3 Average concentrations of boron in different soil types. *(After Ref. 121b).*

ash, and industrial chemicals. Reclamation of high B soils requires about three times as much water as reclamation of saline soils.[121a]

Some authors suppose that the chemistry of soil B, compared to other nutrient elements, is very simple for the following reasons[124]: (1) B does not undergo oxidation–reduction or volatilization reactions in soils, (2) boric acid is a very weak, monobasic acid that acts as a Lewis acid by accepting a hydroxyl ion to form the borate anion, and (3) B-containing minerals are either very insoluble (tourmalines) or very soluble (hydrated B minerals) and these minerals generally do not control solubility of B in soil solution. However, such an approach seems to simplify the matter. For example, B behavior in soil may be governed by redox processes indirectly—as a significant portion of soil B was found to be bound (nonexchangeably) to the redox-sensitive components, such as manganese and iron amorphous minerals.[43] Furthermore, in soil, there are plenty of B-binding agents besides hydroxyl anions from dissociated water molecules. Surprisingly, most of the published works omitted or mentioned in passing B–organic/humus matter interactions (Refs 40,43 for review). Reviewing the literature on soil B chemistry, one finds information on soil B association with various soil components, including adsorbed/exchange forms and those occluded in mineral phases (clays and Fe/Al hydroxides), as well as the common assertion that dissolved B compounds occur as undissociated boric acid and borates in soil solution, which are supposed to be major bioavailable forms of B. Only a few works have considered the necessity of including B-organic forms in their fractionation schemes.[43,45,119,125]

In the mean time, Hingston[126] and Yermiyahu et al.[127] reported higher sorption of B on soil organic matter than on clay; at similar pH levels and B concentration, the sorption was one to two orders of magnitude higher than on clays. Boron deficiency has been observed in soils with high organic matter contents. This deficiency has been shown to be related to the high affinity of organic matter to B.[120,127–130] Sorption/desorption studies of B (as boric acid) on organic matter exhibited strong hysteresis effects and the sorbed B was removed to a partial extent only.[131] The formation of dihydroxy- and/or hydroxy-/carboxy binding of borate to humic substances and/or carbohydrates was proposed to be the main sorption mechanism.[127] Gasper et al.[27b] utilizing the newly developed Fourier transform ion cyclotron resonance mass spectrometry technique, detected and outlined two main groups of B-complexes in peat soluble humified (fulvic) materials: (1) components of the first group could be described as carbohydrate derivatives, and (2) the second group compiles many types of compounds such as substituted aromatic compounds—lignin and its derivatives, and—supposedly—alicyclic and aliphatic (hydroxyl) polycarboxylates, tannin-like molecules and "hydrogen-depleted structures." A bulk of organically bound soil B seems to be inherited from the plant lignin: Kot et al.[43] have found a major portion of the soil B to be bound firmly in refractory fractions of the soil organic matter, such as humin and clay–/iron–aluminum–humus complexes. The important question of mechanisms and forms of B release during decomposition of plant remnants as well as sequestration of B with refractory humus compounds still remains to be considered. It is still unclear if the cycling of B in natural soil–plant system should include free (unbound) dissolved B species in soil water. Boron fate in the rhizosphere has never been investigated, but only touched upon in a few works.

Like other micronutrients, the minimum solution concentration of B required for plant growth is too low to be easily measured directly. This concentration must be low enough not to have an adverse effect on plants. Obviously, the content of available B compounds in the soil water must be controlled by an effective buffer system that could supply trace but continuous quantities of B.[132] Until recently there were not adequate analytical facilities to determine B concentrations at trace level and, especially, to distinguish B chemical species in soil water directly. It is important to stress that most experiments on B behavior in soil and biota—either in laboratory or in the field—have been based on supplying inorganic solutions of boric acid and borates. As Viets[119] marked, soluble boric acid and borate added to soil is weakly adsorbed and remains largely in the water-soluble phase where it can be toxic to sensitive species. Of course, those experiments show that plants are able to take up B (as oxy-B complexes) from the applied inorganic solution, yet such experiments do not reflect the real processes which take place in situ soil–plant system. As Hu and Brown[133] pointed out, the apparent contradiction between experimental results and in-field observations suggests that B uptake is determined by factors that are as yet unknown.

Some authors suppose that most of the soil micronutrients are complexed by organic chelators, which act as a buffer for nutrient supply.[134] The formation constant for the metal–chelate complex, the excess of chelator in solution, and the solution composition govern the activity of the micronutrients. These systems are able to maintain submicromolar concentrations of micronutrients. For the case of B-oxy compounds, such a chelator must be of a specific nature. Asad et al.[132] evaluated a range of possible B chelators and borosilicate glass for their effectiveness in maintaining solution B concentrations for water culture studies, including groups of polyhydric alcohols, sugars, phenolic compounds, and fluoride. They found that the B-specific ion-exchange resin Amberlite IRA-743 containing *N*-methyl-glucamine functional groups appeared to be very promising. Commercial attempts to utilize B complexes with glycine and arginine as controlled release fertilizer for soil and foliar application have been reported to be successful.[135,136] Kot et al.[137] experimented with a collection of the eastern Mediterranean soils and plant litter to evaluate the soils available B potential, sources and turnover. The results showed the soils' high capability to compensate B to the soil water. "Regeneration" of toxic levels of available B following reclamation has been observed for some soils earlier; and this additive B has supposedly been released from previously undissolved sources that are less leachable than adsorbed B.[138] Moreover, the release of B from the water–mobile phase ("soil solution") showed a strong positive correlation with "dissolved" organic matter, indicating B—organic/humus complexes to be a major source of the available B. The measured leaching of B from the plant litter exceeded the evaluated B income with atmospheric precipitation by orders of magnitude; thereby one can suppose that B turnover in regular soil—plant systems is dominated by the semiclosed cycle "plants—litter—soil humus/humin—soil water mobile colloids." Treatment of the soil samples with strong formaldehyde solutions showed some decrease in release of B, indicating the involvement of microbiota in these processes. The dissolved/available B related to colloidal fractions of <0.20 and 0.45 μm, presumably, of organic origin.[137]

1.4 EFFECT OF BORON ON MICROBIOTA AND PLANTS

Maluga[139] noticed that B contents of plant ash are significantly higher than those of soils and the lithosphere—400, 10, and 12 mg/kg, correspondingly, indicating B accumulation by living organisms. The accumulation varies strongly for different groups of organisms (Figure 1.4). Boron is not biomagnified in aquatic food chain, as shown by the fact that B concentrations were higher in filamentous algae and detritus than in invertebrates and fishes.[141] In terrestrial food chains, B may accumulate in plants but not in animals.[142]

Boron essentiality for the heterocystous cyanobacteria, predominant organisms during the Middle Pre-Cambrian Period, indicates that B was an essential element during the early evolution of photosynthetic organisms.[143] Plants needed B in order to grow stems and roots as they left seas and colonized land. Therefore, B was an epochal and indispensable element for terrestrial plants and life development.[144,145]

1.4.1 Bacteria, Archaea, and Fungi

There is as yet but little known about the necessity of B for the growth of microorganisms. Probably, Voicu[146] was the first to report a stimulating effect of boric acid on *Azotobacter*—in the presence of humus (0.1% sodium humate). Subsequent studies have confirmed B essentiality for a variety of bacteria and algae strains.[147,148a] Anderson and

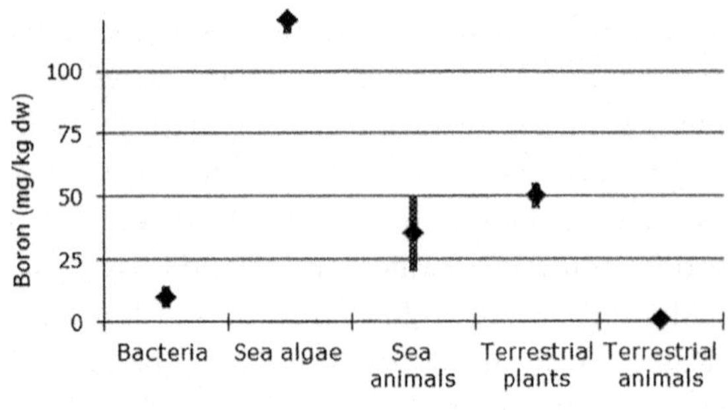

Figure 1.4 Average concentration of boron in plants and animals. *(After Ref. 140).*

Jordan[148b] also found B-stimulated nitrogen fixation in *Azotobacter*, although they admitted that B was not required for the bacterial growth. Among bacteria, B has been reported to be essential for cyanobacteria for proper formation of nitrogen-fixing heterocysts.[149] The B-containing antibiotics tartrolon (A and B) and boromycin synthesizing by bacteria have been discovered.[150,151] Boromycin is notable for being the first natural product found to contain B. It is suggested to have potent anti-HIV activity.[151] Ahmed et al.[152] discovered three strains of B-tolerant bacteria from B-enriched soil. The strains required B for growth and can tolerate more than 450 mM B in feeding broth. The proposed name is *Bacillus boroniphilus* sp. nov.

Nelson and Mele[153] investigation, using growth-based and molecular techniques, implied that B (and NaCl) is more likely to affect rhizosphere microbial community structure indirectly through root exudate quantity and/or quality than directly through microbial toxicity.

Addition of B (as boric acid) to phytoplankton had a significant effect on primary production and carbon assimilation; the effect of B could be species specific.[154] Boron stimulates growth of yeasts (*Saccharomyces cerevisiae*),[155] the organisms without pectic cell walls. A role for B has been demonstrated in the establishment of an effective legume–*Rhizobium* symbiosis.[156] The recent discovery of a B-containing bacterial signal molecule, autoinducer AI-2, revealed B's role in bacterial quorum sensing.[157]

Boron has not been shown to be essential for fungi.[140] The potential for a B requirement in archaea, the third domain of life, has not been evaluated.[158]

1.4.2 Plants

The B requirement for vascular plant growth was first demonstrated a century ago.[159–161] However, its biochemical role is not well understood even now. Several extensive reviews on the topic have been published in recent years.[16,162–166] This time, there is a common agreement that B is implicated in three main processes: (1) maintaining cell wall structure, (2) maintaining membrane function, and (3) supporting metabolic activities. Bolaños et al.[156] marked that as all of the roles clearly established for B are related to its capacity to form diester bridges between *cis*-hydroxyl-containing molecules, and they proposed that the main reason for B essentiality is the stabilization of molecules with *cis*-diol. The work of Teasdale and Richards[167] induced to conclude that B has a primary role in cell wall biosynthesis. Subsequent work has shown that most of the B resides in the pectic fraction of the cell wall, which extensibility is impaired under B deficiency.[168]

A variety of enzymes in microorganisms, plants, animals, and humans interact with B compounds resulting in stimulation, stabilization, or inhibition. Despite these numerous effects of B, until recently no evidence had yet been presented that B is an enzyme constituent or that it has a direct role in plant enzyme activities.[169] Blevins and

Lukaszewski[164] noted that one more site of B action that is not connected with a structural role in cell walls or membranes is auxin metabolism. Boron interaction with auxin has long been postulated, and although the issue remains controversial, it may be crucial to our understanding of the role of B in plants.

As a result of B deficiency, several physiological impairments were reported like affection of sugar transport, cell wall synthesis and structure, lignification, carbohydrate metabolism, RNA metabolism, respiration, indole acetic acid metabolism, phenol metabolism, and membrane integrity.[168] It was recently reported that B deficiency may impair ascorbate metabolism and induce oxygen activation.[164] Boron deficiency inhibits growth, kills growing meristems, inhibits flower development, and causes low fruit and seed set, male sterility, seed formation of damaged embryos and malformed fruit.[115,170] Boron has a critical role in expanding tissues and must be supplied continually throughout the life of the plant, usually through the root. Boron foliar spray application has been proved beneficial in case of B deficiency.[171,172]

Boron is commonly thought to be absorbed from the soil solution by roots mainly as undissociated boric acid.[12,16,133] At physiological pH boric acid is expected to be in the form of an uncharged small molecule with a molecular volume of 71.5 Å, which is similar to urea (75.3 Å) and other small nonelectrolytes. The mechanism of B uptake remains controversial, and there is evidence for active uptake and passive entry into cells. Passive B entry is the most widely accepted mechanism for higher plants.[123,133] Though, there is apparent contradiction between in vitro results, which suggest that B uptake is a passive process, and field results, which demonstrate significant differences among species and genotypes that are difficult to reconcile, but are of fundamental importance to studies of B nutrition. According to findings of Dordas et al.[173] and Stangoulis et al.[174] the permeability of boric acid in the plasma membrane is an order of magnitude lower than those calculated previously, implying active B transport to plants and the need of membrane proteins to satisfy a plant's demand of B, especially under B limitation. Physiological studies using sunflower *Helianthus annuus* plants suggested the existence of energy-dependent high-affinity transport systems that are induced at low B supply and established concentration gradients for B in both processes of uptake and xylem loading.[175] Nable et al.[123] found that B content in all organs of five barley and six wheat cultivars differed dramatically even though all were grown under identical conditions.

Aphalo et al.[176] analyzed the mobility of B within plants and assessed data on the quantitative importance of retranslocation for the B budget of mature conifer forests and as a mechanism for avoiding toxicity. They hypothesized that species with very long-lived leaves are able to redistribute B from old to young tissues. The authors pointed out that when there is B retranslocation through the phloem, it is due to the formation of complexes with sugar alcohols, such as mannitol, sorbitol, and possibly pinitol.[177]

1.5 EFFECT OF BORON ON ANIMALS AND HUMANS

The reported weighted mean B intakes were 1.17 mg/d for men and 0.96 mg/d for women; for vegetarian adults, these intakes were 1.47 mg/d for men and 1.29 mg/d for women.[178] Samman et al.[179] estimated the daily human intake of B from the American diet to be approximately 1 mg/d. At that coffee and milk were the two top B contributors; these products are low in B, yet they make up 12% of the total B intake by virtue of the volume consumed.

The acceptable safe range for B intake is from 1 to 13 mg/d, and a daily intake of up to 20 mg/d can be achieved with a diet high in nuts, dried fruits, and wine.[98] Of selected Australian foods,[180] the most B-enriched are fruits—especially dry fruits (raisins, peaches, apricots, prunes), avocado, and currants, and nuts—all of them, but especially almonds and hazelnuts. Some vegetables have high content of B—first of all, legumes such as red beans, lentils, and chickpeas. Animal foods are relatively poor in B, except honey. In the United Kingdom, The Total Diet Study[181] showed that the highest concentration of B is in the food group nuts, followed by fresh fruit, fruit products, and green vegetables. The main contributors to dietary intake were beverages, fresh fruit, and potatoes.

Boron compounds are widely used in cosmetic products, such as makeup, skin and hair care preparations, deodorants, moisturizing creams, breath fresheners, and shaving creams, in concentrations of up to 5%.[182] Boron compounds can be found in the form of boric acid, borax, and other borates in a wide range of consumer products, including borosilicate glass, soaps and detergents, preservatives, adhesives, porcelain, enamel, leathers, carpets, artificial gemstones, high-contrast photographic materials, wicks, electric condensers, fertilizers, insecticides, and herbicides.[183] A reasonable estimate of B exposure from consumer products is 0.1 mg/d.[184]

While there is no Recommended Dietary Allowance for B, 20 mg/d is the upper limit of intake based on extrapolation from toxicological studies on rats.[185] World Health Organization[98] has introduced a guideline value of 0.5 mg/L for B in water. This guideline value was derived from a no-observed-adverse-effect-level of 9.6 mg/B kg body weight (bw) daily in a developmental study in rats, an uncertainty factor of 60, and allocation of 10% of the total daily intake to drinking water.

1.5.1 Essentiality

A relatively large amount of circumstantial evidence for B essentiality for animals and human has appeared since the 1980s (see extensive reviews Refs 179,186–188). Probably, the major role of B is in membrane-bound processes, early phases of tissue differentiation, or where large amounts of membrane material are required.[163] In humans, inadequate B intake was found to result in decreased activities of several membrane-bound hormones.[189] Listing of signs of B deficiency is difficult because most studies in animals used stressors to enhance the response to changes in dietary B.

Recent research has shown that B is necessary in the early stages of life. This includes the demonstration that lacking B adversely affects reproduction and embryo development in frogs and zebra fish. Studies with rats and mice suggest that low B may affect reproduction in mammals. Culturing of the frog *Xenopus laevis* in a medium containing less than 0.3 μM of B during organogenesis resulted in abnormal development of the gut, craniofacial region and eye, visceral edema, and kinking of the tail (muscular and skeleton). In the zebra fish model, during the early postfertilization period, 45% of B-deprived embryos died, whereas only 2% of B-supplemented embryos died.[190] Studies in rats and mice have not been as definite as those with frogs and zebra fish. However, in one study, preimplantation development of two-cell embryos from both B-deprived and B-supplemented female mice were significantly impaired by culturing of a B-deficient medium.[191] References on B deprivation-induced syndromes and diseases of animals and human are summarized in Table 1.7.

In a pioneering metabolic balance study, Kent and McCance[196] found that the major route of B excretion is through urine. (What seems especially valuable in this work is that the investigators considered the B naturally present in food—along with the B supplemented as boric acid.) Their results showed that the B in natural foods was metabolized in very much the same way as the additions of boric acid.

Low additional doses of B may have a positive effect on the healing of wounds and cerebral functions and by affecting calcium metabolism serving to activate bone and mineral metabolism. According to some authors, B has antiosteoporotic, antiinflammatory, anticoagulating, antineoplastic, and hypolipemic effects (Refs 188,189 for reviews).

1.5.2 Toxicity

Most B compounds are thought to have low toxicity and therefore are not considered an industrial problem. Boron is used in medicine as sodium borate, boric acid, or borax, which is also a common cleaner. Weir and Fisher[197] have studied the chronic toxicity of borax and boric acid in laboratory species and confirm the low order of toxicity for these chemicals.

However, accidental poisoning due to boric acid and borates has been reported. The fatal dose of orally ingested boric acid for an adult is somewhat more than 15 or 20 g and for an infant 5–6 g. The deaths of six babies, 6–11 days of age, were reported as the result of one feeding of milk, which had been diluted with a solution of boric acid (2.5%) in error for sterile water.[198] Boron poisoning causes depression of the circulation, persistent vomiting, and diarrhea, followed by shock and coma. Boric acid intoxication can arise from absorbing toxic quantities from ointments applied to burn areas or to open wounds, but it is not absorbed from intact skin.[199,200]

The most important source of exposure for human populations is ingestion of B from food, primarily from fruits and vegetables. Occupational exposure to B dust and

Table 1.7 Boron Deprivation-induced Syndromes and Diseases of Animals and Humans

Species	Dose/Cofactor	Symptoms/Diseases	References
Frog *Xenopus laevis*	Model dietary deprivation	Increase of necrotic eggs, high frequency of abnormal gastrulation	192
Frog *X. laevis*	<3.5 mg/kg B medium during organogenesis	Abnormal development of the gut, craniofacial region and eye, visceral edema, kinking of the tail (muscular and skeleton)	193
Chicks	Coaffected with vitamin-D_3 deficiency	Growth delay, bone formation affected (calcification)	194,195
Chicks		Hepatic glycolytic pathway (energy substrate metabolism) change	195
Rats		Weakening of immune function	196
Rats		Exacerbation of adjuvant-induced arthritis	194
Rats		Probable role in the maintenance of brain activation	197
Rats	Coaffected with vitamin-D_3 deficiency	Decrease of the apparent absorption and impairment of balance of Ca, Mg, and P	198
Human		Impairment of Ca metabolism	194,196
Human		Decreased brain electrical activity	199
Human		Poorer performance in tasks of motor speed and dexterity, attention, and short-term memory	199

exposure to B in consumer products, such as cosmetics, medicines, and insecticides, are other potentially significant sources.

1.5.3 Toxicity—Inhalation

Boron is absorbed following inhalation exposure, although it is not clear how much is absorbed directly through the mucous membranes of the respiratory tract and how much is cleared by mucociliary activity and swallowed.[201] Low acute inhalation toxicity was observed in those borates tested. In an inhalation study in which rats were exposed to boric acid at actual concentrations of 2.12 mg (0.37 mg B)/L (highest attainable concentration) for 4 h no deaths were observed.

1.5.4 Toxicity—Oral

In the 1870s, it was discovered that borax and boric acid could be used to preserve foods, and for about the next 50 years, borate addition was considered one of the best methods of preserving of fish, meat, and dairy products.[187] Today boric acid (E284) and sodium tetraborate (E285) are permitted preservatives in sturgeons' eggs (caviar) only; the maximum permitted level is 4 g B as $B(OH)_3$/kg.

Boron is readily absorbed from the gastrointestinal tract following oral exposure.[188,189] Studies in laboratory animals conducted by oral exposure have identified the developing fetus and the testes as the two most sensitive targets of B toxicity in multiple species (Ref. 202 for review). The developmental effects on mice and rats that have been reported following B exposure include high prenatal mortality; reduced fetal bw; and malformations and variations of the eyes, central nervous system, cardiovascular system, and axial skeleton.[203] Acute oral toxicity for mouse LD_{50} was 3450 mg/kg bw, for rat LD_{50} was 2660 mg/kg bw.

For human the acute adult quantitative dose response data ranged from 1.4 to 70 mg/kg bw. In cases where ingestion was less than 3.7 mg/kg bw, subjects remained asymptomatic. The human oral lethal dose is regularly quoted as 2–3 g boric acid for infants, 5–6 g boric acid for children and 15–30 g for adults. Symptoms of acute effects may include nausea, vomiting, gastric discomfort, skin flushing, excitation, convulsions, depression, and vascular collapse (Ref. 189 for review).

1.5.5 Toxicity—Intact

Boron is not absorbed across intact skin, but is readily absorbed across damaged skin.[204] The acute dermal toxicity of borates in animals is low, being >2000 mg/kg bw for all borates tested.

1.5.6 Toxicity—Mutagenic Activity

All available in vitro data indicate no mutagenic activity. In addition, the only in vivo study on boric acid (a mouse bone marrow micronucleus study) also indicated no mutagenic activity. Boric acid, disodium tetraborate anhydrous ($Na_2B_4O_7$), disodium tetraborate pentahydrate ($Na_2B_4O_7 \cdot 5H_2O$), and disodium tetraborate decahydrate ($Na_2B_4O_7 \cdot 10H_2O$) have been found nonmutagenic.[202]

1.5.7 Toxicity—Carcinogenicity

No data has been found regarding a possible association between cancer and B exposure in humans. The studies available in animals are inadequate to ascertain whether B has the potential to cause cancer. On the contrary, inhibition of cell proliferation of some human cancer lines by application of boric acid has been reported.[158]

1.5.8 On the Medical Geology of Boron

Naturally occurring and man-accentuated concentrations of major and trace elements display appreciable geographical variations. Adapting to such variations presents a challenge to the health and well-being of humans. Some territories are known for their specific environmental/geochemical background, i.e., deficiency or excess of available forms of some trace and major chemical elements, element misbalances or excessive mobility/bioavailability of certain element compounds. These imbalances may result in so-called "endemic" syndromes and diseases among grazing cattle and humans. The information on B on this topic is scarce and vague, if any reliable data exist at all.[205] Kovalsky[46,47] thought that the productivity of grazing cattle might be correlated to an excess or deficiency of B in the environment. Skukovsky and Kovalsky[206] described a "boric biogeochemical province" in the steppe region of western Siberia and north-eastern Kazakhstan. The authors noted that the increased B content in the soils resulted in B enrichment of the local vegetation and of the blood and tissues of grazing cattle, which, along with—according to the authors—copper deficiency caused "boric enteritis," anemia, and pulmonary diseases.

To my knowledge, the landscape geochemical—medical studies in the former USSR in the 1960—1970s and then in China were based mostly on a statistical correlation between local background concentrations of chemical elements in parent rocks, soils, natural waters, and stream sediments and locally distributed epidemic-like diseases and syndromes ("endemic diseases"). Thornton[207] reviewed national geochemical atlases of geochemistry—health connections based on stream sediments and soil surveys. He noted the advantages and limitations of such surveys to human health studies and emphasized that many of the relations established between the abundance of chemical elements in the environment and human diseases are empirical rather than causal.

More recent etiological investigations of endemically distributed diseases in China, accompanied with biochemical investigations on grazing cattle and humans have given a more profound base on possible B role in some diseases. Peng et al.[208] suggested that a B (along with germanium) deficiency may be a contributing factor in the etiology of Kashin—Beck disease. At the same time, the results of the study regarding the health effect of slightly elevated B in the drinking water of northern France showed a tendency toward a beneficial effect with low-dose B environmental exposure (still less than 1 mg/L) in drinking water.[209]

ACKNOWLEDGMENTS

Financial support provided by Misrad ha-Klita, State of Israel, is highly appreciated.

REFERENCES

1. Weintraub E. Preparation and properties of pure boron. *Trans Am Electrochem Soc* 1909;**16**:165—84.
2. Laubengayer AW, Hurd DT, Newkirk AE, Hoard JL. Boron. I. Preparation and properties of pure crystalline boron. *J Am Chem Soc* 1943;**65**(10):1924—31.
3. Zhai M, Shaw DM. Boron cosmochemistry. Part I: Boron in meteorites. *Meteoritics* 2012;**29**(5):607—15.
4. Perelman AI. *Geochemistry.* Moscow: Vysshaya Shkola; 1979 [in Russian].

5. Garavelli A, Vurro F. Barberiite, NH_4BF_4, a new mineral from Vulcano, Aeolian Islands, Italy. *Am Mineral* 1994;**79**(3–4):381–4.
6. White DE. Magmatic, connate, and metamorphic waters. *Geol Soc Am Bull* 1957;**68**:1659–82.
7. Fujita N, Shinkai S, James TD. Boronic acids in molecular self-assembly. *Chem An Asian J* 2008;**3**(7):1076–91.
8. Greenwood NN, Earnshaw A. *Chemistry of the elements*, vol. 1. Oxford: Pergamon Press; 1984.
9. Johnson AL. Boron. *Annu Rep Sect "A" Inorg Chem* 2005;**101**:34–53.
10. Steinberg H, Brotherton RJ. *Organoboron chemistry*. Interscience Publishers; 1964.
11. Garbett K, Darnall DW, Klotz IM. The effects of bound anions on the reactivity of residues in hemerythrin. *Arch Biochem Biophy* 1971;**142**(2):455–70.
12. Greenwood NN. Boron. In: *Comprehensive inorganic chemistry*, vol. 1; 1973. pp. 665–991.
13. Hunt CD. Boron-binding-biomolecules: a key to understanding the beneficial physiologic effect of dietary boron form prokaryotes to humans. In: Goldbach HE, et al., editors. *Boron in plant and animal nutrition*. NY: Kluwer Academic/Plenum Publishers; 2002. pp. 21–36.
14. Krebs B, Hammerschmidt A, Doch M. Thio- and selenoborates: from rings to clusters and networks. *Phys Chem Glasses Eur J Glass Sci Technol Part B* 2003;**44**(2):132–4.
15. Woods WG. An introduction to boron: history, sources, uses, and chemistry. *Environ Health Perspect* 1994;**102**(Suppl. 7):5–11.
16. Power PP, Woods WG. The chemistry of boron and its speciation in plants. *Plant Soil* 1997;**193**(1):1–13.
17. Dembitsky VM, Smoum R, Al-Quntar AA, Ali HA, Pergament I, Srebnik M. Natural occurrence of boron-containing compounds in plants, algae and microorganisms. *Plant Sci* 2002;**163**(5):931–42.
18. Boron Minerals and Chemicals. *Chemical economic handbook*. Menlo Park, CA: SRA International; 1990.
19. Boron. *U.S. Geological Survey, Mineral Commodity Summaries*; January 2012. http://minerals.usgs.gov/minerals/pubs/commodity/boron/mcs-2012-boron.pdf.
20. Downing RG, Strong PL, Hovanec BM, Northington J. Considerations in the determination of boron at low concentrations. *Biol Trace Elem Res* 1998;**66**(1):3–21.
21. Thellier M, Chevallier A, His I, Jarvis MC, Lovell MA, Ripoll C, et al. Methodological developments for application to the study of physiological boron and to boron neutron capture therapy. *J Trace Microprobe Tech* 2001;**19**(4):623–57.
22. Sah RN, Brown PH. Boron determination – a review of analytical methods. *Microchem J* 1997;**56**(3):285–304.
23. López FJ, Giménez E, Hernández F. Analytical study on the determination of boron in environmental water samples. *Fresenius J Anal Chem* 1993;**346**(10–11):984–7.
24. van Staden JKF, van der Merwe TA. Automated in situ preparation of Azomethine-H and the subsequent determination of boron in fertilizer process and water effluent streams with sequential injection analysis. *Analyst* 2000;**125**(11):2094–9.
25. Gladney ES, Jurney ET, Curtis DB. Nondestructive determination of boron and cadmium in environmental materials by thermal neutron-prompt γ-ray spectrometry. *Anal Chem* 1976;**48**(14):2139–42.
26. Bishop M, Shahid N, Yang J, Barron AR. Determination of the mode and efficacy of the cross-linking of guar by borate using MAS ^{11}B NMR of borate cross-linked guar in combination with solution ^{11}B NMR of model systems. *Dalt Trans* 2004;**17**:2621–34.
27. a. Matsunaga T, Ishii T, Watanabe H. Speciation of water-soluble boron compounds in radish roots by size exclusion HPLC/ICP-MS. *Anal Sci* 1996;**12**:672–6;
b. Gaspar A, Lucio M, Harir M, Schmitt-Koplin Ph. Targeted and non-targeted boron complex formation followed by electrospray Fourier transform ion cyclotron mass spectrometry: a novel approach for identifying boron esters with natural organic matter. *Eur J Mass Spectrom* 2011;**17**:113–23.
28. Seyfried WE, Janecky DR, Mottl MJ. Alteration of the oceanic crust: implications for geochemical cycles of lithium and boron. *Geochim Cosmochim Acta* 1984;**48**(3):557–69.
29. You CF, Spivack AJ, Smith JH, Gieskes JM. Mobilization of boron in convergent margins: implications for the boron geochemical cycle. *Geology* 1993;**21**(3):207–10.

30. Klee RJ, Graedel TE. Elemental cycles: a status report on human or natural dominance. *Annu Revue Environ Resour* 2004;**29**:69–107.
31. Lemarchand D, Gaillardet J, Lewin E, Allègre CJ. The influence of rivers on marine boron isotopes and implications for reconstructing past ocean pH. *Nature* 2000;**408**(6815):951–4.
32. Vengosh A, Kolodny Y, Starinsky A, Chivas AR, McCulloch MT. Coprecipitation and isotopic fractionation of boron in modern biogenic carbonates. *Geochim Cosmochim Acta* 1991;**55**(10):2901–10.
33. Ishikawa T, Nakamura E. Boron isotope systematics of marine sediments. *Earth Planet Sci Lett* 1993;**117**(3):567–80.
34. Kolodny Y, Chaussidon M. Boron isotopes in DSDP cherts: fractionation and diagenesis. *Geochem Soc Spec Publ* 2004;**9**:1–14.
35. Argust P. Distribution of boron in the environment. *Biol Trace Elem Res* 1998;**66**(1):131–43.
36. Fogg TR, Duce RA. Boron in the troposphere: distribution and fluxes. *J Geophys Res* 1985;**90**(D2):3781–96.
37. Koutz FR. *Boron geochemistry of Central American volcanoes* [Doctoral dissertation]. Dartmouth College; 1971.
38. Park H, Schlesinger WH. Global biogeochemical cycle of boron. *Glob Biogeochem Cycles* 2002;**16**(4):1072–82.
39. Anderson DL, Kitto ME, McCarthy L, Zoller WH. Sources and atmospheric distribution of particulate and gas-phase boron. *Atmos Environ* 1994;**28**(8):1401–10.
40. Kot FS. Boron sources, speciation and its potential impact on health. *Rev Environ Sci Biotechnol* 2009;**8**(1):3–28.
41. Spivack AJ, Palmer MR, Edmond JM. The sedimentary cycle of the boron isotopes. *Geochim Cosmochim Acta* 1987;**51**(7):1939–49.
42. Kopf A, Deyhle A. Back to the roots: boron geochemistry of mud volcanoes and its implications for mobilization depth and global B cycling. *Chem Geol* 2002;**192**(3):195–210.
43. Kot FS, Farran R, Kochva M, Shaviv A. Boron in humus and inorganic components of Hamra and Grumosol soils irrigated with reclaimed wastewater. *Soil Res* 2012;**50**(1):30–43.
44. Turekian KK, Wedepohl KH. Distribution of the elements in some major units of the earth's crust. *Geol Soc Am Bull* 1961;**72**(2):175–92.
45. Lemarchand E, Schott J, Gaillardet J. Boron isotopic fractionation related to boron sorption on humic acid and the structure of surface complexes formed. *Geochim Cosmochim Acta* 2005;**69**(14):3519–33.
46. Kovalsky VV. Geochemical ecology and problems of health. *Phil Trans R Soc B Biol Sci* 1979;**288**:185–91.
47. Kovalsky VV. *Geochemical environment and life*. Moscow: Nauka; 1982 [in Russian].
48. Ozol AA. *Sedimentary and volcanogenic-sedimentary genesis of boron ores*. Moscow: Nauka; 1983 [in Russian].
49. Goodarzi F, Swain DJ. Paleoenvironmental and environmental implications of the boron content of coals. *Geol Surv Can Bull* 1994;**471**. 76 pp.
50. Gulyayeva IA, Kaplun VB, Shishenina EP. Distribution of boron among the components of petroleum. *Geochem Int* 1966;**3**:636–41.
51. Shendrikar AD, Faudel GB. Distribution of trace metals during oil shale retorting. *Environ Sci Technol* 1978;**12**(3):332–4.
52. Stollenwerk KG, Runnells DD. Composition of leachate from surface-retorted and unretorted Colorado oil shale. *Environ Sci Technol* 1981;**15**(11):1340–6.
53. Miyata Y, Tokieda T, Amakawa H, Uematsu M, Nozaki Y. Boron isotope variations in the atmosphere. *Tellus B* 2000;**52**(4):1057–65.
54. a. Rose-Koga EF, Sheppard SM, Chaussidon M, Carignan J. Boron isotopic composition of atmospheric precipitations and liquid–vapour fractionations. *Geochim Cosmochim Acta* 2006;**70**(7):1603–15;

    b. Boyd CE, Walley WW. Studies of the biogeochemistry of boron. I. Concentrations in surface waters, rainfall and aquatic plants. *Am Midl Nat* 1972;**80**(1):1–14.

55. Demuth N, Heumann KG. Determination of trace amounts of boron in rainwater by ICP-IDMS and NTI-IDMS and the dependence on meteorological and anthropogenic influences. *J Anal At Spectrosc* 1999;**14**(9):1449–53.
56. Murphy DM, Thomson DS, Mahoney MJ. In situ measurements of organics, meteoritic material, mercury, and other elements in aerosols at 5 to 19 kilometers. *Science* 1998; **282**(5394):1664–9.
57. Gaillardet J, Lemarchand D, Göpel C, Manhès G. Evaporation and sublimation of boric acid: application for boron purification from organic rich solutions. *Geostand Newslett* 2001;**25**(1):67–75.
58. Braker W, Mossman AL. *Matheson gas data book*. East Rutherford, NJ: Matheson; 1980. pp. 476–80.
59. Ellis AJ, Mahon WAJ. *Chemistry and geothermal systems*. Academic Press; 1977.
60. Fyfe WS, Price NJ, Thompson AB. *Fluids in the earth's crust*. Amsterdam: Elsevier Science Publisher; 1978.
61. Chetelat B, Gaillardet J. Boron isotopes in the Seine River, France: a probe of anthropogenic contamination. *Environ Sci Technol* 2005;**39**(8):2486–93.
62. Rose EF, Chaussidon M, France-Lanord C. Fractionation of boron isotopes during erosion processes: the example of Himalayan rivers. *Geochim Cosmochim Acta* 2000;**64**:397–408.
63. Martin JM, Meybeck N. Elemental mass-balance of material carried by major world rivers. *Mar Chem* 1979;**7**:173–206.
64. Neal C, Robson AJ. A summary of river water quality data collected within the Land–Ocean Interaction Study: core data for eastern UK rivers draining to the North Sea. *Sci Total Environ* 2000;**251**:585–665.
65. Wyness AJ, Parkman RH, Neal C. A summary of boron surface water quality data throughout the European Union. *Sci Total Environ* 2003;**314**:255–69.
66. Xing X, Zhang H, Wu G, Wei F. Analysis of boron concentration and pollution status of water in Kuandian County. *Chin J Ecol* 2005;**24**:327–9.
67. Kloppmann W, Pennisi M, Bianchini G, Muti A, Cerbai N, Vengosh A, et al. Boron contamination of water resources in the Mediterranean region: distribution, sources, social impact and remediation: the BOREMED project. In: *Proceeding of International Conference Hydrology of the Mediterranean and Semi Arid Regions, Montpellier*; 2003.
68. Vengosh A, Weinthal E, Kloppmann W. Natural boron contamination in Mediterranean groundwater. *Geotimes* 2004;**49**(5):20–5.
69. Sánchez-Martos F, Pulido-Bosch A. Boron and the origin of salinization in an aquifer in Southeast Spain. *C R Acad Sci Ser IIA–Earth Planet Sci* 1999;**328**(11):751–7.
70. Çöl M, Çöl C. Environmental boron contamination in waters of Hisarcik area in the Kutahya Province of Turkey. *Food Chem Toxicol* 2003;**41**(10):1417–20.
71. Wilson TRS. Salinity and the major elements of sea water. In: Riley JP, Skirrow G, editors. *Chemical oceanography*, vol. 1. London: Academic Press; 1975. pp. 365–413.
72. Holmén K. The global carbon cycle. *Int Geophys* 2000;**72**:282–321.
73. Collins AG. *Geochemistry of oilfield waters*. New York: Elsevier; 1975.
74. Goldschmidt VM, Peters C, Hauptmann H. Zur geochemie des bors. *Nachr Ges Wiss Göttingen, Math-Phys Kl* 1932:402–7.
75. Uppström LR. The boron/chlorinity ratio of deep-sea water from the Pacific Ocean. *Deep Sea Res Oceanogr Abstr* 1974;**21**(2):161–2.
76. Brunskill GJ, Zagorskis I, Pfitzner J. Geochemical mass balance for lithium, boron, and strontium in the Gulf of Papua, Papua New Guinea (Project TROPICS). *Geochim Cosmochim Acta* 2003;**67**(18):3365–83.
77. Mackenzie FT, Garrels RM. Chemical mass balance between rivers and oceans. *Am J Sci* 1966;**264**(7):507–25.
78. Riley JP, Skirrow G, editors. *Chemical oceanography*, vols. 2 & 3. London: Academic Press; 1975.
79. Bassett RL, Buszka PM, Davidson GR, Chong-Diaz D. Identification of groundwater solute sources using boron isotope composition. *Environ Sci Technol* 1985;**29**:2915–22.
80. Gast JA, Thompson TG. Determination of the alkalinity and borate concentration of sea water. *Anal Chem* 1958;**30**:1549–51.

81. Melon MG, Morris VN. An electronic study of the titration of boric acid. *Ind Eng Chem* 1924;**16**:123–6.
82. Noakes JE, Hood DW. Boron-boric acid complexes in sea-water. *Deep Sea Res (1953)* 1961;**8**(2):121–9.
83. Williams PM, Strack PM. Complexes of boric acid with organic cis-diols in seawater. *Limnol Oceanogr* 1966;**11**(3):401–4.
84. Jackson TA. Humic matter in natural water and sediments. *Soil Sci* 1975;**119**(1):56–64.
85. North BB. Primary amines in California coastal waters: utilization by phytoplankton. *Limnol Oceanogr* 1975;**20**(1):20–7.
86. Degens ET, Reuter JH, Shaw KNF. Biochemicalcompounds in offshore California sediments and sea waters. *Geochim Cosmochim Acta* 1964;**28**:45–66.
87. Nissenbaum A, Presley BJ, Kaplan IR. Early diagenesis in a reducing fjord, Saanich Inlet, British Columbia. I. Chemical and isotopic changes in major components of interstitial water. *Geochim Cosmochim Acta* 1972;**36**(9):1007–27.
88. Menzel DW. The distribution of dissolved organic carbon in the Western Indian Ocean. *Deep Sea Res Oceanogr Abstr* 1964;**11**(5):757–65.
89. Druffel ER, Williams PM, Bauer JE, Ertel JR. Cycling of dissolved and particulate organic matter in the open ocean. *J Geophys Res Oceans* 1992;**97**(C10):15639–59.
90. Förstner U, Wittmann GT, Prosi F, van Lierde JH. *Metal pollution in the aquatic environment.* 2nd ed. Berlin: Springer-Verlag; 1987.
91. Frederickson AF, Reynolds RC. Geochemical method for determining paleosalinity. *Clays Clay Miner* 1960;**8**:203–13.
92. Perry EA. Diagenesis and the validity of the boron paleosalinity technique. *Am J Sci* 1972;**272**(2):150–60.
93. Williams LB, Hervig RL, Wieser ME, Hutcheon I. The influence of organic matter on the boron isotope geochemistry of the gulf coast sedimentary basin, USA. *Chem Geol* 2001;**174**(4):445–61.
94. Health Canada. *Guidelines for Canadian drinking water quality.* 6th ed. Ottawa: Minister of supply and services, 96-WHO-196; 1996.
95. Weinthal E, Parag Y, Vengosh A, Muti A, Kloppmann W. The EU drinking water directive: the boron standard and scientific uncertainty. *Eur Environ* 2005;**15**:1–12.
96. SanPiN. *Sanitary regulations and norms of Russian Federation*; 2001. No. 2.1.4.1074–01 [in Russian].
97. EPA, Office of Water. *Drinking water regulations and health advisories.* EPA-822-B-96–002; 1996., http://www.epa.gov/OST/Tools/dwstds.html.
98. Boron. *Environmental health criteria 204.* Geneva: World Health Organization; 1998.
99. Jahiruddin M, Smart R, Wade AJ, Neal C, Cresser MS. Factors regulating the distribution of boron in water in the River Dee catchment in north east Scotland. *Sci Total Environ* 1998;**210**:53–62.
100. Chetelat B, Gaillardet J, Freydier R, Négrel P. Boron isotopes in precipitation: experimental constraints and field evidence from French Guiana. *Earth Planet Sci Lett* 2005;**235**(1):16–30.
101. Lemarchand D, Gaillardet J. Transient features of the erosion of shales in the Mackenzie basin (Canada), evidences from boron isotopes. *Earth Planet Sci Lett* 2006;**245**(1):174–89.
102. a. Gustafsson O, Gschwend PM. Aquatic colloids: concepts, definitions, and current challenges. *Limnol Oceanogr* 1997;**42**(3):519–28;
b. Perelman AI. *Geochemistry of natural waters.* Moscow: Nauka; 1982 [in Russian].
103. Lead JR, Davison W, Hamilton-Taylor J, Buffle J. Characterizing colloidal material in natural waters. *Aquat Geochem* 1997;**3**(3):213–32.
104. Wilkinson MC, Hearn J, Steward PA. The cleaning of polymer colloids. *Adv Colloid Interface Sci* 1999;**81**(2):77–165.
105. Kretzschmar R, Schäfer T. Metal retention and transport on colloidal particles in the environment. *Elements* 2005;**1**(4):205–10.
106. Leenheer JA, Croué JP. Peer reviewed: characterizing aquatic dissolved organic matter. *Environ Sci Technol* 2003;**37**(1):18–26.
107. Chauveheid E, Denis M. The boron–organic carbon correlation in water. *Water Res* 2004;**38**(7):1663–8.

108. Jaquet JM, Gavaud E, Vernet JP. Basic concepts and associated statistical methodology in the geochemical study of lake sediments. *Hydrobiologia* 1982;**91**(1):139–46.
109. Stewart KC, Wilson SA, Severson RC. *Total and water extractable boron in sediments from nine sites of the western United States*. US Geological Survey; 1989.
110. Mun AI, Zhaimina RE. Distribution of boron in modern lake sediments. *Geochem Int* 1962;**11**:1136–44.
111. Hem JD. *Study and interpretation of the chemical characteristics of natural water.* U.S. Geological Survey Water Supplement Paper 2254. 3rd ed. 1985.
112. Barth S. Application of boron isotopes for tracing sources of anthropogenic contamination in groundwater. *Water Res* 1998;**32**:685–90.
113. Vengosh A. The isotopic composition of anthropogenic boron and its potential impact on the environment. *Biol Trace Elem Res* 1998;**66**:145–51.
114. Leyshon AJ, Jame Y-W. Boron toxicity and irrigation management. In: Gupta UC, editor. *Boron and its role in crop production*. Boca Raton (FL): CRC Press; 1993. pp. 207–26.
115. Gupta UC. *Boron and its role in crop production*. Boca Raton (FL): CRC Press; 1993.
116. Vengosh A, Heumann KG, Juraske S, Kasher R. Boron isotope application for tracing sources of contamination in groundwater. *Environ Sci Technol* 1994;**28**(11):1968–74.
117. Kovda VA. *Fundamentals of soil science*, vol. 2. Moscow: Nauka; 1973 [in Russian].
118. Goldschmidt VM. The principles of distribution of chemical elements in minerals and rocks. The seventh Hugo Müller Lecture, delivered before the Chemical Society on March 17, 1937 *J Chem Soc* 1937:655–73.
119. Viets Jr FG. Micronutrient availability, chemistry and availability of micronutrients in soils. *J Agric Food Chem* 1962;**10**(3):174–8.
120. Berger KC, Pratt PF. Advances in secondary and micronutrient fertilization. In: *Fertilizer technology and usage*. Madison (WI): Soil Society of America; 1963. pp. 287–340.
121. a. Keren R, Bingham FT. Boron in water, soils, and plants. In: *Advances in soil science*. New York: Springer; 1985. pp. 229–76;
b. Kabata-Pendias A, Pendias H. *Trace elements in soils and plants*. Boca Raton (FL): CRC Press; 2001.
122. Reisenauer HM, Walsh LM, Hoeft RG. Testing soils for sulphur, boron, molybdenum, and chlorine. *Soil Test Plant Anal* 1973:173–200.
123. Nable RO, Bañuelos GS, Paull JG. Boron toxicity. *Plant Soil* 1997;**193**(1–2):181–98.
124. Goldberg S. Reactions of boron with soils. *Plant Soil* 1997;**193**(1–2):35–48.
125. Raza M, Mermut AR, Schoenau JJ, Malhi SS. Boron fractionation in some Saskatchewan soils. *Can J Soil Sci* 2002;**82**(2):173–9.
126. Hingston FJ. Reactions between boron and clays. *Soil Res* 1964;**2**(1):83–95.
127. Yermiyahu U, Keren R, Chen Y. Effect of composted organic matter on boron uptake by plants. *Soil Sci Soc Am J* 2001;**65**(5):1436–41.
128. Yermiyaho U, Keren R, Chen Y. Boron sorption on composted organic matter. *Soil Sci Soc Am J* 1988;**52**(5):1309–13.
129. Hue NV, Hirunburana N, Fox RL. Boron status of Hawaiian soils as measured by B sorption and plant uptake. *Commun Soil Sci Plant Anal* 1988;**19**(5):517–28.
130. Liu Z. Regularities of content and distribution of boron in soils. *Acta Pedol Sin* 1989;**26**:353–61.
131. Keren R, Communar G. Boron transport in soils as affected by dissolved organic matter in treated sewage effluent. *Soil Sci Soc Am J* 2009;**73**(6):1988–94.
132. Asad A, Bell RW, Dell B, Huang L. Development of a boron buffered solution culture system for controlled studies of plant boron nutrition. *Plant Soil* 1997;**188**(1):21–32.
133. Hu H, Brown PH. Absorption of boron by plant roots. *Plant Soil* 1997;**193**(1–2):49–58.
134. Treeby M, Marschner H, Römheld V. Mobilization of iron and other micronutrient cations from a calcareous soil by plant-borne, microbial, and synthetic metal chelators. *Plant Soil* 1989;**114**(2):217–26.
135. Biomin Boron. http://www.jhbiotech.com/Product%20Labels/WA%20pdf%20files/Biomin%20Boron.pdf.
136. Glycine technology – the answer to zinc and boron applications. http://www.rd2.co.nz/uploads/Glycine%20technology%20-%20Changes.pdf.

137. Kot FS, Fujiwara K, Kochva M, Shaviv A, Sugo T. *On soil boron turnover in soil—plant system with emphasis on organic matter.* Abstracts of BIOGEOMON, 8th International Symposium on Ecosystem Behaviour, July 13—17, 2014. Germany: University of Bayreuth; 2014.
138. Rhoades JD, Ingvalson RD, Hatcher JT. Laboratory determination of leachable soil boron. *Soil Sci Soc Am J* 1970;**34**(6):871—5.
139. Maluga DP. *Biogeochemical methods of prospecting.* New York: Consultants Bureau, Plenum Press; 1964.
140. Bowen JE, Gauch HG. Nonessentiality of boron in fungi and the nature of its toxicity. *Plant Physiol* 1966;**41**(2):319—24.
141. Saiki MK, Jennings MR, Braumbaugh WG. Boron, molybdenum, and selenium in aquatic food chains from the lower San Joaquin river and its tributaries, California. *Arch Environ Contam Toxicol* 2005;**24**:307—19.
142. Underwood EJ. Trace metals in human and animal health. *Int J Food Sci Nutr* 1981;**35**(1):37—48.
143. Bonilla I, Garcia-González M, Mateo P. Boron requirement in cyanobacteria. Its possible role in the early evolution of photosynthetic organisms. *Plant Physiol* 1990;**94**:1554—60.
144. Lewis DH. Boron, lignification and the origin of vascular plants — a unified hypothesis. *New Phytol* 1980;**84**(2):209—29.
145. Pfeiffer CC, Braverman ER. Epochal trace elements and evolution. *Agents Actions* 1982;**12**(3):412—5.
146. Voicu J. Influence de l'humus sur la sensibilité de l'Azotobacter chroococcum vis-à-vis du bore. *C R Acad Sci* 1922;**175**:317—9.
147. Gerretsen FC, Hoop HD. Nitrogen losses during nitrification in solutions and in acid sandy soils. *Can J Microbiol* 1957;**3**(2):359—80.
148. a. Carrano CJ, Schellenberg S, Amin SA, Green DH, Küpper FC. Boron and Marine life: a new look at an enigmatic bioelement. *Mar Biotechnol Int J Focus Mar Genomics, Mol Biol Biotechnol* 2009;**11**(4):431—40;
b. Anderson GR, Jordan JV. Boron: a non-essential growth factor for *Azotobacter chroococcum*. *Soil Sci* 1961;**92**(2):113—6.
149. Mateo P, Bonilla I, Fernandez-Valiente E, Sanchez-Maeso E. Essentiality of boron for dinitrogen fixation in *Anabaena sp. PCC 7119*. *Plant Physiol* 1986;**81**(2):430—3.
150. Irschik H, Schummer D, Gerth K, Hoefle G, Reuchenbach H. The tartrolons, new boron-containing antibiotics from a myxobacterium, *Sorangium cellulosum*. *J Antibiot* 1995;**48**:26—30.
151. Kohno J, Kawahata T, Otake T, Morimoto M, Mori H, Ueba N, et al. Boromycin, an anti-HIV antibiotic. *Biosci Biotechnol Biochem* 1996;**60**:1036—7.
152. Ahmed I, Yokota A, Fujiwara T. *Gracilibacillus boraciitolerans sp. nov.*, a highly boron-tolerant and moderately halotolerant bacterium isolated from soil. *Int J Syst Evol Microbiol* 2007; **57**(4):796—802.
153. Nelson DR, Mele PM. Subtle changes in rhizosphere microbial community structure in response to increased boron and sodium chloride concentrations. *Soil Biol Biochem* 2007;**39**(1):340—51.
154. Subba Rao DV. Effect of boron on primary production of nanoplankton. *Can J Fish Aquat Sci* 1981;**38**:52—8.
155. Bennet A, Rowe RI, Soch N, Eckert CD. Boron stimulates yeasts (*Saccharomyces cerevisiae*) growth. *J Nutr* 1999;**129**(12):2236—8.
156. Bolaños L, Esteban E, de Lorenzo C, Fernandez-Pascual M, de Felipe MR, et al. Essentiality of boron for symbiotic dinitrogen fixation in pea (*Pisum sativum*) rhizobium nodules. *Plant Physiol* 1994;**104**(1):85—90.
157. Chen X, Schauder S, Potier N, van Dorsselaer A, Pelczer I, Bassler B, et al. Structural identification of a bacterial quorum-sensing signal containing boron. *Nature* 2002;**415**:545—9.
158. Eckhert CD, Barranco WT, Kim DH. Boron in animals and humans: prostate cancer a model for understanding boron biology. In: Fangsen X, editor. *Advances in plant and animal boron nutrition.* Springer-Verlag; 2007. pp. 291—7.
159. Agulhon H. Presence et utilité du bore chez les vegetaux. *Ann Inst Pasteur (Paris)* 1910;**24**:321—9.
160. Mazé P. Influences respectives des éléments de la solution minérale sur le dévelopement du maïs. *Ann Inst Pasteur (Paris)* 1914;**28**:21—68.
161. Warington K. The effect of boric acid and borax on the broad bean and certain other plants. *Ann Bot (London)* 1923;**37**:629—72.

162. Loomis WD, Durst RW. Chemistry and biology of boron. *BioFactors (Oxford, England)* 1992;**3**(4):229–39.
163. Goldbach HE, Wimmer M. Boron in plants and animals: is there a role beyond cell-wall structure? *J Plant Nutr Soil Sci* 2007;**170**:39–48.
164. Blevins DG, Lukaszewski KM. Boron in plant structure and function. *Annu Rev Plant Biol* 1998;**49**(1):481–500.
165. Bolaños L, Lukaszewski K, Bonilla I, Blevins D. Why boron? *Plant Physiol Biochem* 2004;**42**(11):907–12.
166. Brown PH, Bellaloui N, Wimmer MA, Bassil ES, Ruiz J, Hu H, et al. Boron in plant biology. *Plant Biol* 2008;**4**(2):205–23.
167. Teasdale RD, Richards DK. Boron deficiency in cultured pine cells quantitative studies of the interaction with Ca and Mg. *Plant Physiol* 1990;**93**(3):1071–7.
168. Parr AJ, Loughman BC. Boron and membrane function in plants. In: Robb DA, Pierpoint WS, editors. *Metals and Micronutrients: Uptake and Utilization by Plants*. London, UK: Academic Press; 1983. pp. 87–107.
169. Cakmak I, Römheld V. Boron deficiency-induced impairments of cellular functions in plants. *Plant Soil* 1997;**193**(1–2):71–83.
170. Dell B, Huang L. Physiological response of plants to low boron. *Plant Soil* 1997;**193**(1–2):103–20.
171. Askew HO, Chittenden E, Thomson RHK. *The use of borax in the control of internal cork of apples*. Government Printer; 1936.
172. Gupta UC, Cutcliffe JA. Effects of methods of boron application on leaf tissue concentration of boron and control of brown-heart in rutabaga. *Can J Plant Sci* 1978;**58**(1):63–8.
173. Dordas C, Chrispeels MJ, Brown PH. Permeability and channel-mediated transport of boric acid across membrane vesicles isolated from squash roots. *Plant Physiol* 2000;**124**(3):1349–62.
174. Stangoulis JC, Reid RJ, Brown PH, Graham RD. Kinetic analysis of boron transport in Chara. *Planta* 2001;**213**(1):142–6.
175. Dannel F, Pfeffer H, Römheld V. Update on boron in higher plants-uptake, primary translocation and compartmentation. *Plant Biol* 2002;**4**(2):193–204.
176. Aphalo PJ, Schoettle AW, Lehto T. Leaf life span and the mobility of "non-mobile" mineral nutrients-the case of boron in conifers. *Silva Fenn* 2002;**36**(3):671–80.
177. Brown PH, Shelp BJ. Boron mobility in plants. *Plant Soil* 1997;**193**(1–2):85–101.
178. Rainey CJ, Nyquist LA, Christensen RE, Strong PL, Culver BD, Coughlin JR. Daily boron intake from the American diet. *J Am Diet Assoc* 1999;**99**:335–40.
179. Samman S, Foster M, Hunter D. The role of boron in human nutrition and metabolism. In: Hosmane NS, editor. *Boron science: new technologies and applications*. Boca Raton, FL: CRC Press; 2011. pp. 73–82.
180. Naghii MR, Wall PML, Verus AP. The nutritional and metabolic effects of boron in humans and animals. *Biol Trace Elem Res* 1998;**66**(1):227–35.
181. MAFF. *1994 total diet study: metals and other elements*. Food Surveillance Information Sheet No. 131; 1997.
182. Beyer KH, Bergfeld WF, Berndt WO, Boutwell RK, Carlton WW, Hoffman DK, et al. Final report on the safety assessment of sodium borate and boric acid. *J Am Coll Toxicol* 1983;**2**(7):87–125.
183. Moore JA. An assessment of boric acid and borax using the IEHR evaluative process for assessing human developmental and reproductive toxicity of agents. Expert Scientific Committee *Reprod Toxicol* 1997;**7**:305–19 [Elmsford, NY].
184. EU Technical Guidance Document 1995. *European Commission: risk assessment of new and existing substances*. Technical Guidance Document Draft; October 1995.
185. DRI. *Dietary reference intakes for vitamin A, vitamin K, arsenic, boron, chromium, copper, iodine, iron, manganese, molybdenum, nickel, silicon, vanadium, and zinc*. A Report of the Panel on Micronutrients... [et al.]. Institute of Medicine (US). Panel on Micronutrients. Food, & Nutrition Board, Institute of Medicine. National Academies Press; 2001.
186. *Trace elements in human nutrition*. Geneva: WHO/FAO/IAEA; 1996.
187. Nielsen FH. Boron in human and animal nutrition. *Plant Soil* 1997;**193**(1):199–208.
188. a. Hunt CD. Dietary boron: progress in establishing essential roles in human physiology. *J Trace Elem Med Biol* 2012;**26**:157–60;

b. Schou JS, Jansen JA, Aggerbeck B. Human pharmacokinetics and safety of boric acid. In: *Disease, metabolism and reproduction in the toxic response to drugs and other chemicals*. Berlin-Heidelberg: Springer; 1984. pp. 232–5.
189. Nielsen FH. The emergence of boron as nutritionally important throughout the life cycle. *Nutrition* 2000;**16**(7–8):512–4.
190. Rowe RI, Eckhert CD. Boron is required for zebrafish embryogenesis. *J Exp Biol* 1999;**202**(12):1649–54.
191. Lanoue L, Taubeneck MW, Muniz J, Hanna LA, Strong PL, Murray FJ, et al. Assessing the effects of low boron diets on embryonic and fetal development in rodents using in vitro and in vivo model systems. *Biol Trace Elem Res* 1998;**66**(1):271–98.
192. Bai Y, Hunt CD. Dietary boron enhances the efficacy of cholecalciferol in broiler chicks. *J Trace Elem Exp Med* 1996;**9**(3):117–32.
193. Penland JG. The importance of boron nutrition for brain and psychological function. *Biol Trace Elem Res* 1998;**66**(1):299–317.
194. Fort DJ, Stover EL, Strong PL, Murray FJ, Keen CL. Chronic feeding of a low boron diet adversely affects reproduction and development in *Xenopus laevis*. *J Nutr* 1999;**129**(11):2055–60.
195. Hegsted M, Keenan MJ, Siver F, Wozniak P. Effect of boron on vitamin D deficient rats. *Biol Trace Elem Res* 1991;**28**(3):243–55.
196. Kent NL, McCance RA. The absorption and excretion of minor elements by man: silver, gold, lithium, boron and vanadium. *Biochem J* 1941;**35**(7):837–44.
197. Weir Jr RJ, Fisher RS. Toxicologic studies on borax and boric acid. *Toxicol Appl Pharmacol* 1972;**23**(3):351–64.
198. Young EG, Smith RP, MacIntosh OC. Boric acid as a poison: report of six accidental deaths in infants. *Can Med Assoc J* 1949;**61**(5):447–50.
199. Goldbloom RB, Goldbloom A. Boric acid poisoning: report of four cases and a review of 109 cases from the world literature. *J Pediatr* 1953;**43**(6):631–43.
200. MacGillivray PC, Fraser MS. Boric acid poisoning in infancy arising from the treatment of napkin rash. *Arch Dis Child* 1953;**28**(142):484–9.
201. Culver BD, Shen PT, Taylor TH, Lee-Feldstein A, Anton-Culver H, Strong PL. The relationship of blood-and urine-boron to boron exposure in borax-workers and usefulness of urine-boron as an exposure marker. *Environ Health Perspect* 1994;**102**(Suppl. 7):133–7.
202. *Opinion on boron compounds*. European Commission Directorate-General for Health and Consumers; 2010. Revision of September, 28 2010.
203. Price CJ, Strong PL, Marr MC, Myers CB, Murray FJ. Developmental toxicity NOAEL and postnatal recovery in rats fed boric acid during gestation. *Fundam Appl Toxicol* 1996;**32**:179–93.
204. Draize JH, Kelley EA. The urinary excretion of boric acid preparations following oral administration and topical applications to intact and damaged skin of rabbits. *Toxicol Appl Pharmacol* 1959;**1**(3):267–76.
205. Lindh U. Biological functions of the elements. In: Sclinus O, editor. *Essentials of medical geology: impacts of the natural environment on public health*. Amsterdam: Elsevier; 2005. pp. 115–60.
206. Skukovsky BA, Kovalsky VV. Western Siberian boric sub-region of biosphere. In: Kovalsky VV, editor. *Biogeochemical zoning and geochemical ecology*. Moscow: Trudy Biogeokhimicheskoy Laboratorii 20; 1985 [in Russian].
207. Thornton I. Specific geoenvironmental topics. Chapter 3. In: *Trace element maps and the regional diseases*. Berlin-Heidelberg: Springer; 1992.
208. Peng X, Lingxia Z, Schrauzer GN, Xiong X. Selenium, boron, and germanium deficiency in the etiology of Kashin-Beck disease. *Biol Trace Elem Res* 2000;**77**:193–7.
209. Yazbeck C, Kloppmann W, Cottier R, Sahuquillo J, Debotte G, Huel G. Health impact evaluation of boron in drinking water: a geographical risk assessment in northern France. *Environ Geochem Health* 2005;**27**:419–27.
210. US Salinity Laboratory Staff. *Diagnosis and improvement of saline and alkali soils*, vol. 6. US Department of Agriculture Handbook; 1954.

CHAPTER 2

The Chemistry of Boron in Water

Victor Kochkodan, Nawaf Bin Darwish, Nidal Hilal
Centre for Water Advanced Technologies and Environmental Research (CWATER), College of Engineering, Swansea University, Swansea, UK

2.1 BORON AND ITS CHEMICAL PROPERTIES

Boron (B) is the fifth element in the periodic table with an atomic mass of 10.81. It is the most electronegative element of Group III, and boron's chemical properties closely resemble those of the nonmetals, particularly silicon.

Pure elemental boron was first isolated simultaneously and independently in 1808 by H. Davy in England, who observed that an electric current sent through a solution of borates produced a brown precipitate on one of the electrodes, and by J. Gay-Lussac and L. Thenard in France, who obtained boron by reducing boric acid with iron at high temperatures.[1]

Elemental boron exists as a solid at room temperature, either as black monoclinic crystals or as a yellow or brown amorphous powder when impure. Amorphous boron can be obtained by the reduction of boron oxide with sodium or potassium fluoroborate with potassium[1]:

$$B_2O_3 + 6Na = 2B + 3Na_2O$$

$$KBF_4 + 3K = 4KF_4 + B$$

Crystalline boron was first prepared when hydrogen and boron bromide vapors at a rather less-than atmospheric pressure were passed over a tantalum filament heated to 1000–1300 °C.[2] At this temperature, the bromide is reduced, and the boron thus becomes deposited on the filament as black hexagonal flakes and needles:

$$2BBR_3 + 3H_2 = 6HBr + 2B$$

Two crystalline modifications of boron, namely, α-rhombohedral boron (Figure 2.1) and β-rhombohedral boron (Figure 2.2) exist at atmospheric pressure. The latter is believed to be thermodynamically stable at high temperatures, whereas α-boron is sometimes called the low-temperature form.[3]

The chemical nature of boron is influenced primarily by its small size (covalent radius of boron of 0.8–1.01 Å) and high ionization energy (344.2 kJ/mol).[1] The high affinity for oxygen is another dominant characteristic of boron, which forms the basis of the extensive chemistry of borates and related oxocomplexes.[2]

Boron Separation Processes
ISBN 978-0-444-63454-2, http://dx.doi.org/10.1016/B978-0-444-63454-2.00002-2

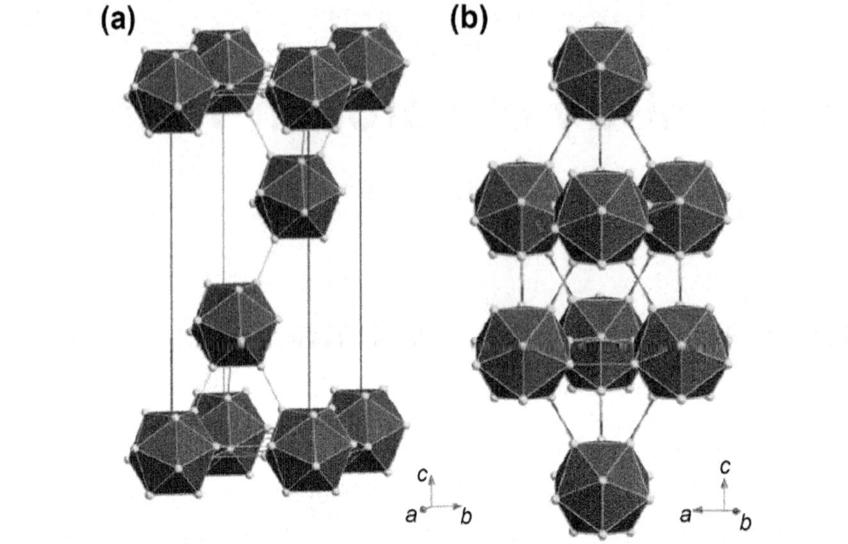

Figure 2.1 Unit cells of α-boron. (a) Hexagonal setting and (b) rhombohedral setting.[3]

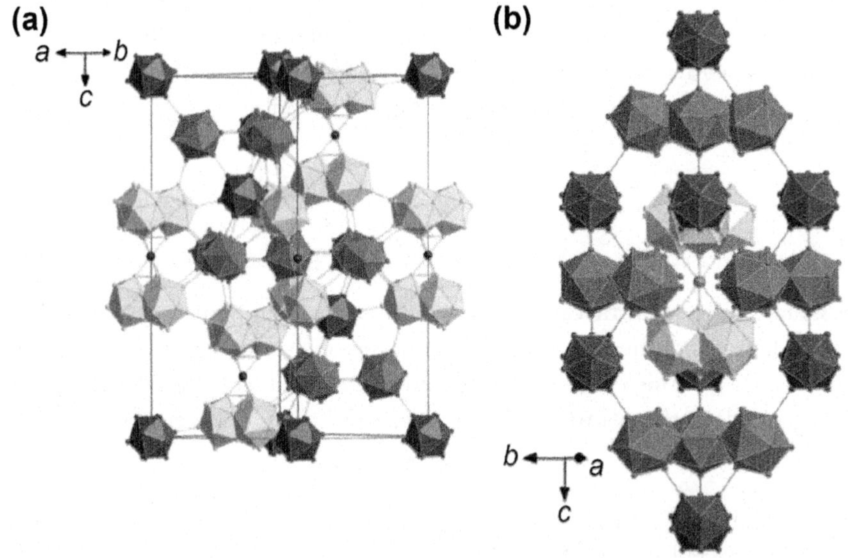

Figure 2.2 Unit cells of β-boron. (a) Hexagonal setting and (b) rhombohedral setting.[3]

The chemical properties of the boron element depend also on its morphology and particle size. Micron sized amorphous boron reacts easily and sometimes intensely, whereas crystalline boron is very inert chemically and resistant to attack even by boiling hydrofluoric or hydrochloric acid. It should be noted that boron is difficult to prepare in a state of high purity owing to its very high melting point of 2079 °C.[4] The boiling point of boron is 2250 °C, and its density is 2.37 g/m^3.[5]

The electron structure of the element is $1s^2\ 2s^2\ 2p^1$ and hence boron can form three or four valence bonds. In its most common compounds, such as oxides, sulfides, nitrides, and halides, boron has the formal oxidation state of +3. In these compounds, the bonds are coplanar, with interbond angles of 120°. The lower oxidation states +1 or 0 are present only in compounds such as higher boranes (e.g., B_5H_9), subvalent halides (e.g., B_4Cl_4), metal borides (e.g., Ti_2B), or in some compounds containing multiple B—B bonds.[2] In naturally occurring compounds, boron usually has a coordination number of either 3 or 4.

Boron salts are generally very water soluble, for example, borax has a water solubility of 25.2 g/L, while boron trifluoride is the least water-soluble boron compound, with a water solubility of 2.4 g/L.[2]

2.2 BORON IN NATURE

Boron element is composed of ^8B, ^{10}B, ^{11}B, ^{12}B, and ^{13}B isotopes. The most stable isotopes are ^{10}B and ^{11}B. The occurrence of these isotopes in nature is 19.1—20.3% and 79.7—80.9%, respectively.[1]

2.2.1 Boron in the Lithosphere

Boron is widely distributed in lithosphere of the earth.[6] It is found in rocks and soils, particularly in clay-rich marine sediments. The concentration of boron in the Earth's crust varies from 1 to 500 mg/kg, depending on the nature of the rock.[7] According to Krauskopf,[8] the average boron in the earth's crust is around 10 mg/kg, representing 0.001% of the elemental composition of the earth. The amount of boron in soils ranges from 2 to 100 mg/kg with an average of 30 mg/kg.[7] Most soils have a low boron content (<10 mg B/kg), while high-boron-content soils (10—100 mg B/kg) are usually associated with volcanic activity.

The total amount of boron stored in the lithosphere is estimated as the continental and oceanic crusts (1018 kg B), coal deposits (1010 kg B), commercial borate deposits (1010 kg B), and biomass (1010 kg B).[9]

It should be noted that boron is never found free in nature, but it invariably occurs as the oxide B_2O_3 in combination with the oxides of other elements to form borates of greater or lesser complexity. There are >200 boron compounds in the Earth, but only 12 are commercially significant.[10] The first known borate mineral to antiquity is sodium tetraborate decahydrate $Na_2B_4O_7 \times 10H_2O$ or borax (Figure 2.3(a)). An early use of borax was to make perborate, the beaching agent once widely used in household detergents. The other important boron-containing minerals are ulexite ($NaCaB_5O_9 \cdot 8H_2O$), colemanite ($Ca_2B_6O_{11} \cdot 5H_2O$) (Figure 2.3(b)), and kernite ($Na_2B_4O_7 \cdot 4H_2O$). Borax, colemanite, ulexite, and kernite provide >90% of the

Figure 2.3 Photographs of boron-containing minerals. Borax (a) and colemanite (b).

world's boron demand.[10] The occurrence of concentrated deposits of borate minerals is intimately connected with past or present volcanic activity and arid climatic conditions are essential for continued preservation of such deposits, which are being exploited in the USA, Turkey, Italy, Spain, Russia, Chile, and some other countries.

2.2.2 Boron in the Aqueous Environment

The majority of the Earth's boron is found in the oceans, with an average concentration of 4.5 mg/L, but it ranges from 0.5 to 9.6 mg/L.[11] For example, the boron content in the Mediterranean sea may be as high as 9.6 mg/L.[9]

The natural borate content of ground water and surface water is usually small and is a result of leaching from rocks and soils containing borates and borosilicates. Concentrations of boron in ground water throughout the world range widely, from <0.3 to >100 mg/L. In the European Union (EU), concentrations of boron change from 0.5 to 1.5 mg/L for southern Europe (Italy, Spain) and up to approximately 0.6 mg/L for northern Europe (Denmark, Germany, UK).[12]

The amount of boron in surface water depends on factors such as the proximity to marine coastal regions, inputs from industrial and municipal effluents, and the geochemical nature of the drainage area. Boron concentrations in surface water range from <0.001 to 2 mg/L in EU, with mean values typically below 0.6 mg/L.[12] A similar boron content has been reported for water bodies within Turkey, Russia, and Pakistan, from 0.01 to 7 mg/L, with most values being <0.5 mg/L. Concentrations of boron in surface waters of North America (Canada, USA) range from 0.02 to 360 mg/L, indicative of boron-rich deposits, up to 0.01 mg/L in Japan and up to 0.3 mg/L in South African surface waters.

A wide variation of boron concentrations in surface water is due to both natural and anthropogenic factors.[13] Natural factors include the weathering of rocks and the leaching of salt deposits. In coastal areas, rain containing sea salt from ocean spray provides another source of boron.[14]

The anthropogenic factors include water pollution with boron-containing wastes from the glass/ceramic industry (the largest market of 56% global borate demand) and metallurgy. Boron may also be discharged to the environment as drainage from disused coal mines and leaching of tips and landfills from the mining industry.[15]

Domestic wastewater effluents may be also extremely enriched in boron, with concentrations varying from several hundred micrograms per liter to several micrograms per liter.[16] By far the most common reason for this enrichment is the presence of sodium perborate that is used as a bleaching agent in detergents and cleaning products.

Boron-containing fertilizers can also be a major source of anthropogenic boron due to their widespread application. Boron is an essential micronutrient for plants and consequently is included in many fertilizers at levels ranging from 0.01 to 0.06 wt.%, most commonly in the form of borax.[17] Wyness et al.[18] found that rivers draining high-intensity agricultural areas of south-eastern England can have average boron concentrations of up to 387 μg/L.

The total distribution of boron contents in the hydrosphere was identified by Argust.[9] The boron content is 1015 kg in the oceans, 1011 kg in ground water, 1011 kg in ice and 108 kg in surface waters.

2.3 PHYSICOCHEMISTRY OF BORON COMPOUNDS IN WATER

In nature, boron is released from rocks and soils through weathering, and subsequently ends up in the aqueous environment as boric acid $B(OH)_3$ or borate ion $B(OH)_4^-$.

2.3.1 Physical Properties of Boric Acid

Boric acid was first prepared from borax by the action of hydrochloric acid[2]:

$$Na_2B_4O_7 \times 10H_2O + 2\,HCl \rightarrow 4\,H_3BO_3\left[\text{or }\; B(OH)_3\right] + 2\,NaCl + 5\,H_2O.$$

It crystallizes from aqueous solutions normally as white, shining, waxy plates of orthoboric acid, H_3BO_3. On heating above 100 °C, the orthoboric acid gradually loses water, and changes to metaboric acid HBO_2:

$$H_3BO_3 \rightarrow HBO_2 + H_2O$$

At higher temperatures, all the water is lost, and anhydrous boric oxide is formed:

$$2H_3BO_3 \rightarrow B_2O_3 + 3H_2O$$

Crystalline boric acid consists of layers of H_3BO_3 molecules held together by hydrogen bonds (Figure 2.4). The dimensions of the unit cell, which contains four molecules of H_3BO_3, are $a_0 = 7.04$ A, $b_0 = 7.05$ A, and $c_0 = 6.56$ A with $\alpha = 92°30'$, $\beta = 101°10'$ and $\gamma = 120°$.[2] The lattice is of the layer type (which explains the ready cleavage into flakes), consisting of sheets of coplanar BO_3 groups. Each oxygen atom,

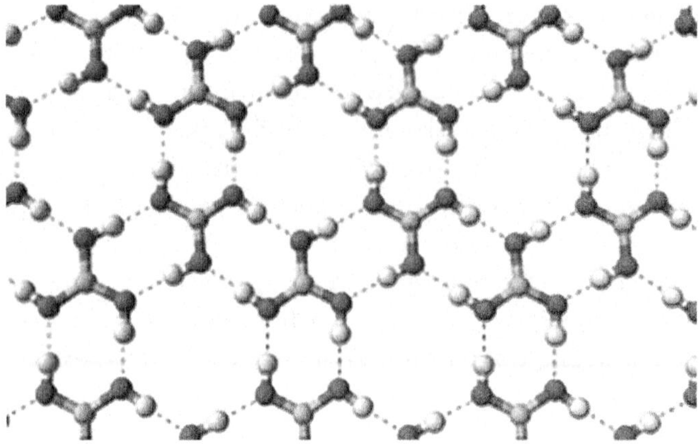

Figure 2.4 A schematic presentation of one sheet of H_3BO_3 lattice. *(Adapted from Ref. 2).*

besides being linked to boron, is joined to two other oxygen atoms by means of hydroxyl bonds. The distance between each pair of sheets is 3.18 A.[2]

The solubility of boric acid in water increases rapidly with temperature at atmospheric pressure (Table 2.1).

Boric acid dissolved in water is uncharged and has a trigonal structure. Figure 2.5 shows an actual size comparison between boron acid molecule and sodium and chloride ions in water.

2.3.2 Dissociation of Boric Acid in Water

Boric acid is very weak and does not dissociate in aqueous solution as a Bronsted acid but it acts as a Lewis acid by accepting a hydroxyl ion to form the tetrahydroxyborate ion, as confirmed by Raman spectroscopy[20]:

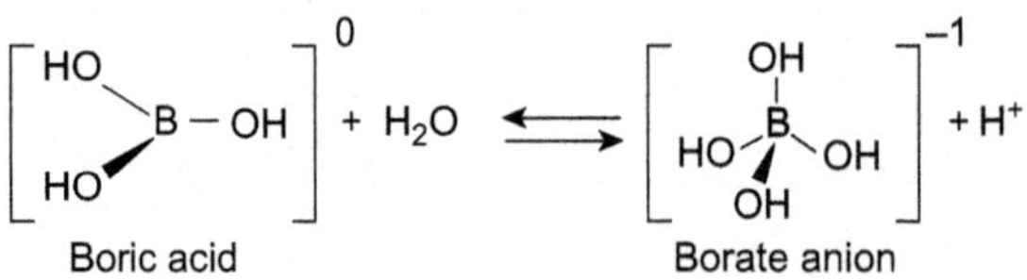

Thus, the dominant forms of inorganic boron in natural aqueous systems are mononuclear species such as boric acid $B(OH)_3$ and borate ion $B(OH)_4^-$. The distribution of these two components depends on the first dissociation constant K_a of boric acid. It was shown that the first dissociation constant is equal to 5.8×10^{-10} mol/L in fresh water at a temperature of 25 °C, while values of 1.8×10^{-13} and 3×10^{-14} have been reported for the second and third dissociation constants of boric acid, respectively. As seen in Table 2.1, K_a increases with an increase of water temperature.[19]

In solutions that are more concentrated than 0.1 M, boric acid acts as a much stronger acid than in diluted solutions, and becomes comparable to acetic acid; the apparent ratio

Table 2.1 Effect of the Temperature on Water Solubility and the First Dissociation Constant K_a of Boric Acid[1,29]

Temperature (°C)	Solubility of $B(OH)_3$ (wt.%)	$K_a \times 10^{10}$
0	2.70	
15	4.17	4.72
20	4.65	5.26
25	5.44	5.79
50	10.24	8.32
75	17.41	
100	27.53	

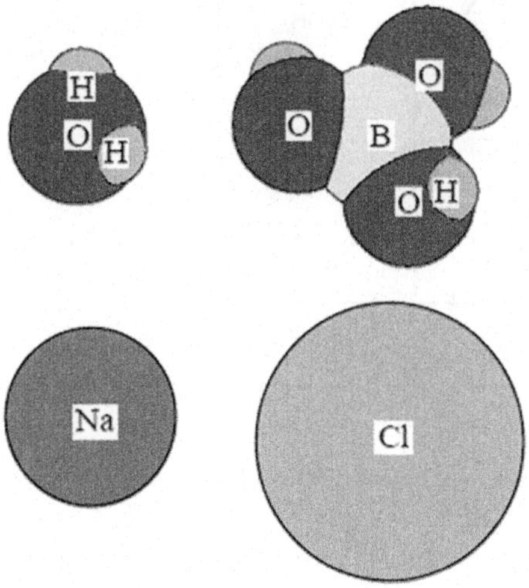

Figure 2.5 Size comparison between boron acid and some other species in aqueous solutions.

of the concentration of borate ions to that of boric acid molecules in the solution progressively increases from 1:1000 at 0.2 M to 1:5 at 3.5 M.[2]

The pK values $(=-\log(K_a))$ of boric acid have been determined to be $pK_a = 8.60$ in artificial seawater at T = 258 °C/salinity of 35 g/L[22] and $pK_a = 9.24$ at 25 °C in fresh waters.[23]

In general, due to a relatively high pK_a, boric acid has limited dissociation at neutral or low pH values. Being a weak acid, the actual pK_a value of boric acid (and distribution of boric acid and borate ion) essentially varies depending on pH, ionic strength, and temperature of the feed solution. The most important parameter, which determines the ratio of boric acid molecules to borate ions in water, is the pH of the medium. As shown in Figure 2.6, the distribution of boron species in seawater changes dramatically with pH at temperatures of 10 and 35 °C. As can be seen in Figure 2.6, the borate monovalent

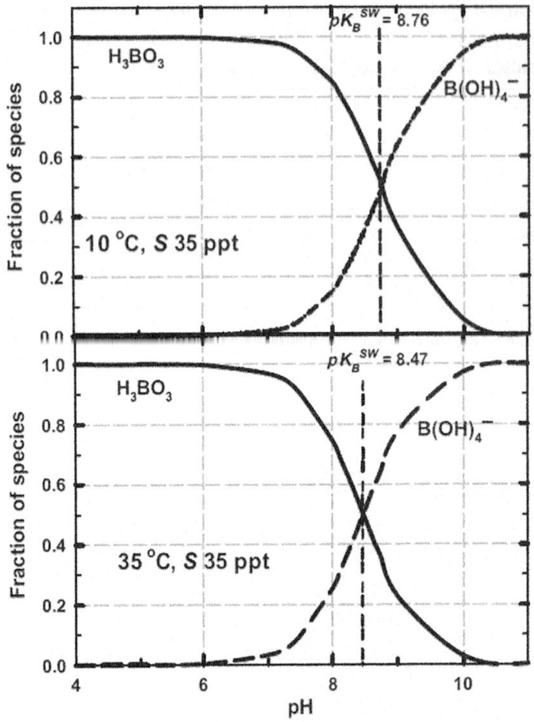

Figure 2.6 Fraction of $B(OH)_3$ and $B(OH)_4^-$ as a function of pH for seawater with a salinity of 35%. Top figure at 10 °C, bottom figure at 35 °C.[21]

anion $B(OH)_4^-$ dominates at higher pH values while nonionized boric acid $B(OH)_3$ dominates at lower pH values.

Dickson[22] investigated the dissociation of boric acid in seawater over a wide range of salinities (5000–45,000 ppm) and temperatures (0–45 °C). Based on the results obtained, Dickson suggested an equation to estimate the dissociation constant of boric acid depending on salinity (S) and temperature of seawater (T):

$$\begin{aligned} \ln K_a = & \left(-8966.90 - 2890.51\mathrm{S}^{0.5} - 77.942\mathrm{S} + 1.726\mathrm{S}^{1.5} - 0.0993\mathrm{S2}\right)/\mathrm{T} \\ & + \left(148.0248 + 137.194\mathrm{S}^{0.5} + 1.62247\mathrm{S}\right) + \left(-24.4344 - 25.085\mathrm{S}^{0.5}\right. \\ & \left.- 0.2474\mathrm{S}\right)\ln T + 0.053105\mathrm{S}^{0.5}\mathrm{T} \end{aligned}$$

Roy et al.[24] have subsequently found that the measurement of the dissociation constant by using the Dickson approach is quite reliable. It was reported that the pK_a of boric acid decreases from 9.23 to 8.60, when the salinity increased from 0 to 40,000 ppm[25] and the pK_a changed from 9.079 to 9.38 as the solution temperature increased from 10 to 50°C.[19]

The distribution of boric acid and borate in seawater at different salinities and temperatures is shown in Figure 2.7, and it is seen that a fraction of borate ion increases with an increase of the salinity and temperature of seawater.

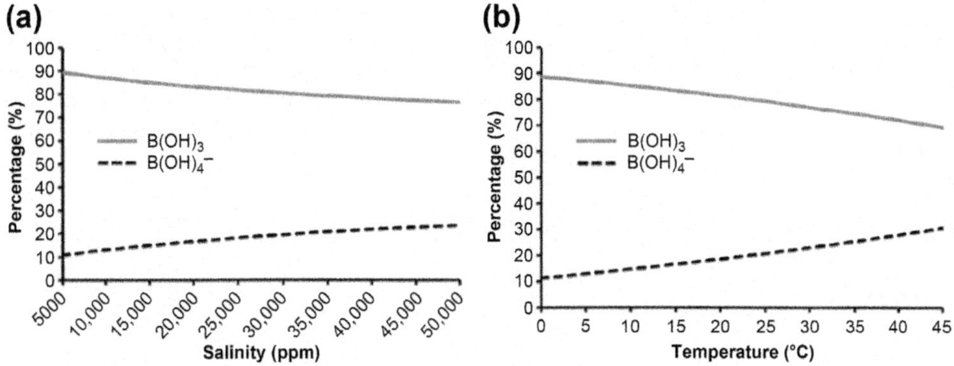

Figure 2.7 The distribution of boric acid molecules and borate ions in seawater at different salinities (a) and temperatures (b).[26]

The dependence of the pK_a of boric acid on pressure has been investigated by Tsuda et al.[27] They found an increase in the pK_a value of up to 2 units as the pressure increased from 0 to 6 kbar.

The kinetics of the boric acid–borate equilibrium at various temperatures and ionic strengths of aqueous solutions was studied by Waton et al.[28] The authors suggested that the temperature dependence of the kinetic rate constant can be fitted by an Arrhenius equation:

$$k = A \exp(-E_a/RT),$$

where A is the preexponential factor or Arrhenius factor, $R = 8.3145$ J/mol × K is the gas constant, T is the absolute temperature in Kelvins, and E_a is the activation energy.

A linear regression of ln k_{+3} versus $1/T$ (Figure 2.8) yields $A = 4.58 \times 10^{10}$ kg/mol × s and $E = 20.8 \pm 5.1$ kJ/mol. Thus, the rate constant k_{+3} can be expressed as

$$k_{+3} = 4.58 \times 10^{10} \exp\left(-20.8 \times 10^3/RT\right),$$

which yields $k_{+3} = 1 \times 10^7$ kg/mol × s at 258°C.

It should be noted that, in contrast to many other acid–base equilibriums, the dissociation of boric acid in aqueous solutions is not diffusion controlled. This is probably due to the substantial structural change that is involved in the conversion from planar $B(OH)_3$ to tetrahedral $B(OH)_4^-$ species.[29] As a result, the rate constant is two to four orders of magnitude smaller than the typical rate constants of diffusion-controlled reactions that are on the order of 10^9–10^{11} kg/mol × s.[30]

Zeebe et al.[31] showed that the theoretically calculated relaxation time for chemical and isotopic equilibrium is approximately 95 and 125 μs, respectively, for typical seawater conditions at temperature $T = 258$ °C and salinity $S = 35\%$. It follows that for

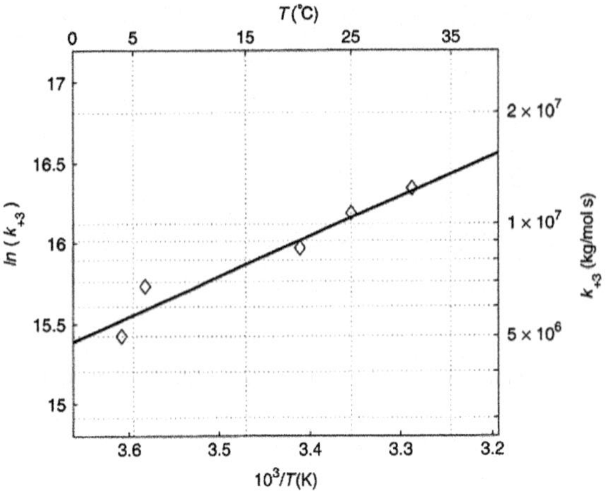

Figure 2.8 Arrhenius plot of the rate constant k_{+3} of boric acid dissociation at ionic strength $I = 0.1$. The solid line represents the best linear fit to the data when ln k_{+3} is plotted versus $10^3/T$ (K).[28]

practical purposes, that is, when time scales of minutes and hours are considered, it can safely be assumed that the dissolved boron species are in equilibrium.

2.3.3 Polyborate Ions in Aqueous Solutions

Depending on the boron concentration in aqueous solutions, various boron-containing species may be found in water. At low boron concentrations (<0.02 M), dissolved boron is mainly present as the mononuclear boron species, $B(OH)_3$ and $B(OH)_4^-$. At higher concentrations (0.025—0.6 M) and with an increase in pH from 6 to 10, water-soluble polyborate ions, such as $B_3O_3(OH)_4^-$, $B_4O_5(OH)_4^-$ and $B_5O_6(OH)_4^-$, are formed.[32]

The formation of these polynuclear ions is attributed to the interaction of boric acid and borate ions in solution according to the following equation[33]:

$$2\,B(OH)_3 + B(OH)_4^- \leftrightarrow B_3O_3(OH)_4^- + 3\,H_2O$$

$$2\,B(OH)_3 + B(OH)_4^- \leftrightarrow B_3(OH)_{10}^-$$

$$2\,B(OH)_3 + 2\,B(OH)_4^- \leftrightarrow B_4O_5(OH)_4^{-2} + 5\,H_2O$$

$$4\,B(OH)_3 + B(OH)_4^- \leftrightarrow B_5O_6(OH)_4^- + 6\,H_2O$$

The structural formulas of some polymeric borate ions and their distribution as a function of pH of the aqueous solutions are presented in Figures 2.9 and 2.10, respectively. An increase in pH usually results in higher nuclearity borates, but at pH > 10, $B(OH)_4^-$ is mainly formed. It should be noted that the formation of polyborates ions in aqueous solutions is negligible at boron concentrations <290 mg/L.[34]

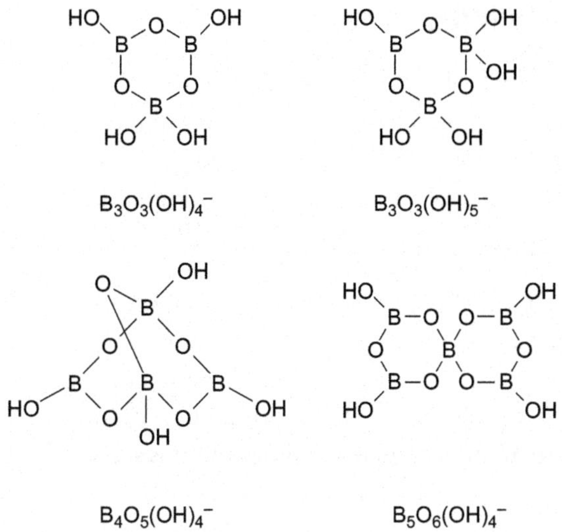

Figure 2.9 Chemical structures of polymeric borate anions in aqueous solutions.[35]

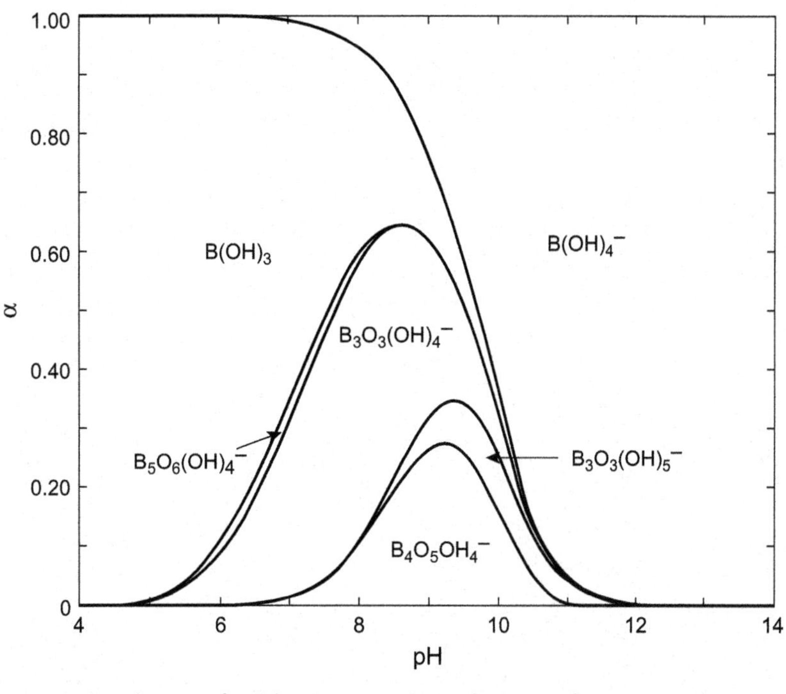

Figure 2.10 Distribution of polyborate species as a function of pH in 0.4 M boric acid.[36]

2.4 COMPLEXATION OF BORON SPECIES IN WATER

2.4.1 Boron Complexes with Alcohols and Polyols

In an aqueous environment, boric acid and borates react with alcohols and multiple hydroxyl-containing compounds (polyols) forming boron esters, neutral *cis*-diol monoborate esters, monoborate complexes, or bis(diol) borate complexes (Figure 2.11).

$$B(OH)_3 + 3\,ROH \leftrightarrow B(OR)_3 + 3H_2O$$

Complexation with polyols increases the acidity of boric acid due to the formation of cyclic borate esters. For example, the dissociation constant of boric acid becomes about 7×10^{-6} in the presence of mannitol, that is, a 10- to 1000-fold increase compared to that without the polyol.[2] Such behavior is not shown on the addition of monofunctional alcohols, nor of glycols, nor of the *trans*-form of cyclopentane diols, but it is shown upon the addition of the *cis*-form of the latter compounds. The phenomenon depends on the formation of monocyclic or dicyclic compounds with the polyol groups, which are more highly dissociated than the boric acid itself, and it follows that the only compounds with two hydroxyl groups suitably placed on the same side of the C—C link can react in this way. Thus, the stability of the borate complex formed is strongly dependent on the type of diol used. If the diol involves the OH groups oriented in such a way that they accurately match the structural parameters required by a tetrahedrally coordinated boron, a strong complex is formed.

The complexation with polyols resulting in an increase in the acidity of boric acid has been used for many years for the quantitative analysis of boric acid. Boric acid can be titrated to a phenolphthalein endpoint, which is not possible in the absence of the polyols.[37]

The complexation process with polyols involves two distinct mechanisms: interaction of boric acid with polyol or borate ion with polyol (Figure 2.12). Contribution of each mechanism in the overall complexation depends on the solution pH, where either boric acid or borate ion is dominantly present.

The equilibrium constants of borate complexes have been investigated in several studies,[38–40] and some of the data are listed in Table 2.2. In general, the stability of the

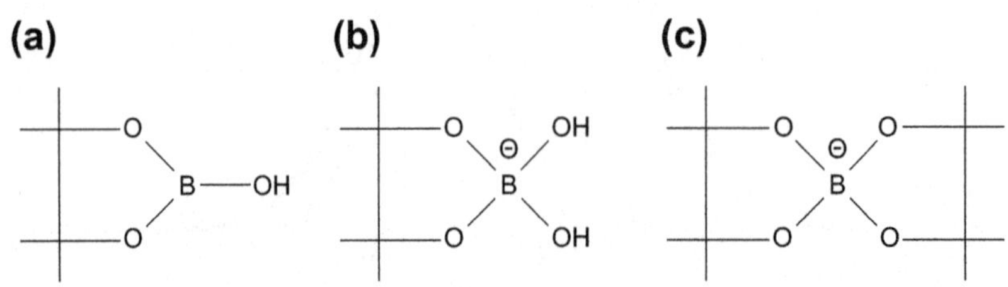

Figure 2.11 Chemical structures of *cis*-diol monoborate esters (a), monoborate complexes (b), or bis(diol)borate complexes (c).[30]

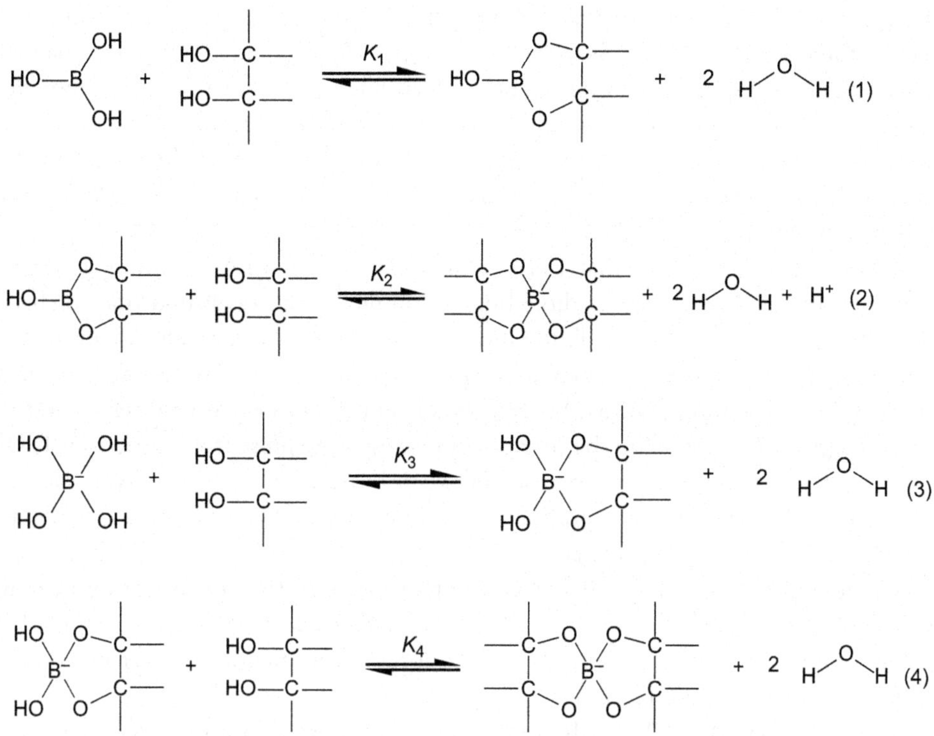

Figure 2.12 General scheme of boron complexation with polyols.[41]

Table 2.2 Equilibrium Constant for Boric Acid Complexes with Polyols[42]

Polyol	K_1	K_2
1,2-ethanediol	2.15	1.15
1,3-propanediol	1.27	0.11
Glycerol	16.0	41.2
Catechol	7.8×10^3	1.42×10^4
D-Mannitol	1.1×10^2	1.37×10^5
D-Glucose	1.50×10^3	7.6×10^2
D-Sorbitol		4.44×10^5
D-Ribose		1.57×10^7

borate complex is strongly dependent on the type of diol used. As has been discussed above, if the orientation of the OH groups in diol molecule is favorable for tetrahedral boron coordination, a stable complex will be formed.

A variety of physical, spectroscopic, and chemical techniques have been utilized for the studies of the formation of borate complexes. It has been shown that the amount of acidification produced upon the addition of polyol is proportional to the extent of ester formation and that the monitoring of the electrical conductivity of the solution may be used to study boron complex formation.[2,32] Later, ^{1}H, ^{11}B, and ^{13}C nuclear magnetic resonance (NMR) spectroscopies have been increasingly applied to the measurement of solution equilibrium in borate complex formation.[43] Thereafter, it was shown that the complex formation between various carbohydrates and borates could be detected by signal broadening and/or chemical shift changes in the ^{13}C NMR or ^{11}B NMR spectra of the parent compounds.[40,44]

Makkee et al.[40] showed that boric acid/borate reacts with mannitol generating anionic mono (1:1) and bis(1:2) diol—borate complexes. It was found that for 0.1 M boric acid and 0.01 M D-mannitol, the monoborate complex is formed and, even at pH values as high as 12, 90% of the boric acid remains uncomplexed. On the other hand, at a mannitol concentration of 0.5 M, boric acid concentration is essentially zero at pH > 8 and the bis(mannitol) ester and the monomannitol ester are present in a ratio of 9:1. Due to the reversible nature of the reactions between boric acid and the polyol, the concentration of the hydrogen ion produced should be kept low to ensure a considerable extent of dichelate complex formation.

Geffen et al.[45] used ^{11}B NMR spectroscopy to analyze the complexation of boron (0.1 M) with mannitol as a function of the pH and found that at pH 7.7 for a boron:mannitol molar ratio of 1:5 and at pH 8.8 for a boron:mannitol molar ratio of 1:2, complexation goes to completion.

Figure 2.13 shows the ^{11}B NMR spectra of main products of boron complexation with mannitol with a chemical shift of −9.6 ppm, which corresponds to the diborate ester (BP_2), and three minor peaks with lower chemical shifts ($\delta = -13.5$, -13.7, and -14.7 ppm).

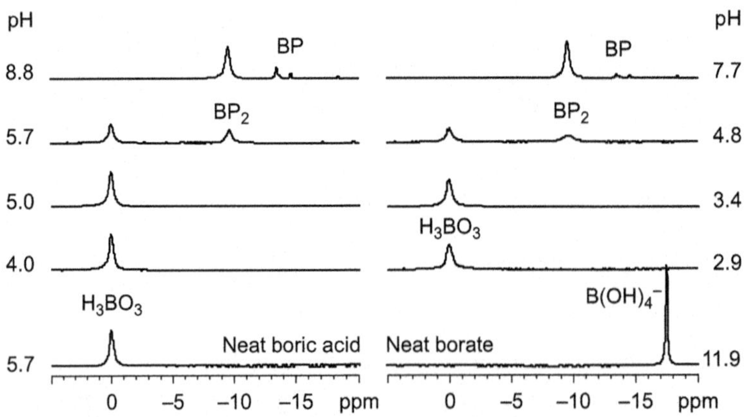

Figure 2.13 ^{11}B NMR spectra of free and mannitol-complexated boron species as a function of the equilibrium pH. Left panel, boron:mannitol ratio of 1:2; right panel, boron:mannitol ratio of 1:5. The boron concentration was 0.1 M. BP is boron:mannitol monoester; BP_2 is boron:mannitol diester. Bottom spectra represent undissociated boric acid (left panel) and fully ionized borate (right panel) controls in the absence of mannitol.[45]

These three peaks were assigned by the authors to the monoborate ester (boron bounds to three different pair ofhydroxyls ofmannitol) in accordance with previous studies.[37,40] It was also shown that the diester is the major product at a 1:5 ratio and a minor product at a 2:1 ratio in which mannitol is the limiting factor for the complexation. Theoretical calculations ofionized boron species products as a function ofpH using Mineql+ program revealed that the reactant concentrations have a strong influence on the ionized boron species: for 32 and 7 mg/L boron contents, an almost complete ionization ofboron species is expected at pH values of approximately 9.5 and 10, respectively.[45]

The similar anionic mono (1:1) and bis (1:2) diol monoborate species are also formed with carbohydrates possessing 1,2-diol systems.[46]

It is interesting to note that natural boron-containing polyol complexes have been isolated from the phloem sap of celery (*Apium graveolens*) and extrafloral nectar of peach (*Prunus persica*).[47] The three-dimensional configuration of these complexes is presented in Figure 2.14.

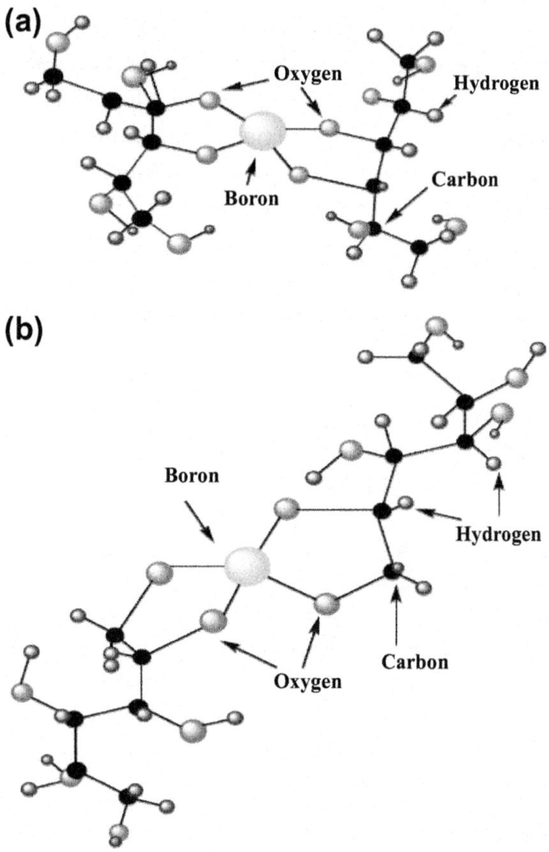

Figure 2.14 Predicted three-dimensional configuration of 3,4,3′,4′ (a) and 1,2,1′,2′ (b) bis-mannitol borate complexes isolated from *Apium graveolens* and *Prunus persica*.[47]

Pozer et al.[48] studied the thermodynamic parameters of aliphatic 1,2-diol—boron complexation by variable-temperature ^{1}H and ^{11}B NMR spectroscopy. The systems studied were $B(OH)_4^-$/1,2-ethanediol; $B(OH)_4^-$/1,2-propanediol; $C_6H_5B(OH)_3^-$/1,2-ethanediol; $CH_3B(OH)_3^-$/1,2-propanediol; and $CH_3B(OH)_3^-$/1,2-dihydroxybenzene. The first four systems have very similar stability constants and thermodynamic parameters. The reactions are all exothermic ($\Delta H° \sim -20$ kJ/mol), and the values of $\Delta S°$ are quite negative ($\Delta S° \sim -60$ J/mol × K). The negative entropy is attributed primarily to a loss of configurational entropy in the ligand on complexation. This assertion was further investigated by studying the complexation of $CH_3B(OH)_3^-$ with the rigid ligand 1,2-dihydroxybenzene. The $CH_3B(OH)_3^-$/1,2-dihydroxybenzene reaction is characterized by a stability constant that is greater by four orders of magnitude than those of the other systems, and this increase is shown to be entirely due to a much more positive value of $\Delta S°$.

Complexation of boric acid with high molecular weight organic alcohols may result in the gelation of polymeric solutions. It has been shown that polyvinyl alcohol undergoes rapid gelation when contacted with boric acid solution, due to the complexation of boron by hydroxyl groups located at different polymeric chains.[49,50] It was considered that the gelation can be avoided by specially designed polymers possessing boron chelating sites such as imino-bis propanediol (IBPD) groups.[51] Poly(vinyl amino-*N*, *N*′-bis-propane diol) (GPVA) polymer containing four hydroxyl groups for boron complexation was synthesized by the reaction of high molecular weight polyvinyl amine with glycidol.[49] During complexation, three hydroxyl groups of four are involved in boron complexation while the fourth one remains unoccupied (Figure 2.15). The experiments showed that the reaction of IBPD functional polymer with boric acid in aqueous solution does not result in observed precipitation during a week.

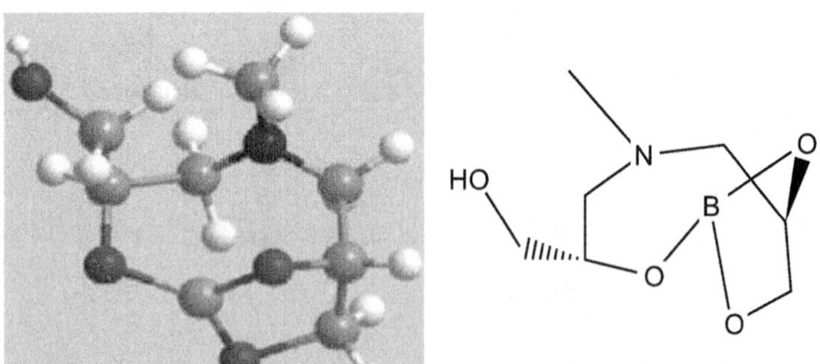

Figure 2.15 Configuration of boron–imino-bis propanediol complex. Atoms are represented as follows: C (gray), N (blue), H (white), O (red), and B (yellow).[51] (For interpretation of the references to color in this figure legend, the reader is referred to the web version of this article.)

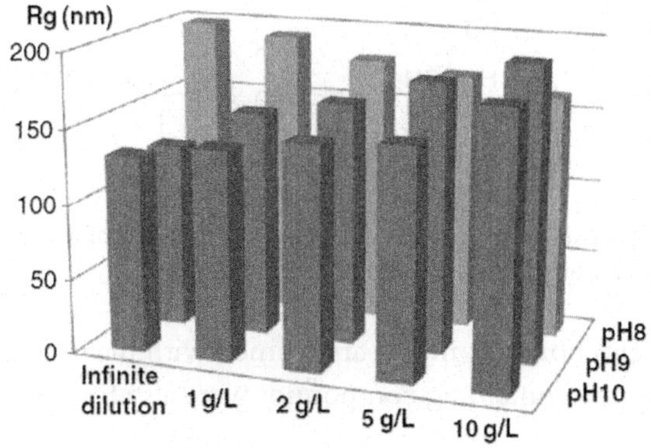

Figure 2.16 Effect of pH and polymer concentration on the gyration radius of GPVA in the presence of boron at a loading of 0.001.[51]

A dynamic and static light scattering method was used for studying the conformation of the polymer macromolecules upon changes in experimental conditions such as pH, polymer concentration, and the presence of boron (Figure 2.16).

As seen in Figure 2.16, the radius of gyration of the polymer chains is proportional to the polymer concentration at pH 9.0 and 10.0, whereas at pH 8.0, the proportionality becomes reversed. This controversy was explained by residual hydroxyl groups retained after complexation of boron with three of the hydroxyl groups. The residual hydroxyl groups would probably form a relatively weak intermolecular boron bridging among the macromolecular chains. Thus, at low polymer concentrations, they may be too far away from each other for the boron bridging to occur.

Smith et al.[52] synthesized three water-soluble polyethylenimine (PEI)-based polymers containing linear alkyl monool, 1,2-diol, and 1,2,3-triol groups for boron complexation. It was suggested that boron acid/borate can interact with the functionalized polymers via ion pairing and/or borate ester formation. Ultrafiltration experiments showed that boron rejection essentially decreased when NaCl was added to the solutions. This finding indicates the presence of significant ion pairing in boron—PEI systems. To suppress the ion-pairing effect of PEI, a high chloride/boron ratio of about 10 was used. It was found that the monool—PEI polymer does not form any borate esters and that boron binding occurs only through the ion-pairing mechanism. For the 1,2-diol-PEI and 100 mg/L boron concentration, almost 43% of the boron was rejected when sodium sulfate was added to the solution. For 1,2,3-triol and the same boron concentration, an even larger rejection of boron (76%) was shown when NaCl was added to the solution. These results indicate that boron interacts with polymers, which contain 1,2-diol and 1,2,3-triol groups, by means of ion pairing and also via borate ester formation.

2.4.2 Boron Complexation with Organic Acids and Enzymes

Besides alcohols and polyols, boric acid reacts with some organic acids in water and forms boron-containing complexes. For example, neutral borate complex, monomalic

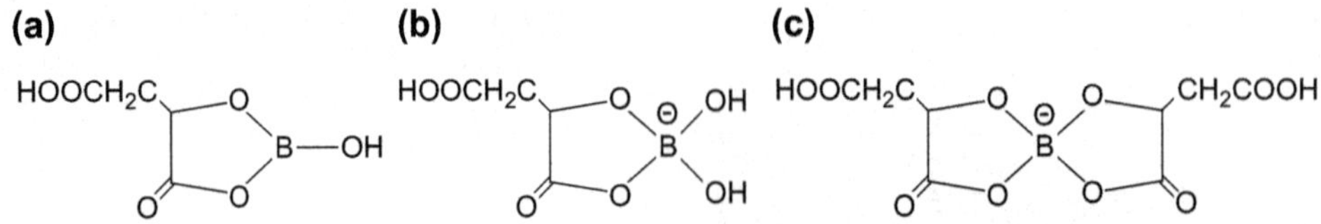

Figure 2.17 Chemical structures of borate complexes with maleic acid. Neutral borate complex (a), monomalic acid borate complex (b), and bis(malic acid) borate complex (c).[55]

acid borate complex, and the bis(malic acid) borate complex are formed with malic acid and its derivatives (Figure 2.17) These boron-containing compounds were also found in apple juice and wine.[53,54]

The complexation of boric acid with salicylic acid, salicyl alcohol, and bis(hydroxymethyl)phenol derivatives has been investigated using ^{11}B NMR spectroscopy by Miyakazi et al.[56] It was shown that boric acid accepts an electron pair through the nucleophilic attack of the salicylic acid (Ac), followed by a condensation reaction, to form the 1:1 monochelate complex (Figure 2.18(a)). The monochelate complex then reacts with the ligand through the condensation reaction to give the 1:2 bischelate complex (Figure 2.18(b)).

The chemical shifts and formation constants of boric acid complexes with salicyl compounds are shown in Table 2.3.

As seen from the data in Table 2.3, bis(hydroxymethyl)phenol derivatives formed quite stable chelate complexes among the salicyl compounds used. It was assumed that boron complexation with bis(hydroxymethyl)phenol compounds was highly favorable from an enthalpy viewpoint.[54]

The 3-D structures of boron acid complexes with salicyl derivatives are presented in Figure 2.19.

As seen in Figure 2.20, where the pH dependence of the equilibrium concentrations of the boron species in a large excess of 3-5-bis(hydroxymethyl)-4-hydroxybenzoil acid is presented, at the pH range of 6–9, the negatively charged 1:2 bischelate complex prevails in the aqueous solution.

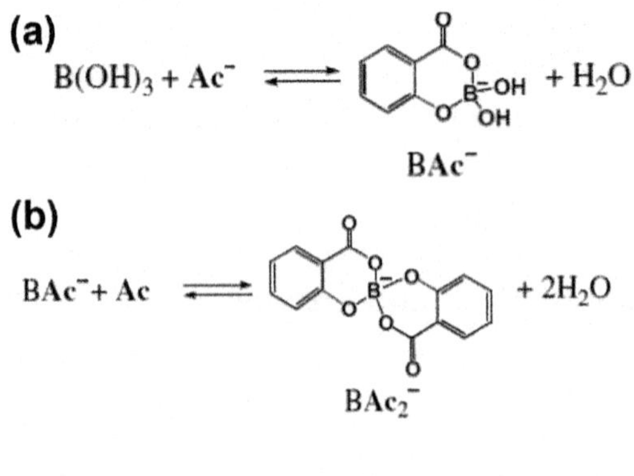

Figure 2.18 Complexation of boric acid with salicylic acid with the formation of a 1:1 monochelate complex (a) and a 1:2 bischelate complex (b).

Table 2.3 ^{11}B NMR Chemical Shifts and Formation Constants of Boric Acid Complexes with Salicyl Derivatives[56]

Ligand	pK_{a1}	pK_{a2}	Complex	δ	log K_f
Salicylic acid (Ac)	2.98	13.61	BAc^-	2.9	1.05 ± 0.01
			BAc_2^-	3.3	2.15 ± 0.03
Salicyl alcohol (Ol)	9.54		BOl^-	1.6	3.60 ± 0.01
			BOl_2^-	1.8	1.50 ± 0.06
2,6-Bis(hydroxymethyl)-*p*-cresol (LMe)	9.44		$BLMe^-$	1.7	4.23 ± 0.03
			$B(LMe)_2^-$	1.9	2.45 ± 0.10
3,5-Bis(hydroxymethyl)-4-hydroxybenzoic acid (LAc)	4.20	8.57	$BLAc^{2-}$	1.8	3.74 ± 0.04
			$B(LAc)_2^{3-}$	2.0	2.42 ± 0.09

Ionic strength is 0.10 mol/L KCl, 25 °C.

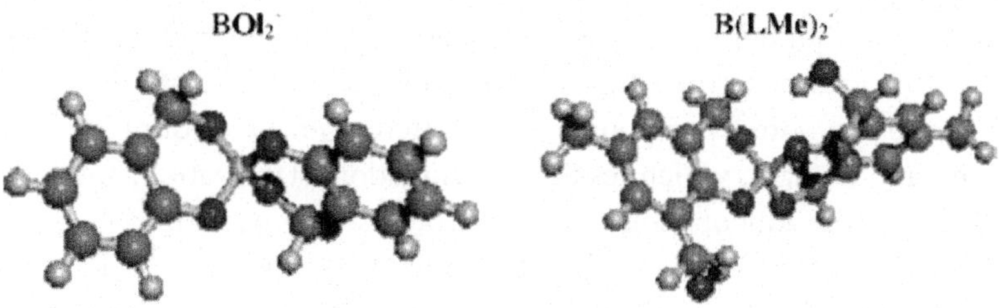

Figure 2.19 Optimized structures of the 1:2 complexes of boric acid with salicyl alcohol (Ol)$^-$ and 2,6-bis(hydroxymethyl)-*p*-cresol (LMe)$^-$.[56]

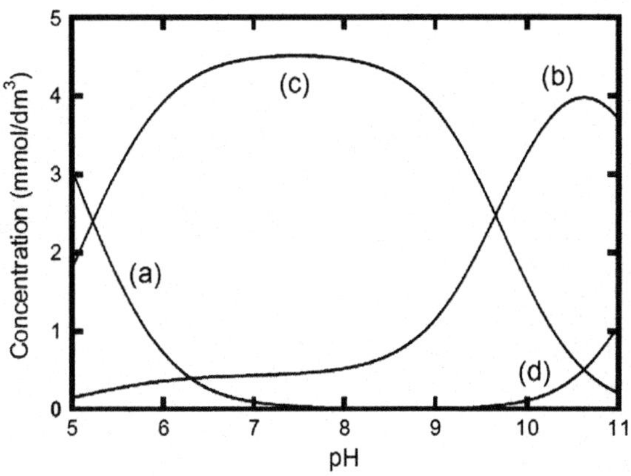

Figure 2.20 pH-dependent equilibrium concentrations of boron species for a solution containing 5 mmol/L of boric acid/borate and 50 mmol/L of LAc. (a) $B(OH)_3$; (b) $B(LAc)^{2-}$; (c) $B(LAc)_2^{3-}$; and (d) $B(OH)_4^-$.[56]

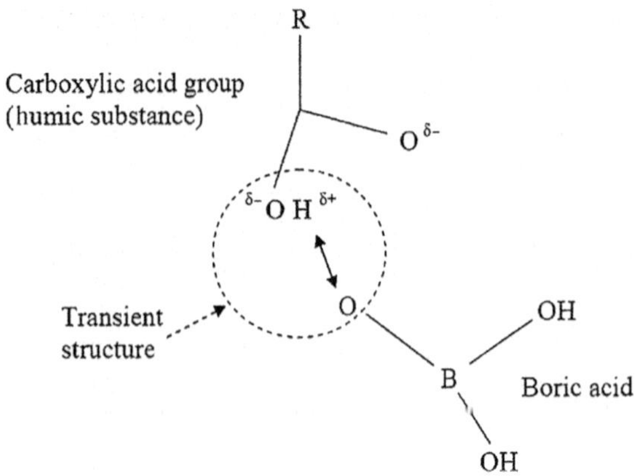

Figure 2.21 Schematic presentation of $B(OH)_3$ complexation with humic substances.[58]

A possible complexation of boron with humic substances in aqueous solutions should also be mentioned. Schmitt-Kopplin et al.[57] postulated that $B(OH)_3$ binds to carboxylate groups (COO^-) within humic acids where it forms a transient hydrogen-bonded structure. A schematic presentation of this complexation mechanism is shown in Figure 2.21.

Complexation of boron compounds with different enzymes of plants, microorganisms, animals, and humans, which results in stimulation, stabilization, and/or inhibition of the ferments in the aqueous solutions, have been also reported.[59]

It was found that the enzyme urease is inhibited by boric acid.[60] This inhibition is attributed to borate occupying the active site, for example, the active site of a serine protease (Figure 2.22).

An another example is the complexation of borate ion with the ribose group of nicotinamide adenine dinucleotide.[59] The charge separated complex is favored over the reduced nicotinamide adenine dinucleotide and leads to the inhibition of this enzyme (Figure 2.23).

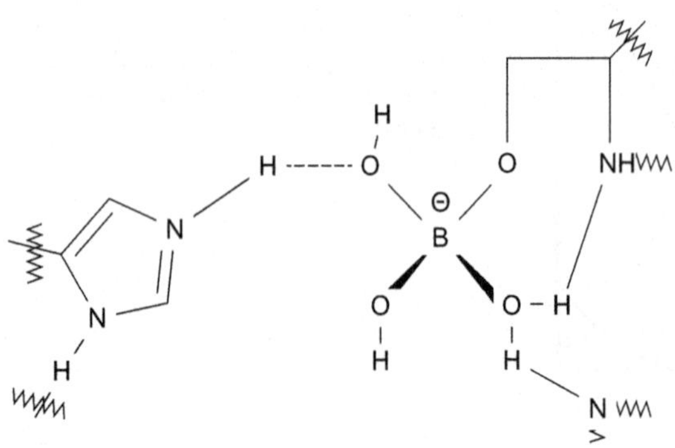

Figure 2.22 Boron inhibition of the active site of a serine protease.[59]

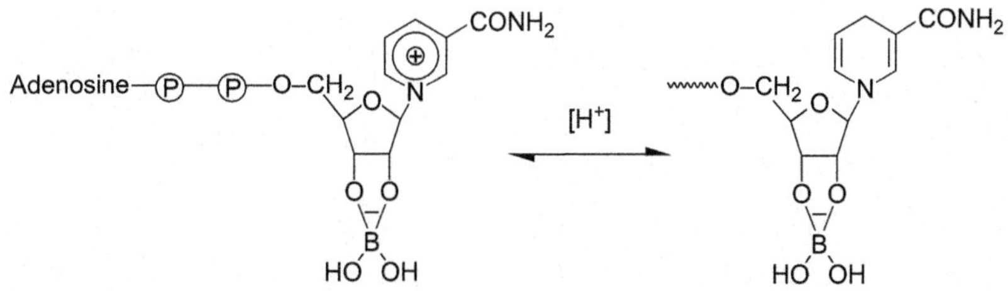

Figure 2.23 Borate inhibition of hydrogenase coenzyme by fibityl group complexing.[59]

It should be noted that boron's complexation ability in aqueous solutions is used as the basis of selective ion exchange technology for boron removal from water.[61] Commercial boron selective resins are primarily classified as macroporous crosslinked polystyrenic resins, functionalized with the *N*-methyl-*D*-glucamine (NMG) group.[62] The functional NMG group includes a tertiary amine and polyol groups. The role of the tertiary amine is to neutralize the proton brought by the formation of tetraborate complex. NMG groups capture boron through a covalent attachment and formation of an internal coordination complex as shown in Figure 2.24.

The complexation ability of boron complexation ability is used for the development of novel boron selective sorbents. Recently, the new sorbent NMG functionalized calix[4] arene-based magnetic sporopollenin sorbent has been synthesized and used for boron removal from aqueous environments.[63] It was shown that boron complexation takes place between the borate ion $B(OH)_4^-$ and hydroxyl groups of the synthesized sorbent (Figure 2.25). It was found that the highest sorption value (84%) was obtained at a pH of 7.5. On the contrary, a lower boron sorption above pH 7.5 may be due to the abundance of OH^- ions in the solution, which compete with $B(OH)_4^-$ ions for the sorption sites.

Also, boron complexation with some specific organic reagents, such as curcumin and carmine, are widely used in spectrophotometric techniques for the determination of boron concentration in water.[64]

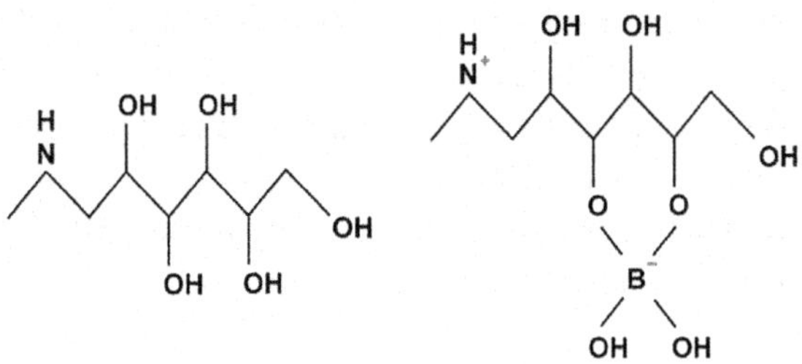

Figure 2.24 Chemical structure of the NMG group (left) and monoborate complex (right).[26]

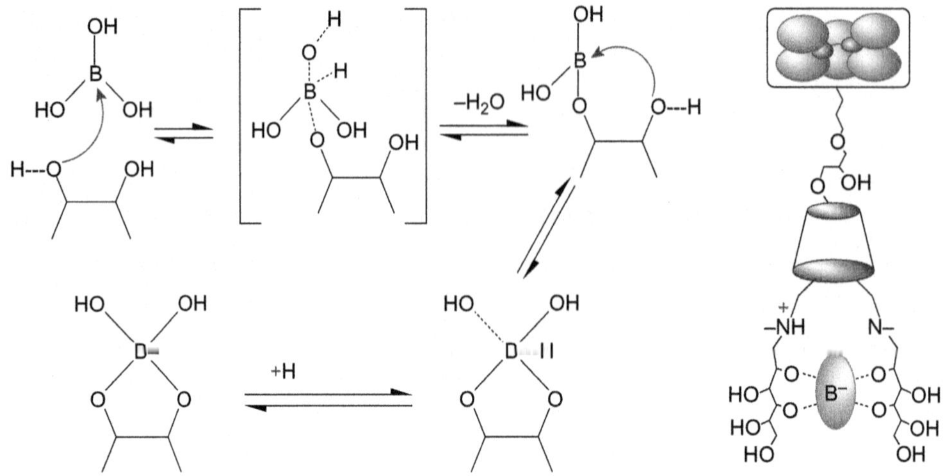

Figure 2.25 Boric acid complexation with vicinal-OH groups and boron sorption by calix[4] arene-based magnetic sporopollenin.[63]

2.5 BORON AND DRINKING WATER REGULATIONS

Boron aquatic chemistry primarily determines the boron concentration in water, including drinking water. It should be noted that for many years the boron content in drinking water was not considered as an important issue regarding a possible impact on human health. In 1958, 1963, and 1971, there was no mention of boron in the World Health Organization (WHO) International Standards for drinking water. Since boron has been shown to induce several harmful effects on animals in laboratory studies,[65,66] provisional guidelines for boron concentration in drinking water by the WHO were first introduced in 1993 at 0.3 mg/L, based on the no-observed-adverse-effect level. This guideline value was increased to 0.5 mg/L in 1998 due to a lack of financially viable boron removal technologies from water. However, extensive data from the UK and the USA on dietary boron intakes showed that the intake from air and food is lower than expected. This led to the increase in the boron intake allocated to drinking water from 10% to 40%,[67] without approaching the tolerable boron daily limit. According to current WHO guidelines on drinking water quality, the recommended boron content in drinking water is established as 2.4 mg/L.[68] It was noted[68] that short- and long-term oral exposures to boric acid or borax in laboratory animals have demonstrated that the male reproductive tract is a consistent target of toxicity. Testicular lesions have been observed in rats, mice, and dogs given boric acid or borax in food or drinking water. Developmental toxicity has been demonstrated experimentally in rats, mice, and rabbits. Negative results in a large number of mutagenicity assays indicate that boric acid and borax are not genotoxic. In long-term studies in mice and rats, boric acid and borax caused no increase in tumor incidence.[68]

Table 2.4 Regional Standards for Boron in Drinking Water[26]

Region	Maximum Boron Concentration (mg/L)	References and Comments
Saudi Arabia	0.5	SASO,[69,70] bottled and unbottled drinking water
United States of America (USA)	—	USEPA,[71] no federal regulations of boron
State of Minnesota	0.6	USEPA[72]
State of New Hampshire	0.63	USEPA[72]
State of Florida	0.63	USEPA[72]
State of Maine	0.63	USEPA[72]
State of Wisconsin	0.9	USEPA[72]
State of California	1	USEPA[72]
European Union (EU) including UK	1	EEA[73]
South Korea	1.4	Ministry of Environment,[74] changed from 0.3 mg/L
Japan	1	NIPH[75]
New Zealand	1.4	Ministry of Health[76]
Australia	4	NHMRC[77]
Canada	5	CDW,[78] it has not changed since 1990
WHO recommendation	0.5	WHO,[79] changed from 0.3 mg/L[80]

As shown in Table 2.4, standards or guideline values for boron concentration in drinking water vary widely around the world, ranging from 0.5 to 5.0 mg/L. In general, the maximum permissible boron concentration in drinking water is commonly determined by considering a range of factors including human and ecological health, social and natural characteristics, and cost of the available water treatment technologies. Because the influence of boron on human health has not been thoroughly elucidated, most of the existing guidelines are still provisional values that are subject to further discovery of boron toxicity on human beings.

2.6 ANALYTICAL METHODS FOR MEASURING BORON CONTENT IN WATER

The most common methods for the determination of boron concentration in water are spectrophotometric and plasma-source spectrometric methods.

2.6.1 Spectrophotometric Methods

They are based on colorimetric reactions of boron with some specific reagents such as curcumin, carmine, and azomethine-H.[64] The optical density of the analyzed solution is usually proportional to the boron concentration in water.

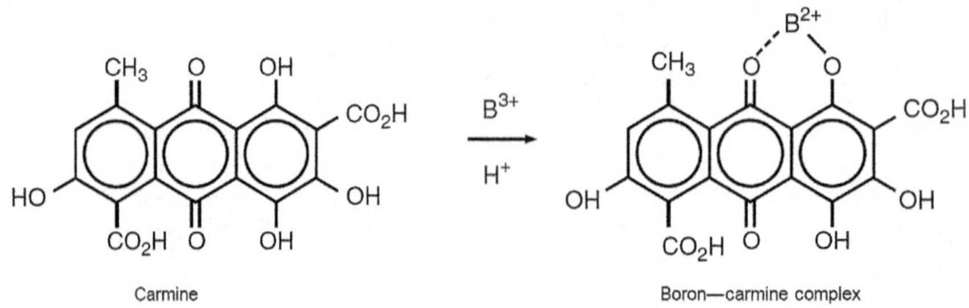

Figure 2.26 Formation of boron-containing complex with carmine indicator.

The curcumin method is based on boron reaction with curcumin (diferuloylmethane—$C_{21}H_{20}O_6$) in acidic solution (pH = 1), when a red-colored rosocyanin complex [$B(C_{21}H_{19}O_6)_2Cl$] is formed (maximal adsorption at $\lambda = 545$ nm).[81]

In the presence of carmine indicator and concentrated sulfuric acid, boron forms a blue-colored complex with the maximal adsorption at $\lambda = 605$ nm (Figure 2.26):

Usually, the curcumin method is recommended for water with boron concentrations between 0.1 and 1.0 mg/L, whereas the carmine method is for boron level in the range of 1−10 mg/L.[82]

The azomethine-H method is based on the boron-catalyzed reaction of 8-aminonaphthyl-1-ol-3,6-pirosulfuric acid with salicylaldehyde (Figure 2.27). At pH = 6 in the presence of dissolved forms of borates, the condensation reaction is completed quickly (within 15 min), and a yellow complex is formed. After complex formation, the solution is adjusted to an acidic pH for optimum color measurement at $\lambda = 410$ nm.

The azomethine-H method does not require concentrated acids, and it is applicable to the determination of borate at concentrations between 0.01 and 1 mg/L. In a comparative evaluation of azomethine-H, carminic acid, and curcumin methods for boron determination in water, the azomethine-H method suffered the least interferences

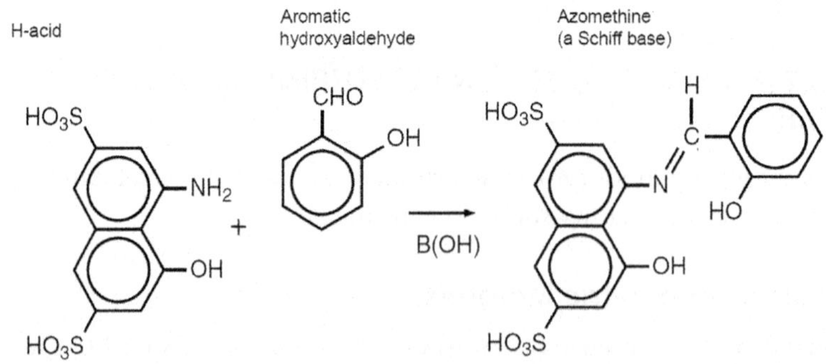

Figure 2.27 A chemical reaction used in the Azomethine-H method of boron determination in water.[83]

and was the most sensitive.[84] It should be noted that the spectrophotometric methods of boron analysis are well suited for field analysis; however, they suffer from numerous interferences and have a low sensitivity and precision.[64]

2.6.2 Atomic Spectrometric Methods

Atomic emission spectrometry (AES) and atomic absorption spectrometry (AAS) generally involve introduction of samples into a flame (usually of acetylene–air or acetylene–N_2), where elements of the sample are atomized. The AAS measurement is based on the principle that free atoms of an element (e.g., boron) in their ground state absorb photons of discrete energy values (a characteristic wavelength) generated by a hollow cathode lamp containing that element.[64] The AES methods measure emission from the atomized and excited species when they fall to the ground state. The AES/AAS determination of boron often requires separation and preconcentration of boron from the sample matrix for acceptable results.[85] Atomic spectrometric methods, such as AES and AAS, revolutionized the determination of a large number of elements, but these methods are not very sensitive for the boron element with serious memory effects of previous samples, and numerous interferences.[86]

2.6.3 Plasma-Source Methods

Introduction of plasmas as ionization sources and the development of plasma-source analytical instruments provided a higher sensitivity and lower detection capability for boron determination than was possible by spectrophotometric and atomic spectrometric methods.[87] The inductively coupled plasma (ICP) method is the most widely used plasma-based techniques for boron analysis.[88,89] The ICP is a type of plasma source formed from electric currents that are caused by electromagnetic induction on a rarefied gas such as argon. Samples are usually prepared in an aqueous phase using steps involving extraction and purification, and are then introduced into the plasma via a nebulizer and spray chamber. ICP has recently been coupled to different types of mass spectrometers to improve method sensitivity and reliability.[64]

Table 2.5 Detection Limits of Analytical Methods of Boron Determination[91]

Methods	Limit of Quantification (μg/L)
ICP–MS	0.15
ICP–AES	5–6
IC	50
Azomethine-H	10–20
Curcumin	100–200
Carmine	1000–2000

ICP–AES, inductively coupled plasma atomic emission spectrometry; ICP–MS, inductively coupled plasma mass spectrometry.

Ion chromatography (IC) has been also used for monitoring boron concentration in water. Boron is usually separated from the sample matrix by an anion exchange resin and converted to tetrafluoroborate BF_4^- by HF treatment. The resulting BF_4^- is determined potentiometrically with a suitable selective electrode.[90]

As can be seen in Table 2.5 the detection limits of boron determination of ICP–AES or ICP–MS methods can be several orders of magnitude lower than those of IC and spectrophotometric methods.

REFERENCES

1. Greenwood NN. Boron. In: *Comprehensive inorganic chemistry.* Oxford: Pergamon Press; 1973. p. 665–993.
2. Kemp PH. *The chemistry of borates.* Ipswich: W.S. Cowell LTD; 1956.
3. Albert B, Hillebrecht H. Boron: elementary challenge for experimenters and theoreticians. *Angew Chem Int Ed* 2009;**48**:8640–68.
4. Emsley J. *The elements.* 2nd ed. Oxford: Oxford University Press; 1991.
5. Krebs RE. *The history and use of our earth's chemical elements: a reference guide.* Greenwood Publishing Group; 2006.
6. Morgan V. *Boron chemistry.* New York: Longman; 1980.
7. Aubert H, Pinta M. *Trace elements in soils.* Amsterdam: Elsevier Scientific; 1997.
8. Krauskopf KB. Geochemistry of micronutrients. In: Mortvedt JJ, Giordano PM, Lindsay WL, editors. Madison: Soil Science Society of America; 1972.
9. Argust P. Distribution of boron in the environment. *Biol Trace Elem Res* 1998;**66**:131–43.
10. Adair R. *Boron.* The Rosen Publishing Group; 2007.
11. Weast RC, Astle MJ, Beyer WH. *CRC handbook of chemistry and physics.* Boca Raton (FL): CRC Press; 1985. p. B-77, B-129.
12. Butterwick L, Oude N, Raymond K. Safety assessment of boron in aquatic and terrestrial environments. *Ecotoxicol Environ Saf* 1989;**17**:339–71.
13. Neal C, Fox KK, Harrow M, Neal M. Boron in the major UK rivers entering the North Sea. *Sci Total Environ* 1998;**210**:41–51.
14. Jahiruddin M, Smart R, Wade AJ, Neal C, Cresser MS. Factors regulating the distribution of boron in water in the River Dee catchment in north east Scotland. *Sci Total Environ* 1998;**210**:53–62.
15. Hebblethwaite RL, Emberson P. Rising from the ashes. *Landsc Des* 1993;**10**:31–4.
16. Fox KK, Daniel M, Morris G, Holt MS. The use of measured boron concentration data from the GREAT-ER UK validation study (1996–1998) to generate predicted regional boron concentrations. *Sci Total Environ* 2000;**251**:305–16.
17. Brady N, Weil R. *The nature and properties of soils.* 14th ed. Upper Saddle River: Prentice Hall; 2008.
18. Wyness AJ, Parkman RH, Neal C. A summary of boron surface water quality data throughout the European Union. *Sci Total Environ* 2003;**314**:255–69.
19. Owen BB. The dissociation constant of boric acid from 10 to 50 C. *J Am Chem Soc* 1934;**56**:1695–7.
20. Jolly WL. *Modem inorganic chemistry.* New York: McGraw-Hill; 1984.
21. Edzwald JK, Haarhoff J. Seawater pretreatment for reverse osmosis: chemistry, contaminants, and coagulation. *Water Res* 2011;**45**:5428–40.
22. Dickson AG. Thermodynamics of the dissociation of boric-acid in synthetic seawater from 273.15-K to 318.15-K. *Deep-Sea Res Part A Oceanogr Res Pap* 1990;**37**:755–66.
23. Dean JA. *Lange's handbook of chemistry.* New York: McGraw-Hill; 1999.
24. Roy RN, Roy LN, Lawson M, Vogel KM, Moore CP, Davis W, et al. Thermodynamics of the dissociation of boric-acid in seawater at S = 35 from 0-degrees-C to 55-degrees-C. *Mar Chem* 1993;**44**:243–8.

25. Choi WW, Chen KY. Evaluation of boron removal by adsorption on solids. *Environ Sci Technol* 1979;**13**:189–96.
26. Hilal N, Kim GJ, Somerfield C. Boron removal from saline water: a comprehensive review. *Desalination* 2011;**273**:23–35.
27. Tsuda M, Shirotani I, Minomura S, Terayama Y. Effect of pressure on dissociation of weak acids in aqueous buffers. *Bull Chem Soc Jpn* 1976;**49**:2952–5.
28. Waton G, Mallo P, Candau SJ. Temperature-jump rate study of the chemical relaxation of aqueous boric-acid solutions. *J Phys Chem* 1984;**88**:3301–5.
29. Mellen RH, Browning DG, Simmons VP. Investigation of chemical sound-absorption in seawater. *J Acoust Soc Am* 1983;**74**:987–93.
30. Eigen M, Hammes G. Elementary steps in enzyme reactions. In: Nord FF, editor. *Advances in enzymology*. New York: Wiley; 1963. p. 1–38.
31. Zeebe RE, Sanyal A, Ortiz JD, Wolf-Gladrow DA. A theoretical study of the kinetics of the boric acid-borate equilibrium in seawater. *Mar Chem* 2001;**73**:113–24.
32. Power PP, Woods WG. The chemistry of boron and its speciation in plants. *Plant Soil* 1997;**193**:1–13.
33. Cotton FA, Wilkinson G. *Advanced inorganic chemistry*. New York: Wiley; 1980.
34. Su CM, Suarez DL. Coordination of adsorbed boron – a FTIR spectroscopic study. *Environ Sci Technol* 1995;**29**:302–11.
35. Salentine CG. High-field B-11 NMR of alkali borates – aqueous polyborate equilibria. *Inorg Chem* 1983;**22**:3920–4.
36. Anderson JL, Eyring EM, Whittaker MP. Temperature jump rate studies of polyborate formation in aqueous boric acid. *J Phys Chem* 1964;**68**:1128–32.
37. Belcher R, Tully GW, Svehla G. A comparative study of various complexing agents (polyols) used in titration of boric acid. *Anal Chim Acta* 1970;**50**:261.
38. Vanduin M, Peters JA, Kieboom APG, Vanbekkum H. The Ph-dependence of the stability of esters of boric-acid and borate in aqueous-medium as studied by B-11 NMR. *Tetrahedron* 1984;**40**:2901–11.
39. Vanduin M, Peters JA, Kieboom APG, Vanbekkum H. Studies on borate esters. 2. Structure and stability of borate esters of polyhydroxycarboxylates and related polyols in aqueous alkaline media as studied by B-11 NMR. *Tetrahedron* 1985;**41**:3411–21.
40. Makkee M, Kieboom APG, Vanbekkum H. Studies on borate esters 3. Borate esters of D-mannitol, D-glucitol, D-fructose and D-glucose in water. *Rec Trav Chim Pays-Bas-J R Neth Chem Soc* 1985; **104**:230–5.
41. Tu KL, Chivas AR, Nghiem LD. Enhanced boron rejection by NF/RO membranes by complexation with polyols: measurement and mechanisms. *Desalination* 2013;**310**:115–21.
42. Sanderson BR. Coordinated compounds of boric acid. In: *Supplement to Mellor's comprehensive treatise on inorganic and theoretical chemistry. Boron-oxygen compounds*, vol. V, part A. New York: Longman; 1980.
43. Lenz RW, Heeschen JP. The application of NMR to structural studies of carbohydrates in aqueous solution. *J Polym Soc* 1961;**51**:247–55.
44. Gorin PAJ, Mazurek M. C-13 resonance spectroscopic studies on formation of borate and diphenylborinate complexes of polyhydroxy compounds. *Can J Chem-Revue Can Chim* 1973;**51**:3277–86.
45. Geffen N, Semiat R, Eisen MS, Balazs Y, Katz I, Dosoretz CG. Boron removal from water by complexation to polyol compounds. *J Membr Sci* 2006;**286**:45–51.
46. Chapelle S, Verchere JF. A B-11 and C-13 NMR determination of the structures of borate complexes of pentoses and related sugars. *Tetrahedron* 1988;**44**:4469–82.
47. Hu HN, Penn SG, Lebrilla CB, Brown PH. Isolation and characterization of soluble boron complexes in higher plants – the mechanism of phloem mobility of boron. *Plant Physiol* 1997;**113**:649–55.
48. Pizer RD, Ricatto PJ, Tihal CA. Thermodynamics of several boron acid complexation reactions studied by variable-temperature H-1 and B-11 NMR-spectroscopy. *Polyhedron* 1993;**12**:2137–42.
49. Sinton SW. Complexation chemistry of sodium-borate with polyvinyl-alcohol) and small diols – a B-11 NMR-study. *Macromolecules* 1987;**20**:2430–41.
50. Kurokawa H, Shibayama M, Ishimaru T, Nomura S, Wu WI. Phase-behavior and sol–gel transition of poly(vinyl alcohol) borate complex in aqueous-solution. *Polymer* 1992;**33**:2182–8.

51. Zerze H, Karagoz B, Ozbelge H, Bicak N, Aydogan N, Yilmaz L. Imino-bis-propane diol functional polymer for efficient boron removal from aqueous solutions via continuous PEUF process. *Desalination* 2013;**310**:158–68.
52. Smith BF, Robison TW, Carlson BJ, Labouriau A, Khalsa GRK, Schroeder NC, et al. Boric acid recovery using polymer filtration: studies with alkyl monool, diol, and triol containing polyethylenimines. *J Appl Polym Sci* 2005;**97**:1590–604.
53. Matsunaga T, Nagata T. In-vivo B-11 NMR observation of plant-tissue. *Anal Sci* 1995;**11**:889–92.
54. Lutz O, Humpfer E, Spraul M. Ascertainment of boric-acid esters in wine by B-11 NMR. *Naturwissenschaften* 1991;**78**:67–9.
55. Dembitsky VM, Smoum R, Al-Quntar AA, Abu Ali H, Pergament I, Srebnik M. Natural occurrence of boron-containing compounds in plants, algae and microorganisms. *Plant Sci* 2002;**163**:931–42.
56. Miyazaki Y, Matsuo H, Fujimori T, Takemura H, Matsuoka S, Okobira T, et al. Interaction of boric acid with salicyl derivatives as an anchor group of boron-selective adsorbents. *Polyhedron* 2008;**27**:2785–90.
57. Schmitt-Kopplin P, Hertkorn N, Garrison AW, Freitag D, Kettrup A. Influence of borate buffers on the electrophoretic behavior of humic substances in capillary zone electrophoresis. *Anal Chem* 1998;**70**:3798–808.
58. Banasiak LJ, Schafer AI. Removal of boron, fluoride and nitrate by electrodialysis in the presence of organic matter. *J Membr Sci* 2009;**334**:101–9.
59. Kliegel W. *Bor in biologie*. New York: Springer-Verlag; 1980.
60. Zaborska W. Competitive inhibitors of free and chitosan immobilized urease. *Acta Biochem Polon* 1995;**42**:115–8.
61. Simonnot MO, Castel C, Nicolai M, Rosin C, Sardin M, Jauffret H. Boron removal from drinking water with a boron selective resin: is the treatment really selective? *Water Res* 2000;**34**:109–16.
62. Kabay N, Guler E, Bryjak M. Boron in seawater and methods for its separation – a review. *Desalination* 2010;**261**:212–7.
63. Kamboh MA, Yilmaz M. Synthesis of *N*-methylglucamine functionalized calix[4]arene based magnetic sporopollenin for the removal of boron from aqueous environment. *Desalination* 2013;**310**:67–74.
64. Sah RN, Brown PH. Boron determination – a review of analytical methods. *Microchem J* 1997; **56**:285–304.
65. Weir RJ, Fisher RS. Toxicologic studies on borax and boric-acid. *Toxicol Appl Pharmacol* 1972;**23**:351–9.
66. Lee IP, Sherins RJ, Dixon RL. Evidence for induction of germinal aplasia in male rats by environmental exposure to boron. *Toxicol Appl Pharmacol* 1978;**45**:577–90.
67. WHO. *Boron in drinking water – background document for development of WHO guidelines for drinking water quality*; 2009.
68. WHO. *Guidelines for drinking water quality.* 4th ed. 2011. Geneva.
69. SASO. *Bottled drinking water.* Saudi Arabian Standards Organization; 2000.
70. SASO. *Unbottled drinking water.* Saudi Arabian Standards Organization; 2000.
71. USEPA. *Edition of the drinking water standards and health advisories*; 2006. Washington, DC.
72. USEPA. *Drinking water health advisory for boron*; 2008. Washington, DC.
73. EEA. *The quality of water intended for human consumption*; 1998. Council Directive 98/83/EC.
74. MOE. *Management of drinking water quality.* Republic of Korea: Ministry of Environment; 2009.
75. NIPH. *Seawater desalination facility on Okinawa*; 2006. Japan.
76. MOH. *Drinking-water standards for New Zealand.* Wellington: Ministry of Health; 2005.
77. NHMRC. *Australian drinking water guidelines.* Canberra: National Health and Medical Research Council; 2004.
78. CDW. *Guidelines for Canadian drinking water quality*; 2008. Ottawa.
79. WHO. *Boron in drinking water*; 2003. Geneva.
80. WHO. *Guidelines for drinking-water quality.* 2nd ed. 1993. Geneva.
81. Dyrssen DW, Uppstrom LR, Novikov YP. Studies on chemistry of determination of boron with curcumin. *Anal Chim Acta* 1972;**60**:139.
82. Greenberg AF, Trusell RR, Clesecrin LS. *Standard method for the examination of water and wastewater.* 16th ed. Washington DC: American Public Health Association; 1985.

83. Fang ZL. Nonequilibrated sample manipulation – the essence of flow-injection analysis. *Microchem J* 1992;**45**:137–42.
84. Lopez FJ, Gimenez E, Hernandez F. Analytical study on the determination of boron in environmental water samples. *Fresenius J Anal Chem* 1993;**346**:984–7.
85. Botelho GMA, Curtius AJ, Campos RC. Determination of boron by electrothermal atomic-absorption spectrometry – testing different modifiers, atomization surfaces and potential interferents. *J Anal At Spectrom* 1994;**9**:1263–7.
86. Papaspyrou M, Feinendegen LE, Mohl C, Schwuger MJ. Determination of boron in cell-suspensions using electrothermal atomic-absorption spectrometry. *J Anal At Spectrom* 1994;**9**:791–5.
87. Farhat A, Ahmad F, Arafat H. Analytical techniques for boron quantification supporting desalination processes: a review. *Desalination* 2013;**310**:9–17.
88. Barth S. Comparison of NTIMS and ICP-OES methods for the determination of boron concentrations in natural fresh and saline waters. *Fresenius J Anal Chem* 1997;**358**:854–5.
89. Vogl J, Rosner M, Pritzkow W. Development and validation of a single collector SF-ICPMS procedure for the determination of boron isotope ratios in water and food samples. *J Anal At Spectrom* 2011;**26**:861–9.
90. Katagiri J, Yoshioka T, Mizoguchi T. Basic study on the determination of total boron by conversion to tetrafluoroborate ion (BF_4^-)followed by ion chromatography. *Anal Chim Acta* 2006;**570**:65–72.
91. Tu KL, Nghiem LD, Chivas AR. Boron removal by reverse osmosis membranes in seawater desalination applications. *Sep Purif Technol* 2010;**75**:87–101.

CHAPTER 3

Risk Assessment of Borates in Occupational Settings

Yalçın Duydu[1], Nurşen Başaran[2], Hermann M. Bolt[3]
[1]Ankara University, Faculty of Pharmacy, Department of Toxicology, Tandoğan, Ankara, Turkey
[2]Hacettepe University, Faculty of Pharmacy, Department of Toxicology, Sıhhiye, Ankara, Turkey
[3]Leibniz Research Centre for Working Environment and Human Factors (IfADo), Dortmund, Germany

3.1 INTRODUCTION

Human health risk assessment for reproductive effects of boric acid and sodium borates is the primary scope of this chapter. Animal experiments and published occupational and environmental exposure data will form the main frame of this assessment. Exposure to sodium perborates (perboric acid and sodium salts) is not within the scope of this assessment.

Boron is a naturally occurring element that is widespread at relatively low concentrations.[1,2] Boron in the environment is chemically bound to oxygen, usually as alkali or alkaline earth borates, or as boric acid (EPA, 2004, syf 3—5). Boric acid is a weak acid with a pK_a of 9.2. In aqueous solutions at physiological pH, boric acid and borate salts exist primarily in the form of an undissociated acid (H_3BO_3).[1,2] Boron oxide also reacts with water to form boric acid. Weir and Fisher found that borax and boric acids have similar toxicity when dose is calculated as boron.[3] Therefore, the toxicity associated with these compounds is expected to be similar when based on boron equivalents. Many studies have been reported on the toxicity of boron in laboratory animals. The studies have been carried out with different boron compounds. Therefore, comparing the equivalent doses of boron (B) is the best way to evaluate the results of different studies. Table 3.1 shows the molecular formulae and conversion factors for equivalent doses of boron.

3.2 TOXICOKINETICS

The toxicokinetics (TKs) of boric acid and sodium borates are similar in rats and humans with respect to absorption, distribution, and metabolism.[4] Boric acid and sodium borates are readily absorbed, and >90% of an orally administered dose is excreted unchanged in the urine of both animals and humans.[5] The elimination half-life of an orally administered dose is <24 h in humans and animals.[6] Boric acid is not metabolized in the body, and sodium borates are converted to boric acid at physiological pH.[2] Absorption of

Boron Separation Processes
ISBN 978-0-444-63454-2, http://dx.doi.org/10.1016/B978-0-444-63454-2.00003-4

Table 3.1 Molecular Formulae of Boric Acid/Borates and Conversion Factors for Equivalent Doses of Boron

	Boron	Boric Acid	Boron Oxide	Borax	Borax Pentahydrate	Anhydrous Borax
CAS registry number	7440-42-8	10043-35-3	1303-86-2	1303-96-4	12179-04-3	1330-43-4
Molecular formula	B	H_3BO_3	B_2O_3	$Na_2B_4O_7 \cdot 10H_2O$	$Na_2B_4O_7 \cdot 5H_2O$	$Na_2B_4O_7$
Molecular weight	10.81	61.83	69.62	381.43	291.35	201.27
Boron content (%)	100	17.48	31.06	11.34	14.85	21.49
Conversion factor	–	0.175	0.311	0.113	0.148	0.215

boron from the respiratory tract via inhalation is another major route of boron exposure, especially in occupational settings.[7] Dermal absorption of boric acid and sodium borates through intact skin is negligible.[2]

3.3 HEALTH RISK ASSESSMENT

Risk assessment is an integrated process consisting of four steps: (1) "Hazard identification" is the first step to determine whether exposure to a chemical has the potential to harm human health, and to identify the specific chemical hazard. (2) The second step is "dose–response assessment". This step determines the relationship between the quantity of exposure to a hazard and the severity of adverse effects. (3) "Exposure assessment" is the third step within the risk assessment process that determines the actual levels of exposure and absorption of the toxicant. (4) The last step of the risk assessment process is risk "characterization." This step combines information from the previous three steps and uses this to interpret the nature and magnitude of risk (Figure 3.1).

3.3.1 Hazard Identification

There are many studies on acute, subchronic, chronic, reproductive toxicity, and genotoxicity of boric acid and borates. Some of the key studies are summarized below, in order to identify the critical effect and critical dose level for risk assessment.

3.3.1.1 Animal Experiments

3.3.1.1.1 Acute Toxicity

3.3.1.1.1.1 Acute Oral Toxicity Most acute toxicity figures in humans are based on historical data of the medicinal use of boric acid in antiseptic ointments.[8] Weir and

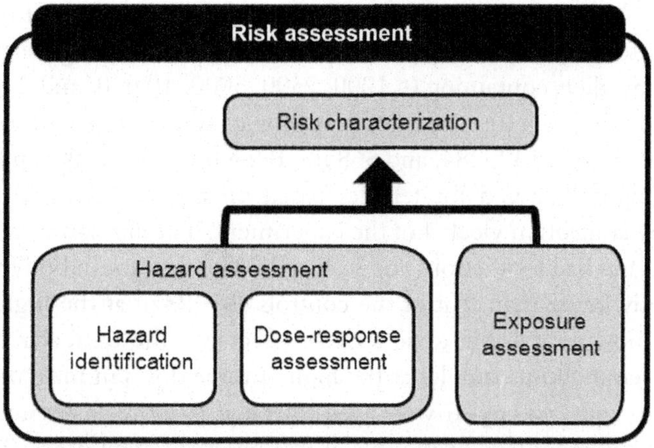

Figure 3.1 The integrated process of risk assessment.

Fisher[3] provided experimental data on acute toxicity of boric acid and borax. Accordingly, acute oral toxicities values (LD_{50}) of boric acid and borax in rats were 3000–4000 mg/kg bw and 4500–6000 mg/kg bw, respectively. Similar signs of toxicity for both boric acid and borax were reported: central nervous system depression, ataxia, convulsion, and finally death occurred. Acute oral toxicity (LD_{50}) of borax pentahydrate in rats was 2403–4207 mg/kg bw, and the clinical signs of toxicity were similar to those for borax and boric acid.[9]

3.3.1.1.1.2 Acute Dermal Toxicity Boric acid is apparently not absorbed through intact skin.[10,11] The LD_{50} of boric acid and borax pentahydrate in rabbits is reported to be >2000 mg/kg bw.[9]

3.3.1.1.1.3 Acute Inhalation Toxicity A pilot study performed by Bio/dynamics Inc.[9] is the only study on acute inhalation toxicity (LC_{50}) of boric acid. This study was performed in five male and five female Sprague–Dawley rats. These rats were exposed for 4 h to an airborne concentration of 0.16 mg/L of boric acid. No mortalities were reported during the exposure period and the subsequent 14-day observation period. The LC_{50} (4 h) of boric acid was therefore >0.16 mg/L.

3.3.1.1.1.4 Irritation Boric acid and sodium tetraborates are generally not skin irritants. Significant adverse effects on the human eye have not been reported in occupational settings. Boric acid is not an eye irritant in rabbits.[12]

3.3.1.1.2 Subchronic Toxicity

The NTP Technical Report on the "Toxicology and Carcinogenesis Study of Boric Acid in B6C3F1 Mice" is the most detailed study on subchronic toxicity, chronic toxicity, and carcinogenicity.[13] The subchronic toxicity (13-week) study as part of this whole study was performed to evaluate the cumulative toxic effects of repeated exposure to boric acid. In this study, diets containing 0, 1200, 2500, 5000, 10,000, and 20,000 ppm boric acid were fed to groups of 10 mice of each sex for 13 weeks. These doses approximately correspond to 0, 34, 71, 142, 284, and 568 mg B/kg bw/d for males and 0, 47, 98, 196, 392, and 784 mg B/kg bw/d for females based on reported average values for feed consumption by controls in week 4 of the experiment. The animals were thin, hunched, and dehydrated and had foot lesions and scaly tails. The final mean body weights of mice were significantly lower than that of the controls (10–23%) at the highest three doses. However, the most remarkable toxic adverse effect was the testicular degeneration or atrophy of the seminiferous tubules at the highest three doses in male mice. The lowest two doses (1200 and 2500 ppm) were well tolerated by both males and females. Mild extramedullary hematopoiesis of the spleen was observed at all doses. The lowest dose tested (1200 ppm) was the lowest observed adverse effect level (LOAEL) of this study. A NOAEL was not identified in this study.[13]

3.3.1.1.3 Chronic Toxicity and Carcinogenicity

3.3.1.1.3.1 Chronic Toxicity In a chronic toxicity study (2-year study), Sprague–Dawley rats were fed diets containing 0, 117, 350, and 1170 ppm B as borax or boric acid for 2 years (~0, 5.9, 17.5, and 58.5 mg B/kg bw/day). Adverse effects and histopathological changes were not reported for the two lowest dose groups. The animals allocated to the high-dose (1170 ppm) group, however, had decreased food consumption, retarded growth, bleeding eyes, postural abnormalities, and scaly tails. Testicular atrophy was reported in all male rats receiving the highest dose. This study identified an LOAEL of 1170 ppm (58.5 mg B/kg bw/day) and an NOAEL of 350 ppm (17.5 mg B/kg bw/day) for testicular effects.[3]

3.3.1.1.3.2 Carcinogenicity Histopathologic evidence of carcinogenicity was not reported in the above-mentioned chronic toxicity study by Weir and Fisher.[3] This study, however, was designed to assess the systemic toxicity of boric acid. Therefore, the NTP has considered it as an adequate but incomplete carcinogenesis study conducted in one species (Sprague–Dawley rats) and decided to complete the test for carcinogenicity of boric acid by conducting another carcinogenesis study in B6C3F1 mice.[13] Accordingly, male (50/group) and female (50/group) B6C3F1 mice were fed a diet containing 0, 2500, or 5000 ppm boric acid for 103 weeks. The average amount of boric acid consumed was approximately 400–500 mg/kg bw/day (70–87.5 mg B/kg bw/day) or 1100–1200 mg/kg bw/day (192.5–210 mg B/kg bw/day) for low- or high-dose mice. Boric acid-mediated testicular atrophy was observed in male rats allocated to the high dose group as reported in the other chronic and subchronic studies. Dose-related significant increase in neoplasms was not reported. The survival rate was significantly lower in males when compared with that in the control group. The decrease in survival of treated male mice may have reduced the sensitivity of this study. Nevertheless, the overall conclusion of the NTP was that there was no evidence of carcinogenicity of boric acid in male or female B6C3F1 mice.

3.3.1.1.4 Genotoxicity

3.3.1.1.4.1 In vitro Studies Boric acid was not mutagenic in bacterial reverse mutation assays. Boric acid-induced gene mutation was not seen in *Salmonella typhimurium* strains TA 1535, TA 1537, TA 98, and TA 100 with or without rat or hamster liver S-9 activating system.[13] Besides, boric acid was not mutagenic in mammalian cell gene mutation assays. It did not induce mutations in L5178Y mouse lymphoma cells, with or without metabolic activation.[13] Boric acid-mediated induction in chromosome aberrations and increase in the frequency of sister chromatid exchange in Chinese hamster ovary cells, with or without metabolic activation, was not observed.[13–15] Boron-mediated induction in unscheduled DNA synthesis was also not identified at concentrations between 5 and 5000 μg/ml.[2]

3.3.1.1.4.2 In vivo Studies Boric acid-mediated in vivo effects on the micronucleus formation in bone marrow erythrocytes were studied in Swiss Webster. Boric acid was administered orally at dose levels of 900, 1800, and 3500 mg/kg bw/day for 2 days. A dose-dependent increase in the frequency of micronucleus formation was not observed.[2]

3.3.1.1.5 Reproductive Toxicity

Boric acid and borates occur naturally. Many studies have been published on the presence of boric acid or inorganic boron compounds in ground waters, surface waters, plants, and soil. Besides, boric acid and borates are widely used in industrial, agricultural, and household applications. These factors triggered toxicity studies of boric acid and inorganic boron compounds, particularly in 1964–2014. Subchronic and chronic toxicity studies have indicated that the testes are a primary target for boron compounds in male animals. Therefore, such studies have gained more importance in assessing the safety of the human boron exposures, particularly in occupational settings.

In 1972, Weir and Fisher examined the reprotoxic effects of boric acid and borax in a multigeneration (three generation) study by using Sprague–Dawley rats.[3] The rats (weighing 110–150 g) were randomized in groups of eight males and 16 females and received borax or boric acid at 0 (control), 117, 350, and 1170 ppm boron in the diet (~0, 5.9, 17.5, or 58.5 mg B/kg bw/day). The overall fertility indices ((number of pregnancies/number of matings) × 100) for boric acid and borax in the low- and medium-dose groups were significantly higher than those of the controls. Besides, adverse effects on reproduction or gross abnormalities in the examined organs were not observed in low-dose (117 ppm B) and medium-dose (350 ppm B) groups. The animals receiving the highest boric acid or borax doses (1170 ppm B) were found to be sterile. The lack of viable sperm in the atrophied testes of all males and decreased ovulation in the majority of the females were reported as the major adverse effects at the highest dose group. According to the results of this study, 17.5 mg B/kg bw/day was identified as an NOAEL for the above-mentioned reproductive adverse effects (critical toxic effects) in male rats. The animal experiments conducted to determine the boron-mediated adverse effects have consistently pointed out to the similar adverse effects on fertility at relatively high doses and rats were the most sensitive species for these effects. These studies have been compiled in Table 3.2.

3.3.1.1.6 Developmental Toxicity

Many studies have been conducted on boron-mediated developmental toxicity in different species of experimental animals (rat, rabbit, and mice). Heindel et al.[18] studied boron-mediated developmental toxicity in time-mated Sprague—Dawley rats. The rats were exposed to boric acid (in the diet) at 0, 13.7, 28, and 58 mg B/kg bw/day. Decreased mean fetal body weights per litter were reported even at the lowest dose (13.7 mg B/kg bw/day). Remarkable skeletal malformations were reported at a dose level of 28 mg B/kg bw/day and above. It was not possible to identify an NOAEL for the developmental effects of boron in this study. However, 13.7 mg B/kg bw/day was considered to be an LOAEL for the rats, based on the decrease in fetal body weight. Another study was conducted by Price et al.[19] in order to determine an NOAEL for the developmental effects of boron exposure in time-mated Sprague—Dawley rats. The rats were exposed to boric acid (in the diet) at 3.3, 6.3, 9.6, 13.3, and 25 mg B/kg bw/day. Developmental effects were not reported at doses of 3.3, 6.3, and 9.6 mg B/kg bw/day. The mean fetal body weight/litter was decreased at 13.3 mg B/kg bw/day and above. Additionally, increased incidence of short rib XIII and wavy rib were also reported at this dose level. The decrease in the mean fetal body weight/litter and skeletal abnormalities were reported to be more pronounced at higher dose levels. According to the results of this study, the NOAEL for the developmental effects of boron was 9.6 mg B/kg bw/day in rats.

Allen et al.[20] published a study addressing the benchmark-dose (BMD) approach as the point of departure (POD) for the boron-mediated developmental toxic effects by combining the results of two existing rat developmental toxicity studies.[18,19] The reduction in fetal body weight and skeletal malformations were considered as the critical end points of interest in this study. However, it was reported that the decreased fetal body weight provided the best basis for BMD calculations. Accordingly, the calculated $BMDL_{05}$ (BMD Lower Confidence Limit; the 95% lower confidence limit of the BMD for 5% reduction of fetal body weight) for the developmental effects of boron was 10.3 mg B/kg bw/day, which was very similar to the NOAEL (9.6 mg B/kg bw/day).

Table 3.2 Studies Supporting Boron-Mediated Adverse Effects on Fertility in Animals

Experimental Animal	Dose and Duration of Exposure	Adverse Effects	NOAEL (mg B/kg bw/day)	References
Rats (male/female)	• 0-117-350-1170 ppm as boron equivalent in the diet (0-5.9-17.5—58.5 mg B/kg bw/day). • Multigeneration study	• The animals were found to be sterile at 58.5 mg B/kg bw/day • Lack of viable sperm • Atrophy of testes • Decreased ovulation	17.5	Weir and Fisher, 1972[3]
Rats (male)	• 0-250 (44)-500 (88)-1000 (175)-2000 (350) mg boric acid (mg B)/kg/bw • 1 Day (acute)	• Adverse effects on spermiation and sperm quality	88	Linder et al., 1990[16]
Mice (male/female)	• 0-152 (27)-636 (111)-1262 (220) mg Boric acid (mg B)/kg bw/day • Continuous breeding study	• Reduced body weights • Increased incidence of abnormal sperm • Decreased sperm concentration and motility • Seminiferous tubule degeneration	27	Fail et al., 1991[17]

When the related animal studies on the developmental toxicity of boron are reviewed, rats appeared as the most sensitive experimental species[20,21] (Table 3.3). The studies of importance for risk assessment are compiled in Table 3.3.

3.3.1.1.7 Endocrine Toxicity

The male reproductive system is controlled by the interaction of hormones from the hypothalamus and anterior pituitary with hormones from reproductive tissues (Figure 3.2). Gonadotropin releasing hormone (GnRH), follicle stimulating hormone (FSH), luteinizing hormone (LH), inhibin, and testosterone are the major reproductive hormones of males and interact with each other. When required, GnRH is released from the hypothalamus and exerts its effect on the anterior pituitary. This interaction stimulates the secretion of FSH and LH. FSH and LH enter the testis and stimulate the Sertoli cells to facilitate spermatogenesis and testosterone secretion from the Leydig cells. Consequently, spermatogenesis is regulated by Sertoli cell products and testosterone. The male reproductive cycle is under the control of a negative feedback system. High levels of testosterone inhibit the release of GnRH, FSH, and LH. Additionally, the Sertoli cells produce the hormone inhibin when the rate of sperm production is sufficient. This hormone inhibits the secretion of FSH and causes the sperm production to slow down.

Animal experiments have revealed that a high dose of boric acid led to inhibited spermiation and testicular atrophy in rats. Depressed testosterone concentrations and elevated FSH and LH levels in blood have been reported in laboratory animals treated with high doses of boric acid.[22,23] These results might suggest that endocrine mechanisms might be responsible for the testicular effects of boric acid. However, it appears that hormonal changes are secondary to germ cell changes.[23]

Unfavorable effects of boric acid on the endocrine control of the male reproductive system generally occur at doses higher than required for reproductive toxicity in rats (NOAEL: 17.5 mg B/kg bw/day). Therefore, the NOAEL determined for the reproductive toxicity also covers the endocrine control of male reproductive system.

3.3.1.2 Human Health Hazard Assessment

Epidemiological studies on boron-mediated adverse effects on reproductive health are very few, and adverse effects on reproductive health were not identified in the available studies with one exception. In 1972, Tarasenko et al.[24] reported testicular atrophy and sterility to occur in boric acid production workers in Russia. The study was conducted in 28 Russian workers exposed for >10 years to high levels of boron salts (22–80 mg/m^3). Adverse effects on sperm quality parameters (sperm count and motility) were reported for six workers. This is the only study stating adverse effects on reproductive health in male workers exposed to a very high level of inorganic borates. However, limited information on smoking habits, diet, other chemical exposure, and lack of information on the method of semen analysis compromises the value of this study.

Table 3.3 Summary of Developmental Toxicity Studies for Boric Acid

Experimental Animal	Dose	Adverse Effects	NOAEL (mg B/kg bw/day)	References
Rats	• 0-13.7-28-58 mg B/kg bw/day as boric acid in the diet	• Decreased mean fetal body weights/litter • Skeletal malformations	13.7 (LOAEL)	Heindel et al., 1992[18]
Rats	• 3.3-6.3-9.6-13.3-25 mg B/kg bw/day as boric acid in the diet	• Decreased mean fetal body weights/litter • Increased incidence of short rib XIII • Increased incidence of wavy rib	9.6	Price et al., 1996[19]
Rats	• The analysis was performed by using the results of the above-mentioned two studies	• Decreased mean fetal body weights/litter • Skeletal malformations	10.3 ($BMDL_{05}$)	Allen et al., 1996[20]
Mice	• 0-43-79-175 mg B/kg bw/day as boric acid in the diet	• Decreased fetal body weight • Increased incidence of short rib XIII	43	Heindel et al., 1992[18]
Rabbits	• 0-11-22-44 mg B/kg bw/day as boric acid, by gavage	• Reduced live litter size • Increased malformed fetuses/litter	22	Price et al., 1996[21]

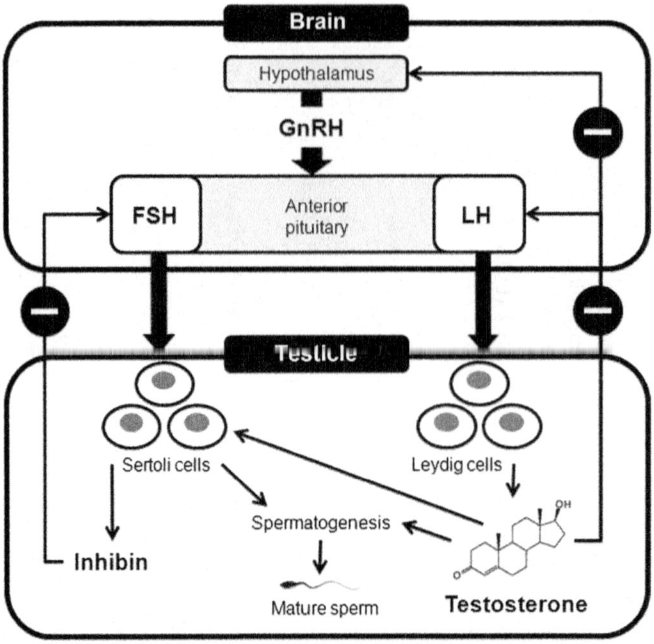

Figure 3.2 Endocrine control of the male reproductive system. (−) Represents the negative feedback action of hormones, GnRH, FSH, and LH.

Reproductive effects of inorganic borates on male employees employed in borax mining and production facilities in the United States have been studied by Whorton et al.[25,26] Of the 753 eligible employees, 542 participated in the study. The live births were used as the indicator of fertility, and the standardized birth ratio (SBR) was used to determine the rate of fertility. An SBR >100 denotes an excess of births, while lower values show a deficit. The SBR for the employees was calculated as 113. Accordingly, the employees have had slightly more children than expected. However, an excess of female offspring (52.7%) was determined in this study. According to the authors, this excess was not due to a decline in male offspring. Consequently, exposure-related adverse effects on infertility and SBR could not be identified.

The health impact of boron in drinking water was investigated in a study in Northern France by Yazbeck et al.[27] Boron-related effects on birth rates, mortality rates, and sex ratios were evaluated in zones of different boron contents in drinking water. Three zones were identified depending on the different levels of boron content in drinking water: "zone I" with no boron background (0.00−0.09 mg B/L), "zone II" with low boron levels (0.10−0.29 mg B/L), and "zone III" with higher levels (≥0.30 mg B/L). Blood boron concentrations were determined in 180 healthy individuals, to assess the relation between boron concentrations in blood and boron concentrations in drinking water of the three zones. The mean blood boron concentration of individuals living in zone III (138.78 ng/g) was slightly (but not significantly) higher than the mean blood boron concentration of individuals living in zones I and II (122.94 ng/g). The SBR in zone III was significantly higher than that of zone I and that of the French general population ($p < 0.0001$). Interestingly, the mortality rate in zone III was less than that of zone I and of the French general population ($p < 0.001$). The male-to-female sex ratios were also compared between the three zones, and the female offspring was slightly higher in zone III than in zones II and I (49.1%, 48.9%, and 48.8%, respectively). These differences, however, did not reach the level of statistical significance.

The first extensive study on boron exposure in boron mining and processing workers was conducted in Kuandian County, Liaoning Province, P.R. China. The results of this study were published between 2006 and 2010.[28–30] This first (preliminary) study explored the relationship between boron exposure and adverse effects on reproduction without providing data on the daily boron exposure levels and boron concentrations in biological fluids of the workers.[26] To this end, the workers ($n = 936$) and the control group ($n = 251$) were compared in terms of lifestyles, smoking, alcohol use, food and beverage consumption, boron exposure in workplace, and general/reproductive health. The control group consisted of office workers, taxi drivers, coal mine workers, train station workers, and farmers. The reproductive health outcomes, including delayed pregnancy, multiple births, spontaneous miscarriages, induced abortions, stillbirths, and sex ratios were investigated in this study. The researchers stated delays in pregnancy and an excess of female offspring (male-to-female birth ratio) in boron-exposed workers when compared with the control group. However, both findings were statistically not significant.

More detailed data on the daily boron exposure levels and boron concentrations in biological fluids of workers living in Kuandian County were reported by the same team in 2008.[29] The major aim of this study was to identify an accurate biomarker of boron exposure that could be used in worker populations. However, the daily boron exposure levels and the boron concentrations in biological fluids (urine, blood, and semen) reported in this study are also valuable for assessing the safe levels of boron exposure in occupational settings. These results will be discussed later in more detail under "Studies on Occupational Boron Exposure."

Some studies have also been performed on the sex ratios at birth related to boron exposures. Available studies have pointed to an excess of female offspring in highly boron-exposed areas.[25–28,31] However, none of these studies have identified a statistically significant relation. The study conducted by Robbins et al.[32] has brought a new approach to this issue by investigating a paternal influence of these shifts. To this end, sperm samples were assayed for Y- or X-bearing sperm cells using the fluorescence in situ hybridization method.[33] This evaluation was performed in 146 participants. The boron worker group (exposed group) was composed of 63 workers employed in boron mines or processing plants. The results obtained from the boron worker group were compared with those of two groups named "community comparison group" (moderate exposure group, $n = 39$) and "control group" (low-exposure group, $n = 44$). Interestingly, the mean Y:X ratio determined in semen samples of boron-exposed workers was the lowest, and the ratio of Y- to X-bearing sperm in the semen samples of the three exposure groups was significantly different. Moreover, boron in biological fluids (urine, blood, and semen) was significantly correlated with the Y:X ratio in sperm samples of 146 participants. Within each comparison group, however, the correlation between boron concentrations in biological fluids and Y:X ratio were statistically not significant. The authors concluded that environmental or occupational boron exposures were associated with a decrease in Y-bearing versus X-bearing sperm. Besides, the shifts in sex ratios at birth toward females was explained with the lower Y:X ratios in the sperm of boron-exposed men.

These results, however, are in contradiction with the results of animal experiments. Boron-mediated shifts in the sex ratio at birth toward females were not identified in a multigeneration study in rats[3] and a continuous breeding study in mice.[17] Also, the results of the recent studies investigating the association between exposure to endocrine-disrupting chemicals and the proportion of Y-bearing sperm are controversial. Tiido et al.[34] investigated the correlation between the proportion of Y-bearing sperm and persistent organohalogen pollutants (POPs) exposure biomarkers (PCB-153 and *p,p′*-DDE) in subjects from Greenland, Poland, Ukraine, and Sweden. Positive associations between the proportion of Y-bearing sperm and POP exposure biomarkers were determined in Swedish fishermen. However, this was not corroborated in three other study populations. A similar study was conducted by Kvist et al.[35] The authors explored whether perfluorooctanoic acid (PFOA) and perfluorooctanesulfonate (PFOS) affect the sperm sex chromosome ratio in subjects from Greenland, Poland, and Ukraine. A positive association between exposure to PFOA/PFOS and the Y:X ratio in men was not identified. Under this circumstances, the Y:X ratios in sperm cannot be considered as a determinant of reproductive toxicity. Also, the seemingly positive results were criticized by Scialli et al.[36] because of methodological shortcomings. These authors concluded that their study conducted in boron mining and processing area of China provided no evidence for boron-mediated reprotoxic effects in humans.

Adverse effects of chronic boron exposure on the semen parameters of men living in the same boron mining and processing area in Kuandian County were also reported by Robbins et al.[30] The semen quality parameters (total sperm count, sperm concentration, motility, and morphology) and sperm DNA integrity measures (aneuploidy, DNA strand breaks, and apoptosis) were compared between boron workers ($n = 66$), the "community comparison group ($n = 59$)," and the "control comparison group ($n = 67$)." Statistically significant differences were not reported. The most remarkable finding was the statistically significant positive correlation between the percentage of sperm cells with normal morphology and boron concentrations in blood or urine. However, this was statistically not significant after controlling for age, smoking status, abstinence interval, alcohol intake, pesticide exposure, and Mg blood concentrations. The authors also reported some additional correlations between increased boron concentrations in blood/urine and decreased specific sperm defects (sperm head, neck, and midpiece defects). However, these correlations were again statistically not significant. Consequently, boron-mediated unfavorable effects on sperm quality parameters and DNA integrity of sperm cells could not be identified in occupationally boron exposed Chinese workers.

3.3.1.2.1 Studies on Reproductive and Developmental Effects of Boron in Turkey

Turkey has the largest boron deposits worldwide; the total boron deposits of Turkey (851 million tons on the basis of B_2O_3) corresponds to about 72.2% of the world's boron deposits.[37] The major boron deposits in Turkey are in Bigadiç, Emet, Kırka, and Kestelek, which are located south of the Marmara Sea. Especially, the water boron contents in some villages of Bigadiç are extremely high compared to the upper limit suggested by the "WHO-Guidelines for Drinking Water Quality".[38] The limit of boron in drinking water recommended by this guideline is 2.4 mg/L. Therefore, studies on the adverse effects of boron on reproduction in humans are important for the populations living in such surroundings. In this regard, Bigadiç and its surroundings are the most preferred areas for studies on boron-mediated adverse effects in humans. Osmanca and İskele are two small villages located in the vicinity of Bigadiç, being of special importance in this regard. According to published studies, boron contents of drinking waters varied between 1.72 and 3.97 mg B/L and 1.42–29.0 mg B/L in the surroundings of Osmanca and İskele villages, respectively.[31,39] Due to these characteristics, these are the preferred areas for epidemiological studies.

The first study on boron-mediated effects on the primary infertility in humans was conducted in the vicinity of Bigadiç by Sayli et al.[31] Two regions were defined in this study and compared in terms of primary infertility. Region I and region II were designated as high boron exposure area (2.05–29 mg B/L in drinking water) and low boron exposure area (0.03–0.40 mg B/L in drinking water), respectively. The study was designed as a retrospective cohort study. The authors compared the reproductive history between families living in regions I and II by identifying married adults (probands) who were able to provide information about their family pedigrees covering three generations. Birth of a living child was considered as an evidence of fertility. Primary infertility was retrospectively evaluated up to three generations of 159 probands (1068 families) living in region I and up to three generations of 154 probands (610 families) living in region II. In region I (the high boron exposure area), 95.7% of marriages produced at least one child, and primary infertility was reported as 2.34%. In region II (low-exposure area), 95.6% of the marriages produced at least one child, and the primary infertility rate was reported as 2.62%. The most interesting finding of this study was the differences in the sex ratios calculated for region I and region II. The sex ratios were 0.89 (189 males/212 females) and 1.04 (247 males/237 females) for region I and region II, respectively. According to this study, more females than males were born in region I. This difference, however, was statistically not significant. The study finally concluded that primary infertility rate was statistically not different between high-exposure (region I) and low-exposure (region II) areas.

Some additional studies providing additional information on boron-mediated reprotoxic effects in humans were conducted by the same author in the same region between 1998 and 2003.[31,40–42] Boron-mediated reprotoxic effects in humans have not been identified. These studies have been summarized in Table 3.4.

The most recent comprehensive study on the reproductive toxicity in occupationally and/or environmentally boron-exposed workers has been conducted by Duydu et al.[43,44] in the Bandırma Boric Acid Production Plant, which is located on the south coast of the Marmara Sea. Two hundred and four workers (male) were enrolled in this study. Forty-nine of them were controls (<48.5 ng B/g blood). The rest of the workers were allocated according to their blood boron concentrations, to groups of low ($n = 72$, 48.5–100 ng B/g blood), medium ($n = 44$, 100–150 ng B/g blood), and high ($n = 39$, >150 ng B/g blood) exposure. The external and internal boron exposure was characterized by determining the boron concentrations in biological samples (blood, urine, and semen), in workplace air, in food, and in water sources. The mean calculated daily boron exposure of the highly exposed group was 14.45 ± 6.57 (range: 3.32–35.62) mg/day. Unfavorable effects of boron exposure on the reproductive toxicity indicators (concentration; motility; morphology of sperm cells; and blood levels of FSH, LH, and testosterone) have not been identified in this study.[43,44]

DNA damage in sperm cells of workers employed in the boric acid production plant was also determined within this study. The DNA integrity of sperm cells was determined using the Comet assay using both alkaline and neutral conditions. A boron-dependent increase in the DNA damage of sperm cells was not identified between the control, low-, medium-, and high-exposed groups.

Interestingly, under the occupational exposure conditions in the Bandırma boric acid production plant, the DNA damage (tail % intensity in neutral Comet assay) in sperm cells was decreasing versus increasing blood boron concentrations in boron-exposed workers. This was statistically significant. Also, this study reported statistically significant correlations between sperm motility/morphology parameters (motility %, motile sperm counts, normal morphology %, neck/mid peace defects %, and tail defects %) and DNA integrity (tail % intensity) under neutral Comet assay conditions. In contrast to the animal experiments, these data pointed to boron-mediated protective effects.[45,46] The results of the epidemiological studies conducted in China and Turkey are compiled in Table 3.5.

Table 3.4 Studies on Boron-Exposure Related Primary Infertility in Boron-Rich Areas of Turkey

Study Method	Study Area	Study Population	End Points	References
Pedigree technique; Probands have provided information about their family pedigrees covering three generations.	*Boron-rich area:* Villages of Balıkesir, Eskişehir, and Kütahya. Boron concentration in drinking water: 0.7–29 mg/L *Control area:* Ankara and Turkey in general.	*Boron-rich area:* 5934 Marriages were evaluated by means of the information from 927 probands. *Control area:* 51 Families were evaluated in Ankara. 49856 Families were evaluated throughout Turkey.	Primary infertility was the major end point investigated. Evidence on primary infertility was not identified.	Sayli 1998.[31]
The marital state and childbearing were assessed through probands.	*Boron rich area:* Villages of Balıkesir, Eskişehir and Kütahya. *Control area:* Çamlıdere, Ankara, Balya Balıkesir, general population.	The study deals with the marriages of male and female sibs. 2197 Subjects were participated and 12981 marriages were evaluated within this study.	Evidence on primary infertility was not identified.	Sayli 2001.[41]
Questionnaire-based study	Bandırma, boric acid production plant.	The study was performed in two stages. At the first stage, 191 workers were studied. At the second stage, 712 workers were studied. Control group: 91 workers employed in a sulfuric acid plant.	Evidence on primary infertility was not identified.	Sayli 2003.[42]

Table 3.5 Comparison of the Results of Two Major Epidemiological Studies on Reprotoxic Effects of Boron Exposure in Occupational Settings

Epidemiological Studies		External Exposure	Internal Exposure Parameters			Unfavorable Effects on the Reproductive Toxicity Indicators						
Studies	Exposure groups	DBI[c] (mg/day)	Urine boron (mg B/L)	Blood boron (ppb)	Semen boron (ppb)	Sperm concentration parameter	Sperm motility parameter	Sperm morphology parameter	FSH	LH	Test[e]	Comet (DNA in tail, %)
Chinese study	Control comparison	2.3[a] (3.0) $n=15$	2.0 (0.9) $n=67$	47.9 (24.1) $n=67$	214.0 (113.9) $n=67$	NI[b] $n=67$	NI $n=67$	NI $n=67$	—	—	—	NI $n=44$
Xing 2008[29]	Community comparison	4.3 (3.1) $n=15$	5.5 (15.6) $n=59$	96.1 (92.1) $n=59$	310.6 (245.3) $n=59$	NI $n=59$	NI $n=59$	NI $n=59$	—	—	—	NI $n=39$
Robbins 2010[30]	Boron workers	41.2 (37.4) $n=15$	16.7 (31.4) $n=66$	499.2 (790.6) $n=66$	785.6 (605.7) $n=66$	NI $n=66$	NI $n=66$	NI $n=66$	—	—	—	NI $n=63$
Turkish study	Control group	4.68 (1.63) $n=49$	2.59 (1.32) $n=49$	<48.5 (LOQ)[d] $n=49$	807.92 (1625.58) $n=49$	NI $n=48$	NI $n=48$	NI $n=47$	NI $n=49$	NI $n=49$	NI $n=49$	NI $n=48$
Duydu 2011,[43] 2012[45]	Low exposure	7.39 (3.97) $n=72$	5.01 (2.07) $n=72$	72.94 (15.43) $n=72$	1427.07 (1939.03) $n=72$	NI $n=68$	NI $n=68$	NI $n=65$	NI $n=72$	NI $n=72$	NI $n=72$	NI $n=68$
	Medium exposure	11.02 (4.61) $n=44$	7.03 (2.37) $n=44$	121.68 (15.62) $n=44$	1482.19 (1410.71) $n=44$	NI $n=43$	NI $n=43$	NI $n=42$	NI $n=44$	NI $n=44$	NI $n=44$	NI $n=43$
	High exposure	14.45 (6.57) $n=39$	9.83 (5.13) $n=39$	223.89 (69.49) $n=39$	1875.68 (2255.07) $n=39$	NI $n=39$	NI $n=39$	NI $n=37$	NI $n=39$	NI $n=39$	NI $n=39$	NI $n=39$

[a]Mean (SD).
[b]Not identified.
[c]Total daily boron intake.
[d]Limit of quantitation.
[e]Testosterone.

Semen parameters in boron-exposed populations living in Osmanca and İskele villages were studied by Korkmaz et al.[39] The daily boron exposures of exposed ($n = 34$) and control subjects ($n = 34$) were 6.5 and 1.4 mg/day, respectively. Sperm concentration, morphology, and motility parameters were compared between exposed and control groups, to evaluate fertility effects of boron in male volunteers. However, boron-mediated unfavorable effects on the above-mentioned reproductive toxicity indicators were not identified.

3.3.2 Dose–Response Assessment

Generally, there are no adequate data for a dose–response assessment in humans. The key data are obtained from animal experiments. In animal experiments, however, dose–response studies were generally conducted at relatively high doses, compared with realistic exposure levels in humans. Therefore, the high doses used in animal experiments must be extrapolated to lower doses and translated to humans. Accordingly, the dose–response assessment includes two steps: The first step is to assess the dose response relationships, and the second step consists of extrapolation from high doses to low doses.

As was discussed in detail under the "hazard assessment" section, boron was neither genotoxic nor carcinogenic in cancer bioassays.[13] Toxic effects of boron on reproduction and development were identified in multiple experimental animal toxicity studies at high doses.[3,18,19,47,48] In this context, reproductive and developmental effects are considered as the critical toxic effects of boron. These effects have a threshold, that is, a dose below which no such adverse response will occur. This type of assessment is referred to as "nonlinear dose–response assessment".

3.3.2.1 Nonlinear Dose Response Assessment

NOAEL and BMDL are the most widely used points of departure for extrapolation to lower doses in nonlinear dose response assessment. The LOAEL can also be used (with additional uncertainty factors, UFs) as POD in the absence of an experimentally determined NOAEL. The NOAEL represents the highest experimental dose for which no adverse health effects have been documented. However, the NOAEL approach has its limitations; it does not take into account the slope of the dose–response curve, and the NOAEL must be a dose that had been tested experimentally.

The BMD approach overcomes such limitations by taking into account the slope of the dose–response curve. Therefore, its use in nonlinear dose response assessments in international scientific organizations (World Health Organization (WHO), United States Environmental Protection Agency (USEPA), etc.) and in regulations is of increasing importance. The BMD is a statistical lower confidence limit for a dose that produces a predetermined change in a response related to an adverse effect (called the benchmark response) compared to the background.[49]

The first step in nonlinear dose response assessment is to identify the NOAEL for the most sensitive end point (critical NOAEL). The NOAEL for the testicular effects (most sensitive end point) of boron in male rats was 17.5 mg B/kg bw/day (Table 3.2). This NOAEL can be used to extrapolate to a safe daily boron exposure level for men. However, the developmental toxic effects of boron are manifested at lower levels than are the reprotoxic effects. The NOAEL and $BMDL_{05}$ for the developmental effects of boron in rats were 9.6 mg B/kg bw/day and 10.3 mg B/kg bw/day as shown in Table 3.3. However, some international scientific organizations (WHO, USEPA etc.) prefer to use the BMDL (if available) as the POD, due to the above-mentioned limitations of the NOAEL approach. Therefore, 10.3 mg B/kg bw/day ($BMDL_{05}$) and 17.5 mg B/kg bw/day (NOAEL) can be used in computing the reference dose (RfD) for boron-mediated reproductive and developmental effects, respectively.

3.3.2.2 Reference Dose

The RfD is defined as an estimate of the daily exposure to the human population, including sensitive subgroups such as children, that is likely to be without deleterious effect even if continued exposure occurs over a lifetime. RfDs are used for health effects that are thought to have a threshold for producing effects. It can be derived from an NOAEL, LOAEL, or BMD, with UFs generally applied to reflect limitations of the data used and generally expressed in milligrams per kilogram body weight per day.[50] Tolerable daily intake (TDI; the common term in WHO publications) is also used as a term having the same meaning as RfD. Accordingly, the RfD or TDI levels are computed by the following formula:

$$\mathrm{RfD(TDI)} = \mathrm{NOAEL(BMD)/UF}.$$

3.3.2.3 Uncertainty Factors

The UFs are used to extrapolate the doses from experimental animals to an average human and from average human to potentially sensitive subpopulations.[51] The standard default UF of 100 allows derivation of a safe human dose from an animal NOAEL in a chronic toxicity test.[51,52] This default value comprises interspecies differences (10-fold) and interindividual variability (10-fold) in humans. These two factors of 10 are default values and are composed of interspecies differences in toxicodynamics (TDs, 10^{04})/TKs (10^{06}) and interindividual variability in TDs (10^{05})/toxicokinetics (10^{05}). These default UFs specified for the TD and TK differences can be replaced with data-derived values (chemical-specific adjustment factors) when adequate data are available.[53,54]

Valid study results are also available on the TKs of boron in animal and humans.[55,56] Accordingly, the UF of 100 is not an appropriate choice in computing an RfD for boron. Therefore, international scientific organizations and some authors have computed a data derived composite UF reflecting the TDs and TKs of boron in both animals and humans. However, the RfD of boron computed by these scientific international organizations and authors are unfortunately inconsistent with each other. In other words, the authors and international scientific organizations are deriving different UFs by using the same existing experimental data set. As a natural consequence of this discrepancy, different UFs are leading to completely different RfDs or TDI levels for boron as shown in Table 3.6. This discrepancy is still a matter of discussion in determining a safe daily intake level of boron.

3.3.2.4 Interpretation Based on Blood Boron Concentrations

In chemical risk assessment, the comparison of NOAEL or BMD levels with human exposure levels is a common procedure and termed as the margin of exposure (MOE) assessment. This procedure requires external exposure levels (oral doses), and the majority of available chemical risk assessments are based on this approach. However, the available data provide blood boron concentrations measured at the NOAEL for the critical effects of boron. Therefore, a direct comparison of measured blood boron concentrations in animals to measured blood boron concentrations in humans is possible

Table 3.6 The Proposed Composite UFs by Different Scientific Organizations and Authors are Leading to Completely Different Daily Boron Intake Levels Considered to be Safe

The Authors and International Scientific Organizations	Composite UF[b]	POD[c] (mg B/kg-bw/day)	POD/UF (mg B/kg-bw/day)
Based on the Developmental Effects in Rats as the Critical End Point			
Murray, F.J., 1995[57]	32	9.6 (NOAEL)	0.3 (RfD)[d]
Dourson et al., 1998[4]	60	9.6 (NOAEL)	0.16 (RfD)
Hasegawa et al., 2013[58]	78	10.3 ($BMDL_{05}$)	0.13 (TDI)[e]
ECETOC, 1995[9]	30	9.6 (NOAEL)	0.32 (TDI)
EGVM, 2003[59]	60	9.6 (NOAEL)	0.16 (UL)[f]
US EPA, 2004[2]	66	10.3 ($BMDL_{05}$)	0.16 (RfD)
EFSA, 2004[60]	60	9.6 (NOAEL)	0.16 (UL)[g]
EU[a], 2007 (general population)[61]	30	9.6 (NOAEL)	0.32 (DNEL)[h]
EU[a], 2007 (workers)[61]	15	9.6 (NOAEL)	0.64 (DNEL)
WHO, 2011[38]	60	10.3 ($BMDL_{05}$)	0.17 (TDI)
Based on the Reproductive Effects in Rats as the Critical End Point			
Hasegawa et al., 2013[58]	68	17.5 (NOAEL)	0.26 (TDI)
ECETOC, 1995[9]	30	17 (NOAEL)	0.57 (TDI)

[a]European Union Risk Assessment Report.
[b]UFs.
[c]Point of departure.
[d]RfD.
[e]Tolerable daily intake.
[f]Safe UL.
[g]Tolerable upper intake level.
[h]Derived no-effect levels.

Table 3.7 The Blood Boron Concentrations Measured in Animals at the NOAEL and the Blood Boron Concentrations Measured in Humans under Environmental or Occupational Exposure Conditions

Experimental or Epidemiological Studies	NOAEL[a] (rat)	Blood B (ng/g)
Experimental Study		
Price et al., 1996[19]	9.6[b]	1270
Weir and Fisher, 1972[3]	17.5[c]	2300[d]
Environmental Studies		
Yazbeck et al., 2005[27]		159.1[e]
Harari et al., 2012[62]		430[e,f,g]
Occupational Studies		
Culver et al., 1994[7]		260[e]
Xing et al., 2008[29]		499.2[e]
Duydu et al., 2011[43]		223.89[e]

[a]NOAEL is expressed in milligrams per kilogram body weight per day.
[b]NOAEL for the developmental effects in rats.
[c]NOAEL for the reproductive effects in rats.
[d]Extrapolated from the published data by Price et al., 1997.
[e]The highest mean blood boron concentration reported in that study.
[f]Geometric mean.
[g]Mean blood B concentration in pregnant women.

to determine the safe MOE. In this way, the interspecies and interindividual UFs related to TK parameters can be reduced or even eliminated. This issue will be discussed in the section "risk characterization". The blood boron concentrations in humans and the blood boron concentrations in animals at the NOAEL are compiled in Table 3.7.

3.3.3 Exposure Assessment

3.3.3.1 Dietary Boron Intake

Humans are exposed to boron primarily through food and water. Total boron intake data based on food and water point to 1–3 mg B/day and 0.2–0.6 mg B/day, respectively.[59,63] Rainey et al.[64,65] reported very detailed results on daily boron intakes (DBI) in the USA by using the Continuing Survey of Food Intakes by Individuals (CSFIIs) as shown in Table 3.8.

Other possible contributors to the daily boron intake are dietary supplements, cosmetics, and consumer products. The Expert Group on Vitamins and Minerals (EGVM) reported a Maximum Estimated Intake level of 14 mg B/day by taking the all-possible boron sources into account (2.6 (food) + 0.6 (water) + 10 (supplements) + 0.47 (cosmetics and consumer products)) for the population living in the UK.[59] However, the contribution of boron in the drinking water to the overall daily boron intake can be much higher in boron-rich areas of countries such as Turkey, Chile, Argentina, and China.

Table 3.8 Dietary Boron Intake Levels

Studies	n (Respondents)	Gender	Mean (mg B/day)	Percentiles	
				5th	95th
CSFII[a] 1989–1991[64]	11,009	Male	1.17	0.43	2.42
		Female	1.01	0.40	2.18
CSFII 1994–1996[65]	15,267	Male	1.28	0.53	2.40
		Female	1.01	0.41	1.87

[a]CSFIIs.

3.3.3.2 Studies on Environmental Boron Exposure

Some studies on environmental boron exposure provide data on boron concentrations in drinking water sources and boron concentrations in blood and/or urine samples of humans. The study conducted in Northern France by Yazbeck et al.[27] reported relatively higher blood boron concentrations in men living in municipalities with >0.30 mg/L boron in drinking water, when compared with those living in municipalities with boron water levels of <0.3 mg/L. However, the difference of mean blood boron concentrations in men was statistically not significant. This statistical nonsignificance might be a reflection of the relatively low level of daily exposure via drinking water. These levels of boron concentrations in drinking waters are low compared with the boron concentrations in drinking water sources close to boron deposits. Higher boron concentrations in drinking waters were reported in studies conducted in Turkey and Chile.

In Turkey, environmental boron exposure via drinking water (1.4–6.5 mg B/L) was studied in persons ($n = 34$) living in the Osmanca and İskele villages of Balıkesir, by Korkmaz et al.[39] Control subjects ($n = 34$) were selected from distant rural areas with boron concentrations <0.01 mg/L in drinking water. The calculated mean daily boron exposure for the exposed and control groups was 6.5 and 1.4 mg/day, respectively, and the difference was statistically significant ($p < 0.05$).

Chile is among the countries with natural deposits of boron as well as Turkey, Russia, USA, and China. High levels of boron concentrations in drinking waters are reported in the Arica region of Northern Chile. The association between boron concentration in drinking water and boron concentration in urine samples of 109 healthy male volunteers was studied by Cortes et al. (2011). The correlation between boron levels in urine and in drinking water was statistically significant (Spearman correlation, $p = 0.64$).

The transfer of boron from exposed mothers to fetuses and to breast-fed infants was investigated in the areas in northern Argentina and Chile with up to 5–10 mg B/L in drinking water by Harari et al.[62] This study provides data also on the relation between boron concentrations in drinking water and blood/urine/breast milk boron concentrations. The boron-exposed persons were recruited from San Antonio de los Cobres (Northwestern Argentina, $n = 11$) and from Arica (Chile, $n = 24$). Control samples were collected in Santiago (Chile, $n = 11$). The maternal boron concentrations in biological fluids of controls were significantly lower than in the samples from both San Antonio de los Cobres and Arica. This reflects the environmental boron exposure via drinking water in these cities. The overall results of studies on environmental boron exposure are compiled in Table 3.9.

Human beings in daily life are mostly exposed to boron via drinking water and food (fruits and vegetables). Therefore, oral exposure is the most important route of boron exposure. In occupational settings, however, exposure to borate dust via inhalation is the primary route of exposure. Dermal absorption of boron across intact skin in humans and laboratory animals has not been reported.[10,11] Apparently, boron absorption from intact skin is negligible. Accordingly, both oral and inhalational routes of boron exposure should be taken into consideration in estimating the daily boron intake levels for the workers employed in boron mining areas and/or production facilities of boron compounds.

3.3.3.3 Studies on Occupational Boron Exposure

The available studies on boron exposure in occupational settings generally provide more detailed data on the daily boron intake, compared to environmental boron exposure studies.

The first comprehensive study on the relation between daily boron exposure levels and boron concentrations in biological fluids (blood and urine) was published by Culver et al.[7] The study was performed in the workers employed in a borax packaging and shipping facility of a borax and boric acid production plant. The workers were selected from low- ($n = 4$), medium- ($n = 5$), and high- ($n = 5$) exposure areas. The total daily boron exposure was calculated as the sum of the dietary (food + fluid) intake plus inhaled amount of the boron during the work shift. Fourteen workers were studied throughout full shifts for five consecutive days. A statistically significant relation was reported between the total daily boron intake levels and blood/urine boron concentrations. The results of this study are summarized in Table 3.10.

Another comprehensive study was performed in Kuandian City, People's Republic of China. The sampling procedure was performed in two sampling periods. The first part of the study was completed in 2003 by taking samples from 60 boron-exposed workers employed in boron mines or processing plants and from nine workers (controls) living in an area where environmental boron levels were low. The total daily boron exposure (food, fluid, and inhalation) of exposed and control groups was 37.0 ± 77.5 and 1.2 ± 0.8 mg/day, respectively, in this pilot study.[29] The main study was conducted in 2004 in the same area. Seventy-four boron-exposed workers were enrolled in this study. Two separate groups were generated, representing areas of high (community comparison, $n = 61$) and low (control comparison, $n = 68$) environmental boron. Boron concentrations were determined in blood, urine, and semen samples for the full sample size

Table 3.9 Studies on Environmental Boron Exposure

Environmental Boron Exposure Studies		External Boron Exposure		Internal Boron Exposure			
Studies	Exposure Groups	Boron Concentration in Drinking Water	DBI[a]	n	Boron Concentration in Urine	Boron Concentration in Blood	Boron Concentration in Breast Milk
Yazbeck et al., 2005[27]	Zone I + II	<0.3 mg/L	—	143	—	123.0 ng/g	—
	Zone III	≥3 mg/L	—	36	—	159.1 ng/g	—
Korkmaz et al., 2011[39]	Control Group	<0.01 mg/L	1.4 mg/day	34	—	—	—
	Exposed Group	1.4–6.5 mg/L	6.5 mg/day	34	—	—	—
Cortes et al., 2011[66]	Low exposure	<3 mg/L (0.22–3)	—	53	3.41 ± 2.13 mg/L (0.45–13.50)	—	—
	Medium/high exposure	>3 mg/L (3.04–11.3)	—	56	7.90 ± 4.31 mg/L (1.21–17.40)	—	—
Harari et al., 2012[62]	Control (Santiago)	0.19^b mg/L (0.15–0.26)	—	11	1.15^b mg/L (0.19–3.2)	$35^{b,c}$ μg/L (21–66)	38^b μg/L (18–180)
	Exposed (San Antonio de los Cobres)	5.2^b mg/L (4.8–6.0)	—	11	15.7^b mg/L (11.0–23.3)	430^b μg/L (210–1500)	255^b μg/L (140–360)
	Exposed (Arica)	7.9^b mg/L (4.2–10.53)	—	24	12.8^b mg/L (6.3–25.0)	$380^{b,c}$ μg/L (125–1360)	270^b μg/L (140–695)

Mean ± SD, range in brackets.
[a]Calculated daily boron intake.
[b]Geometric mean.
[c]Plasma boron concentration.

Table 3.10 Boron Concentrations in Blood and Urine Samples as a Function of the Daily Boron Exposure Levels of Workers Employed in Production Facilities of Boron Compounds or in Boron-Mining Facilities

Occupational Boron Exposure Studies		External Boron Exposure		Internal Boron Exposure	
Studies	Exposure Groups	Mean Inhaled Boron (mg/day)[c]	Mean Total Daily Boron Intake (mg/day)[d]	Mean Blood Boron Levels (ng/g)	Mean Urine Boron Levels (mg/g Creatinine)
Culver et al., 1994[7]	Low exposure[a] $n = 4$	4.70 ± 1.69	6	90	NP
	Medium exposure[a] $n = 5$	16.18 ± 11.64	19.3	NP	NP
	High exposure[a] $n = 5$	24.77 ± 15.35	27.9	260	10.72[e]
Xing et al., 2008[29]	*Samples collected in 2003*				
	Control comparison[a] $n = 9$	NP	1.2 ± 0.8 (0.4–3.2) $n = 15$	22.1 ± 6.7 (14.0–33.2)	1.1 ± 0.8 (0.2–1.7)
	Boron workers[a] $n = 60$	NP	37.0 ± 77.5 (2.3–469.3) $n = 15$	204.8 ± 356.8 (27.1–2003.5)	16.0 ± 37.1 (0.7–235.7)

	Samples collected in 2004				
	Control comparison[a] $n = 68$	NP	2.3 ± 3.0 (0.4–12.7) $n = 15$	48.0 ± 23.9 (8.2–113.0)	1.7 ± 0.6 (0.9–3.2)
	Community comparison[a] $n = 61$	NP	4.3 ± 3.1 (0.7–14.0) $n = 15$	96.5 ± 90.8 (3.3–536)	4.8 ± 6.5 (0.8–50.7)
	Boron workers[a] $n = 74$	NP	41.2 ± 37.4 (11.6–111.4) $n = 15$	499.2 ± 790.6 (20.4–3568.9)	19.3 ± 30.9 (1.4–168.1)
Duydu et al., 2011[43]	Control[b] $n = 49$	0.23 ± 0.79 (<LOQ − 4.09)	4.68 ± 1.63 (0.2–7.54)	<48.5 (LOQ)	2.59 ± 1.32 (0.78 ± 5.62)
	Low exposure $n = 72$	1.15 ± 3.14 (<LOQ − 22.16)	7.39 ± 3.97 (2.56–24.72)	72.94 ± 15.43 (48.46 ± 99.91)	5.01 ± 2.07 (2.38 ± 13.54)
	Medium exposure[b] $n = 44$	1.47 ± 2.69 (<LOQ − 11.34)	11.02 ± 4.61 (2.56–20.88)	121.68 ± 15.62 (100.51 ± 146.07)	7.03 ± 2.37 (2.57 ± 13.43)
	High exposure[b] $n = 39$	2.58 ± 4.96 (<LOQ − 26.9)	14.45 ± 6.57 (3.32–35.62)	223.89 ± 69.49 (152.82 ± 454.02)	9.83 ± 5.13 (3.34 ± 32.68)

[a]Personal air sampling was performed by using the Institute of Occupational Medicine (IOM) sampler.
[b]Personal air sampling was performed by using a closed-face cassette (37 mm).
[c]The amount of inhaled boron was calculated on the basis of the respiratory volume of 10-m^3/8 h shift.
[d]Total daily boron intake by food + fluid + inhalation.
[e]Average postshift concentration for the high-exposure group, NP: numerical value was not provided in the related paper.

of 203 men. The daily boron intake by food and fluids was determined by collecting duplicate food and fluid samples from 15 men from each of the exposure groups. The total daily boron exposure was calculated as the sum of boron in 24-h duplicate food and fluid intakes plus inhaled amount of boron at the workplace for 8-hr work shift. The total daily boron exposure (food, fluid, and inhalation) of exposed, community comparison, and control groups were 41.2 ± 37.4, 4.3 ± 3.1, and 2.3 ± 3.0 mg/day, respectively, in the main study.[29] A statistically significant relation between total daily boron exposure (food + fluid + inhalation) and boron concentrations in blood/urine was also reported in this study and thus confirmed the results by Culver et al.[7] The results of this study are summarized in Table 3.10.

The most recent comprehensive study on the exposure assessment of boron in occupational settings was published by Duydu et al.[43–46] The study was conducted in the boric acid production plant, Bandırma, Turkey. Two hundred and four workers were enrolled in this study. These workers were classified as control, low-exposure, medium-exposure, and high-exposure groups according to their blood boron concentrations. The results of this study corroborate those of the former studies by confirming a statistically significant relation between total daily boron exposure (food + fluid + inhalation) and boron concentrations in blood/urine. The results of this study are summarized in Table 3.10.

3.3.3.4 Occupational Exposure Limits

The occupational exposure limits (OELs) are set to ensure an adequate margin between harmful and safe exposure levels. The proposed OELs are expressed in different terms by different scientific organizations. For instance, American Conference of Governmental Industrial Hygienists (ACGIH) is used "Threshold Limit Values (TLV)." TLV is a level to which it is believed a worker can be exposed day after day for a working lifetime without adverse health effects. The TLV is a recommendation made by the ACGIH and has no regulatory status. The Occupational Safety and Health Administration (OSHA) sets enforceable permissible exposure limits (PELs) to protect workers against the health effects of exposure to hazardous substances. OSHA PELs are based on an 8-h time-weighted average exposure. In Germany, the Deutsche Forschungsgeminschaft uses the term of "Maximum Workplace Concentration (Maximale Arbeitsplatz Konzentration, MAK)" for expressing the exposure limits in the workplaces. The MAK value is defined as the maximum concentration of a chemical substance (as gas, vapor, or particulate matter) in the workplace air that generally does not have known adverse effects on the health of the employee nor cause unreasonable annoyance (e.g., by a nauseous odor) even when the person is repeatedly exposed during long periods, usually for 8 h daily but assuming on average a 40 h working week. As a rule, the MAK value is given as an average concentration for a period of up to one working day or shift. The MAK values are set to promote the protection of health at the workplace.[67] The OELs of different organizations are compiled in Table 3.11.

Table 3.11 Occupational Exposure Limits of Boric Acid and Sodium Tetraborates

Boric Acid and Sodium Borates	ACGIH[a]	OSHA	MAK[a]
Boric acid, H_3BO_3	2 mg/m^3	–	10 mg/m^3
Disodium tetraborate anhydrous, $Na_2B_4O_7$	2 mg/m^3	–	0.75 mg B/m^3
Disodium tetraborate pentahydrate, $Na_2B_4O_7 \cdot 5H_2O$	2 mg/m^3	–	0.75 mg B/m^3
Disodium tetraborate decahydrate, $Na_2B_4O_7 \cdot 10H_2O$	2 mg/m^3	–	0.75 mg B/m^3

[a]Based on inhalable dust measurements.

3.3.4 Risk Characterization

Risk characterization, as was previously outlined, is the stage of risk assessment that integrates information from hazard identification, dose–response assessment, and exposure assessment.

3.3.4.1 Hazard Assessment

According to the globally harmonized system of classification and labeling of chemicals/ the regulation on classification, labeling and packaging of substances and mixtures acute toxicity hazard category of boric acid and sodium borates corresponds to "category 5" ($LD_{50} > 2000$ mg/kg-bw). Inorganic borates are generally nonirritant to the skin and the eye. According to the results of the relevant studies, boric acid is not genotoxic, mutagenic, and carcinogenic as was described before under the related sections. Therefore, these hazards are not of primary concern to the relevant regulatory bodies in regard to boron exposure. On the other hand, decreased fetal body weight and testicular toxicity were identified as the critical toxic effects of boron exposure in animal experiments. The NOAELs of boron for developmental and reproductive effects in rats were 9.6 and 17.5 mg/kg-bw/day, respectively. The BMD was also computed by Allen et al.[20] for the developmental effects of boron. This analysis was based on the decreased fetal body weight and gave a $BMDL_{05}$ of 10.3 mg B/kg-bw/day. These results are of particular interest, since these PODs can be used to derive safe daily boron intake levels (RfD, TLV, etc.). Boron-mediated unfavorable effects on the endocrine control of male rats were also reported. However, these effects occurred at relatively higher doses than those that caused testicular toxicity in rats.

The results of the published environmental and/or occupational epidemiology studies did not support the experimental findings. Boron-mediated unfavorable effects equal to the experimental animals have not been reported in well-established epidemiological studies so far.

3.3.4.2 Dose–Response Assessment

The reproductive and developmental effects are considered to have a threshold. The relation between the dose and threshold toxic effects is translated to safe exposure levels by the components of nonlinear dose response assessment. The NOAEL, LOAEL, and

BMD are the most often referred components of this type of assessment to derive an RfD (or TLV) by dividing the related POD (NOAEL, LOAEL, or BMD) by the composite UF. The NOAELs for the developmental/reproductive effects and $BMDL_{05}$ for the developmental effects of boron exposure are available to calculate a safe daily boron intake level. To this end, many studies have been carried out by different authors and scientific organizations to establish a safe daily boron intake level for many years. However, these analyses were concluded with completely different RfDs being derived. The main reason for this inconsistency was the differences of the assumptions being made by the authors in computing the composite UFs. The proposed composite UFs varied between 15 and 78 as shown in Table 3.6.

The WHO has adopted the composite UF of 60 in drinking water quality guideline.[38] This composite UF was calculated on the basis of the data published by Dourson et al.[4] Accordingly, the default UF of 10 was used for the interspecies uncertainty due to insufficient data.

$$\text{Interspecies uncertainty} = \text{TK} \times \text{TD},$$

$$\text{Interspecies uncertainty} = 10^{06}(\text{default value}) \times 10^{04}(\text{default value}),$$

$$\text{Interspecies uncertainty} = 4 \times 2.5,$$

$$\text{Interspecies uncertainty} = \underline{10}.$$

Pregnant females are considered to be the most sensitive subpopulation. Therefore, the TK portion of the UF for interindividual variability was computed by using the glomerular filtration rate (GFR) in healthy pregnant females. The mean GFR in pooled data were 144 ± 32 ml/min for healthy pregnant women in their last trimester. To cover 95% of the population, 2 standard deviations (SDs) from the mean were subtracted from the mean GFR. Afterward the mean GFR was divided by the difference of the subtraction as shown below. The default value of 3.2 was used for the TD variation in pregnant woman due to insufficient data.

$$\text{Chemical specific value} = 144(\text{mean GFR}) - [2 \times 32(\text{SD of the mean})],$$

$$\text{Chemical specific value} = 144 - 64 = 80 \text{ ml/min},$$

$$\text{Chemical specific value} = 144/80 = 1.8,$$

$$\text{Interindividual uncertainty} = \text{TK} \times \text{TD},$$

$$\text{Interindividual uncertainty} = 1.8(\text{chemical specific value}) \times 10^{05}(\text{default value}),$$

$$\text{Interindividual uncertainty} = 1.8(\text{chemical specific value}) \times 3.2(\text{default value}),$$

$$\text{Interindividual uncertainty} = 5.76 \text{ rounded to } \underline{6}.$$

Consequently, the composite UF was calculated as follows[4]:

$$\text{Composite UF} = \text{Interspecies uncertainty} \times \text{Interindividual uncertainty},$$

$$\text{Composite UF} = 10 \times 6 = \underline{60}.$$

In spite of the inconsistency between the proposed UFs, the composite UF of 60 is used in estimating safe daily boron intake levels by the WHO,[38] EGVM,[59] and European Food Safety Authority (EFSA).[60] This UF is more conservative than the UFs proposed by European Centre for Ecotoxicology and Toxicology of Chemicals (ECETOC),[9] EU Risk Assessment Report,[61] and Murray et al.[57] as can be noticed in Table 3.6. Therefore, dividing the $BMDL_{05}$ (for the developmental effects of boron) with the composite UF of 60 provides a very conservative TDI for the general population.

$$\text{TDI} = \text{BMDL}_{05}/\text{UF},$$

$$\text{TDI} = 10.3/60,$$

$$\text{TDI} = 0.17\ \text{mg B/kg} - \text{bw/day}.$$

According to the WHO,[38] a daily boron exposure level <0.17 mg/kg-bw/day should be considered as safe for the general population. The body weight proposed by WHO[38] for an adult is 60 kg for estimating a safe daily boron intake level for a person. The daily safe level of boric acid intake can be calculated by dividing the boron concentration with the conversion factor of boric acid (Table 3.1) as the follows:

$$\text{TDI} = 0.17 \times 60 = 10.3\ \text{mg B/day},$$

$$\text{TDI} = 10.3/0.175 = 58.9\ \text{mg boric acid/day}.$$

The safe daily consumption over a lifetime is termed by the EGVM as the "Safe Upper Level (UL)." The "Safe UL" for the daily consumption of boron over a lifetime has been derived by dividing the NOAEL with the composite UF of 60 as follows:

$$\text{Safe UL} = \text{NOAEL for the developmental effects/composite UF},$$

$$\text{Safe UL} = 9.6/60,$$

$$\text{Safe UL} = 0.16\ \text{mg B/kg} - \text{bw/day}.$$

These safe daily boron intake levels proposed by different scientific organizations cover the general population. However, the primary scope of this chapter is to assess the safe daily intake levels for workers. It is widely accepted that the interindividual variations between the workers are lower than those of the general population. Therefore, a lower UF can be used in estimating safe daily intake levels for workers without compromising on safety. Accordingly, the EU Risk Assessment Report[61] and the European Chemical Agency (ECHA)[68] propose a substantially lower UF for the workers.

The default interspecies TK UFs of 4 was reduced to three in the EU Risk Assessment Report.[61] The main reason for this reduction is based on the comparison of the renal clearance between rats and humans. The renal clearance of boric acid in humans was three times slower than in rats.[56] The TD component of the interspecies uncertainty has been disregarded due to the similarity of the boric acid metabolism between rats and humans. The interindividual UFs of 5 is used as a default for the workers in the risk characterization process of ECHA.[61,68] Consequently, the composite UF for the workers proposed by the EU Risk Assessment Report can be calculated as follows:

$$\begin{aligned}
\text{Interspecies uncertainty} &= \text{TK} \times \text{TD},\\
\text{Interspecies uncertainty} &= 10^{06}(\text{default value}) \times 10^{04}(\text{default value}),\\
\text{Interspecies uncertainty} &= 3 \times 1,\\
\text{Interspecies uncertainty} &= \underline{3},
\end{aligned}$$

Interindividual uncertainty = $\underline{5}$ (default value for the workers),[68]

$$\begin{aligned}
\text{Composite UF} &= \text{Interspecies uncertainty} \times \text{Interindividual uncertainty},\\
\text{Composite UF} &= 3 \times 5 = \underline{15}.
\end{aligned}$$

The composite UF of 15 introduced a higher safe daily boron intake level for the workers when compared with the general population as shown below (the safe daily exposure levels are expressed in terms of derived no-effect levels (DNEL) in ECHA):[61,68]

$$\begin{aligned}
\text{DNEL}_{\text{worker}} &= \text{NOAEL/UF},\\
\text{DNEL}_{\text{worker}} &= 9.6/15,\\
\text{DNEL}_{\text{worker}} &= 0.64\ \text{mg B/kg} - \text{bw/day},\\
\text{DNEL}_{\text{worker}} &= 0.64 \times 70(\text{default body weight for the workers}),\\
\text{DNEL}_{\text{worker}} &= 44.8\ \text{mg B/day}.
\end{aligned}$$

According to the EU Risk Assessment Report,[61] the daily boron exposure level <0.64 mg B/kg-bw/day should be considered as safes for workers. The body weight proposed by the ECHA for the workers was 70 kg while estimating a safe daily boron exposure level for a worker.[68] The daily safe level of boric acid intake can be calculated by dividing the boron concentration with the conversion factor of boric acid (Table 3.1) as follows:

$$\begin{aligned}
\text{DNEL}_{\text{worker}} &= 0.64/0.175 = 3.66\ \text{mg boric acid/kg} - \text{bw/day},\\
\text{DNEL}_{\text{worker}} &= 3.66 \times 70 = 256.2\ \text{mg boric acid/day}.
\end{aligned}$$

However, the $DNEL_{worker}$ derived by the above-mentioned assumptions does not cover the general population as was stated in the related guideline of the ECHA.[68]

3.3.4.3 Risk Characterization for the General Population

Although the primary scope of this chapter was to discuss the safety of the occupational exposure levels of boron, the safety of the daily boron intake levels of the general population has also been briefly discussed below.

As was stated before, the EGVM estimated a daily boron intake level of 3.7 mg/day (from food, water, and consumer products) for the population living in the UK. This level of daily boron intake is lower than the estimated "Safe UL" and provides a margin of about 6 mg B/day for the supplemental intake for other sources of boron exposure.[59]

$$\text{Safe UL} = 0.16\ \text{mg B/kg} - \text{bw/day},$$

$$\text{Safe UL} = 0.16 \times 60(\text{default human body weight for the general population}),$$

$$\text{Safe UL} = 9.6\ \text{mg B/day},$$

$$\text{Available margin} = 9.6 - 3.7 = 5.9\ \text{mg B/day (for other sources of boron exposure)}.$$

Detailed results on the daily boron intake have also been reported for the population living in the USA.[64,65] The CSFII cover 11,009 and 15,267 respondents surveyed between 1989–1991 and 1994–1996, respectively. The 95th percentile boron intakes were lower than the RfD (0.16 mg/kg-bw/day) derived by the USEPA for boron (Table 3.6). The results of these surveys show that a reasonable margin of safety between the estimated daily boron intake levels and the RfD exists for the general population. Generally, there is an agreement that the estimated safe daily intake levels (RfD, TDI, UL, and DNEL) provide an adequate safety margin in order to protect the general population from the developmental and testicular effects of boron.[2,9,59]

There are very few published studies regarding the environmental boron exposure of the general population. Yazbeck et al.[27] conducted a study in Northern France to evaluate the health impact of boron in drinking water. However, estimation of the daily boron intake levels has unfortunately not included in the study. Nevertheless, the study provides a mean blood boron concentration for the highly boron-exposed individuals via drinking water (Table 3.9). The mean blood boron concentration of the high-exposure group was lower than the mean blood boron concentration determined in rats exposed to the NOAEL for the developmental effects of boron (Table 3.7). The comparison of internal doses between experimental animals and humans is a quite new approach that is named "internal dose-based risk assessment" (this approach will be discussed later). The study reported that the birth rate in high-exposure zone (zone III) was significantly higher ($p < 10^{-4}$) than that in the low-exposure zone (zone I).

Additionally, the mortality rate in zone III was significantly lower than for the population living in zone I ($p < 0.001$). The female offspring was slightly higher in zone III than in zone I. However, this difference was statistically insignificant. The authors concluded that unfavorable effects of boron exposure were not determined in this study. The available correlations between water boron levels and birth or mortality rates may indicate the beneficial effects of boron under the exposure conditions in Northern France.

3.3.4.4 Risk Characterization for the Population Living in Boron-Rich Areas

The high boron content in drinking water constitutes the dominant source of boron exposure for the population living in boron-rich areas.[31,39,62,66] Therefore, it is essential to monitor the boron concentrations in drinking water sources of boron-rich areas for an accurate estimation of the daily boron intake. Especially Turkey and Chile possess some unique areas with drinking water sources containing extremely high levels of boron. The boron standards in drinking water are 2.4 and 1 mg/L in the WHO Guidelines (2009) and EU Drinking Water Directive (98/93/EC) (1998), respectively. However, the boron content in drinking water is one order of magnitude higher than the WHO standard in İskele village of Bigadiç (29 mg/L), Turkey.[31] Nevertheless, boron-mediated unfavorable effects on the fertility have not been reported in a series of studies by Sayli et al.[31,40–42] Unfortunately, the daily boron intake levels of the study population have not been provided by the authors.

Korkmaz et al.[39] conducted a study on the effects of chronic boron exposure on human semen in the population living in Osmanca and İskele village. The study population ($n = 34$) was exposed to high levels of boron from the drinking water (1.4–6.5 mg/L). The daily boron intake for the highly exposed members of this population would be $6.5 \times 2 = 13$ mg/day when considering the default daily water consumption of 2 L. This level of daily boron intake is higher than the TDI (10.3 mg/day) and Safe UL (9.6 mg/day) even without taking into account the other sources of boron exposure. Nevertheless, boron-mediated unfavorable effects on sperm morphology, concentration, and motility parameters have not been observed in this population when compared with the control group ($n = 34$).

The procedures used to derive a safe daily boron intake level by the different scientific organizations (WHO, EFSA, EGVM, etc.) are based on the NOAEL for the decreased fetal body weight (9.6 mg B/kg-bw/day) in order to protect the entire population. However, the study published by Korkmaz et al.[39] was based on the reproductive parameters in males. Actually, the threshold for the critical effects in males is higher than the developmental effects. The NOAEL for the reprotoxic effects of boron is 17.5 mg/kg-bw/day in rats. Accordingly, a TDI can be derived for a 60-kg male by using the UF used by the WHO as follows:

$$\begin{aligned} TDI_{Men} &= NOAEL/UF, \\ TDI_{Men} &= 17.5/60, \\ TDI_{Men} &= 0.29\ \text{mg B/kg} - \text{bw/day}, \\ TDI_{Men} &= 0.29 \times 60(\text{body weight}), \\ TDI_{Men} &= 17.5\ \text{mg B/day}. \end{aligned}$$

The daily boron intake by drinking water for the highly exposed members of this population is lower than this calculated TDI_{Men} and provides a margin for other sources of boron exposure as shown below:

$$\begin{aligned} \text{Available margin} &= 17.5 - 13 \\ &= 4.5\ \text{mg B/day(for other sources of boron exposure)}. \end{aligned}$$

In this context, the lack of unfavorable effects on the semen parameters in the study conducted by Korkmaz et al.[39] is not a surprise. The results of this study are in concordance with the animal experiments.

Harari et al.[62] conducted a study on the early life exposure to lithium and boron from drinking water in Northern Argentina and Chile. The main objective of this study was to investigate the transfer of lithium and boron from exposed mothers to the fetuses and breast-fed infants. In this context, concentrations of boron in whole blood, urine, plasma, and breast milk were determined in voluntarily participating mothers living in San Antonio de los Cobres (Northern Argentina, $n = 11$), Arica (Chile, $n = 24$), and Santiago (Chile, $n = 11$). The drinking water in San Antonio de los Cobres, Arica, and Santiago (control group) contains mean boron concentrations of 5.2 mg/L, 7.9 mg/L, and 0.19 mg/L, respectively (Table 3.9). The mean daily boron intake levels for the voluntary mothers living in San Antonio de los Cobres and Arica will be 10.4 and 15.8 mg/day when considering the default daily water consumption of 2 L. These daily boron intake levels are again higher than the TDI and Safe UL even without taking into account the other sources of boron exposure. However, boron-mediated unfavorable effects on the infants have not been reported. The women in Arica had infants with the largest size at birth and a similar mortality rate as in Santiago (control location).[62] These results do not support the results of animal experiments.

3.3.4.5 Risk Characterization for Workers

The recently published studies on occupational boron exposure were conducted in China and Turkey.[29,43,44] In an earlier study, the total daily boron exposure in occupational settings was also assessed by Culver et al.[7] These are the most comprehensive studies to provide data on occupational boron exposure. Therefore, the exposure assessment and risk characterization for the workers will be based primarily on these three studies (Table 3.10). The mean total daily boron exposure (food + fluids + inhalation) of 41.2 mg/day reported for the Chinese workers is the highest boron intake ever reported in occupational settings.[29,30] This extreme level of daily boron exposure represents the worst-case exposure scenario. Such high levels of daily boron exposure do not seem to be possible in western countries that implemented the standard workplace regulations. In spite of these extreme exposure conditions, impairment in semen quality parameters (which are the most sensitive end points of male reproductive toxicity) and increased damage in sperm DNA have not been identified in Chinese workers. On the contrary, a positive correlation was reported between percent normal morphology of sperm cells and levels of boron in postwork shift urine/blood. However, the statistical significance disappeared when the confounders (smoking, alcohol consumption, and pesticide exposure) were taken into account.

The mean daily boron intake level of 41.2 mg/day reported for the Chinese workers was about four times higher than the TDI (10.3 mg B/day) proposed by the WHO.[38] In this context, this is an unacceptable level of daily boron exposure. However, the worker population is assumed to be much more homogeneous than the general population. The EU Risk Assessment Report and ECHA take this assumption into account and apply a lower UF in the extrapolation of the safe doses from animals to the worker population. The $DNEL_{worker}$ of 44.8 mg/day was mentioned above under the "dose–response assessment" section.[61] It appears that even the highest mean total daily boron exposure ever reported (41.2 mg/day) is still lower than the DNEL. On the other hand, the DNEL proposed in the EU Risk Assessment Report[61] was derived from the NOAEL instead of from the BMD for the developmental effects. This margin will provide an additional safety for the male workers. Therefore, the lack of the boron-mediated toxicity on the semen quality parameters of the Chinese workers is not surprising.

The most recent study on the reprotoxic effects of boron in occupationally exposed workers was published by Duydu et al.[43–46] The mean total daily boron exposure (food + fluids + inhalation) was 14.45 mg/day for the highly exposed workers employed in the Bandırma boric acid production plant (Table 3.10). The potential effects of boron on the male reproductive system were assessed by means of the most sensitive reproductive toxicity indicators (semen quality parameters, FSH, LH, and testosterone); however, boron-mediated unfavorable effects have not been identified in workers under the exposure conditions described by Duydu et al.[43,44] Culver at al[7] reported a mean total daily boron exposure of 27.9 mg/day for the workers employed in borax packaging and shipping facilities. The mean total daily boron intake levels calculated in these two studies were much lower than the daily intake levels reported in the Chinese study.

Therefore, these two studies represent a reasonable worst-case exposure scenario for the workers employed in packaging, production, and shipping facilities of boron compounds. The mean levels of total daily boron exposure calculated in both studies are higher than the TDI (10.3 mg B/day) level; however, they are lower than the DNEL (44.8 mg/day) proposed in the EU Risk Assessment Report.[61] Therefore, the lack of the boron-mediated reprotoxic effects in workers employed in Bandırma boric acid production plant is not surprising.

3.3.4.6 Assessment of Blood Boron Concentrations

The traditional risk assessment procedures for noncancer end points are based on the external exposure. The identified POD (NOAEL, BMD, etc.) in animals is scaled to humans by using a series of UFs. In this context, the default MOE between animal NOAEL and human exposure level has been considered as 100. The MOE greater than or equal to 100 is considered to be safe. The default MOE of 100 corresponds to the default interspecies (×10) and interindividual (×10) UFs.

$$\text{MOE} = \frac{\text{NOAEL(or other related POD)}}{\text{Human exposure}}.$$

However, the default MOE could be reduced when applied to comparative blood concentrations.[69–71] Direct comparison of blood concentrations between humans and animals at the POD for the critical effect will reduce or even eliminate the TK components of the interspecies and interindividual UF.

$$\text{MOE} = \frac{\text{Blood concentration of the parent compound at the POD for the critical effect in animal}}{\text{Blood concentration of the parent compound in human}}.$$

Aylward et al. proposed to reduce the interspecies and interindividual TK UF to 1 when the blood concentration of the parent compound is used as the exposure metric.[71] This approach reduces the target MOE to 10 (or lower). However, this is a general approach and does not represent the required target MOE for boron exposure. Especially, the TD components of the UF should be reevaluated to calculate a chemical-specific target MOE for boron. However, such studies are not available. Nevertheless, the MOEs derived by comparative blood boron concentrations have been calculated and included in Table 3.12.

In spite of the high mean blood boron concentrations, birth and mortality rates were not adversely affected in the population studied by Yazbeck et al.[27] Harari et al.[62] reported really extreme boron concentrations in pregnant women; however, the sizes of infants at birth are larger than those of the control infants. This observation is very important in view of the reduced fetal body weight considered as the critical effect in rat for the developmental toxicity of boron. On the other hand, the occupational exposure studies represent a worst-case exposure scenario with extreme levels of blood boron

Table 3.12 The Mean Blood Boron Concentrations of Rats at the NOAELs for the Critical Effects and the Mean Blood Boron Concentrations of Humans Reported in Related Studies. The Calculated MOEs are Based on Comparative Blood Concentrations

	Experimental Studies		Environmental Exposure Studies		Occupational Exposure Studies		
Parameters	Price et al., 1997[19]	Weir and Fisher, 1972[3]	Yazbeck et al., 2005[27]	Harari et al., 2012[62]	Culver et al., 1994[7]	Xing et al., 2008[29]	Duydu et al., 2011[43]
$NOAEL_{dev.}$	9.6	—	—	—	—	—	—
$NOAEL_{rep.}$	—	17.5	—	—	—	—	—
Blood B, (ng/g)	1270	2300	159.1[a]	430[a,b,c]	260[a]	499.2[a]	223.89[a]
$MOE_{dev.}$	—	—	7.98	2.95	4.88	2.54	5.67
$MOE_{rep.}$	—	—	14.45	5.35	8.84	4.61	10.27

NOAEL is expressed in milligrams per kilogram body weight per day in rats.
[a]The highest mean blood boron concentration reported in that study.
[b]Geometric mean.
[c]Mean blood boron concentration in pregnant women.

concentrations. Nevertheless, the semen quality of workers was not adversely affected by these high blood boron concentrations (Table 3.5) as was reported by Xing et al.[29] and Duydu et al.[43] This information is again very important, as the impairment of spermiation was considered to be the critical effect in rats for the reproductive toxicity of boron. As shown in Table 3.12, the blood boron concentrations reported in environmental and occupational exposure studies are all lower than the blood boron concentrations of the animals at the NOAEL for the critical effects.

3.4 CONCLUSION

Boric acid and sodium borates are toxic to reproduction and development in experimental animals at high doses. The NOAELs for the developmental and reproductive adverse effects of boron are 9.6 mg/kg-bw/day and 17.5 mg/kg-bw/day, respectively. Impaired spermiation and decreased fetal body weight are the most sensitive end points in rats. However, none of these adverse effects of boron have been proven for humans in environmental and occupational epidemiology studies.[7,27,29,43,62] Therefore, the results of numerous independent human studies raise doubts about the relevance in humans of the developmental and reproductive effects of boric acid observed in animal experiments.

In essence, the main reason for the absence of the unfavorable effects of boron in published epidemiological studies seems to be the low level of exposure. As discussed above, in many studies, the exposure level of the population is lower than the limit considered as safe by the relevant exposure guidance values (TDI, RfD, DNEL, etc.). On the other hand, blood boron concentrations are also available in recently published occupational epidemiology studies. The mean blood boron level of the high-exposure group in China was lower than the mean blood boron level at the NOAEL for male fertility and for developmental toxicity in rats by a factor of 4.6 and 2.5, respectively. Similarly, compared to the blood boron level at the NOAEL for male fertility and developmental toxicity in rats, the mean blood boron level of the high-exposure group in Turkey was lower by factors of 10.3 and 5.7, respectively.[72] The extreme mean blood boron concentration in pregnant women reported in Northern Argentina was still lower than the NOAEL for reproductive and developmental toxicity in rats by factor of 5.4 and 3, respectively. These comparisons show that there is no contradiction between human and experimental reproductive toxicity data for boric acid and sodium borates. It clearly appears that human boron exposure, even in the highest exposure cohorts, are too low to reach the blood concentrations that would be required to exert adverse effects on reproduction.[72]

ABBREVIATIONS

ACGIH American Conference of Governmental Industrial Hygienists
B Boron
BMD Benchmark dose
$BMDL_{05}$ Benchmark dose lower confidence limit
CHO Chinese hamster ovary
CLP The Regulation on classification, labeling and packaging of substances and mixtures
CNS Central Nervous System
CSFII Continuing Survey of Food Intakes by Individuals
DNA Deoxyribonucleic acid
DNEL Derived no effect level
ECETOC European Centre for Ecotoxicology and Toxicology of Chemicals
ECHA European Chemical Agency
EDCs Endocrine disrupting chemicals
EFSA European Food Safety Authority
EGVM Expert Group on Vitamins and Minerals
EPA Environmental Protection Agency
FISH Fluorescence in situ hybridization
FSH Follicle stimulating hormone
GFR Glomerular filtration rate
GHS Globally Harmonized System of classification and labeling of chemicals
GnRH Gonadotropin releasing hormone
LC_{50} The chemical in air or in water that kills 50% of the test animals in a given time (lethal concentration).
LD_{50} The amount of a chemical, which causes the death of 50% (one half) of a group of test animals (lethal dose).
LH Luteinizing hormone
LOAEL Lowest observed adverse effect level

LOQ	Limit of quantitation
MAK	Maximum Workplace Concentration (Maximale Arbeitsplatz Konzentration)
MOE	Margin of exposure
NOAEL	No observed adverse effect level
NTP	National Toxicology Program
OEL	Occupational exposure limits
OSHA	Occupational Safety and Health Administration
PEL	Permissible exposure limits
POD	Point of departure
RfD	Reference dose
SBR	Standardized birth ratio
TD	Toxicodynamic
TDI	Tolerable daily intake
TK	Toxicokinatic
TLV	Threshold Limit Values
TWA	Time weighted average
UDS	Unscheduled DNA synthesis
UF	Uncertainty factor
UK	United Kingdom
UL	Upper limit
US	United States
WHO	World Health Organization

REFERENCES

1. Woods WG. An introduction to boron: history, sources, uses and chemistry. *Environ Health Perspect* 1994;**102**(Suppl. 7):5–11.
2. US EPA. *Toxicological review of boron and compounds*. U.S. Environmental Protection Agency; 2004. EPA 635/04/052.
3. Weir RJ, Fisher RS. Toxicologic studies on borax and boric acid. *Toxicol Appl Pharmacol* 1972;**23**:351–64.
4. Dourson M, Maier A, Meek B, Renwick A, Ohanian E, Poirier K. Boron tolerable intake, re-evaluation of toxicokinetics for data-derived uncertainty factors. *Biol Trace Elem Res* 1998;**66**:453–63.
5. Schou JS, Jansen JA, Aggerbeck B. Human pharmacokinetics and safety of boric acid. *Arch Toxicol* 1984;**7**:232–5.
6. Jansen JA, Schou JS, Aggarbeck B. Gastrointestinal absorption and in vitro release of boric acid from water-emulsifying ointments. *Food Chem Toxicol* 1984;**22**:49–53.
7. Culver BD, Shen PT, Taylor TH, Lee-Feldstein A, Anton-Culver H, Strong PL. The relationship of blood- and urine-boron to boron exposure in borax-workers and the usefulness of urine-boron as an exposure marker. *Environ Health Perspect* 1994;**102**:133–7.
8. Pfeiffer CC, Hallman LF, Gersh I. Boric acid ointment. A study of possible intoxication in the treatment of burns. *J Amer Med Assoc* 1945;**128**(4):266–74.
9. ECETOC. *Reproductive and general toxicology of some inorganic borates and risk assessment for human beings*. European Centre for Ecotoxicology and Toxicology of Chemicals; 1995. ECETOC Technical Report No. 63.
10. Draize JH, Kelly EA. The urinary excretion of boric acid preparations following oral administration and topical applications to intact and damaged skin of rabbits. *Toxicol Appl Pharmacol* 1959;**1**:267–76.
11. Friis-Hansen B, Aggerbeck B, Jansen JA. Unaffected blood boron levels in newborn infants treated with a boric acid ointment. *Food Chem Toxicol* 1982;**20**:451–4.
12. HERA. *Human and environmental risk assessment on ingredients of household cleaning products*. "Substance: boric acid"; 2005.

13. NTP. *Toxicology and carcinogenesis study of boric acid in B6C3F1 mice.* National Toxicology Program; 1987. Technical Report Series No. 324, NIH publication No. 88-2580.
14. Turkez H, Geyikoglu F, Tatar A, Keles S, Ozkan A. Effects of some boron compounds on peripheral human blood. *J Biosci* 2007;**62**(11/12):889–96.
15. Arslan M, Topaktas M, Rencuzogullari E. The effects of boric acid on sister chromatid exchanges and chromosome aberrations in cultured human lymphocytes. *Cytotechnology* 2008;**56**:91–6.
16. Linder RE, Strader LF, Rehnberg GL. Effect of acute exposure to boric acid on the male reproductive system of the rat. *J Toxicol Environ Health* 1990;**31**:133–46.
17. Fail PA, George JD, Seely JC. Reproductive toxicity of boric acid in CD-1 Swiss mice: assessment using the continuous breeding protocol. *Fund Appl Toxicol* 1991;**17**:225–39.
18. Heindel JJ, Price CJ, Field EA, Marr MC, Myers CB, Morrissey RE, et al. Developmental toxicity of boric acid in mice and rats. *Fund Appl Toxicol* 1992;**18**:266–77.
19. Price CJ, Strong PL, Marr MC, Myers CB, Murray FJ. Developmental toxicity NOAEL and postnatal recovery in rats fed boric acid during gestation. *Fund Appl Toxicol* 1996;**32**:179–93.
20. Allen BC, Strong PL, Price CJ, Hubbard SA, Datson GP. Benchmark dose analysis of developmental toxicity in rats exposed to boric acid. *Fund Appl Toxicol* 1996;**32**:194–204.
21. Price CJ, Marr MC, Myers CB, Seely JC, Heindel JJ, Schwetz BA. The developmental toxicity of boric acid in rabbits. *Fund Appl Toxicol* 1996;**34**:176–87.
22. Ku WW, Chapin RE, Wine RN, Gladen BC. Testicular toxicity of boric acid (BA): relationship of dose to lesion development and recovery in the F344 rat. *Reprod Toxicol* 1993;**7**:305–19.
23. Fail PA, Chapin RE, Price CJ, Heindel JJ. General, reproductive, developmental, and endocrine toxicity of boronated compounds. *Reprod Toxicol* 1998;**12**:1–18.
24. Tarasenko NY, Kasparov AA, Strongina OM. Effects of boric acid on sexual function in males. *Gig Tr Prof Zabol* 1972;**16**(11):13–6.
25. Whorton MD, Haas JL, Trent L, Wong O. Reproductive effects of sodium borates on male employees: birth rate assessment. *Occup Environ Med* 1994;**51**(11):761–7.
26. Whorton D, Haas J, Trent L. Reproductive effects of inorganic borates on male employees: birth rate assessment. *Environ Health Perspec* 1994;**102**(7):129–32.
27. Yazbeck C, Kloppmann W, Cottier R, Sahuquillo J, Debotte G, Huel G. Health impact evaluation of boron in drinking water: a geographical risk assessment in Northern France. *Environ Geochem Health* 2005;**27**(5–6):419–27.
28. Chang BL, Robbins WA, Wei F, Xun L, Wu G, Li N, et al. Boron workers in China, exploring work and lifestyle factors related to boron exposure. *AAOHN J* 2006;**54**(10):435–43.
29. Xing X, Wu G, Wei F, Liu P, Wei H, Wang C, et al. Biomarkers of environmental and workplace boron exposure. *J Occup Environ Hyg* 2008;**5**:141–7.
30. Robbins WA, Xun L, Jia J, Kennedy N, Elashoff DA, Ping L. Chronic boron exposure and human semen parameters. *Reprod Toxicol* 2010;**29**:184–90.
31. Sayli BS, Tuccar E, Elhan AH. An assessment of fertility in boron-exposed Turkish subpopulations. *Reprod Toxicol* 1998;**12**(3):297–304.
32. Robbins WA, Wei F, Elashoff DA, Wu G, Xun L, Jia J. Y:X sperm ratio in boron-exposed men. *J Androl* 2008;**29**(1):115–21.
33. Robbins WA, Baulch JE, Moore D, Weier HU, Blakey D, Wyrobek AJ. Three-probe fluorescence in situ hybridization to assess chromosome X, Y, and 8 aneuploidy in sperm of 14 men from two healthy groups: evidence for a paternal age effect on sperm aneuploidy. *Reprod Fertil Dev* 1995;**7**:799–809.
34. Tiido T, Rignell-Hydbom A, Jonsson BAG, Giwercman YL, Pedersen HS, Wojtyniak B, et al. Impact of PCB and p,p′-DDE contaminants on human sperm Y:X chromosome ratio: studies in three European populations and the Inuit population in Greenland. *Environ Health Perspect* 2006;**114**(5):718–24.
35. Kvist L, Giwercman YL, Jönsson BA, Lindh CH, Bonde JP, Toft G, et al. Serum levels of perfluorinated compounds and sperm Y:X chromosome ratio in two European populations and in Inuit from Greenland. *Reprod Toxicol* 2012;**34**(4):644–50.

36. Scialli AR, Bonde JP, Brüske-Hohlfeld I, Culver BD, Li Y, Sullivan FM. An overview of male reproductive studies of boron with an emphasis on studies of highly exposed Chinese workers. *Reprod Toxicol* 2010;**29**:10–24.
37. ETI MINE Works General Management, http://en.etimaden.gov.tr/about-boron-62s.htm, [accessed 15.08.13].
38. WHO. *Guidelines for drinking water quality.* 3rd ed. 2011.
39. Korkmaz M, Yenigun M, Bakirdere S, Ataman OY, Keskin S, Muezzinoglu T, et al. Effects of chronic boron exposure on semen profile. *Biol Trace Elem Res* 2011;**143**(2):738–50.
40. Sayli BS. An assessment of fertility in boron-exposed Turkish subpopulations, 2. Evidence that boron has no effect on human reproduction. *Biol Trace Elem Res* 1998;**66**(1–3):409–22.
41. Sayli BS. Assessment of fertility and infertility in boron-exposed Turkish subpopulations, 3. Evaluation of fertility among sibs and in "borate families". *Biol Trace Elem Res* 2001;**81**(3):255–67.
42. Sayli BS. Low frequency of fertility among workers in a borate processing facility. *Biol Trace Elem Res* 2003;**93**(1–3):19–28.
43. Duydu Y, Basaran N, Ustundag A, Aydin S, Undeger U, Ataman OY, et al. Reproductive toxicity parameters and biological monitoring in occupationally and environmentally boron-exposed persons in Bandırma, Turkey. *Arch Toxicol* 2011;**85**:589–600.
44. Duydu Y, Basaran N, Bolt HM. Exposure assessment of boron in Bandırma boric acid production plant. *J Trace Elem Med Biol* 2012;**26**:161–4.
45. Duydu Y, Basaran N, Ustundag A, Aydin S, Undeger U, Ataman OY, et al. Assessment of DNA integrity (COMET assay) in sperm cells of boron-exposed workers. *Arch Toxicol* 2012;**86**:27–35.
46. Basaran N, Duydu Y, Bolt HM. Reproductive toxicity in boron exposed workers in Bandırma, Turkey. *J Trace Elem Med Biol* 2012;**26**:165–7.
47. Ku WW, Chapin RR, Moseman RF, Brink RE, Pierce KD, Adams KY. Tissue disposition of boron in male Fisher rats. *Toxicol Appl Pharmacol* 1991;**111**:145–51.
48. Ku WW, Chapin RE. Mechanism of the testicular toxicity of boric acid in rats: in vivo and in vitro studies. *Environ Health Perspect* 1994;**102**:99–105.
49. EPA. *The use of benchmark dose approach in health risk assessment*; 1995. EPA/630/R-94/007.
50. IPCS. *Risk assessment terminology.* Harmonization Project Document No. 1. WHO; 2004.
51. Renwick AG. Toxicokinetics in infants and children in relation to the ADI and TDI. *Food Addit Contam* 1998;**15**:17–35.
52. IGHRC. *The Interdepartmental group on health risks from chemicals, "uncertainty factors: their use in human health risk assessment by UK Government".* Published by the Institute for Environment and Health; 2003.
53. Renwick AG, Walton K. The use of surrogate endpoints to assess potential toxicity in humans. *Toxicol Lett* 2001;**120**:97–110.
54. WHO. *Chemical-specific adjustment factors for interspecies differences and human variability: guidance document for use of data in dose/concentration–response assessment.* IPCS Harmonization; 2005. Project Document No. 2.
55. Jansen JA, Andersen J, Schou JS. Boric acid single dose pharmacokinetics after intravenous administration to man. *Arch Toxicol* 1984;**55**:64–7.
56. Pahl MV, Culver BD, Strong PL, Murray FJ, Vaziri ND. The effects of pregnancy on renal clearance of boron in humans: a study based on normal dietary intake of boron. *Toxicol Sci* 2001;**60**:252–6.
57. Murray FJ. A human health risk assessment of boron (boric acid and borax) in drinking water. *Regul Toxicol Pharmacol* 1995;**22**:221–30.
58. Hasegawa R, Hirata-Koizumi M, Dourson ML, Parker A, Ono A, Hirose A. Safety assessment of boron by application of new uncertainty factors and their subdivision. *Regul Toxicol Pharmacol* 2013;**65**:108–14.
59. EGVM. *Safe upper levels for vitamins and minerals.* Expert Group on Vitamins and Minerals; 2003.
60. EFSA. European Food Safety Authority, "Opinion of the scientific panel on dietetic products, nutrition and allergies on a request from the commission related to the tolerable upper intake level of boron (sodium borate and boric acid)". *EFSA J* 2004;**80**:1–22.
61. European Union Risk Assessment Report. *Disodium tetraborate, anhydrous boric acid, boric acid, crude natural (1)*; 2007 [Substance Evaluation Report].

62. Harari F, Ronco AM, Concha G, Lianos M, Grander M, Castro F, et al. Early-life exposure to lithium and boron from drinking water. *Reprod Toxicol* 2012;**34**:552–60.
63. Meacham S, Karakas S, Wallace A, Altun F. Boron in human health: evidence for dietary recommendations and public policies. *Open Min Process J* 2010;**3**:36–53.
64. Rainey CJ, Nyquist LA, Christensen RE, Strong PL, Culver BD, Coughlin JR. Daily boron intake from the American diet. *J Am Diet Assoc* 1999;**99**:335–40.
65. Rainey CJ, Nyquist LA, Coughlin JR, Downing G. Dietary boron intake in the United States: CSFII 1994–1996. *J Food Comp Anal* 2002;**15**:237–50.
66. Cortes S, Reynaga-Delgado E, Sancha AM, Ferreccio C. Boron exposure assessment using drinking water and urine in the North of Chile. *Sci Total Environ* 2011;**410-411**:96–101.
67. DFG. *Deutsche Forschungsgemeinschaft, "List of MAK and BAT values"*. Report No. 48. WILEY-VCH Verlag GmbH & Co. KGaA; 2012.
68. ECHA. *Guidance on information requirements and chemical safety assessment.* European Chemical Agency; 2008. "Chapter R8: Characterization of dose [concentration]-response for human health".
69. Hays SM, Becker RA, Leing HW, Aylward LL, Pyatt DW. Biomonitoring equivalents: a screening approach for interpreting biomonitoring results from a public health risk perspective. *Regul Toxicol Pharmacol* 2007;**47**:96–109.
70 Hays SM, Aylward LL, LaKind JS, Bartels MJ, Barton HA, Boogaard PJ, et al. Guidelines for the derivation of biomonitoring equivalents: Report from the biomonitoring equivalents expert workshop. *Regul Toxicol Pharmacol* 2008;**51**:4–15.
71. Aylward LL, Becker RA, Kirman CR, Hays SM. Assessment of margin of exposure based on biomarkers in blood: an exploratory analysis. *Regul Toxicol Pharmacol* 2011;**61**:44–52.
72. Bolt HM, Basaran N, Duydu Y. Human environmental and occupational exposures to boric acid: reconciliation with experimental reproductive toxicity data. *J Toxicol Environ Health A* 2012; **75**(8–10):508–14.

CHAPTER 4

Ion Exchange Borate Kinetics

Jidong Lou, Gary L. Foutch
Oklahoma State University Chemical Engineering Department, Engineering North Stillwater, OK, USA

4.1 INTRODUCTION

Boron sorption from an aqueous solution onto an ion exchange resin involves several mechanisms, including dissociation and ionization of boric acid molecules both in solution and within the resin phase, complex formation, and protonation of *N*-methylglucamine function groups in boron-specific (chelating) resins. Boric acid ionization at the resin surface or within the resin matrix consumes counterions as they are released from the resin and distorts the ion exchange mechanism. In turn, a lack of counterion diffusion out of the resin retards ionic boron carrier species diffusion into the resin phase as a result of the principle of no net current. Protonation of the amino group and boron complex formation turn the *N*-methylglucamine function group into a zwitterion form, resulting in not only a different sorption mechanism but also issues of instability and decreasing apparent capacity.

Boron sorption on a strong base, OH form anion resins or boron-specific (chelating) resins from aqueous solutions is a pH-dependent process. The concentration of ionic boron carriers increases with increasing solution pH. At pH values >11, boron carriers shift to their ionic form, and sorption may be treated as an ordinary ion exchange process. At a pH <6, molecular boric acid is the dominant boron carrier, and sorption on ion exchange resin more closely follows an adsorption process. The functional group *N*-methylglucamine of a boron-specific (chelating) resin is an aminopolyol consisting of an *N*-methylamino end (tertiary amine) and a diol side chain "sugar tail". As boron loads, resin stability and function are subject to the environmental pH.

Boron sorption on boron-specific (chelating) resins is a relatively new research topic. The chelating mechanism, equilibrium isotherm, and kinetics are in need of systematic experimental and theoretical studies. The purpose of this communication is to present our current thoughts on both kinetic mechanisms for reference, interpolating experimental phenomenon, and results and modeling.

4.2 BORATE IONIC CHEMISTRY

4.2.1 Boric Acid in Aqueous Solution

The boron atom has the gas phase electron configuration $1s^2$, $2s^2$, $2p^1$ and is placed in Group III of the Periodic Classification. With its chief oxidation state being +3, boron

Boron Separation Processes
ISBN 978-0-444-63454-2, http://dx.doi.org/10.1016/B978-0-444-63454-2.00004-6

may be considered a typical metalloid. The elemental form of boron is unstable in nature and has the tendency to form anionic rather than cationic complexes. The chemistry of an aqueous boron solution features the existence of a series of polyborate anionic species along with boric acid and single negative charge monomeric borates, and by the fact that boric acid undergoes hydration before ionization. The formation of different borate groups depends on the pH, total boron concentration, and temperature. At the acidic and basic extremes, the primary species in solution are $B(OH)_3$ and monomeric borate ion. However, one can conceive of, at least, three possible structures for a mononuclear borate ion of single negative charge, BO_2^-, $H_2BO_3^-$ (or $BO(OH)_2^-$), and tetrahedral $B(OH)_4^-$. The tetrahedral $B(OH)_4^-$ has been confirmed to be the only mononuclear borate ion present at the highest pH, largely due to $B(OH)_3$ acting as a Lewis acid. Boric acid is very weak, with a pK_a value of 9.24. The ionization of $B(OH)_3$ in aqueous solution to form $B(OH)_4^-$ (the conjugate base of $B(OH)_3$) is spontaneous.[1] The two representations of ionization are presented in Eqns (4.1) and (4.2):

$$B(OH)_3 + H_2O \overset{K_a}{\Longleftrightarrow} B(OH)_4^- + H^+ \tag{4.1}$$

or

$$B(OH)_3 + OH^- \overset{K_1}{\Longleftrightarrow} B(OH)_4^- \tag{4.2}$$

The two processes are coupled to the water autoprotolytic equilibrium and cannot be distinguished thermodynamically.[2]

In a dilute boric acid aqueous solution (<0.025 M), only mononuclear species $B(OH)_3$ and $B(OH)_4^-$ are essentially present.[3,4] Notably, detailed research on dilute boric acid solutions can be found in Manov et al. (1944)[5] and Owen (1938).[6] As the concentration increases or temperature decreases, the possibility of forming polynuclear borates increases. Although there is no unanimous agreement about the structure and forms of polyborates in solution, the evidence from data obtained by studies of electrical conductance, hydrogen-ion concentration (electromotive force (EMF) or pH), X-ray diffraction, temperature jump technique, nuclear magnetic resonance, and Raman spectroscopy has indicated the formation of polynuclear species in solution.

There was considerably less agreement as to the nature of the polyanions present. The problem is complicated by the facts that (1) the equilibrium formation constants K_3 are not large, (2) more than one polyanion is probably present, and (3) the polyanions under one set of conditions are not necessarily the same as those under an alternative set (Ross and Edwards, 1967). In the 1950s, there did seem to be a general agreement that the polyborate with three boron atoms and one negative charge is most important at higher total boron concentrations $B \geq 0.025$ M.[5]

Without sufficient details to establish molecular formulas, Edwards used B_3^- to represent a single negative trimer of $B(OH)_3$, assuming the formation

$$2B(OH)_3 + B(OH)_4^- \overset{K_3}{\Longleftrightarrow} B_3^- \tag{4.3}$$

and suggested that a six-member ring containing alternating boron and oxygen atoms and a chain configuration are two possible structures for the trimer. The estimated value of the equilibrium constant for the reaction is about 3.2×10^{-8}, which is relatively small compared to values reported from other studies.[3,7]

Ingri and coworkers[3,8–10] used an EMF (pH) measurement method with a hydrogen electrode to measure the equilibrium concentrations of boric acid solutions in the range from 0.01 up to 0.6 M at 25 °C. Their data in the pH range 5–9 give evidence of $B_3O_3(OH)_4^-$ and $B_3O_3(OH)_5^{-2}$, at a total boron concentration $B \geq 0.025$ M, and lead to the postulations.

$$3B(OH)_3 \overset{K_{3,1}}{\Longleftrightarrow} B_3O_3(OH)_4^- + H^+ + 2H_2O, \quad \log K_{3,1} = -6.84 \pm 0.10 \tag{4.4}$$

$$3B(OH)_3 \overset{K_{3,2}}{\Longleftrightarrow} B_3O_3(OH)_4^{2-} + 2H^+ + H_2O, \quad \log K_{3,2} = -15.44 \pm 0.20 \tag{4.5}$$

At the total boron concentration ≥0.4 M, data indicate that other polyboronic species exist among trimers, tetramer, and pentamer, which were assumed with molecular formulas $B_4O_5(OH)_3^-$, $B_4O_5(OH)_4^{-2}$, and $B_5O_6(OH)_4^-$. Among these polyborates postulated by Ingri and coworkers, univalent triborate $B_3O_3(OH)_4^-$ and pentaborate $B_5O_6(OH)_4^-$, divalent tetraborate $B_4O_5(OH)_4^{-2}$ have been identified in concentrated boron solutions with the pH range 6–10 by Raman Spectroscopy,[7] which was known to be ideally suited for the derivation of structural formation about ions in solution. Divalent triborate $B_3O_3(OH)_5^{-2}$ was indicated only by Zhou's report.[11] The convincing polyborate structures are six-membered rings containing alternate boron and oxygen atoms as Edwards first suggested in 1953.[7]

Summarizing the previous research about borate formation, Edwards and Ross[12] postulated important structural features and classifications:

- All borates, except for some gas species, have triangular-planar, tetrahedral, or triangular and tetrahedral structures.
- The fundamental unit of the polyanionic structure is a trimeric ring containing both trigonal and tetrahedral boron atoms.
- Boron atoms exist in threefold and fourfold coordination.
- The ring must contain one or two tetrahedral boron atoms to be stable, though it may contain zero, one, two, or possibly three tetrahedral boron atoms.
- The formation of higher polyanions, for example, the tetramer and pentamer, results from the fusion of two or more rings at the tetrahedral boron atoms.

- Long-chain polyborates may be formed from the rings by repeated dehydration.
- Some borates exist with more than one-dimension linkage.

In highly concentrated boron aqueous solution (5.24 mol/kg of boron), Raman spectroscopy, and atmospheric pressure chemical ionization/mass spectrometry give evidence of ring-structured long-chain polyborates such as $B_9O_{10}(OH)_9^{-2}$, $B_{10}O_{12}(OH)_8^{-2}$, $B_{11}O_{14}(OH)_7^{-2}$, and $B_{12}O_{16}(OH)_6^{-2}$.[7]

4.2.2 Physicochemical Properties

The physicochemical properties reviewed here include boric acid solubility in water, the heat of solution, the osmotic and activity coefficients, and the equilibrium constants.

In commercial usage, the name boric acid is usually associated with orthoboric acid $B(OH)_3$. Orthoboric acid precipitates from aqueous solution as white crystals. Its normal melting point is 170.9 °C. The solubility of boric acid is moderate and increases rapidly with increasing temperature. Table 4.1 lists the experimental solubility of boric acid in water at various temperatures. The presence of inorganic salts may enhance or depress the solubility of boric acid in water. Adding potassium chloride, potassium sulfate, or sodium sulfate increases solubility, while the solubility is decreased by adding lithium chloride or sodium chloride.

The heat of the solution is dependent upon the concentration. For solutions in the range of 0.03–0.9 M, the molar heat of solution may be estimated by the empirical relation:

$$\Delta H = \left(22062 - 222 \cdot m + 979 \cdot e^{-123 \cdot m}\right) \quad (\mathrm{J/mol}). \tag{4.6}$$

The osmotic and activity coefficients of some simple borates in aqueous solution at 25 °C are determined by the following relationships[13]:

$$\pi = 2\, m_R \pi_R / nm, \tag{4.7}$$

Table 4.1 Solubility of Boric Acid in Water at Various Temperatures

Temperature °C	Molality
0	0.4304
10	0.5776
20	0.8154
30	1.0678
40	1.4108
50	1.8670
60	2.3961
70	2.7067
80	3.8424
90	4.9151
100	6.5119

where the m_R and π_R are the molality and osmotic coefficient of sodium chloride, respectively; m is the molality of the borate in equilibrium with the reference solution; n is the empirical number, order that $\pi \rightarrow 1$, as $m \rightarrow 0$, unity for boric acid and six for tetraborate $B(OH)_4^-$. The term nm in Eqn (4.7) should be replaced by a summation over all the solutes in solution. With the above estimation of the osmotic coefficient, the mean activity coefficient $\gamma_\pm$ is calculated by Eqn (4.8)

$$-\ln\gamma_\pm = 1 - \pi + 2\int_0^{m^{0.5}} (1-\pi)/m^{0.5} dm^{0.5}. \tag{4.8}$$

For an estimation of the dissociation constant of boric acid varying with temperature in dilute solution, within a temperature range of 10–50 °C, Owen's relationship[6]

$$pK_a = 9.023 + 8 \times 10^{-5}(76.7 - t)^2 \tag{4.9}$$

or Manov's relationship,[5] where T is the absolute temperature,

$$pK_a = 2237.94/T + 0.016883T - 3.305 \tag{4.10}$$

are suggested. For boric acid equilibria with a wider range of temperature (50–290 °C), the relationship in Eqn (4.11) reported by Mesmer et al.[14] is suggested.

$$\log K_a = 1573.21/T + 28.6059 + 0.012078T - 13.2258 \log T + f(I), \tag{4.11}$$

where T is the absolute temperature, I is the ionic strength, $f(I) = (0.325 - 0.00033T)$ $I - 0.0912I^{3/2}$.

4.2.3 Ionization Mechanisms and Ionic Equilibrium

Boric acid is a very weakly dissociated acid in aqueous solution. Boric acid in solution acts as a Lewis acid (adds an OH^-) rather than a Brönsted acid (donates H^+). In aqueous solutions, the ionization of boric acid takes place through simple boron–oxygen bond breaking and formation, which can be very fast and spontaneous. For instance, the specific rate constant k_f of ionization

$$B(OH)_3 + OH^- \xrightarrow{k_f} B(OH)_4^- \tag{4.12}$$

was estimated to be around a value of 10^{10}/M s[15]

The stoichiometry of boric acid ionization, expressed commonly in the literature

$$B(OH)_3 + 2H_2O \overset{K_a}{\Longleftrightarrow} B(OH)_4^- + H_3O^+ \tag{4.13}$$

or

$$B(OH)_3 + H_2O \overset{K_a}{\Longleftrightarrow} B(OH)_4^- + H^+ \tag{4.14}$$

or

$$B(OH)_3 + OH^- \overset{K_1}{\Longleftrightarrow} B(OH)_4^- \tag{4.15}$$

reflects the different mechanisms and pathways. According to Ross and Edwards (1967), there are four possible mechanisms that are feasible pathways for boric acid ionization in aqueous solution,

1. $B(OH)_3 + N \leftrightarrow BO(OH)_2^- + HN^+; \quad BO(OH)_2^- + H_2O \leftrightarrow B(OH)_4^-$ (4.16)

2. $B(OH)_3 + H_2O \leftrightarrow B(OH)_3(OH_2); \quad B(OH)_3(OH_2) + N \leftrightarrow B(OH)_4^- + HN^+$ (4.17)

3. $B(OH)_3 + H_2O + N \leftrightarrow B(OH)_4^- + HN^+$ (4.18)

4. $H_2O + N \leftrightarrow OH^- + NH^+; \quad B(OH)_3 + OH^- \leftrightarrow B(OH)_4^-$ (4.19)

where the symbol N denotes any Brönsted base (often being water). The second consisted of boric acid undergoing hydration before ionization. However, other mechanisms may also exist and cannot be ruled out (Ross and Edwards, 1967). The equilibrium constant related to the concentration for different mechanisms can be generally expressed as follows for comparison:

$$K_1 = \frac{\left[B(OH)_4^-\right]}{\left[B(OH)_3\right] \cdot \left[(OH)^-\right]} \tag{4.20}$$

or

$$K_a = \frac{\left[B(OH)_4^-\right] \cdot \left[(H_3O)^+\right]}{\left[B(OH)_3\right]} = K_w \frac{\left[B(OH)_4^-\right]}{\left[B(OH)_3\right] \cdot \left[(OH)^-\right]} = K_w K_1, \tag{4.21}$$

where K_w represents the water dissociation constant at a given temperature.

When both the $B(OH)_3$ and its conjugate base $B(OH)_4^-$ are present in an aqueous solution, the result is polymerization and the formation of water (Ross and Edwards, 1967), which have been described as being similar to the neutralization reaction (Edwards et al. 1961). Therefore, the equilibrium relations for the formation of polyborates, for instance, $B_3O_3(OH)_4^-$ and $B_5O_6(OH)_4^-$, is better expressed as

$$2B(OH)_3 + B(OH)_4^- \overset{K_3}{\Longleftrightarrow} B_3O_3(OH)_4^- + 3H_2O \tag{4.22}$$

$$4B(OH)_3 + B(OH)_4^- \overset{K_5}{\Longleftrightarrow} B_5O_6(OH)_4^- + 6H_2O \tag{4.23}$$

which were also suggested by Momii and Nachtrieb[16] and Anderson et al.[15]

A generally hypothetical expression for the formation of polyborates may be suggested as

$$xB(OH)_3 + yB(OH)_4^- \overset{K_{(x+y),y}}{\Longleftrightarrow} B_{x+y}O_z(OH)_{3x+4y-2z}^{-y} + zH_2O \tag{4.24}$$

Correspondingly, the equilibrium quotient $K_{(x+y),y}$ for the formation of polyborate with the total boron number $(x + y)$ and charge number y in Eqn (4.24) is

$$K_{(x+y),y} = \frac{\left[B_{x+y}O_z(OH)_{3x+4y-2z}^{-y}\right]}{\left[B(OH)_3\right]^x \cdot \left[B(OH)_4^-\right]^y}. \tag{4.25}$$

At temperatures from 50 to 290 °C, a set of equilibrium quotients for the ionization of boric acid and the formation of some possible polynuclear species was postulated by Mesmer et al. (1972), where subscripts m and n in $K_{m,n}$ denote the number of boron atoms and charges (in the formed borate or polyborates), respectively.

$$\log K_{1,1} = \frac{1573.21}{T} + 28.6059 + 0.012078 \cdot T - 13.2258 \cdot \log T, \tag{4.26}$$

$$\log K_{3,1} = \frac{3339.5}{T} + 8.084 + 1.497 \cdot \log T, \tag{4.27}$$

$$\log K_{4,2} = \frac{12820}{T} - 134.56 + 42.105 \cdot \log T, \tag{4.28}$$

$$\log K_{5,3} = \frac{10499}{T} - 118.115 + 36.237 \cdot \log T, \tag{4.29}$$

This is a convenient relation for the quantitative estimation of equilibrium quotients in the formation of related borate or polyborate species with various operation temperatures.

4.3 SORPTION MECHANISM OF BORON ON ION EXCHANGER

Ion exchange is a promising technology to remove boron from various aqueous solutions to below parts per million (ppm) levels. The general methods and procedures for determining or removing boron from aqueous solutions were outlined by Korkisch[17] as follows:

- A solution is passed through strong acid cation exchange resins to remove metals.
- The acidic solution generated by the cation exchange column is then passed through a weakly basic anion exchanger to remove strong acids.
- The effluent solution from the weakly basic bed is analyzed colorimetrically for boron or boric acid traces.

- This solution is passed through a strong anion exchange resin to remove boron or boric acid.

Applications reported include seawater osmosis permeate,[18–20] wastewater of the Kizildere geothermal power plant,[21] drinking water,[22] and sewage effluent.[23] In addition, ion exchange technology is applied to boron thermal regeneration systems of nuclear power plants for controlling the boron concentration in primary coolant,[24–26] and for enriching boron isotope ^{10}B.[27–30] Strong base OH^- form anion resin and boron selective resin (boron-chelating resin) are two types of ion exchange resins used in these applications, with different mechanisms governing the boron sorption process.

4.3.1 On Strong Base Anion Exchange Resins

Boron sorption on a strong base OH^- form anion exchange resin is essentially the ionization process of boric acid inside the resin along with the consumption of the released counterion. Free counterions OH^- result in a very high pH inside the resin bead, well above 10, and such an environment leads to all the boron carriers shifting to the tetrahedral form $B(OH)_4^-$. For applications with a total boron concentration <0.025 M, the boric acid molecule is the dominant boron carrier. During the uptake process, a concentration gradient between the solution and resin phases drives molecules near the resin surface where polarization exists by the attraction of the electrical field inside the resin and the resin surface. At the surface and immediately inside the resin, the $B(OH)_3$ molecule forms tetrahedral $B(OH)_4^-$ through the reaction with counterions OH^- released from the resin. The generated tetrahedral $B(OH)_4^-$ ion replaces OH^- ions to maintain electroneutrality inside the resin. No actual ion exchange occurs during the sorption of boric acid molecules. Therefore, the boric acid uptake process on a strong base OH form anion exchange resin can be written in a general form as

$$\overline{R^+} + \overline{OH^-} + B(OH)_3 \rightarrow \overline{R^+} + \overline{B(OH)_4^-} \tag{4.30}$$

The quantities with bars refer to the inside of the resin phase. This equation is valid as long as boric acid is the dominant boron carrier in solution. If we assume that the molecular form boron carrier and ionic species boron carrier are comparable, the postulation in Eqn (4.30) will hold without significant deviation because of the following reasons:

1. Donnan potential effect of ion exchanger: the nonionic boric acid molecule is not impacted by Donnan exclusion and may be taken up in preference to ionic species.
2. Electroneutrality ($C_{B(OH)_4^-} = C_{H^+} - C_{OH^-}$) and no electric current ($J_{B(OH)_4^-} + J_{(OH)^-} = 0$) restrictions in the diffusion film outside the resin surface limit the concentration of tetrahedral $B(OH)_4^-$ because of a very low coion H^+ concentration

(from the last section, comparable concentration between the molecular form and ionic form occurs at a pH close to pK 9.25 for boric acid) and transport of tetrahedral $B(OH)_4^-$ because of the consumption of OH ions inside the resin.

3. The local equilibrium through the dissociation relation $B(OH)_3 + 2H_2O \overset{K_a}{\Longleftrightarrow} B(OH)_4^- + (H_3O)^+$ tends to further decrease the concentration of tetrahedral $B(OH)_4^-$ ions in the film during the uptake process of boric acid on the resin.
4. As boric acid shifts to tetrahedral $B(OH)_4^-$ inside the resin, the boundary conditions of the film are changed correspondingly thereby decreasing the concentration gradient of tetrahedral $B(OH)_4^-$ and so does the driving force of the present tetrahedral $B(OH)_4^-$ in the film transporting to the resin phase.

The uptake of boric acid and borate ions changes the conditions inside the resin by increasing the total boron concentration and reducing the pH. Deviation from Eqn (4.30) could occur at very late stages of resin conversion or with high total boron concentrations in the solution. When boric acid is saturated near resin exchange sites, the tendency of boron polymerization to form polyborates increases, resulting in a relatively high apparent resin capacity. However, when the total boron concentration in solution is low, the liquid side concentration gradient is quite small and the film resistance controls the molecular diffusion of boric acid through the film. Consequently, the interface concentration of boric acid is very low. There are not enough boric acid molecules involved in the polymerization to form polyborates inside the resin. The tetrahedral $B(OH)_4^-$ sorption to resin exchange sites becomes the dominant borate species uptake mechanism, resulting in a relatively low apparent resin capacity. This was shown experimentally by Tomizawa.[31,32]

4.3.2 On Boron-Selective Resins

Boron-selective resins have existed since the 1950s.[33] The principle is that polyhydric compounds possess a high affinity for boric acid or tetraborate to form boron complexes. These resins, like other gel-type resins, consist of a copolymer matrix, divinylbenzene, crosslinked macroporous polystyrene. Instead of fixed charge groups within a matrix, these resins are impregnated with saccharide derivatives (sugar alcohol, commonly amino sugar) as functional groups to provide active sites for boron sorption. Commercially available boron selective resins contain the functional group *N*-methylglucamine, $R-CH_2-N(CH_3)-C_6H_8(OH)_5$ (i.e., Amberlite™ IRA743; Diaion CRB02; Purolite, S108). The functional group derives from sorbitol, consisting of a diol side chain "sugar tail" and an *N*-methylamino end (tertiary amine), which serves in structure as an anchor attached on the polymer matrix of a resin (Figure 4.1).

The type of boron-selective resin above is cataloged as a weak-base anion resin due to the nature of the tertiary amine in the attached functional group and the apparent capacity is subject to the change in pH. Instead of counterion exchange, the boron-specific

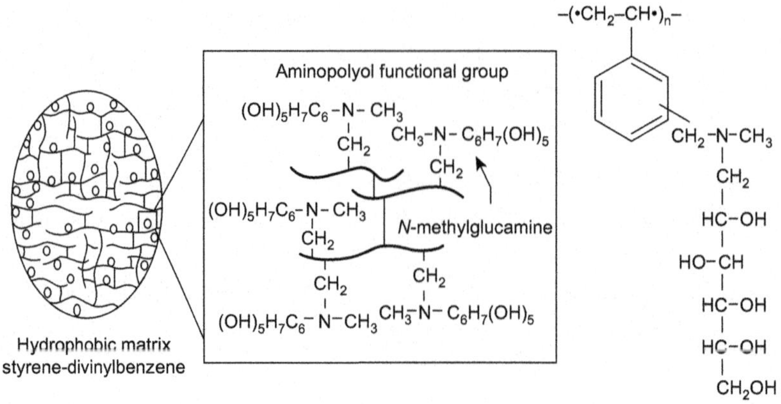

Figure 4.1 Schematic structure of boron-selective-resin-attached (left: Ref. 34) *N*-methylglucamine functional group (see right for details).

resins apply the principle in the formation of stronger boron complexes with the functional group attached on the matrix as chelating resin. The uptake of boron as tetraborate, $B(OH)_4^-$, as illustrated in Figure 4.2, involves the ionization of boric acid, protonation of the amine group, deprotonation of the polyol sugar tail, and the formation of more acidic boroester complexes, either borate monoester or borate diester, or both. There are no positively or negatively charged ionic groups, thus no electropotential inside the resin initially, until the protonization of the *N*-methylamino group and the formation of boron complexes. When using $\overline{R}$ as the functional group inside the resin

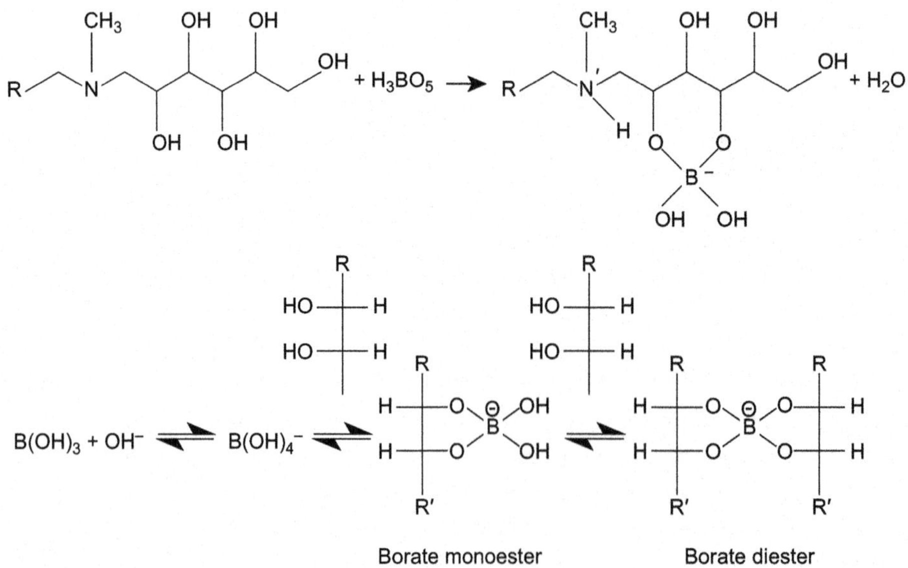

Figure 4.2 Schematic diagram of the formation of borate esters: top (Ref. 35) and bottom (Ref. 36).

and B^- as boron complexes, the process of complex formation can be represented in a general form as

$$\overline{R} + B(OH)_3 \rightarrow \overline{R^+B^-} \quad \text{or} \quad \overline{R} + B(OH)_4^- + H^+ \rightarrow \overline{R^+B^-} \tag{4.31}$$

This is a chemisorption process, without ion exchange involved, and chemical potential is the dominant driving force. The *N*-methylglucamine functional group becomes the zwitterion form (possesses acid–base properties) after complexation, and thus, its stability depends on the pH. In addition to the stability of complexes, conditions in bulk solution such as concentration and pH play more important roles on sorption pattern and performance such as rate, rate-controlled step, and apparent capacity (detailed discussion in the next section).

The formation of boroester complexes can be rapidly reversible and pH dependent.[37] In an aqueous medium, the stability of the boroester formed on *N*-methylglucamine functional groups depends on both the pH and the type of diols. To pH dependence the highest complex stability appears at the pH of its isoelectric point, at which the concentration of its zwitterion form is a maximum. On the other hand, the tolerable strain on O–B–O bonds in the resultant boroester complexes limits only certain diols that can react with boric acid or borates to form boroesters. According to Power and Woods,[37] the strongest boroester complexes are formed when a diol that contains OH groups matches accurately with the structural parameters required by a tetrahedrally coordinated boron. Typically, polyols that contain adjacent *cis* hydroxyl groups are most likely to react with boron compounds to form boroester complexes,[38] and the resultant furanose configuration (five-membered ring structure) is more stable than that with six-membered rings because of the higher strain on the reactive OH group from the furanose configuration.[39]

At low total boron concentrations, there are no significant polyborates formed. Either boric acid or tetraborate ion, or both, are the dominant boron carriers. The formation of boroester complexes inside the resin between boric acid or tetraborate ion and the functional polyol in the boron-specific resin can be represented as follows:

$$B(OH)_3 + H_2L \overset{\beta_{0,1}}{\Longleftrightarrow} \{BL(OH)_2\}^- + (H_3O)^+ \tag{4.32}$$

or

$$B(OH)_4^- + H_2L \overset{\beta_{1,1}}{\Longleftrightarrow} \{BL(OH)_2\}^- + 2H_2O \tag{4.33}$$

and

$$\{BL(OH)_2\}^- + H_2L \overset{\beta_{1,2}}{\Longleftrightarrow} \{BL_2\}^- + 2H_2O \tag{4.34}$$

In these equations, H_2L represents a polyol or diol group (diol side chain "sugar tail"); β is the corresponding association constant of the reaction; the subscripts are the number of boron and the number of polyol groups, respectively; $\{BL(OH)_2\}^-$ and $\{BL_2\}^-$ are borate-monoester and borate–diester complexes, respectively.

In a strong acidic environment, for example, pH $< pK_a$ of a boron–sorbitol 1:1 complex (between 4.5 and 5.0), only the reaction in Eqn (4.34) leads to a stable boroester complex because there is essentially no reaction in Eqn (4.33) taking place and the reaction indicated in Eqn (4.32) has a tendency toward the left side. The protonation of a tertiary amine is an acid–base reaction with a pK_a value of 9.65.[40] At pH values >9.65, deprotonation occurs so that the instability issue is raised for *N*-methylglucamine function groups in a boron-specific chelating resin. At pH close to 7 (the isoelectric point of the functional group with the formed boron complex), the concentration of $\{BL(OH)_2\}^-$ reaches a maximum and so likely does the resin apparent capacity.

Polyborates can be generated in solution, inside the resin, or in both phases through polymerization between boric acid and tetraborate under certain conditions (see the previous section). The possibility of the formation of some stable polyborate–polyol complexes should be anticipated, and the conditions under which they occur recognized.[41]

4.4 SORPTION EQUILIBRIUM AND KINETICS

4.4.1 Equilibrium

Adsorption is the adhesion of atoms, ions, or molecules (sorbate) from a gas, liquid, or dissolved solid to the active site of a solid surface (sorbent). This process creates a film of the adsorbate on the surface of the adsorbent and/or fills the pours within the matrix. A quantitative relationship of the retention or release of a sorbate on sorbent has been termed the sorption isotherm. Depending on the type of binding or interaction and binding energy, adsorption is classified as physisorption (e.g., weak van der Waals forces) or chemisorption (e.g., covalent bonding, electrostatic attraction). Using A as the solute in solution, S^* as the active site on a solid surface, an adsorption process can be expressed in general as

$$A + S^* \leftrightarrow AS^* \tag{4.35a}$$

The sorption isotherm is usually expressed as the concentration of the sorbate retained on a sorbent q as a function of the sorbate concentration C remaining in the solution, referencing certain experimental conditions—particularly temperature, that is,

$$q = f(C). \tag{4.35b}$$

For binary ion exchange, the process characterized by counterion exchange between the resin phase and solution is commonly represented by

$$z_B\overline{A}^{z_A} + z_A B^{z_B} \leftrightarrow z_B A^{z_A} + z_A \overline{B}^{z_B}, \tag{4.36a}$$

where z_i is the electrochemical valence of the ith ionic species, and $\overline{A}$ and $\overline{B}$ are ions in the resin phase. An isotherm is another way to describe the ionic composition of the ion exchanger as a function of the experimental conditions. As a rule, the equivalent ionic fraction $\overline{x_A}$ of the counterion A in the ion exchanger is plotted as a function of the equivalent ionic fraction x_A in the solution, while other variables are kept constant.[42] Accordingly, for the process expressed in Eqn (4.36a), the equivalent ionic fraction x_A is defined by

$$x_A \equiv z_A m_A \cdot (z_A m_A + z_B m_B)^{-1}, \tag{4.36b}$$

where m_i is the concentration of solute i in molality. Quantitative expressions in molality, molarity, and equivalent fraction, of ion–exchange equilibrium, for example, in Eqn (4.36a), is similar to the law of mass action, such as the separation factor α_B^A, which is defined by

$$\alpha_B^A \equiv \overline{m_A} m_B \cdot (\overline{m_B} m_A)^{-1} = \overline{C_A} C_B \cdot \left(\overline{C_B} C_A\right)^{-1} = \overline{x_A} x_B \cdot (\overline{x_B} x_A)^{-1} \tag{4.36c}$$

or selectivity coefficient K_B^A, which is a general expression as a function of the activities a_i of two components that depend only on temperature, as defined by

$$K_B^A(T) \equiv \left(\overline{a}_A^{z_B} \cdot a_B^{z_A}\right) \cdot \left(\overline{a}_B^{z_A} \cdot a_A^{z_B}\right)^{-1}, \tag{4.36d}$$

where m_i and C_i are the concentrations of solute i in molality and molarity, respectively, a_i is the activity of solute i in the liquid phase, quantities with bars refer to the inside of the resin phase. For an ideal system such as a dilute solution, molality or molarity expression may be used as follows:

$$\text{Molality} \quad K_B^A \equiv \frac{\overline{m}_A^{|z_B|} \cdot m_B^{|z_A|}}{\overline{m}_B^{|z_A|} m_A^{|z_B|}}, \tag{4.36e}$$

$$\text{Molarity} \quad K'^A_B \equiv \frac{\overline{C}_A^{|z_B|} \cdot C_B^{|z_A|}}{\overline{C}_B^{|z_A|} C_A^{|z_B|}}. \tag{4.36f}$$

The selectivity coefficient is more convenient for theoretical studies.[42]

Considering the sorption of boron on a strong base, OH form anion resin, the counterion OH^- released from the resin phase is consumed at, or very near, the resin surface by boric acid from solution to form tetraborate $B(OH)_4^-$ in replacing OH^- ion inside the resin. With a very high pH environment inside the resin, all boron carrier species shift spontaneously to the tetraborate form as soon as diffusion into the resin phase occurs. Inside the resin, only two counterion pairs, tetraborate $B(OH)_4^-$ and OH^-, participate in interdiffusion as ordinary ion exchange. Recall Eqn (4.30) for the representation of the process.

$$\overline{R^+} + \overline{OH^-} + B(OH)_3 \rightarrow \overline{R^+} + \overline{B(OH)_4^-}$$

The corresponding equilibrium relationship may be described by the molarity selectivity coefficient.

$$K_{\mathrm{B(OH)_3}}^{\mathrm{OH}} = \left(\overline{C}_{\mathrm{OH}}^{1} C_{\mathrm{B(OH)_3}}^{0}\right) \cdot \left(\overline{C}_{\mathrm{B(OH)_4}}^{1} C_{\mathrm{OH}}^{1}\right)^{-1} = \overline{C}_{\mathrm{OH}} \cdot \left(\overline{C}_{\mathrm{B(OH)_4}} C_{\mathrm{OH}}\right)^{-1}. \tag{4.37}$$

As sorption of boron on a weak-base boron-specific (chelating) resin, conversion of the resin likely consists of two mechanisms. One is the boron carriers, boric acid molecule or borate ions chelated on diols of the function group *N*-methylglucamine through the formation of boroester complexes. There are a series of reactions involved in the sorption process, including boric acid dissociation, protonation of the free-base form ($-CH_2-NCH_3-CH_2-$), and formation of boroester complexes. The other is the uptake of either the boric acid molecule or borate ions on the protonated amino group, depending on the conditions inside the resin and the *N*-methylglucamine function group itself. Such a process can be written in the general form.

$$\overline{\mathrm{R}} + \mathrm{B(OH)_3} \rightarrow \overline{\mathrm{R^+B^-}} \quad \mathrm{or/and} \quad \overline{\mathrm{R^+}} + \mathrm{B(OH)_4^-} \rightarrow \overline{\mathrm{R^+B^-}} \tag{4.38}$$

Therefore, sorption of boron from aqueous solutions on boron-specific (chelating) resins may be characterized as competitive sorption of different sorbates for different sites inside the resin, which is more likely the nature of an (chemical) adsorption process than that of an ion exchange process. Theory-governed adsorption processes, such as ideal, or modified, Langmuir and Freundlich isotherms, could be adapted to interpret the boron sorption isotherm.

4.4.1.1 Models and Correlations of Sorption Isotherm

There are various sorption isotherm theories, models, empirical, and mechanistic correlations reported in the past. In addition to the consideration of modeling and experiment measurement, a detailed review on physical bases and classification of sorption isotherm is presented by Limousin et al.[43] Sorption isotherms that may apply to boron sorption on ion exchange resins are briefly reviewed here for convenience.

4.4.1.2 Langmuir Isotherm

With homogeneous surfaces and negligible interaction among adsorbed molecules, the ideal Langmuir isotherm for single solute is

$$q = q_{max} K_L C \cdot (1 + K_L C)^{-1} \tag{4.39a}$$

with a linear expression

$$\frac{1}{q} = \frac{1}{q_{max} K_L} \cdot \frac{1}{C} + \frac{1}{q_{max}}, \tag{4.39b}$$

where q and q_{max} represent the sorbate concentration adsorbed and the maximum monolayer capacity saturated on the surface of a sorbent, C is the sorbate molarity concentration in solution, K_L is the Langmuir constant, reflecting the binding energy. Extending Langmuir isotherm to a multicomponent system is expressed as

$$q_i = q_{max} K_i C_i \cdot \left(1 + \sum_1^{i=n} K_i C_i\right)^{-1}, \tag{4.39c}$$

where K_i is the Langmuir constant of the ith species, C_i is the sorbate concentration of the ith species, q_i and q_{max} represent the ith sorbate concentration and the maximum monolayer capacity saturated on the surface of a sorbent, respectively.

The ideal Langmuir isotherm is limited to applications of low sorbate concentrations. Deviations can be pronounced even at low sorbate concentrations due to the heterogeneity of sites on a sorbent and significant sorbate–sorbate interaction near the saturated region of a sorbent.[44]

4.4.1.3 Two-Site Langmuir Model and Langmuir–Freundlich Model

Two-site Langmuir and Langmuir–Freundlich models are modifications of the ideal Langmuir isotherm to reduce the deviation duo to heterogeneity of sites on a sorbent. The protonated amino group on *N*-methylglucamine with a positive charge can provide the other active site for sorption of either boric acid molecule or negatively charged borate ion, along with the side chelating diol group. Two-site Langmuir or Langmuir–Freundlich model may be more adequate to interpret the isotherm in this situation.

Two-Site Langmuir Model

$$q = q_{max,1} K_{L,1} C \cdot \left(1 + K_{L,1} C\right)^{-1} + q_{max,2} K_{L,2} C \cdot \left(1 + K_{L,2} C\right)^{-1}. \tag{4.40}$$

Langmuir–Freundlich Model (Sip's Equation)

$$q = q_{max} K_{L-F} C^n \cdot \left(1 + K_{L-F} C^n\right)^{-1}, \tag{4.41}$$

In Eqn (4.40) and (4.41), $K_{L,i}$ and $q_{max,i}$ are the Langmuir constants and total capacities corresponding to different types of sites, K_{L-F} and n are the Langmuir–Freundlich isotherm constant and heterogeneity coefficient, respectively. The value of n varies fractionally from 0 to 1. As the value of n is set to 1, the Langmuir–Freundlich isotherm reduces to the Langmuir isotherm for homogeneous material, while <1 for heterogeneous materials.

For applications of the Langmuir–Freundlich isotherm model, refer to Ho et al. (2002) for sorption isotherm studies of divalent metal ions on peat, Turiel et al.[45] for the assessment of the crossreactivity and binding site characterization of a propazine-imprinted polymer, and Jeppu and Clement[46] for the simulation of pH-dependent adsorption effects. However, the Langmuir–Freundlich isotherm model lacks thermodynamic consistence and should be treated with caution.[44]

4.4.1.4 Equilibrium of Boron–Diol Complexation

The association constants of complexations in Eqns (4.32)–(4.34) are expressed, respectively, as follows:

$$\beta_{0,1} = \left[(BL(OH)_2)^-\right]\cdot\left[(H_3O)^+\right]\cdot\left\{\left[B(OH)_3\right]\cdot\left[H_2L\right]\right\}^{-1}, \quad (4.42a)$$

$$\beta_{1,1} = \left[(BL(OH)_2)^-\right]\cdot\left\{\left[B(OH)_4^-\right]\cdot\left[H_2L\right]\right\}^{-1}, \quad (4.42b)$$

$$\beta_{1,2} = \left[BL_2^-\right]\cdot\left\{\left[(BL(OH)_2)^-\right]\cdot\left[H_2L\right]\right\}^{-1}. \quad (4.42c)$$

The association constants, $\beta_{0,1}$ and $\beta_{1,1}$, are not independent and can be related with the aid of the dissociation relation (or ionization) of boric acid,

$$B(OH)_3 + 2H_2O \overset{K_a}{\leftrightarrow} B(OH)_4^- + H_3O^+ \quad K_a = \left[B(OH)_4^-\right]\cdot\frac{\left[(H_3O)^+\right]}{\left[B(OH)_3\right]} \quad (4.43)$$

resulting in

$$\beta_{0,1} = K_a\beta_{1,1}. \quad (4.44)$$

There are about nine orders of magnitude difference between $\beta_{0,1}$ and $\beta_{1,1}$. Obviously, with conditions of the coexisting comparable boric acid molecule and tetraborate form, tetraborate is profoundly favored over the boric acid molecule involved in the formation of boron–ester complexes. Undissociated boric acid molecules inside the resin, if any, are likely to compete for sites of protonated amino groups. At pH >9.24, contribution to boron sorption of tetraborate is dominant; thus, a film diffusion-controlled process is likely.

From Eqns (4.42b) and (4.42c), one may have the concentration relationship of complete boron chelate form (boron diester) BL_2 with the diol H_2L, and tetraborate $B(OH)_4^-$

$$\left[BL_2\right] \propto \left[H_2L\right]^2 \quad \text{and} \quad \left[BL_2\right] \propto \left[B(OH)_4^-\right]. \quad (4.45)$$

Boron diester is likely to form at the early stage of sorption and under the conditions leading to a very high tetraborate concentration, which is likely to decrease the capacity of the boron-specific (chelating) resin.

4.4.2 Sorption Kinetics

Sorption kinetics determines the change in concentration with time and position. In fixed-bed ion exchange or adsorption, the kinetics consists of two parts: local sorption rate within the particle and fluid layer surrounding the particle, and through the bed space. With a set of proper assumptions, negligible radial concentration gradient, uniform bed porosity, and particle size, and plug flow, the sorption history through the bed space can be described by a mathematically simplified material balance equation (good for dilute concentration, trace system),

$$\frac{u_s}{\varepsilon}\frac{\partial C_i}{\partial x} + \frac{\partial C_i}{\partial t} + \frac{(1-\varepsilon)}{\varepsilon}\frac{\partial \overline{q_i}}{\partial t} = 0 \tag{4.46a}$$

in which C_i is the bulk liquid concentration of the ith sorbate, $\overline{q}_i$ is the average concentration of the ith sorbate in the sorbent phase, u_s is the superficial velocity, ε is the porosity of a packed bed, and x is the longitudinal coordinate of a packed bed.

For a particle with a spherical geometry, the time dependence of concentration inside the particle is given by

$$\text{Material balance (Fick's second law)} \quad \frac{\partial q_i}{\partial t} = \frac{1}{r^2}\frac{\partial}{\partial r}\left(r^2 D_{p,i}\frac{\partial q_i}{\partial r}\right), \tag{4.47a}$$

$$\text{Equilibrium isotherm} \quad q_i = f(C_i), \tag{4.47b}$$

$$\text{Average concentration inside the particle} \quad \overline{q}_i = \frac{3}{R_p^3}\int_0^{R_p} q_i r^2\, dr. \tag{4.47c}$$

In the liquid layer (diffusion film), Fick's first law is applied to develop the flux equations of nonionic species.

$$J_i = -D_i \cdot \nabla C_i. \tag{4.48}$$

The Nernst—Planck Equation is applied to describe the flux of ionic species.

$$J_i = -D_i\left[\frac{\partial C_i}{\partial r} + z_i C_i \frac{F}{RT}\frac{\partial \varphi}{\partial r}\right]. \tag{4.49}$$

The condition of no accumulation at the interface between liquid phase and resin particle phase leads to the time dependence of the concentration inside the resin particle interrelated with the flux through the diffusion film

$$\frac{\partial q_i}{\partial t} = -J_i a_s, \tag{4.50}$$

where D_p is the diffusion coefficient within the resin bead, C_i refers to the molarity concentration of the ith species in the liquid phase, D_i is the self-diffusion coefficient of the ith species in the liquid, F is the Faraday constant, r is the radial coordinate of particles, R and T are the gas constant and absolute temperature, respectively, φ is the electric potential, q_i is the concentration of the ith species in the sorbent phase, z_i represents the electrochemical valence of the ith species, and a_s is the specific surface area of the resin particle, $a_s d_p = 6$ for spherical particles with a diameter d_p.

4.4.2.1 Resistance to Boron Mass Transfer and Sorption Control Step

The overall rate of sorption of boron from aqueous solutions on either a strong base OH form anion resins or *N*-methylglucamine functional weak-base chelating resins is likely controlled by mass transfer resistance. Chemical reactions, such as boric acid dissociation and ionization, boron polymerization, protonation of the amino group, and borate-diol esterification, involved in the sorption process, take place rapidly compared to diffusion so that local equilibrium is valid. Either separately or together, diffusion inside the resin and within a Nernst film layer adherent to the resin surface could be the rate-determining step.

For boron sorption from aqueous solutions on strong base OH form anion resins, boric acid converts to tetraborate $B(OH)_4^-$ at the interface and right inside the resin. Counterion OH^- liberated from the resin cannot make headway into the Nernst film but is consumed in the ionization of molecular boric acid. Conditions of no net current and electroneutrality limit borate ions in the film. Maintaining the dissociation relationship of boric acid increases interaction and consequently reduces the mobility of boric acid. Diffusion inside the resin is by interdiffusion of counterions OH^- ion originally inside the resin and $B(OH)_4^-$ as in ordinary ion exchange. Therefore, the rate-controlling step is Nernst film diffusion of molecular boric acid and borate ions.

Introducing the functional group *N*-methylglucamine results in a strong affinity for borate ion and boric acid, and thus tends to form boron complexes with a negative charge. However, the gain in selectivity must be paid for by a loss in the ion exchange rate.[42] A consequence of such an association greatly reduces the mobility of the incoming borate ion inside the boron-chelating resin. Particle diffusion resistance could be dominant after a period of loading. Besides, the resin will be saturated with ionic species after regeneration and for similar reasons to those for strong base OH form anion resins, diffusion resistance of boron carriers in the Nernst film is significant. Thus, the rate-controlling step could be combined film and particle diffusion for the sorption of boron on boron-specific (chelating) resins.

4.4.2.2 Determination of Apparent Diffusion Coefficient of Boron in the Film

Because of dissociation and association as boron carriers get transported through the film, the total concentration of counterions is no longer constant, and the flux — and thus material balance (Fick's second law) — must be modified to include such an effect. With the assumption of no coion flux, the only coion hydrogen leads to

$$J_H = -D_H\left[\frac{\partial C_H}{\partial r} + z_H C_H \frac{F}{RT}\frac{\partial \varphi}{\partial r}\right] = 0, \tag{4.51a}$$

$$\frac{F}{RT}\frac{\partial \varphi}{\partial r} = -\frac{1}{C_H}\frac{\partial C_H}{\partial r}. \tag{4.51b}$$

Substituting the relation in Eqn (4.51b) into the tetraborate (indicated as $B1$) flux expression, one has

$$J_{B1} = -D_{B1}\left[\frac{\partial C_{B1}}{\partial r} + z_{B1} C_{B1} \frac{F}{RT}\frac{\partial \varphi}{\partial r}\right] = -D_{B1}\left[\frac{\partial C_{B1}}{\partial r} + \frac{C_{B1}}{C_H}\frac{\partial C_H}{\partial r}\right]. \tag{4.52}$$

From the equilibrium relation of boric acid dissociation, we have

$$K_a = C_H \cdot C_{B1} \cdot \frac{1}{C_B}. \tag{4.53}$$

Differentiation of Eqn (4.53) and solving for the tetraborate concentration gradient in particle radial direction give

$$\frac{\partial C_{B1}}{\partial r} = \frac{K_a}{C_H}\frac{\partial C_B}{\partial r} - \frac{C_{B1}}{C_H}\frac{\partial C_H}{\partial r}. \tag{4.54}$$

Substituting the relation in Eqn (4.54) into Eqn (4.52) gives the flux of tetraborate with the concentration gradient of the boric acid molecule

$$J_{B1} = -D_{B1}\frac{K_a}{C_H}\frac{\partial C_B}{\partial r}. \tag{4.55}$$

The total boron flux in the film is

$$J_{BT} = J_B + J_{B1} = -D_B\frac{\partial C_B}{\partial r} - D_{B1}\frac{K_a}{C_H}\frac{\partial C_B}{\partial r} = -\left[D_B + D_{B1}\frac{K_a}{C_H}\right]\cdot\frac{\partial C_B}{\partial r}. \tag{4.56}$$

The apparent diffusivity D_e for the total boron in the film, combining the relationship of water dissociation, $K_w = C_H \cdot C_{OH}$, and Eqn (4.21), is then

$$D_e = D_B + D_{B1}K_a\frac{1}{C_H} \quad \text{or} \tag{4.57a}$$

$$D_e = D_{BT} = D_B + D_{B1}C_{OH}\frac{K_a}{K_w} = D_B + D_{B1}C_{OH}K_1, \tag{4.57b}$$

where subscripts $B1$, B, and BT represent tetraborate ion $B(OH)_4^-$, boric acid molecule, and the total boron, respectively, K_a and K_1 are the equilibrium constants of boric acid dissociation and ionization, respectively.

Equations (51.57a) and (51.57b) reflect the effects caused by the dissociation and association of the species involved in diffusion within the film and hydrogen concentration, and are thus pH dependent as well. At pH = 9, the ratio of K_a/C_H is within zero order of magnitude; therefore, the contribution of the boric acid molecule and tetraborate to boron sorption is comparable, and the contribution of the ionic form tetraborate will increase as the pH value increases or as the concentration of hydrogen decreases.

4.4.2.3 Apparent Diffusion Coefficient in the Resin Particle

For a system with binary counterion interdiffusion within the resin phase, the formulation of the problem was given by Helfferich and Plesset[47] with the precondition of constant fixed ionic groups and the coion concentration. With the system of counterion hydroxyl OH^- ions and borate ions $B(OH)_4^-$ (indicated as $B1$), the flux inside the resin for either counterion is

$$J_i = -\left[D_{OH}D_{B1}\left(z_{OH}^2 q_{OH} + z_{B1}^2 q_{B1}\right)\cdot\left(D_{OH}z_{OH}^2 q_{OH} + D_{B1}z_{B1}^2 q_{B1}\right)^{-1}\right]\cdot\nabla q_i, \quad (4.58)$$

where D_i and z_i are the self-diffusivity and electrochemical valence of the ith species, q_i refers to the concentration of the ith species in the resin phase, and the apparent diffusivity D_e for either counterion is

$$D_e = \left[D_{OH}D_{B1}\left(z_{OH}^2 q_{OH} + z_{B1}^2 q_{B1}\right)\cdot\left(D_{OH}z_{OH}^2 q_{OH} + D_{B1}z_{B1}^2 q_{B1}\right)^{-1}\right] \quad (4.59)$$

as long as the protonated condition of the amino group is valid. If the pH inside the resin is above the pK_a value of the protonated amino group, deprotonation occurs, and the precondition breaks down; thus, this relation is no longer valid.

4.4.2.4 Interfacial Concentrations

The interface is the boundary between the liquid phase and the resin particle phase. Interfacial concentrations consist of the concentration C^* at the liquid side and the concentration q^* at the resin particle side. Their relationships must reflect the equilibrium isotherm defined in the model. Consider the two limitations of diffusion control. If the process is under liquid side film diffusion control, the distribution of the concentration in the resin phase can be treated as being uniform, and correspondingly, the interfacial concentration at the resin side q^* equals the average concentration in the resin phase $\bar{q}$. In the other limitation, the resistance on the liquid side is negligible. The interfacial concentration on the liquid side C^* can be set as the liquid bulk concentration C^0. By an applied sorption isotherm, they are

$$\text{Film} - \text{diffusion control} \quad q^* = \bar{q}; \quad \text{then} \quad C^* = F(q^*), \quad (4.60a)$$

$$\text{Particle} - \text{diffusion control} \quad C^* = C^0; \quad \text{then} \quad q^* = f(C^*). \quad (4.60b)$$

If resistance, both on the liquid side and on the resin side, needs to be considered, there is no explicit formulation to specify the surface concentration on each side. Interfacial concentrations on both sides are implicit functions of the average concentration inside the resin phase $\bar{q}$.

4.4.2.5 Simplified Rate Laws and Semiempirical Models

In practical applications, various mathematically simplified rate expressions have been suggested and widely used. For a list of semiempirical models and simplified rate laws and brief discussion refer to Helfferich.[42] Most of them are apparently based on the original work of Lagergren in 1898, including linear driving force approximation[48,49] and pseudo-first-order (Lagergren kinetic equation) approximation, which are commonly applied to interpret the kinetic performance of ion exchanger or other sorbents:

$$\text{Linear Driving Force} \quad \frac{d\bar{q}}{dt} = k_p(q^* - \bar{q}) = k_L(C - C^*) \quad \text{with} \quad k_p = \frac{15D_p}{R_p^2}, \tag{4.61}$$

$$\text{Pseudo} - \text{first} - \text{order} \quad \frac{dq}{dt} = k_1(q_e - q), \tag{4.62}$$

where q^* and $\bar{q}$ represent the resin surface concentration and the average concentration inside the resin, respectively, C and C^* are the concentrations in the bulk solution and on the interface of the liquid side, R_p is the radius of the resin particle, k_p and D_p are the mass transfer coefficient and diffusion coefficient in the resin particle, respectively, k_L is the mass transfer coefficient in the liquid phase, q and q_e are adsorbed concentrations on the sorbent with respect to time and the adsorbed concentration on the sorbent at equilibrium, respectively, k_1 is the pseudo-first-order rate constant.

The linear driving force and the pseudo-first-order approximations have similar expressions. The former implies that the uptake history is proportional to the dynamic change of the driving force and the latter indicates that the sorption rate is proportional to the ultimate driving force at operation conditions. In the case of infinite solution volume or negligible concentration change in the bulk liquid through an uptake process, the two approximations are essentially equivalent.

These approximations are good for reversible, linear, or near linear sorption systems. For an irreversible sorption system, Vermeulen's equation[50] is a superior approximation[49]:

$$\text{Vermeulen's equation} \quad \frac{d\bar{q}}{dt} = \frac{\pi^2 D_p}{R_p^2}\frac{\left(q^{*2} - \bar{q}^2\right)}{2\bar{q}}, \tag{4.63}$$

where q^* and $\bar{q}$ represent the resin surface concentration and average concentration inside the resin, R_p is the radius of the resin particle, D_p is the diffusion coefficient in the resin particle.

A linear driving force approximation is particularly useful in the quantitative treatment of a packed bed and has found widespread application.[44] The mass transfer coefficient on the liquid side k_L defined in Eqn (4.61), interrelated diffusivity D_e, and film thickness δ ($k_L = D_e/\delta$) can be correlated as a lumped parameter, Sherwood number (as a function of the Reynolds number and Schmidt number), from mass transfer experiments. A list of correlations accounting for ion exchange data for calculating the Sherwood number is summarized by Chowdiah et al.[51] The ratio of k_p to k_L may be used as a measure of sorption mechanisms. As $k_p/k_L \gg 1$, the sorption process is the liquid side film controlling process, as $k_p/k_L \ll 1$, it is the particle diffusion controlling process, and while near equal is a combined film and particle resistance controlling process.

4.4.3 Comments

The optimum performance of an ion exchange process must be a balance of selection, capacity, exchange rate, stability, and regeneration. Considering applications for boron removal from aqueous solutions on boron-specific (chelating) resins, the solution pH is an essential operational parameter. A pH range from the pK_a of a sorbitol—boron 1:1 complex to the pK_a of the protonated amino group of the *N*-methylglucamine functional group may be suitable; the pH value range from the isoelectric point (of the zwitterionic form, its numerical value is the average of the two pK_as, mentioned above) of the functional group, *N*-methylglucamine to the pK_a of boric acid can be even better, and the pH value around the isoelectric point could be the best in terms of boron monoester stability; thus, uptake effectiveness and apparent capacity are maximized. Within the pH range discussed, the higher the pH, the higher the tendency of boron monoester to lose hydrogen and to form boron diester. Apparently, the function group, *N*-methylglucamine provides two sorption sites, which is the effect of an already bound ligand. The thermodynamics and kinetics of the other site are not yet known.

REFERENCES

1. Ross VF, Edwards JO. The structural chemistry of the borates. *The Chemistry of Boron and its Compounds* 1967:155—207.
2. Corti H, Crovetto R, Refnandez-Prini R. Properties of the borate ion in dilute aqueous solutions. *J Chem Soc Faraday Trans 1* 1980;**76**:2179—86.
3. Ingri N, Lagerstrom G, Frydman M, Sillen LG. Equilibrium studies of polyanions II. Polyborates in $NaClO_4$ medium. *Acta Chem Scand* 1957;**11**:1034—58.
4. Edwards JO. Detection of anionic complexes by pH measurements. I. Polymeric borates. *J Am Chem Soc* 1953;**75**(24):6151—4.
5. Manov GG, DeLollis NJ, Acree SF. Ionization constant of boric acid and the pH of certain borax-chloride buffer solutions from 0 to 60 °C. Res Paper RP1609 *Part J Res Natl Bureau Stand* 1944;**33**:287—308.
6. Owen BB. The dissociation constant of boric acid from 10 to 50°. *J Am Chem Soc* 1934;**56**:1695—7.

7. Maya L. Identification of polyborate and fluoropolyborate ions in solution by Raman spectroscopy. *Inorg Chem* 1976;**15**(9):2179–84.
8. Ingri N. Equilibrium studies of polyanions. 8. On the first equilibrium steps in the hydrolysis of boric acid, a comparison between equilibria in 0.1 M and 3.0 M $NaClO_4$. *Acta Chem Scand* 1962;**16**:439–48.
9. Ingri N. Equilibrium studies of polyanions 10. On the first equilibrium steps in the acidification of $B(OH)_4^-$, an application of the self-medium method. *Acta Chem Scand* 1963a;**17**:573–80.
10. Ingri N. Equilibrium studies of polyanions 11. Polyborates in 3.0 M Na(Br), 3.0 M Li(Br), and 3.0 M K(Br), a comparison with data obtained in 3.0 M $NaClO_4$. *Acta Chem Scand* 1963b;**17**:581–9.
11. Zhou Y, Fang C, Fang Y, Zhu F. Polyborates in aqueous borate solution: a Raman and DFT theory investigation. *Spectrochim Acta Part A* 2011;**83**:82–7.
12. Edwards JO, Ross V. Structure principles of the hydrated polyborates. *J Inorg Nucl Chem* 1960; **15**(3–4):329–37.
13. Platford RF. Osmotic and activity coefficients of some simple borates in aqueous solution at 25°. *Can J Chem* 1969;**47**:2271–3.
14. Mesmer RE, Baes Jr CF, Sweeton FH. Acidity measurements at elevated temperatures. VI. Boric acid equilibria. *Inorg Chem* 1972;**11**(3):537–43.
15. Anderson JL, Eyring EM, Whittaker MP. Temperature jump rate studies of polyborate formation in aqueous boric acid. *J Phys Chem* 1964;**68**:1128–32.
16. Momii RK, Nachtrieb NH. Nuclear magnetic resonance study of borate-polyborate equilibria in aqueous solution. *Inorg Chem* 1967;**6**(6):1189–92.
17. Korkisch J. Their application to inorganic analytical chemistry. In: *Handbook of ion exchange resins*, vol. VI. Boca Raton, FL: CRC Press; 1989.
18. Jacob C. Boron removal by ion exchange technology. *Desalination* 2007;**205**:47–52.
19. Nadav N. Boron removal from seawater reverse osmosis permeate utilizing selective ion exchange resin. *Desalination* 1999;**124**:131–5.
20. Nadav N, Priel M, Glueckstern P. Boron removal from the permeate of a large SWRO plant in Eilat. *Desalination* 2005;**185**:121–9.
21. Badruk M, Kabay N, Demircioglu M, Mordogan H, Ipekoglu U. Removal of boron from wastewater of geothermal power plant by selective ion-exchange resins. I. Batch sorption-elution studies. *Sep Sci Technol* 1999;**34**(13):2553–69.
22. Simonnot MO, Castel C, Nicolai M, Rosin C, Sardin M, Jauffret H. Boron removal from drinking water with a boron selective resin: is the treatment really selective? *Wat Res* 2000;**34**(1):109–16.
23. Vengosh A, Heumann KG, Juraske S, Kasher R. Boron isotope application for tracing sources of contamination in groundwater. *Environ Sci Technol* 1994;**28**:1968–74.
24. Lou J, Foutch GL, Na JW. Kinetics of boron sorption and desorption in boron thermal regeneration system. *Sep Sci Technol* 2000;**35**(14):2259–77.
25. Peterka F. Selection of anion-exchange resins for boron thermal-regeneration systems. *J Chrom A* 1980;**201**:359–70.
26. van der Schoot MR. US Patent 4,017,358 1977.
27. Byrne RH, Yao W, Klochko K, Tossell JA, Kaufman AJ. Experimental evaluation of the isotopic exchange equilibrium $^{10}B(OH)_3 + {}^{11}B(OH)_4^- = {}^{11}B(OH)_3 + {}^{10}B(OH)_4^-$ in aqueous solution. *Deep-Sea Res* 2005;**53**:684–8.
28. Kakihana H, Kotaka M, Satoh S, Nomura M, Okamoto M. Fundamental studies on the ion-exchange separation of boron isotopes. *Bull Chem Soc Jpn* 1977;**50**(1):158–63.
29. Sonoda A, Makita Y, Ooi K, Takagi N, Hirotsu T. pH-dependence of the fractionation of boron isotopes with *N*-methyl-*D*-Glucamine resin in aqueous solution systems. *Bull Chem Soc Jpn* 2000:1131–3.
30. Yoneda Y, Uchijima T, Makishima S. Separation of boron isotopes by ion exchange. *J Phys Chem* 1959;**63**(12):2057–8.
31. Tomizawa T. Studies on the absorption of boric acid on anion exchange resin I. Absorption characteristics of boric acid on strong base anion exchange resins. *Denki Kagaku* 1979;**47**(10):602–7.
32. Tomizawa T. Studies on the absorption of boric acid on anion exchange resin II. Effects of temperature on the absorption characteristics of boric acid on strong base anion exchange resins. *Denki Kagaku* 1981;**49**(6):339–42.

33. Lyman W, Preuss A. U.S. Patent 2,813,838. November 19, 1957.
34. Kabay N, Sarp S, Yuksel M, Kitis M, Koseoglu H, Ararc O, et al. Removal of boron from SWRO permeate by boron selective ion exchange resins containing *N*-methylglucamine groups. *Desalination* 2008;**223**:49–56.
35. Hubicki Z, Kolodynska D. Selective removal of heavy metal ions from waters and waste waters using ion exchange methods. In: Kilislioglu A, editor. *Ion exchange technologies*. InTech; 2012.
36. Labouriau A, Smith BF, Khalsa GRK, Robison TW. Boric acid binding studies with diol containing polyethylenimines as determined by ^{11}B NMR spectroscopy. *J Appl Poly Sci* 2006;**102**:4411–8.
37. Power PP, Woods WG. The chemistry of boron and its speciation in plants. *Plant Soil* 1997;**193**:1–13.
38. Hunt CD. Specific boron-binding biomolecules: a key to understanding the beneficial physiologic effects of dietary boron from prokaryotes to humans. In: Goldbach Heiner, et al., editors. *Boron in plant and animal nutrition*. New York: Kluwer Academic/Plenum Publishers; 2002.
39. Loomis WD, Durst RW. Chemistry and biology of boron. *Bio Factors* 1992;**3**:229–39.
40. García-Soto M, del Mar de la Fuente, Camacho EM. Boron removal by processes of chemosorption. *Solvent Extr Ion Exch* 2005;**23**:741–57.
41. Gaspar A, Harir M, Lucio M, Hertkorn N, Schmitt-Kopplin P. Targeted borate complex formation as followed with electrospray ionization fourier transform ion cyclotron mass spectrometry: monomolecular model system and polyborate formation. *Spectrom* 2008;**22**:3119–29.
42. Helfferich FG. *Ion exchange*. New York: McGraw Hill Book Company; 1962.
43. Limousin G, Gaudet J-P, Charlet L, Szenknect S, Barthes V, Krimissa M. Sorption isotherm: a review on physical bases, modeling and measurement. *Appl Geochem* 2007;**22**:249–75.
44. Ruthven DM. *Principles of adsorption and adsorption processes*. New York: John Wiley & Sons Inc.; 1984.
45. Turiel E, Perez-Condea C, Martin-Esteban A. Assessment of the cross-reactivity and binding sites characterisation of a propazine-imprinted polymer using the Langmuir–Freundlich isotherm. *Analyst* 2003;**128**:137–41.
46. Jeppu GP, Clement TP. A modified Langmuir–Freundlich isotherm model for simulating pH-dependent adsorption effects. *J Contaminant Hydrol* 2012;**129–130**:46–53.
47. Helfferich F, Plesset MS. Ion exchange kinetics. A nonlinear diffusion problem. *J Chem Phys* 1958;**28**(3):418–24.
48. Glueckauf E, Coates JI. Theory of chromatography. Part IV: the influence of incomplete equilibrium on the front boundary of chromatogram and on the effectiveness of separation. *J Chem Soc* 1947:1315–21.
49. Glueckauf E. Theory of chromatography. Part 10. Formulae for diffusion into spheres and their application to chromatography. *Trans Faraday Soc* 1955;**51**:1540–51.
50. Vermeulen T. Theory for irreversible and constant-pattern solid diffusion. *Ind Eng Chem* 1953;**45**(8):1664–70.
51. Chowdiah VN, Foutch GL, Lee G-C. Binary liquid-phase mass transport in mixed-bed ion exchange at low solute concentration. *Ind Eng Chem Res* 2003;**42**:1485–94.
52. Ho YS, Porter JF, McKay G. Equilibrium isotherm studies for the sorption of divalent metal ions onto peat: copper, nickel and lead single component systems. *Water Air Soil Pollut* 2002;**141**:1–33.
53. Ristic MDj, Rajakovic LjV. Boron removal by anion exchangers impregnated with citric and tartaric acids. *Sep Sci Technol* 1996;**31**(20):2805–14.

CHAPTER 5

Separation and Recovery of Boron From Various Resources Using Chelate Adsorbents

Kazuharu Yoshizuka, Syouhei Nishihama
Department of Chemical Engineering, The University of Kitakyushu, Kitakyushu, Japan

5.1 INTRODUCTION

Boron is an essential material in fiber glass insulation, borosilicate glass, metal welding, and pharmaceutical manufacturing, and is a dopant for semiconductors.[1,2] Boron is generally obtained by the refinement of boron ore, mainly colemanite, and borax.[2] Boron is also contained in natural waters and is an essential element for life on Earth. The amount of boron required for life is however very limited, and ingestion of an excess amount of boron is a well-known cause of nervous system disease. The World Health Organization currently recommends limiting the concentration of boron in drinking water to 2.4 mg/L.[3] Removal of boron from industrial waste[4,5] and ceramic or opto-electronic industries[6,7] has therefore been investigated. Boron removal technology is also required to utilize natural water sources, with the most important application being desalination of seawater to obtain fresh water.[8] Geothermal water, which has recently attracted attention as a source of renewable energy, can also be regarded as an alternative source of boron.[9–11] Separation and recovery technologies for boron, especially from aqueous solutions, are therefore required.

Several separation technologies, such as adsorption with inorganic adsorbents,[12,13] ion exchange,[1,11,12,14–16] solvent extraction,[16,17] and reverse osmosis,[18] have been applied for recovering boron from aqueous solutions. Kabay et al. reviewed the technologies for boron removal from seawater.[19] Ion exchange is one of the most promising technologies for the separation and recovery of boron from aqueous solutions, combined with reverse osmosis to obtain fresh water from seawater. The glucamine group is well known as an effective chelating group in boron adsorbents. The Mitsubishi Chemical Corporation also provides Diaion® CRB series adsorbent,[20] and these are commonly used as an adsorbent for boron.[11] Recently, the adsorption of boron by cellulose fibers containing glucamine groups has also been developed, and these fibers were found to possess faster adsorption kinetics for boron than conventional ion exchange resins.[21]

Boron Separation Processes
ISBN 978-0-444-63454-2, http://dx.doi.org/10.1016/B978-0-444-63454-2.00005-8

In this chapter, we describe the several organic chelate resins having glucamine-based functional groups, such as chelate resin (Diaion® CRB series) and chelate fiber (Chelest Fiber® series), to compare the performance. Furthermore, boron recoveries from various boron resources such as geothermal waters and salt lake brines are described using these chelating adsorbents.

5.2 REMOVAL TECHNOLOGY OF BORON FROM AQUEOUS SOLUTION

Taking into account the increasing concentration of boron in surface waters and the need for treatment of seawater, which contains large amounts of this element, the current research focuses on the effective technologies for removal of boron. The high incidence of boron in natural and wastewaters determines the importance of research in developing the effective process of boron removal from aqueous solution. However, there is the great difficulty that boron appears in water as several chemical species with different concentrations.[23] The methods commonly used in water purification, such as sedimentation, coagulation, or adsorption on clays are not effective in the case of boron compounds. Water treatment by biological or chemical compounds is not effective as they remove only small amounts of boron or they do not remove it at all.[22,23] The processes of evaporation, crystallization, or solvent extraction, suited only for solutions with high concentrations, are more useful in the production of boric acid than in the process of removing traces of boron from water.[24]

The technology most commonly used for the removal of boron from aqueous solutions, the application of chelating resin, seems to be one of the most effective methods.[25] The chelating resins containing ligands having three or more hydroxyl groups in the *cis*-position show a high selectivity to boron, while these groups are not reactive to ordinary metals and other elements. Selective sorption of these resins is due to reactions that are characteristic for boron. Molecules of polyoxide compounds tend to bond through the formation of boric acid esters of boron or borate anion complexes with a proton as a counterion.[9,22,26,27] The results obtained suggest that the presence of a tertiary amine group is critical for boron chelating.[27–30] Most synthesized resins were formed by modification with the *N*-methyl-*D*-glucamine (NMDG) copolymer of styrene and divinylbenzene (DVB). The functional groups of this resin capture boron through a covalent attachment and form a coordination complex as is shown in Figure 5.1.[31]

Several kinds of chelate resins are commercially available such as Amberlite IRS 743 (Rohm & Haas Corporation, USA), BSR1 (Dow Chemical, USA), Purolite S-108 (Purolite, USA), Diaion CRB 03, and Diaion CRB 05 (Mitsubishi Chemical Corporation, Japan). These materials are able to remove boron selectively to the effectiveness of 93–98%, even with highly mineralized solutions.[20] Some of them were tested even in a pilot plant.[32]

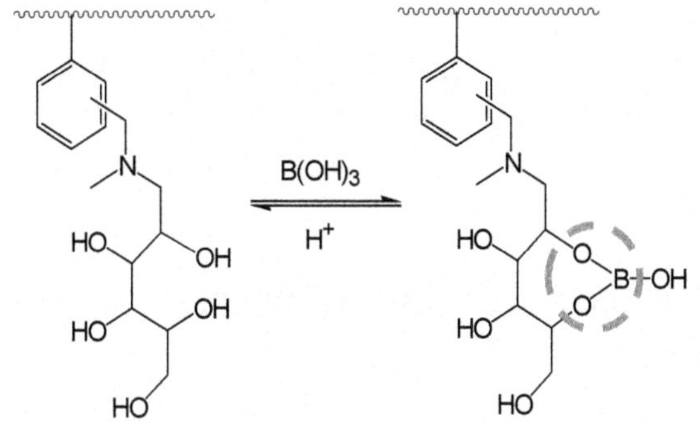

Figure 5.1 Binding mechanism of boron by NMDG-type chelating resin.[14]

Studies on boron-selective resins with different functionalities to NMDG have gained the attention of researchers for last two decades. Bicek et al. synthesized polymer matrices modified with different compounds. They produced a resin based on crosslinked polystyrene, modified with glycidyl groups and grafted with sorbitol.[28] Other materials, obtained by the same authors were NMDG-bearing resins with matrices of crosslinked glycidyl methacrylate (GMA) with *N,N'*-tetrallyl piperazinium dichloride or terpolymer of GMA/methyl methacrylate (MMA)/DVB.[27,33] The surface of the polymer matrix GMA/MMA/DVB was modified by ethylene diamine and glycidol.[29] In another study, they modified the GMA/MMA/DVB substrate with 2-hydroxyethylamono propylene glycol for getting a B-selective sorbent.[31,34]

Liu et al. obtained a hybrid gel prepared with tetraethoxysilane (TEOS), (3-glycidoxypropyl)trimethoxysilane (GPTMS), and precursor synthesized from GPTMS and NMDG that had a good mechanical strength and affinity to boron.[35] The silica support modified with NMDG was also used in the study.[36] Li et al. described the process of synthesis of a boron adsorbent by grafting NMDG onto the sillica-polyallylamine composites.[29] This material had a high capacity toward boron even in seawater spiked with high concentrations of other ions. A novel silica-supported NMDG adsorbent (Si-MG) was synthesized by Xu et al., by anchoring the NMDG modified (3-glycidoxypropyl) trimethoxysilane to the silica surface.[37]

A number of studies using natural polymers have also been carried out. The idea of using natural materials was initiated by Orlando et al.[38] They noted that waste natural materials could be used with a good efficiency as anion-exchange resins. Besides, the wastes were mostly built of polysugars, which means that they have chelating properties for boron. These compounds also contain amino groups, and these depend on the nature of biopolymer nitrogen content in the range from 0.2% to 6%. Inukai et al. synthesized NMDG derivatives of cellulose.[21] Bryjak et al. tried to use pretreated sawdust from several species of trees for the removal of boron.[39] Sabarudin et al. obtained chitosan resin derivatized with NMDG using chitosan that was crosslinked with ethyleneglycol diglycidyl ether.[40] The prepared materials had a good affinity to boron and removed B from aqueous solutions faster than did any of the commercially available resins. Wei et al. synthesized the modified chitosan adsorbent with NMDG groups using atom transfer radical polymerization and showed that their resins had a boron uptake much higher than any commercial resins.[41]

5.3 ADSORPTION BEHAVIOR OF BORON BY CHELATE RESINS AND CHELATING FIBERS

Diaion® CRB 03 and CRB 05 were supplied by the Mitsubishi Chemical Co. (Tokyo, Japan), and Chelest Fiber® GRY-HW was supplied by Chelest Co. (Osaka, Japan). CRB 03 and CRB 05 are polystyrene-based resins, and Chelest Fiber is a cellulose-based fiber. The functional group is *N*-methylglucamine in both the resin and the fiber. Characteristics of the adsorbents based on their catalog are summarized in Table 5.1.

The aqueous solutions were prepared by the addition of appropriate concentrations of HCl or NaOH to a boric acid solution to control the equilibrium pH. In the case of single boron system, the feed concentration of boron was set to 5 mmol/L to investigate the pH dependence of the adsorption and was varied from 1 to 24 mmol/L to measure adsorption isotherms. In the case of multicomponent system, for the investigation of the pH dependence on the adsorption, feed solution containing LiCl, KCl, and $MgCl_2$, together with boron, was used ($[B]_{ini} = [Li]_{ini} = [Mg]_{ini} = [K]_{ini} = 5$ mmol/L).

Table 5.1 Characteristics of CRB 03, CRB 05, and Chelest Fiber

	CRB 03	CRB 05		Chelest Fiber
Support	Crosslinked polystyrene		Support	Cellulose-based fiber
Ligand	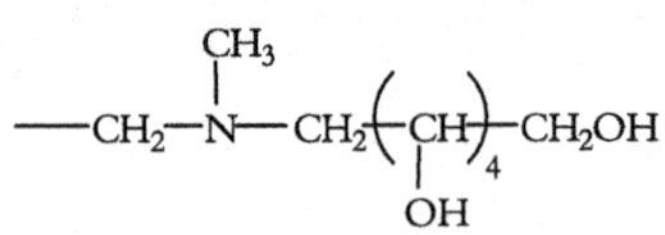 (*N*-methylglucamine)		Ligand	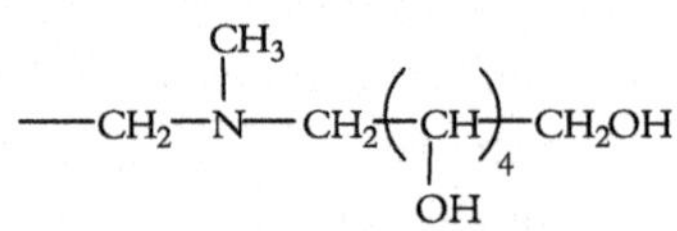 (*N*-methylglucamine)
Density	670 g/L-resin	760 g/L-resin	True specific gravity	1.5
Water content	45–55%	43–53%	Water content	<45%
Effective size	0.35–0.55 mm	>0.35 mm	Length of fiber	~0.5 mm
			Diameter of fiber	~100 μm

Batchwise adsorption was carried out by bringing into contact 20 mg of each adsorbent and 10 mL of aqueous solution. The suspended mixture was shaken using a mechanical shaker at 150 rpm at 25 °C for >24 h. The aqueous solution was then filtered, and the concentrations of boron and other metal ions in the aqueous solution were determined by inductively coupled plasma atomic emission spectrometer (ICP–AES) or atomic adsorption spectrometer (AAS), and the pH was measured by a pH meter. The amount of each component (M) adsorbed (q) was calculated by

$$q = \frac{\left([\mathrm{M}]_{\mathrm{ini}} - [\mathrm{M}]\right)\cdot L}{w}, \tag{5.1}$$

where $[\mathrm{M}]_{\mathrm{ini}}$ and [M] are the initial and equilibrium concentrations of the component, L is the volume of aqueous solution, and w is the weight of the adsorbent.

The adsorption rate of boron for the three adsorbents was first investigated by batchwise adsorption at initial pH values ($\mathrm{pH_{ini}}$) of 3, 7, and 10. Figure 5.2 shows the time course variations for boron adsorption at $\mathrm{pH_{ini}} = 7$. The adsorption of boron reached equilibrium within 24 h in the cases of CRB 03 and CRB 05 and 8 h in the case of Chelest Fiber. Similar results were obtained with an aqueous solution at $\mathrm{pH_{ini}}$ 3 and 10. Kabay et al. reported that the kinetics of boron adsorption using an adsorbent containing glucamine groups (Diaion® CRB 02 and Dowex® XUS 43,594.00) can be expressed using a pseudo-second-order model, as given by Eqn (5.2).[42]

$$\frac{dq}{dt} = k\left(q_{\mathrm{eq}} - q_{\mathrm{t}}\right)^2, \tag{5.2}$$

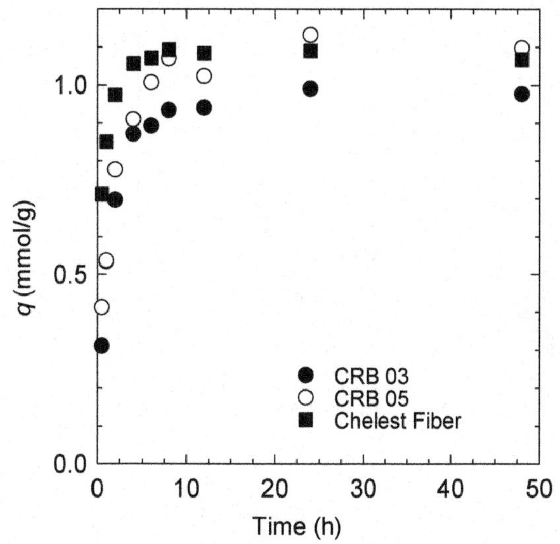

Figure 5.2 Time course variation for boron adsorption with CRB 03, CRB 05, and Chelest Fiber at $\mathrm{pH_{ini}} = 7$.

where k is the pseudo-second-order rate constant (grams per millimoles per hour) and q_{eq} and q_t are the amounts of boron adsorbed (millimoles per gram) at equilibrium and at time t (hours), respectively. By varying the variables in Eqn (5.2), a linear relationship for the pseudo-second-order model can be obtained as Eqn (5.3):

$$\frac{t}{q_t} = \frac{1}{kq_{eq}^2} + \frac{1}{q_{eq}}t. \tag{5.3}$$

Figure 5.3 shows the linearized relationships for boron adsorption by each adsorbent at initial pH_{ini} values 3, 7, and 10. Linear relationships between t and t/q_t were obtained in all systems, indicating that the adsorption behavior of boron by the adsorbents agreed with pseudo-second-order kinetics. The rate constants (k), calculated from Eqn (5.4), are summarized in Table 5.2, together with equations of the regression lines and correlation coefficients (r^2). The rate constant of Chelest Fiber was much higher than those of CRB 03 and CRB 05. Inukai et al. reported that the faster rate of adsorption by fiber-type adsorbents compared with resin-type adsorbents is caused by the circumstances of the functional group. That is, the functional groups in the fiber-type adsorbent exist on the surface of the fiber, while those of the resin-type adsorbent exist within micropores. In addition, cellulose fiber as a raw material of Chelest Fiber possesses hydroxyl group, which allows the adsorbate to be feasibly contacted with the functional groups due to their hydrophilicity.[21]

Figure 5.4 shows the effect of the equilibrium pH (pH_{eq}) on the amount of boron adsorbed by the adsorbents. The boron adsorption behaviors of the adsorbents were

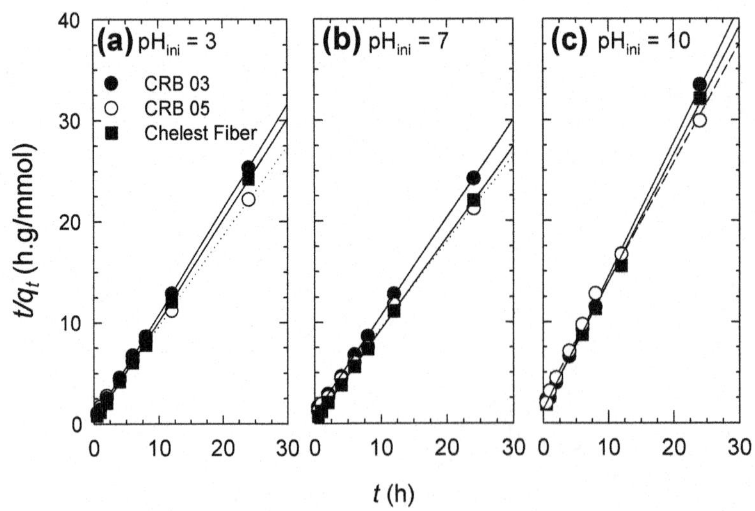

Figure 5.3 Pseudo-second-order kinetics for the adsorption of boron on CRB 03, CRB 05, and Chelest Fiber at pH_{ini} = (a) 3, (b) 7, and (c) 10.

Table 5.2 Regression Lines and Correlation Coefficients Obtained from Figure 5.2 and Rate Constants for the Adsorbents at $pH_{ini} = 3, 7$, and 10

Adsorbent	pH_{ini}	Equation	r^2	k (g/mmol·h)
CRB 03	3	$y = 1.04x + 0.386$	1.00	2.80
	7	$y = 0.971x + 0.921$	1.00	1.02
	10	$y = 1.33x + 1.21$	0.999	1.46
CRB 05	3	$y = 0.888x + 0.797$	0.999	0.989
	7	$y = 0.853x + 0.923$	0.999	0.788
	10	$y = 1.18x + 2.24$	0.995	0.622
Chelest fiber	3	$y = 1.00x + 0.102$	1.00	9.80
	7	$y = 0.907x + 0.199$	1.00	4.13
	10	$y = 1.26x + 1.35$	0.998	1.18

similar, and the maximum values occurred around neutral pH. The complexation of boric acid with the hydroxyl group in the glucamine group liberates a proton. The adsorption amount of boron in the acidic range therefore increases with increasing pH. The decrease in the amount adsorbed with increasing pH in the alkali range was due to the hydrolysis of boric acid.[43] The glucamine-based adsorbents are therefore effective for the removal of boron at neutral pH. Figure 5.5 shows the effect of pH_{eq} on the adsorption amount of boron, magnesium, lithium, and potassium from their multicomponents system by Chelest Fiber. Similar adsorption behavior of boron, compared with the single boron system, is observed in the case of multicomponents system. No adsorption of other metal ions is progressed in the acidic and neutral pH ranges, while adsorption of

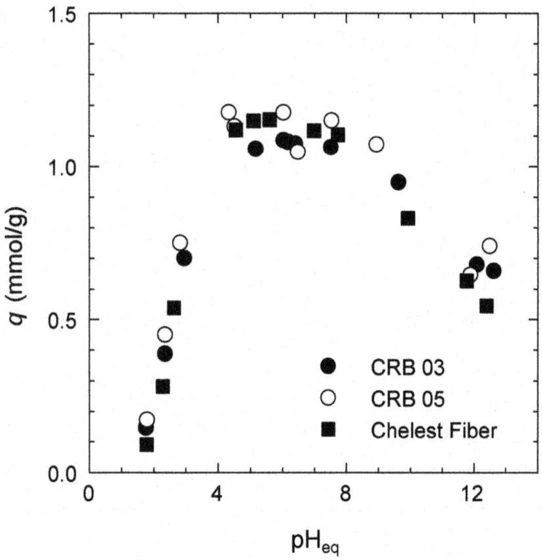

Figure 5.4 Effect of equilibrium pH on the amount of boron adsorbed by CRB 03, CRB 05, and Chelest Fiber in a single boron system. $[B]_{ini} = 5$ mmol/L.

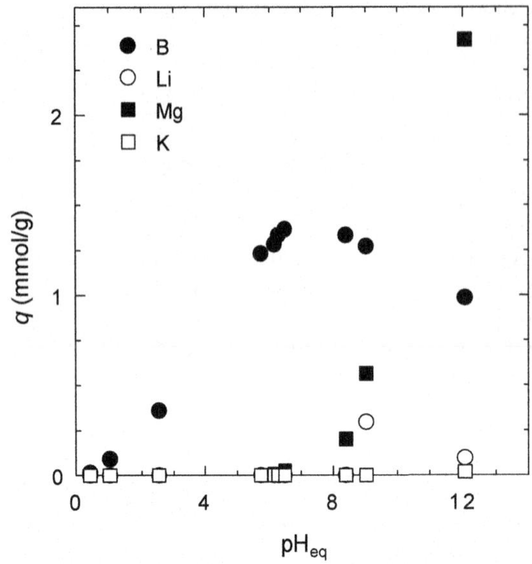

Figure 5.5 Effect of equilibrium pH on the adsorption amount of boron, lithium, magnesium, and potassium by Chelest Fiber in the multicomponent system. $[B]_{ini} = [Li]_{ini} = [Mg]_{ini} = [K]_{ini} = 5$ mmol/L.

magnesium is progressed in the alkali range. These results indicate that the glucamine group possesses a high selectivity for boron especially in acidic and neutral pH ranges.

The adsorption isotherms for boron by the adsorbents were then investigated at $pH_{eq} = 6.05 \pm 0.05$ in the cases of CRB 03 and CRB 05 and at $pH_{eq} = 6.60 \pm 0.05$ in the case of Chelest Fiber. Figure 5.6(a) shows the adsorption isotherms. The adsorption isotherms of all adsorbents fit well with the linear relationship of the Langmuir mechanism, as shown in Figure 5.6(b). The maximum amounts of boron adsorbed (q_{max}) obtained from the Langmuir adsorption isotherms were 0.989 (CRB 03), 1.18 (CRB 05), and 1.15 mmol/g (Chelest Fiber), respectively. CRB 05 and Chelest Fiber had higher maximum adsorption capacities for boron than CRB 03.

5.4 CHROMATOGRAPHIC SEPARATION OF BORON FROM AQUEOUS SOLUTION

CRB 05 and Chelest Fiber were therefore applied to a chromatographic operation for the adsorptive separation of boron.

CRB 05 (weight: 0.6 g, wet volume: 1.6 mL) or Chelest Fiber (weight: 0.6 g, wet volume: 1.8 mL) was packed into a column tube (10 cm in length and 8 mm in inner diameter), sandwiched between quartz wool. The aqueous feed solution, containing approximately 800 mg/L of boron at pH 6.5, was then fed upward through the column using a dual-plunger pump. The flow rate of the feed solution was changed to compare the breakthrough curves. The flow rates and space velocities (SVs) used in the experiments are summarized in Table 5.3. The effluent was collected with a fraction collector.

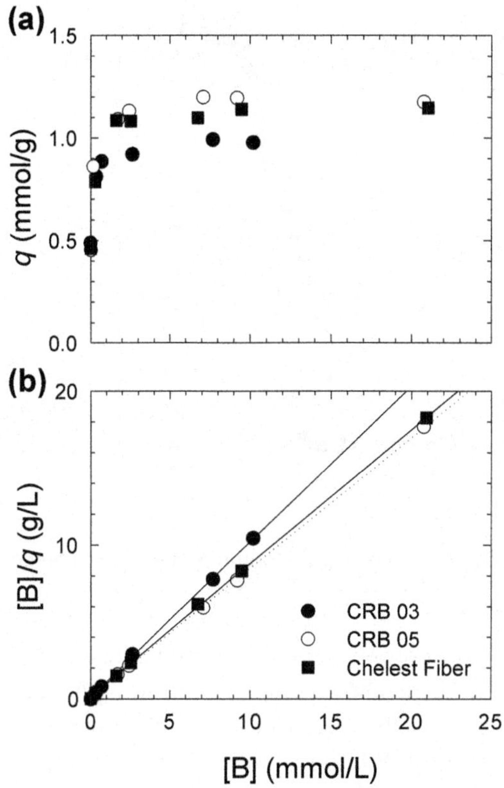

Figure 5.6 (a) Adsorption isotherms for boron with adsorbents at equilibrium pH $= 6.05 \pm 0.05$ (CRB 03 and CRB 05) and 9.60 ± 0.05 (Chelest Fiber). (b) Linearized Langmuir plots of the adsorption isotherms. CRB 03: $y = 1.01\,x + 0.0744$, $r^2 = 1.00$; CRB 05: $y = 0.846\,x + 0.0376$, $r^2 = 1.00$; Chelest Fiber: $y = 0.869\,x + 0.113$, $r^2 = 1.00$.

The concentrations of boron collected by the fraction collector were determined by ICP–AES or AAS. The bed volume (BV) of the effluent was calculated by

$$\text{Bed Volume} = \frac{v \cdot t}{V}, \tag{5.4}$$

where v is the flow rate of the solution, t is the time of the feed solution application, and V is the wet volume of the adsorbent in the column.

Table 5.3 SVs used for the Chromatographic Operation

	SV (1/h)	
Flow rate (mL/min)	**CRB 05**	**Chelest Fiber**
0.25	9.4	8.3
0.50	17.6	16.7
0.75	28.1	25.0
1.00	37.5	33.3

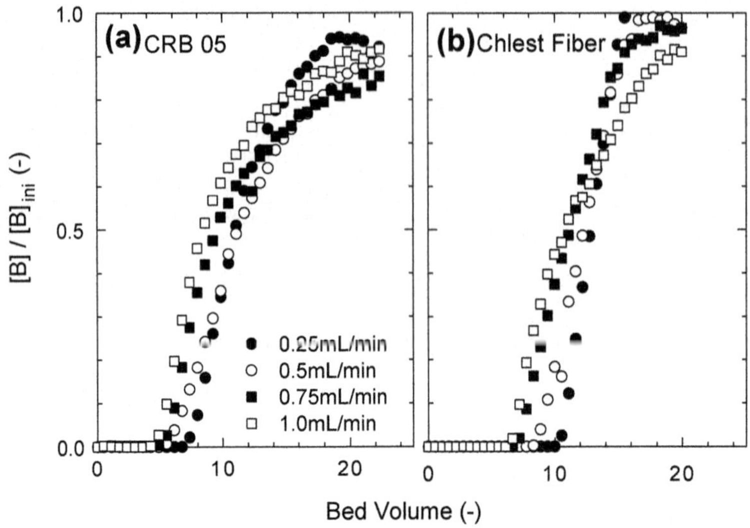

Figure 5.7 Breakthrough curves for boron adsorption from a pure boron solution at different flow rates by (a) CRB 05 and (b) Chelest Fiber.

Figure 5.7 shows the breakthrough curves for boron with CRB 05 and Chelest Fiber at different flow rates. In the case of CRB 05, the breakthrough of boron was quite fast, and the breakthrough curve was not sharp, even at the lowest flow rate (0.25 mL/min, SV = 9.4/h). In the case of Chelest Fiber, the breakthrough curves were quite sharp up to a flow rate of 0.75 mL/min (SV = 25/h), while the bed volume at the breakthrough point was lower. The sharp breakthrough curves indicate that a narrow adsorption band could be obtained even when the SV was as high as 25/h. These results correspond to the fast adsorption kinetics obtained by batchwise adsorption, with Chelest Fiber having the potential for use at high SVs.

Figure 5.8 shows the breakthrough and elution curves for boron at SV = 8.6 h^{-1} from a pure boron solution containing 800 mg/L of boron at pH6.5. Effective adsorption could be achieved, similar to Figure 5.7, and quantitative elution could also be achieved by 2 mol/L HCl solution with an elution yield of 99.5%. The elution is progressed by the reverse reaction of the complexation of boric acid with the hydroxyl group in the glucamine group, and thus, the quantitative elution can be conducted with the acid solution due to prevention of dissociation of hydroxyl group.[43]

5.5 BORON REMOVAL FROM GEOTHERMAL WATER

The three kinds of adsorbents, CRB 03, CRB 05, and Chelest Fiber, are applied for the selective recovery of boron from the geothermal water sampled at Obama Hot Spring, Kyushu, Japan. The components of the geothermal water are listed in Table 5.4.[44]

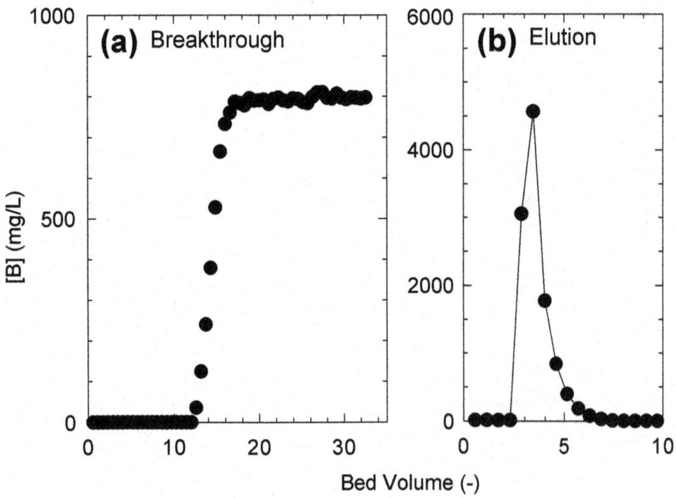

Figure 5.8 (a) Breakthrough and (b) elution curves for boron from a pure boron solution with Chelest Fiber.

CRB 03 (weight: 0.6 g, wet volume: 1.8 mL), CRB 05 (weight: 0.6 g, wet volume: 1.6 mL), or Chelest Fiber (weight: 0.6 g, wet volume: 1.8 mL) was packed into a column tube (10 cm in length and 5 mm in inner diameter), sandwiched between quartz wool. After breakthrough, deionized (DI) water was fed into the column to wash out the excess feed solution remaining in the column, and the boron adsorbed on the column was then eluted with 0.1 mol/L HCl solution.

The adsorption—elution processing of boron was conducted six times with the same column. In the case of CRB 03, the flow rate was 0.25 mL/min (SV = 16.7/h), in the case of CRB 05, the flow rate was 0.25 mL/min (SV = 18.8/h), and in the case of Chelest Fiber, the flow rate was 0.25 mL/min (SV = 16.7/h). The breakthrough point was reached for BV = 80 (CRB 03), and BV = 120 (CRB 05 and Chelest Fiber). When the elution of boron from the resin was performed with HCl solution (0.1 mol/L), it was completed after BV = 15 with a high elution efficiency. The breakthrough curves for CRB 03, CRB 05, and Chelest Fiber and the elution curve of boron are given in Figure 5.9, and the elution yields were 85.4 ± 2.7% (CRB 03), 88.2 ± 1.0% (CRB 05), and 99.8 ± 0.5% (Chelest Fiber). CRB 03, CRB 05, and Chelest Fiber can be used for repeated processing, for application to the adsorption of boron from hot spring water.

Table 5.4 Components of the Geothermal Water in Obama Hot Spring in Kyushu, Japan

Component	B	Na^+	Mg^{2+}	K^+	Ca^{2+}	pH
Concentration (mg/L)	14.8	207	148	156	141	8.3

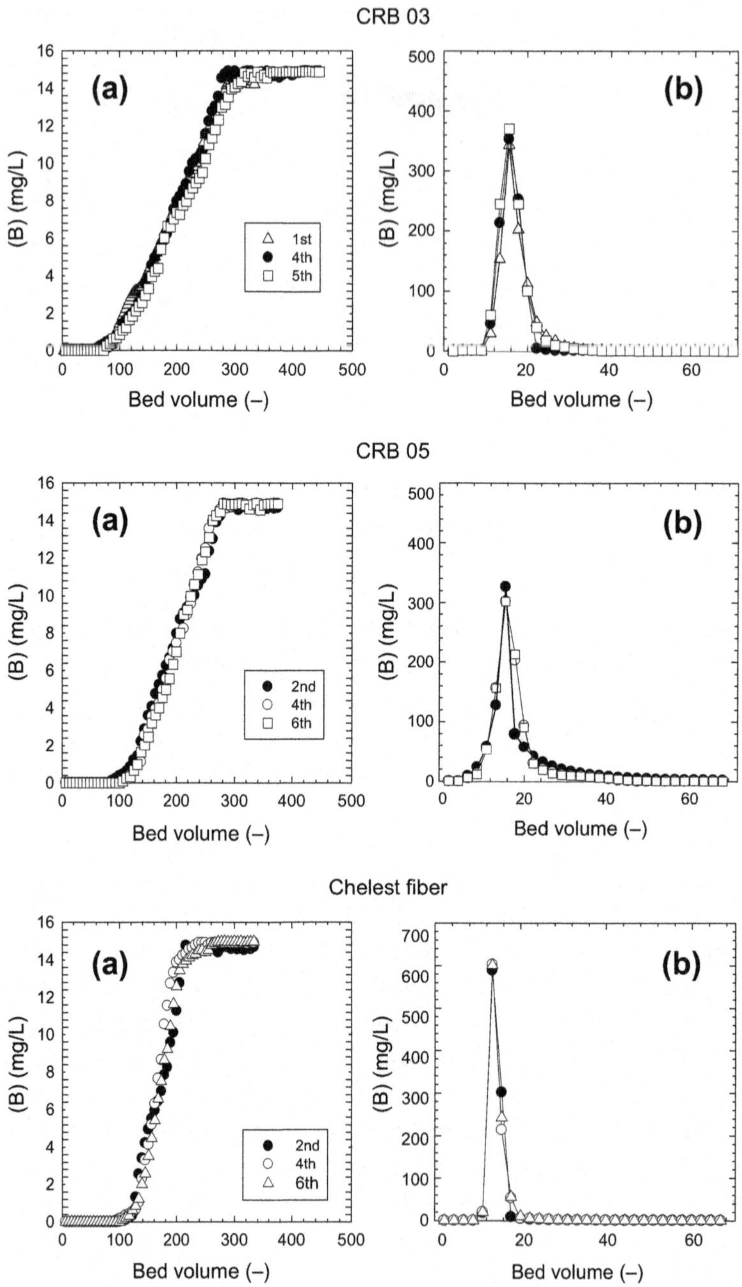

Figure 5.9 (a) Breakthrough and (b) elution curves for boron from geothermal water with CRB 03, CRB 05, and Chelest Fiber.

5.6 BORON RECOVERY FROM SALT LAKE BRINE

Chelest Fiber was applied for the selective recovery of boron from the brine of a salt lake.[45] Some salt lake brines have recently attracted attention as lithium resources. Boron should be removed from brine before producing the lithium product to avoid serious contamination, and solvent extraction is currently employed for this purpose.[46,47] In this work, the actual brine from the Salt Lake of Uyuni in Bolivia, kindly supplied by Corporación Minera de Bolivia, was used as the feed solution after filtering the suspended solids from the brine. The Chelest Fiber (weight: 1.9 g, wet volume: 3.5 mL) was packed into the column tube, and the brine was fed upward to the column at a flow rate of 0.5 mL/min (SV = 8.3/h). The composition and pH of the brine are summarized in Table 5.5. After breakthrough, DI water was fed into the column to wash out the excess feed solution remaining in the column, and the boron adsorbed on the column was then eluted with 2.0 mol/L HCl solution.

Figure 5.10 shows the breakthrough and elution curves of boron and coexisting components of salt lake brine, as listed in Table 5.5. Here, the breakthrough curves of the

Table 5.5 Composition and pH of Brine of Salt Lake of Uyuni in Bolivia

Component	B	Li^+	Mg^{2+}	K^+	Na^+	pH
Concentration (mg/L)	850	1410	19,140	23,320	60,560	7.0

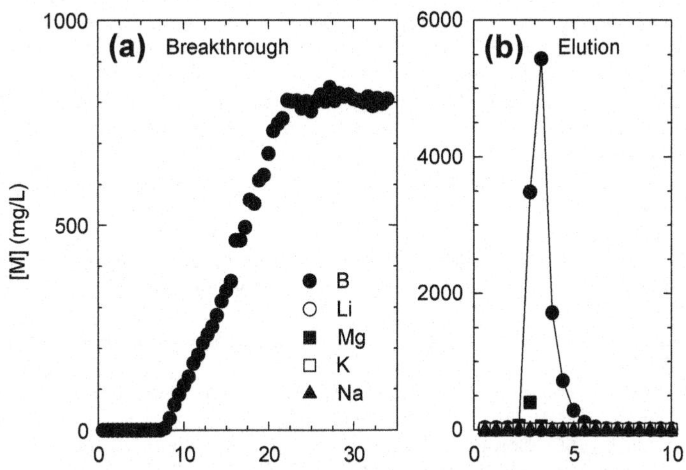

Figure 5.10 (a) Breakthrough and (b) elution curves of boron and coexisting components of salt lake brine with Chelest Fiber. The breakthrough curves for lithium, magnesium, potassium, and sodium are omitted.

coexisting components are omitted from the figure, due to large differences in the concentration scale. The breakthrough curve for boron had a slightly lower slope than that from the pure boron solution. In addition, the elution yield of boron decreased to 81.7%. The coexisting components were however broken through immediately, though breakthrough curves of the coexisting components are omitted from the figure. In addition, almost no coexisting components are observed in the elution step, indicating that selective separation of boron could be achieved with the Chelest Fiber. The Chelest Fiber is, therefore, considered to have potential as an adsorbent for boron from salt lake brine which is a multicomponent solution containing very high concentrations of other elements.

REFERENCES

1. Öztürk N, Köse TE. Boron removal from aqueous solutions by ion-exchange resin: batch studies. *Desalination* 2008;**227**:233–40.
2. Japan Oil, Gas and Metals National Corporation (JOGMEC). *Kobutsu material flow*; 2009.
3. World Health Organization. *Guidelines for drinking-water Quality Fourth Edition* [Chapter 12] Chemical Fact Sheets. Geneva: Switzerland; 2011.
4. Özdemir M, Kıpçak İ. Boron recovery from borax sludge, boron industrial waste, by solid–liquid extraction. *Ind Eng Chem Res* 2003;**42**:5256–60.
5. Sinirkaya M, Kocakerim MM, Boncukçuoğlu R, Küçük Ö, Öncel S. Recovery of boron from tincal wastes. *Ind Eng Chem Res* 2005;**44**:427–33.
6. Chong MF, Lee KP, Chieng HJ, Ramli IISB. Removal of boron from ceramic industry wastewater by adsorption–flocculation mechanism using palm oil mill boiler (POMB) bottom ash and polymer. *Water Res* 2009;**43**:3326–34.
7. Kentjono L, Liu JC, Chang WC, Irawan C. Removal of boron and iodine from optoelectronic wastewater using Mg-Al(NO_3) layered double hydroxide. *Desalination* 2010;**262**:280–3.
8. Kabay N, Bryjak M, Schlosser S, Kitis M, Avlonitis S, Matejka Z, et al. Adsorption-membrane filtration (AMF) hybrid process for boron removal from seawater: an overview. *Desalination* 2008;**223**:38–48.
9. Badruk M, Kabay N, Demircioglu M, Mordogan H, Ipekoglu U. Removal of boron from wastewater of geothermal power plant by selective ion-exchange resins. I. Batch sorption–elution studies. *Sep Sci Technol* 1999;**34**:2553–69.
10. Badruk M, Kabay N, Demircioglu M, Mordogan H, Ipekoglu U. Removal of boron from wastewater of geothermal power plant by selective ion-exchange resins. II. Column sorption–elution studies. *Sep Sci Technol* 1999;**34**:2981–95.
11. Koseoglu P, Yoshizuka K, Nishihama S, Yuksel U, Kabay N. Removal of boron and arsenic from geothermal water in Kyushu Island, Japan, by using selective ion exchange resins. *Solv Extr Ion Exch* 2011;**29**:440–57.
12. Ooi K, Kanoh H, Sonoda A, Hirotsu T. Screening of adsorbents for boron in brine. *J Ion Exch* 1996;**7**:166–73.
13. Yan CY, Yi WT. Preparation, characterization, and boron adsorption behavior of gluconate-intercalated hydrotalcite. *AIChE* 2010;**29**:450–6.
14. Köse TE, Öztürk N. Boron removal from aqueous solutions by ion-exchange resin: column sorption–elution studies. *J Hazard Mater* 2008;**152**:744–9.
15. Yan C, Yi W, Ma P, Deng X, Li F. Removal of boron from refined brine by using selective ion exchange resins. *J Hazard Mater* 2008;**154**:564–71.
16. Tsuboi I, Kunugita E, Komasawa I. Recovery and purification of boron from coal fly ash. *J Chem Eng Jpn* 1990;**23**:480–5.

17. Matsumoto M, Kondo K, Hirata M, Kokubu S, Hano T, Takada T. Recovery of boric acid from wastewater by solvent extraction. *Sep Sci Technol* 1997;**32**:983–91.
18. Magara Y, Aizawa T, Kunitake S, Itoh M, Kohki M, Kawasaki M, et al. The behavior of inorganic constituents and disinfection by products in reverse osmosis water desalination process. *Water Sci Technol* 1996;**34**:141–8.
19. Kabay N, Güler E, Bryjak M. Boron in seawater and methods for its separation—a review. *Desalination* 2010;**261**:212–7.
20. Mitsubishi Chemical Corporation. *Diaion® manual of ion exchange resins and synthetic adsorbent*, vol. 2. Tokyo: Mitsubishi Chemical Corporation; 2010.
21. Inukai Y, Tanaka Y, Matsuda T, Mihara N, Yamada K, Nambu N, et al. Removal of boron(III) by *N*-methylglucamine-type cellulose derivatives with higher adsorption rate. *Anal Chim Acta* 2004;**511**:261–5.
22. Melnyk L, Goncharuk V, Butnyk I, Tsapiuk E. Boron removal from natural and wastewaters using combined sorption membrane process. *Desalination* 2005;**185**:147–57.
23. Kabay N, Yilmaz I, Bryjak M, Yuksel M. Removal of boron from aqueous solutions by hybrid ion exchange-membrane process. *Desalination* 2006;**198**:158–65.
24. Sahin S. A mathematical relationship for the explanation of ion exchange for boron removal. *Desalination* 2002;**143**:35–43.
25. Simonnat MO, Castel C, Nicolai M, Rosin C, Sardin M, Jauffret H. Boron removal from drinking water with a boron selective resin: is the treatment really selective? *Water Res* 2000;**34**:109–16.
26. Kabay N, Yilmaz I, Yamac S, Samatya S, Yuksel M, Yuksel U, et al. Removal and recovery of boron from geothermal wastewater by selective ion exchange resins. I. Laboratory tests. *React Funct Polym* 2004;**60**:163–70.
27. Bicak N, Bulutcu N, Senkal BF, Gazi M. Modification of crosslinked glycidyl methacrylate-based polymers for boron-specific column extraction. *React Funct Polym* 2001;**47**:175–84.
28. Bicak N, Senkal BF. Sorbitol-modified poly(*N*-glycidyl styrene sulfonamide) for removal of boron. *J Appl Polym Sci* 1998;**68**:2113–9.
29. Li X, Liu R, Wu S, Liu J, Cai S, Chen D. Efficient removal of boron acid by *N*-methyl-D-glucamine functionalized silica-polyallylamine composites and its adsorption mechanism. *J Coll Interface Sci* 2011;**361**:232–7.
30. Senkal BF, Bicak N. Polymer supported iminodipropylene glycol functions for removal of boron. *React Funct Polym* 2003;**55**:27–33.
31. Bicak N, Ozbelge O, Yilmaz L, Senkal BF. Crosslinked polymer gels for boron extraction derived from *N*-glucidol-*N*-methyl-2-hydroxypropyl methacrylate. *Macromol Chem Phys* 2000;**201**:577–84.
32. Chillón Arias MF, Valero i Bru, Prats Rico D, Pedro V Galvañ. Comparison of ion exchange resins used in reduction of boron in desalinated water for human consumption. Desalination 278: 244–249.
33. Gazi M, Bicak N. Selective boron extraction by polymer supported 2-hydroxyethylamono propylene glycol functions. *Reac Funct Polym* 2007;**67**:936–42.
34. Bicak N, Gazi M, Senkal BF. Polymer supported amino bis-(cis-propan 2,3 diol) functions for removal of trace boron from water. React Func Polym 65;143–148.
35. Liu H, Ye X, Li Q, Kim T, Qing B, Guo M, et al. Boron adsorption using a new boron-selective hybrid gel and the commercial resin D564. *Coll Surf A: Physicochem Eng Asp* 2009;**341**:118–26.
36. Kaftan Ö, Acikel M, Eroglu AE, Shahwan T, Artok L, Ni C. Synthesis, characterization and application of a novel sorbent, glucamine-modified MCM-41, for the removal/preconcentration of boron from waters. *Anal Chim Acta* 2005;**547**:31–41.
37. Xu L, Liu Y, Hu H, Wu Z, Chen Q. Synthesis, characterization and application of a novel silica based adsorbent for boron removal. *Desalination* 2012;**294**:1–7.
38. Orlando US, Okuda T, Nishijima W. Chemical properties of anion exchangers prepared from waste natural materials. *React Funct Polym* 2003;**55**:311–8.
39. Bryjak M, Koltuniewicz A, Trochimczuk A. *Report for NATO science programme, development of an innovative treatment process for geothermal wastewater*; 2004.

40. Sabarudin A, Oshita K, Oshima M, Motomizu S. Synthesis of cross-linked chitosan possessing *N*-methyl-D-glucamine moiety (CCTS-NMDG) for adsorption/concentration of boron in water samples and its accurate measurement by ICP–MS and ICP–AES. *Talanta* 2005;**66**:136–44.
41. Wei Y-T, Zheng Y-M, Chen JP. Design and fabrication of an innovative and environmental friendly adsorbent for boron removal. *Water Res* 2011;**45**:2297–305.
42. Kabay N, Sarp S, Yuksel M, Arar Ö, Bryjak M. Removal of boron from seawater by selective ion exchange resins. *React Funct Polym* 2007;**67**:1643–50.
43. Qi T, Sonoda A, Makita Y, Kanoh H, Ooi K, Hirotsu T. Synthesis and borate uptake of two novel chelating resins. *Ind Eng Chem Res* 2002;**41**:133–8.
44. Shao S, Nishihama S, Yoshizuka K. Separation and recovery of boron from water using ion exchange methods. In: *Proc. The 12th international symposium on East Asian resources recycling technology*; 2013. pp. 581–4.
45. Nishihama S, Sumiyoshi Y, Ookubo T, Yoshizuka K. Adsorption of boron using glucamine-based chelate adsorbents. *Desalination* 2012;**310**:81–6.
46. Risacher F, Fritz B. Quaternary geochemical evolution of the Salars of Uyuni and Coipasa, Central Altiplano, Bolivia. *Chem Geol* 1991;**90**:211–31.
47. Garrett D. *Handbook of lithium and natural calcium chloride, their deposits, processing, uses and properties.* Amsterdam: Elsevier; 2004.

CHAPTER 6

Adsorption of Boron by Minerals, Clays, and Soils

Tomasz Koźlecki, Izabela Polowczyk
Faculty of Chemistry, Wrocław University of Technology, Wrocław, Poland

6.1 INTRODUCTION

Boron is a vital microelement, necessary for the growth of plants, albeit there is only a narrow gap between desirable and toxic concentrations.[1–4] The tolerance varies between different species. For example, avocado, lemons, and wheat are sensitive, while cotton, tomato, and oat are relatively tolerant.[5]

Although boron is a low-abundance element, there are many sources of possible contamination, due to anthropogenic activity and natural processes. According to the Canadian Water Quality Guidelines for the Protection of Aquatic Life, large amounts of boron are present in sediments and sedimentary rock, particularly in clay-rich marine ones, mainly due to the high concentration of B in seawater.[6] Boron is also slowly released into the environment by natural weathering processes. Due to the widespread occurrence of clay-rich sedimentary rocks on the Earth's land surfaces, the majority of boron mobilized into soils and the aquatic environment by weathering probably originates from this source. Natural weathering releases more boron into the environment than industrial sources do. It is very important to estimate the sorption and release of boron by minerals, soils, and soil components (e.g., humic acids). This knowledge is important to provide a sufficient amount of boron and, on the other hand, prevent excessive accumulation of boron in the environment.

6.2 ADSORPTION OF BORON ON MINERALS AND CLAYS

Goldberg and Glaubig investigated the adsorption of boron on various crystalline and amorphous iron and aluminum oxide minerals, including gibbsite, α-alumina, alon, aluminum oxide C, pseudoboehmite, amorphous aluminum oxide, hematite, goethite, 34% goethite/maghemite, 64% goethite/maghemite, and amorphous iron oxide.[7] The experimental results have been interpreted by means of a constant capacitance model. It has been shown that sorption increased at low pH, reached a maximum at pH values between 7 and 9, then decreased at a high pH. Adsorption maxima were usually quite broad, particularly for aluminum oxide C (~2.5 pH units). In the case of amorphous

Boron Separation Processes
ISBN 978-0-444-63454-2, http://dx.doi.org/10.1016/B978-0-444-63454-2.00006-X

aluminum oxide, the maximum was slightly shifted toward lower pH values. Generally, amorphous materials adsorbed much more boron than crystalline ones did, while gibbsite and α-alumina did not show any adsorption (Table 6.1).

In order to apply a constant capacitance model, three surface reactions are defined, as shown in Table 6.1, together with equilibrium constant equations and calculated average values. One can see that the values of the protonation–dissociation B surface complexation constants for both Al and Fe are not significantly different.

Seki et al. examined the adsorption of boron on Al_2O_3-based materials, namely, Siral and Pural 30.[8] The former contains 72% Al_2O_3 as boehmite, and the latter is composed of 28% SiO_2 and 72% Al_2O_3. A 2^3 full factorial design was used to estimate the effect of individual variables and their interactions for boron removal. Three coded variables have been used: adsorbent type (Siral = −1, Pural 30 = +1), pH (5.70 = −1,9.50 = +1), and temperature (298 K = −1, 318 K = +1). Two runs have been performed for each variable set, and the results have been averaged. Statistical analysis shows that the type of adsorbent has a positive effect on adsorption, while temperature and pH had a negative effect. Siral 30 was found to be a more efficient adsorbent than Pural. Adsorption data (Figure 6.1) have been fitted using three isotherms:

- Langmuir:

$$Q_{eq} = \frac{K_L Q_{max} C_{eq}}{1 + K_L C_{eq}}$$

where Q_{eq} is the amount of boron adsorbed (millimoles per gram), K_L is the constant related to adsorption energy, Q_{max} is the amount of adsorption corresponding to

Table 6.1 Sorption of Boron on Various Aluminum and Iron Oxides.

Oxide	Specific Surface Area (m²/g)	Maximum Sorption (mmol/kg)
Al-oxides		
Gibbsite	2	nd[a]
α-Alumina	0.9	nd
Alon	69.6	19.10
Aluminum oxide C	96.9	6.75
Pseudoboehmite	227	46.56
Amorphous aluminum oxide	163	55.58
Fe-oxides		
Hematite	16.4	1.89
Goethite	31.2	3.92
34% Goethite/maghemite	55.7	7.70
64% Goethite/maghemite	74.2	6.48
Amorphous iron oxide	112	35.00

[a]nd, not detected.
Adapted from Ref. [7].

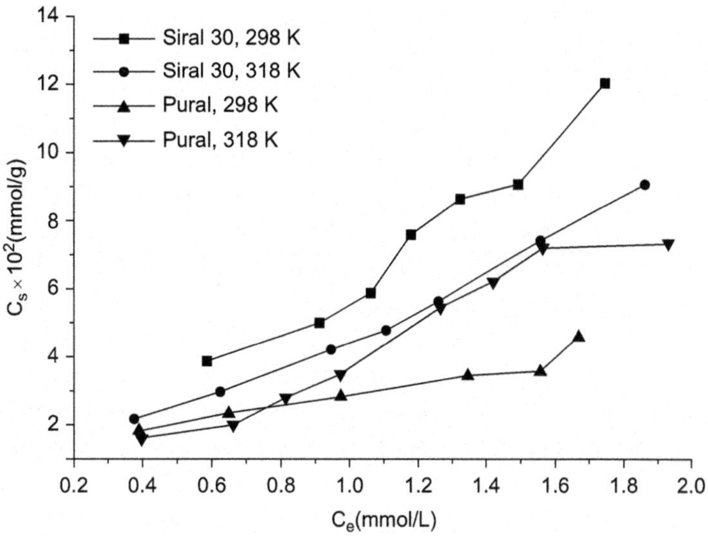

Figure 6.1 Adsorption isotherms for boron sorption on Siral 30 and Pural.[8]

monolayer coverage (millimoles per gram), and C_{eq} is the equilibrium concentration (millimoles per cubic decimeter);

- Freundlich:

$$Q_{eq} = K_F C_{eq}^n,$$

where K_F is the relative adsorption capacity (millimoles per gram) and n is a constant;

- Dubinin–Radushkevich:

$$\ln Q_{eq} = \ln Q_{max} - k\left[RT\ln\left(1+\frac{1}{C_{eq}}\right)\right]^2$$

where R is the gas constant (8.314 J/mol K), T is the temperature (kelvins), and k is a constant, used to calculate adsorption energy (moles squared per kilojoules squared). The expression in square brackets is called the Polanyi potential.

Calculations performed by Seki et al.[8] showed that the results can be satisfactorily fitted by both Freundlich and Dubinin–Radushkevich equations. The latter should be converted into a linear form:

$$\ln Q_{eq} = \ln K_F + n \ln C_{eq}$$

Albeit the standard enthalpy of sorption on Pural was negative ($\Delta H = -25.651$ kJ/mol), the process was exothermic; for Siral 30, a positive value was obtained ($\Delta H = 0.788$ kJ/mol). The value of Gibbs' free energy, $\Delta G°$ increased from 7.628 to 9.860 kJ/mol for Pural and from 7.854 to 8.319 kJ/mol for Siral 30, respectively, when the temperature increased from 298 to 318 K. This indicates the nonspontaneous nature of the process.

Peak et al. studied the adsorption of H_3BO_3 and $B(OH)_4^-$ on hydrous ferric oxide (HFO) using attenuated total reflection infrared spectroscopy (ATR-IR).[9] Three possible adsorption mechanisms have been postulated, as shown in Figure 6.2. The first mechanism (Figure 6.2(a)) assumes reactions of trigonal boric acid with FeOH groups on the HFO surface to produce trigonal and tetrahedral complexes. Initially, an outer-sphere surface complex is formed as a result of the Lewis acidity of the boron metal center, followed by ligand exchange giving trigonal or tetrahedral inner-sphere surface complexes of boron. The former result from the water ligand displacement, while for the latter, the proton from the surface hydroxyl leaves instead. The second mechanism (Figure 6.2(b)) is the direct formation of an inner-sphere surface complex via a Lewis acid–base mechanism. The third (Figure 6.2(c)) assumes the formation of some un-identified outer-sphere intermediate, giving rise to inner-sphere complexation between the hydroxyl and borate. Because the surface complex postulated in the first mechanism can be observed in the ATR-IR spectrum at 1395/cm at pH 6.5, pathway (a) in Figure 6.2 was found to be most likely.

Recently, Demetriou and Pashalidis examined the adsorption of boron on iron oxide, FeO(OH), in aqueous solutions, as a function of various parameters: pH, ionic strength,

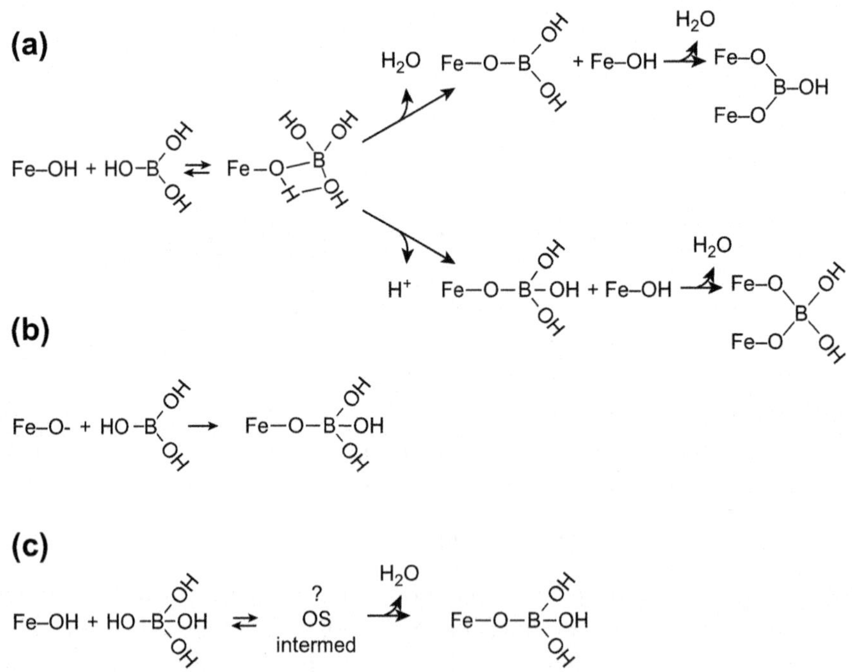

Figure 6.2 Possible reaction mechanisms of boric acid at the HFO surface.[9]

temperature, boron concentration, and amount of the adsorbent.[10] They concluded that the driving force of adsorption is a Lewis acid/base reaction of H_3BO_3 at the surface of the iron oxide with replacement of a water molecule by boric acid (Figure 6.3). At a low concentration of H_3BO_3, the bidentate complex was predominant, while at higher concentrations, the monodentate complex was favored. The process was found to be exothermic and spontaneous ($\Delta H^\circ = -15.5$ kJ/mol, $\Delta G^\circ = -4.9$ kJ/mol at 298 K, -3.3 kJ/mol at 343 K), an energetic effect practically did not depend on ionic strength; thus, the formation of inner-sphere complexes was postulated. The adsorption isotherm was the Langmuir one; the sorption capacity was determined as 0.03 mol/kg. The adsorption was strongly pH dependent, with the maximum at pH 8. This value is close to the point of zero charge value of iron oxide and slightly lower than the pK_a of boric acid, indicating that the optimum conditions for boron removal occur when there is no surface charge and H_3BO_3 is the predominant species in solution. Similar results have been presented by the authors in another paper; the conclusions were supported with Raman measurements.

Goli et al. examined the interactions between boron and synthetic goethite, α-FeOOH.[11] Measurements were carried out in sodium nitrate solution, as a function of pH, ionic strength, goethite, and boron concentration. Experimental results were supported by charge distribution multisite surface complexation (CD-MUSIC) calculations. This method is composed of models: MUSIC, describing the effects of surface heterogeneity,[12] and CD.[13] To model the adsorption process, the set or surface reactions have been proposed. These reactions are summarized in Table 6.2. All calculations agreed well with experimental results: Figure 6.4 shows calculated and measured adsorption isotherms, and Figure 6.5 depicts the calculated surface speciation of boron, according to the above model.

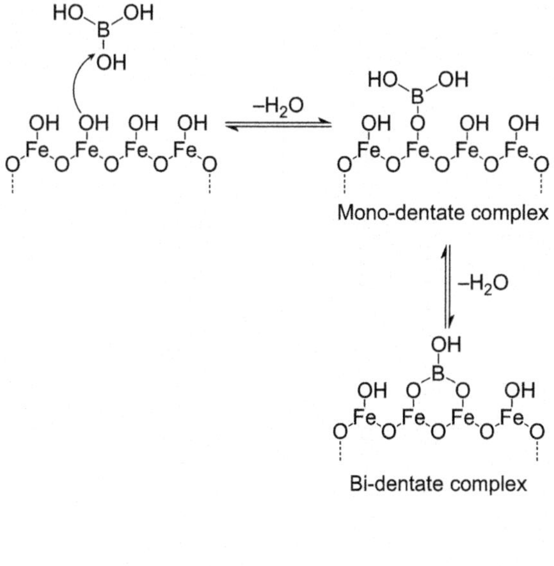

Figure 6.3 Proposed mechanism for the adsorption of boric acid on iron oxide surface, at pH 8.[10]

Table 6.2 Summary of Reactions Used to Model Adsorption of Boron on Goethite Using CD-MUSIC Model [11]

Reaction	Description
$\equiv SH + B(OH)^0_{3(aq)} \Leftrightarrow \equiv SH_{1-q}B(OH)_3 + qH^+_{(aq)}$	The interaction of boron with goethite surface
$B(OH)^0_3 + H_2O_{(l)} \Leftrightarrow \equiv B(OH)^-_{4(aq)} + H^+_{(aq)}$	Dissociation of boric acid; log $K_a = 9.24$
$\equiv SH + B(OH)^-_{4(aq)} \Leftrightarrow \equiv SH_{1-q}B(OH)_3 + (q-1)H^+_{(aq)} + H_2O_{(l)}$	The combination of above reactions
$2\equiv FeOH^{-1/2} + B(OH)^0_{3(aq)} \Leftrightarrow \equiv (FeO)_2^{-1+\Delta z0}BOH^{\Delta zl} + 2H_2O_{(l)}$	The interaction of neutral boric acid molecule with two singly coordinated surface groups; $\Delta z0$ and $\Delta z1$ are changes in the charge of 0- and 1-plane as a result of $B(OH)^0_{3(aq)}$ adsorption; $\Delta z0 = -\Delta z1$
$2\equiv FeOH^{-1/2} + B(OH)^0_{3(aq)} \Leftrightarrow \equiv (FeO)_2^{-1+\Delta z0}B(OH)_2^{-1+\Delta zl} + H_2O_{(l)} + H^+_{(aq)}$	High pH formation of tetrahedral borate surface complex $\Delta z0 + \Delta z1 = -1$
$\equiv FeOH^{-1/2} + H^+(aq) + B(OH)^0_{3(aq)} \Leftrightarrow \equiv FeOH_2^{+1/2} \quad B(OH)^0_{3(aq)}$	Formation of outer-sphere complex is defined by the interaction of boric acid with positively charged surface groups at low pH

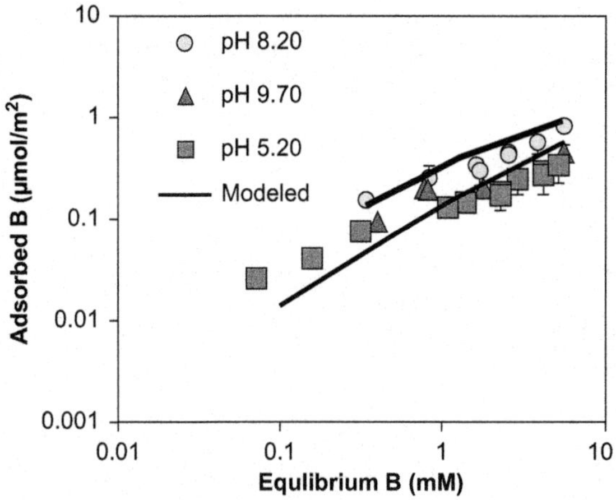

Figure 6.4 The log of boron adsorption isotherms on goethite (10 g/dm^3) in 0.10 M $NaNO_3$ at different pH values, calculated using the CD-MUSIC model.[11]

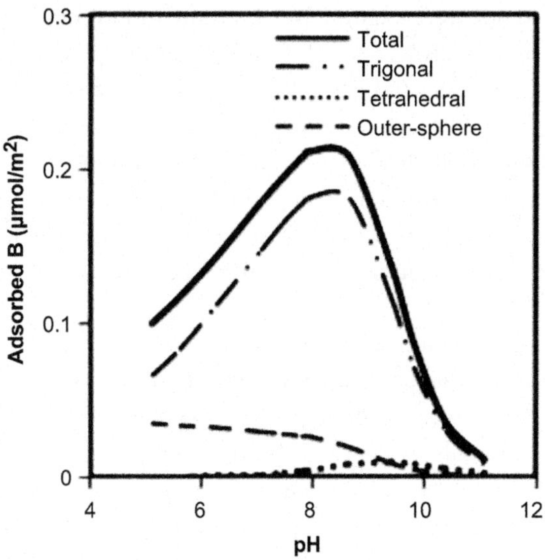

Figure 6.5 The calculated surface speciation of boron in a single-ion system (10 g/dm^3, surface area = 98.0 m^2/g) with 0.80 mM $B(OH)_{3(aq)}$ in 0.10 M $NaNO_3$ as a function of pH. Calculations performed using the CD-MUSIC model.[11]

Singh and Mattigod[14] modeled the adsorption of boron on kaolinite in the presence of potassium or calcium perchlorate as an electrolyte, with an ionic strength equal to 0.09 mol/dm pretreated clay particles, 0.2–2 μm in size, have been used, the temperature was set at 25 ± 2 °C. The generalized triple-layer surface complexation model ($TL_{(g)}$-SCM)[15] has been used to interpret the results. Adsorbed ions were assumed to form surface complexes at 0 or the β plane of the electrical double layer (Figure 6.6); these species were analogous to inner and outer-sphere complexes.

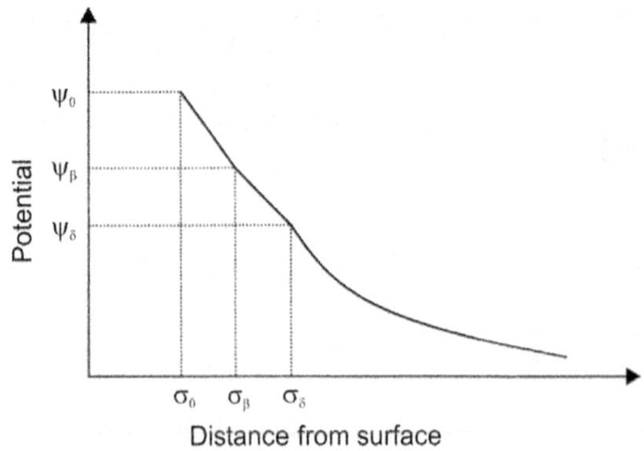

Figure 6.6 Surface charges (σ) and surface potentials (ψ) in the triple-layer model.

Phenomenological studies led to the set of reactions, involving the following components: SOH,[1] XOH,[2] OH^-, $B(OH)_3$, $B(OH)_4^-$, K^+, Ca^{2+}, and ClO_4^-, as shown in Table 6.3, together with the corresponding equilibrium constants. The adsorption isotherm was determined as well. Two models were examined: Langmuir and Freundlich models. The correlation coefficients of the models were 0.9956 and 0.9558, respectively; thus, the latter equation can be assumed to be valid. The isotherm parameters were identical for both potassium and calcium perchlorate solutions.

Maximum boron adsorption was observed at pH 8.5–9, which corresponds to the pK_a of boric acid. The amount of boron adsorbed was dependent on the initial B concentration, reaching approximately 60 $nmol/m^2$ in 0.1 M $KClO_4$ and 65 $nmol/m^2$ in 0.033 M $Ca(ClO_4)_2$ for an initial boron concentration of 10 mg/dm^3. These results were consistent with $TL_{(g)}$-SCM modeling.

In 2005, Goldberg analyzed an inconsistency in the triple-layer model description of ionic strength-dependent boron adsorption.[16] Several minerals were used in this study: iron oxide, goethite, the aluminum oxide, gibbsite, kaolinite, montmorillonite, and two arid-zone soils. Except for gibbsite, all minerals exhibited an inner-sphere adsorption mechanism, as proven by ionic strength dependence of sorption. The triple-layer model is able to describe changes in boron adsorption with changing ionic strength. For several materials, the ionic strength dependence of model boron adsorption is opposite to that observed in the experimental data; that is, the use of an inner-sphere adsorption mechanism in the triple-layer model resulted in an ionic strength dependence typical for the formation of outer-sphere surface complexes and vice versa.

Keren and Sparks investigated the boron adsorption on pyrophillite, $Al_2Si_4O_{10}(OH)_2$, as a function of pH and ionic strength.[17] Sodium nitrate has been used

[1] S denotes aluminol site.
[2] X denote silanol site.

Table 6.3 Equations for Surface-Complexation Reactions of Boron on Kaolinite [14]

Reaction	Equilibrium Constant
Surface Protolysis Reactions	
$SOH_2^+ = SOH + H^+$	$K_{a1}^* = \frac{[SOH][H^+]}{[SOH_2^+]}\exp(-F^*\psi_0)$
$SOH = SO^- + H^+$	$K_{a2}^* = \frac{[SO^-][H^+]}{[SOH]}\exp(-F^*\psi_0)$
Electrolyte Surface-Complexation Reactions	
$SOH + K^+ = [SO^-\ K^+] + H^+$	$K_K^* = \frac{[SO^-K^+][H^+]}{[SOH][K^+]}\exp(F^*(\psi_\beta - \psi_0))$
$SOH + Ca^{2+} = [SO^-\ Ca^{2+}]^+ + H^+$	$K_{Ca}^* = \frac{[SO^-Ca^{2+}][H^+]}{[SOH][Ca^{2+}]}\exp(F^*\psi_0)$
$SOH + H^+ + ClO_4^- = SOH_2^+\ ClO_4^-$	$K_{ClO_4^-}^* = \frac{[SOH_2^+ClO_4^-]}{[SOH][H^+[ClO_4^-]]}\exp(F^*(\psi_0 - \psi_\beta))$
Boron Surface-Complexation Reactions	
$SOH + B(OH)_3 + H^+ + K^+ = [SB(OH)_3]^-\ K^+ + 2H_2O$	$K_1^* = \frac{[(SB(OH)_3)^-K^+]}{[SOH][B(OH)_3[Ca^{2+}]]}\exp(F^*(\psi_\beta - \psi_0))$
$2SOH + B(OH)_4^- + H^+ = [(SO)_2B(OH)] + 3H_2O$	$K_2^* = \frac{[(SO)_2B(OH)]}{[SOH]^2[B(OH)_4^-][H^+]}$
$SOH + B(OH)_4^- + K^+ = [SOB(OH)_3]^-\ K^+ + H_2O$	$K_3^* = \frac{[(SOB(OH)_3)^-K^+]}{[SOH][B(OH)_4^-][K^+]}\exp(F^*(\psi_\beta - \psi_0))$
$SOH + B(OH)_3 + Ca^{2+} = [SB(OH_3)]^-\ (CaOH)^+ + 2H_2O$	$K_4^* = \frac{[(SB(OH)_3)^-(CaOH)^+]}{[SOH][B(OH)_3][Ca^{2+}]}\exp(F^*(\psi_\beta - \psi_0))$
$2SOH + B(OH)_4^- + CaOH^+ = [SO_2B(OH)_2]^-\ (CaOH)^+ + 2H_2O$	$K_5^* = \frac{[(SO_2B(OH)_2)^-(CaOH)^+]}{[SOH]^2[B(OH)_4^-][CaOH^+]}\exp(F^*(\psi_\beta - \psi_0))$
$XOH + B(OH)_4^- + CaOH^+ = [XO^-Ca^{2+}]^+B(OH)_4^- + H_2O$	$K_6^* = \frac{[(XO^-\ Ca^{2+})^+B(OH)_4^-]}{[XOH][B(OH)_4^-][CaOH^+]}\exp(F^*(\psi_0 - \psi_\beta))$

S, silanol sites; X = aluminol; $F^* = \frac{F}{RT}$; F = Faraday constant (96485 C/mol^{-1}), R = gas constant (8.314 J/mol K), T = temperature (K).

as an electrolyte; the ionic strength was 0.01 and 0.1 mol/dm^3 for pH 5 and 0.005, 0.01, and 0.1 mol/dm^3 for pH 7 and 9. The model assuming that $B(OH)_3$, $B(OH)_4^-$, and OH^- compete for the same adsorption sites has been used to estimate boron adsorption:

$$Q_{BT} = \frac{Q_{max}(K_{HB}[HB] + K_B[B])}{1 + K_{HB}[HB] + K_B[B] + K_{OH}[OH]}$$

where Q_{BT} is the amount of B adsorbed (moles per gram), Q_{max} is the apparent maximum B adsorption (moles per gram), K_{HB}, K_B, and K_{OH} are the absorption constants for $B(OH)_3$, $B(OH)_4^-$, and OH^-, respectively, and [HB], [B], and [OH] are the concentrations of $B(OH)_3$, $B(OH)_4^-$, and OH^-, respectively. It has been found that boron adsorption increases with both increasing pH and ionic strength. At pH 5, the effect of ionic strength on the amount of adsorbed boron is relatively low. This was explained by the low clay edge surface electric potential of −47 mV. This value was found to reduce sharply with distance from the edge surfaces. This phenomenon, together with the low concentration of $B(OH)_4^-$ and the low affinity coefficient for $B(OH)_4^-$, explains the low amount of adsorbed B and the small effect of ionic strength. A subsequent paper by Keren et al. presented the results of pressure-jump relaxation measurements of the adsorption–desorption of borate ions on the edge surfaces of pyrophyllite.[18] This mineral is a typical 2:1 clay, with only minor deviations from the dioctahedral structure. All experiments were carried out at pH 9, and the ionic strength was 0.01 mol/dm^3 (sodium nitrate solution). At this pH 9, 37% of the total boron in solution exists as $B(OH)_4^-$, while for the total adsorbed B, the fraction of the adsorbed $B(OH)_4^-$ is assumed to be 99%, independently of boron concentration. This high fraction is related to the absence of repulsive forces associated with the negative electrical field since the permanent electrical charge can be neglected. The sorption capacity can then be defined as

$$Q_B = \frac{Q_{max} K_B[B]}{1 + K_{HB}[HB] + K_B[B] + K_{OH}[OH]}$$

Analysis of figures from the reference shown the adsorption is described by the Langmuir equation (the authors did not provide appropriate calculations). The maximum sorption capacity calculated was $Q_B = 16.4$ μmol/g. The amount of adsorbed B on pyrophyllite at pH 9 is much higher than that found for montmorillonite, even though the specific surface area of the former is approximately 20 times smaller than that of the latter mineral (43.7 and 800 m^2/g, respectively). Keren et al. reasoned out that the dioctahedral structure of pyrophyllite consists of essentially neutral tetrahedral–octahedral–tetrahedral layers held together by van der Waals forces. Therefore, no potential barrier originating from the planar surfaces exists for borate ions approaching the adsorption sites on the edge surfaces. Thus, the B adsorption reaction on the edge surfaces of the clay and their intrinsic adsorption–desorption coefficients can be evaluated. Two sets of reversible reactions contribute to the adsorption process, as shown in Figure 6.7. This conclusion is strongly supported by relaxation experiments.

Kim and Kirkpatrick investigated the adsorption sites of boron on boehmite AlO(OH), silica gel, and illite $(K_{0.82}Na_{0.04}Ca_{0.01})(Si_{3.12}Al_{0.88})(Al_{1.98}Mg_{0.02})O_{10}(OH)_2$, using solid-state ^{11}B magic-angle spinning nuclear magnetic resonance (MAS NMR).[19] Moreover, boehmite was found to be a possible model for Fe(III) oxyhydroxides. This technique is sensitive enough to investigate adsorption at concentrations down to 0.03 weight%. The NMR signal could be easily resolved for both trigonal and tetrahedral boron exchanged onto investigated minerals, and the resonances were assigned, based on both chemical shift and quadrupole coupling constants. Example spectra, obtained for boehmite at various pH values, are depicted in Figure 6.8. One can see the peak related to the resonance for trigonal boron with an intensity at δ 7−20, exhibiting a well-resolved quadrupolar splitting, along with a narrow, symmetrical resonance at δ 0.7−1.6, characteristic for tetragonal boron. The behavior of minerals differs substantially. For boehmite, both trigonal boron (B(3)) and tetragonal boron (B(4)) occur predominantly as inner-sphere complexes formed by a ligand exchange reaction with surface aluminol sites. The B(3)/[B(3) + B(4)] ratio of approximately 0.87 remains almost constant with pH from 3 to 11, as well as solution concentration, or with washing. Kim and Kirkpatrick explained this effect by the formation of a boroaluminate structure that is stable over a wide range of conditions. For silica gel, B(3) and B(4) occur as outer-sphere complexes or as residual precipitate from the unremoved solution. The B(3)/B(4) ratio decreases with increasing pH paralleling the speciation in solution, but the relative abundance of B(4) is greater than in solution. A small fraction of the B(4) occurs as inner-sphere complexes with B(4)−O−Si linkages formed by ligand exchange reaction with silanol sites. The comparison of B(3) and B(4) relative intensities for boehmite and silica gel is shown in Figure 6.9. Illite has higher boron surface atomic densities than the others, and boron occurs as outer-sphere B(3) and B(4), like for silica gel, and as inner-sphere B(3) and B(4), like for boehmite. Outer-sphere B(3) and B(4) appear to dominate at pH < 5, whereas inner-sphere B(3) and B(4) are dominant at pH values >11.

Investigation of the adsorption of boron on bentonite, sepiolite, and illite, both raw and treated with nonylammonium chloride, has been carried out by Karahan et al.[20] The

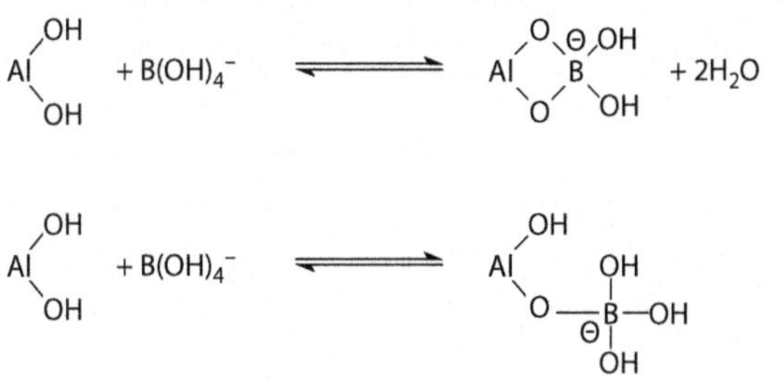

Figure 6.7 Mechanism of adsorption of boron on the edge surfaces of pyrophyllite.[18]

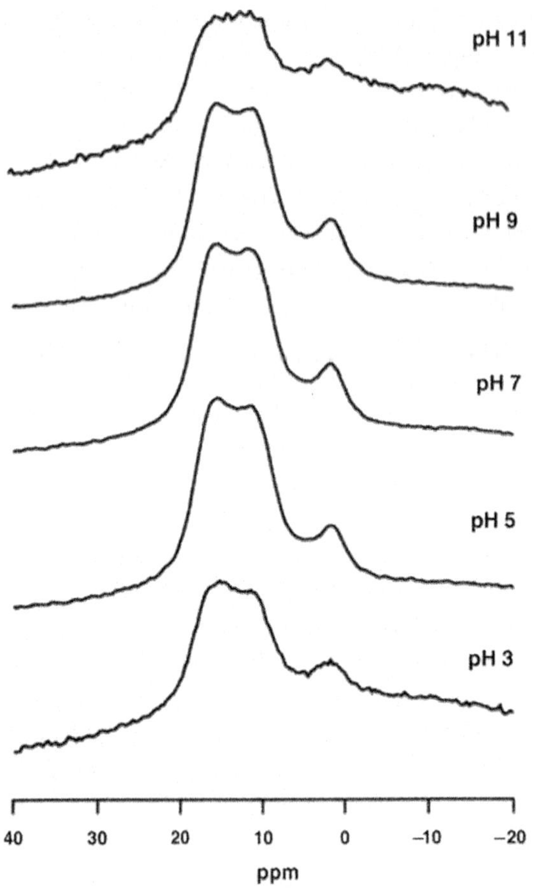

Figure 6.8 ^{11}B MAS NMR spectra of B-exchanged boehmite collected after the reaction in 0.01 M boric acid solution at the indicated pH for 24 h.[19]

authors varied the pH up to 10 and the concentration of boron was between 4 and 50 mg/dm^3. Modification decreased the surface area, and increased the median pore size. In the case of bentonite and illite, treatment with nonylammonium chloride increased the sorption of boron. The adsorption was pH dependent, maximum values were attained at pH 7.90, 7.82, 8.58, 9.03, 9.14, and 9.54 for bentonite, illite, sepiolite, modified bentonite, modified illite, and modified sepiolite, respectively. Experimental results could be fitted with both Freundlich and Dubinin–Radushkevich equations, while the Langmuir equation gave poor results. Calculations of adsorption energy, using constant k from the latter equations gave values between −4.310 and −8.450 kJ/mol; these numbers indicate that the mechanism of the process is ion exchange. Fourier transform infrared (FTIR) measurements have also been performed to get an insight into the nature of the complexes. For bentonite, an absorption band in the range of 3200–3600/cm exists, due to the stretching band of the OH groups. After adsorption,

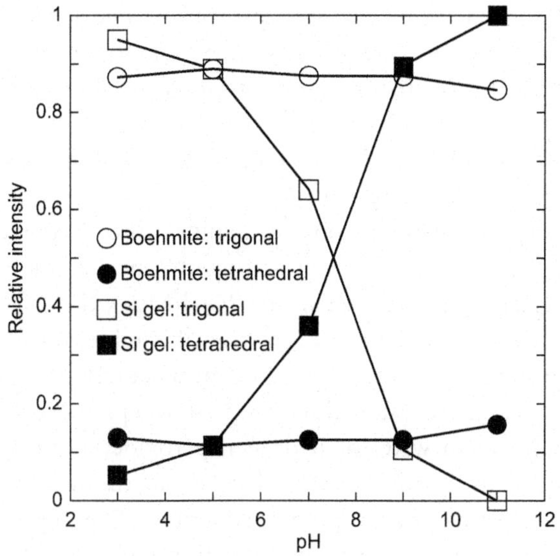

Figure 6.9 Relative intensities of B(3) and B(4) in B-exchanged boehmite and silica gel as functions of pH obtained from the curve fitting of the ^{11}B MAS NMR spectra.[19]

bentonite displays two bands, at 3629 and 3430/cm; their intensities increased compared to initial values. This indicates that hydrogen bonding may occur between the hydroxyl group of bound water and boron species. In the case of illite, the absorption bands at 3550 and 3428/cm were assigned to the lattice OH and bound water stretching vibrations. After boron adsorption, they shifted to 3552 and 3422/cm, because similar hydrogen bonding may occur. In the case of sepiolite (Figure 6.10), zeolitic water gives a

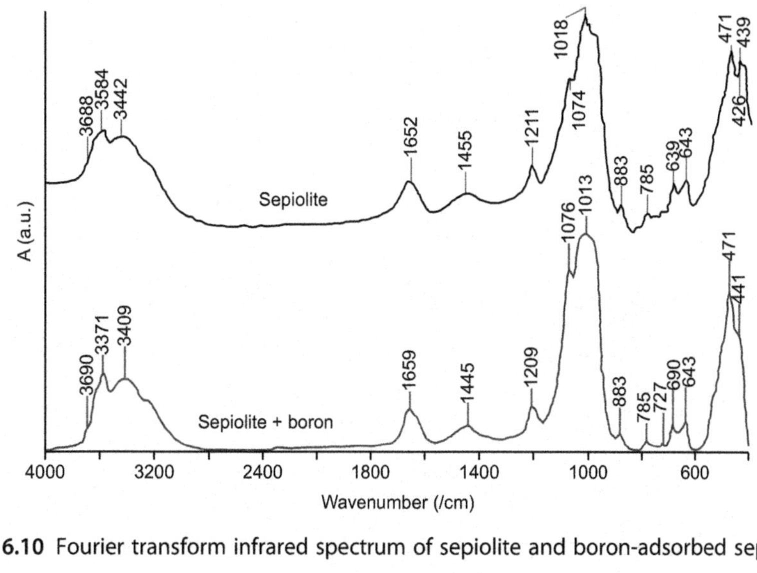

Figure 6.10 Fourier transform infrared spectrum of sepiolite and boron-adsorbed sepiolite.[20]

broad adsorption band at 3442/cm, shifting to 3409/cm after adsorption. The authors concluded that the entrance of boron species into channels may affect the structure of residual zeolitic water. Two possible association modes were proposed as shown in Figure 6.11. Another stretching absorption of bound water can be observed at 3688/cm (weak) and 3584/cm, shifting after sorption to 3690 and 3571/cm. Such small shifts may indicate the presence of the structure presented in Figure 6.11(b).

Kehal et al. researched on the adsorption of boron on vermiculite from Palabora Mining Co, South Africa.[21] According to Del Rey-Perez-Caballero and Poncelet, this is a mineral of the average formula $[Si_{3.02}Al_{0.79}Ti_{0.05}Fe^{III}{}_{0.14}][Mg_{2.50},Fe^{II}{}_{0.38}\text{-}Fe^{III}{}_{0.09}(X)_{0.03}]O_{10}(OH,F)_2\ Ba_{0.29}K_{0.14}Ca_{0.08}$; the structure is a trioctahedral vermiculite intermediate between mica (biotite) and vermiculite, with a low mica content. It was modified by thermal shock at 700 °C or chemical exfoliation (80 °C in the presence of hydrogen peroxide) and ultrasonic treatment in water or hydrogen peroxide; all these operations resulted in a size reduction of the particles and a decrease in the density, ultrasound treatment being the most effective way for size reduction. The boron adsorption was found to be highly dependent on the modification process. The adsorption isotherms were fitted by means of the Freundlich model with a moderate-to-good determination coefficient $R^2 \geq 0.94$ (Figure 6.12). One can see that exfoliation and size reduction under ultrasound in the presence of H_2O_2 increased the adsorption uptake compared to raw vermiculite (0.015 mmol/g for raw material, about 0.028 and 0.032 mmol/g for thermal shock worked-up and exfoliated material, respectively, to 0.107 and 0.151 mmol/g for ultrasound treatment in water and hydrogen peroxide, respectively).

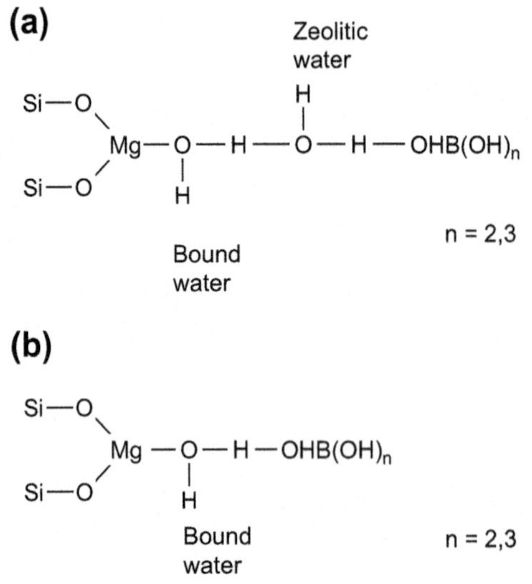

Figure 6.11 Possible association of boron species and zeolites.[20]

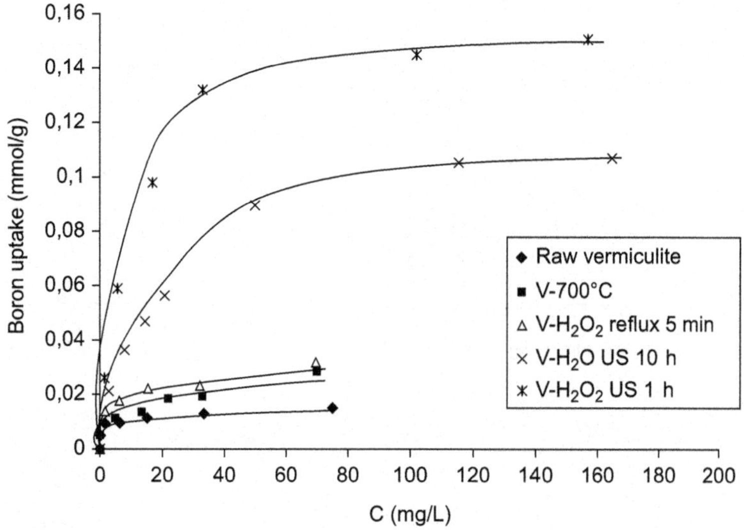

Figure 6.12 Adsorption isotherms of boron at room temperature by raw and modified vermiculites at pH = 9.26 (6 g of vermiculite in 200 mL of solution, t = 5 h).[21]

Recently, Tabelin et al. investigated the adsorption of boron from rocks excavated during the construction of a road tunnel on Hokkaido island, Japan.[22] They identified pumiceous tuff material (three samples, all loamy sand), partly weathered volcanic ash (two samples, one loamy sand, and one clay loam), and coastal marine sediment (six samples, two sandy loam, two loamy sand, and two conglomeratic loamy sand). It has been found that adsorption is best described by the Freundlich equation, except for one pumiceous tuff sample, for which the linear equation should be applied (Figure 6.13).

Sasaki et al. studied the sorption of borate onto calcined dolomite.[23] This reaction was carried out at 700—900 °C, either under an inert atmosphere or in air. A sequential decarbonation occurred with an increase in the calcination temperature, first decomposition of $CaMg(CO_3)_2$ to MgO and $CaCO_3$ up to 700 °C, followed by the conversion of $CaCO_3$ to CaO between 700 and 900 °C. The surface area and pore volume of the samples increased with increasing calcination temperature; this effect was much stronger when the process was carried out in air. The sorption capacity decreased significantly when dolomite was calcined at a higher temperature. Adsorption isotherm plots were provided, but the article missed the quantitative interpretation of data.

Kentjono et al. used synthetically manufactured Mg—Al (NO_3)-layered double hydroxide (LDH) to remove boron and iodine from optoelectronic wastewater.[24] The formula of this compound was $[Mg_{0.66}Al_{0.34}(OH)_2](NO_3)_{0.34}\cdot 0.52H_2O$. It was found to be a very effective sorbent for both boron and iodine; a higher affinity was attained for the latter. It has been determined that the adsorption equilibrium is reached after 2 h of the process. The adsorption is described by the Langmuir isotherm (determination

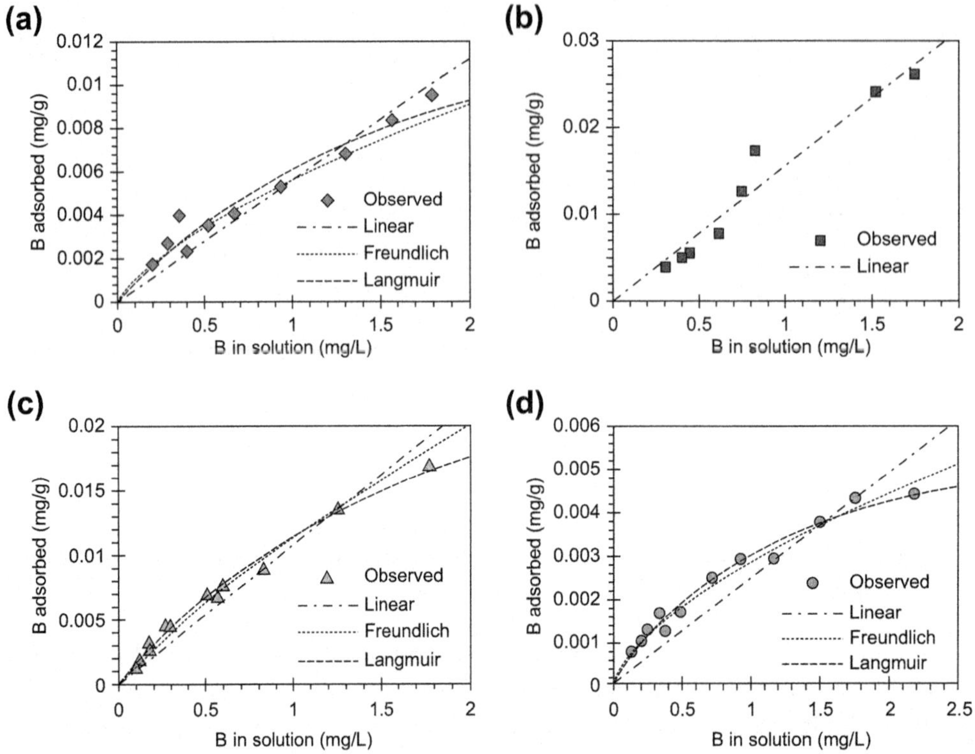

Figure 6.13 Adsorption characteristics of B onto natural geologic materials fitted with linear, Freundlich, and Langmuir isotherms. (a) Partly weathered volcanic ash/loamy sand, (b) partly weathered volcanic ash/clay loam, (c) coastal marine sediment/sandy loam (sample S-1), (d) coastal marine sediment/sandy loam (sample S-5).[22]

coefficient R^2 was generally at least 0.98), and was found to be slightly pH dependent. The maximum sorption capacity was 37.90 mg/g at pH 9, while at pH 8, it was 34.01 mg/g, and at pH 10–12, it decreased from 31.00 to 29.99 mg/g. Also, the temperature dependence of the adsorption isotherm was examined at the temperature range 295–323 K; the pH was kept constant at 9.0 ± 0.2. The temperature only slightly affected the boron adsorption capacity, and the Langmuir constant, decreased with increasing temperature, from 0.0706 dm^3/g at 295 K to 0.0452 dm^3/g at 323 K. The thermodynamics of adsorption was determined based on these experiments. It has been found that the process is weakly exothermic, with $\Delta H^\circ = -3.22$ kcal/mol. The electrokinetic potential was measured at pH 9 after the adsorption process, and the dependence is shown in Figure 6.14. Due to the zeta potential shifting toward negative values when the sorption was high, the conclusion has been drawn that the adsorption occurs on external surfaces. The authors proposed a mechanism of boron removal to be a combination of ion exchange and adsorption. The very low value of enthalpy suggests

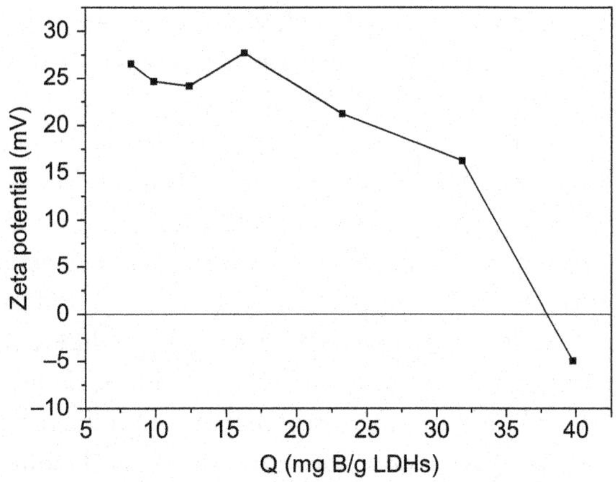

Figure 6.14 Zeta potential of layered double hydroxide (LDH) after treating industrial wastewater at pH 9.0 ± 0.2.[24]

that the process is some combination of the above mechanisms, because for physisorption, $\Delta H°$ usually ranges between 1.19 and 9.56 kcal/mol, and for ion exchange, it is <2 kcal/mol. Moreover, boron is weakly adsorbed on LDH.

6.3 ADSORPTION OF BORON ON SOILS AND HUMIC ACIDS

Soil is a complex system composed of various amounts of clays, minerals, organic matter, water, and air. Typical soil contains about 25% air, 25% water, and 50% solids. The latter has approximately 5% organic matter and 95% inorganic matter. Peat soils may contain up to 95% organic material, while other soils contain as little as 1% organic matter.[25] According to Goldberg,[26] there are numerous soil components, able to adsorb boron: aluminum and iron oxides, magnesium hydroxide, clay minerals, calcium carbonate, and organic matter. Boron adsorption reactions may be described empirically using the Langmuir isotherm, the Freundlich isotherm, or the phenomenological Keren model.[27]

Goldberg analyzed boron adsorption on soils and soil minerals using the constant capacitance model.[28] Both trigonal, $B(OH)_3$, and tetrahedral, $B(OH)_4$, surface complexes were taken into account, and the results were consistent with experimental spectroscopic data. Numerous minerals were investigated: iron oxides—hematite, goethite, and amorphous iron oxide; aluminum oxides—alon, aluminum oxide C, pseudoboehmite, and amorphous aluminum oxide; kaolinites—KGa-1, well crystallized, KGa-2, poorly crystallized, and Georgia kaolinite; 2:1 clays—SAz-1 montmorillonite, STx-1 montmorillonite, SWy-1 montmorillonite, IMt-1 illite, Fithian illite, and Morris illite, and 14 types of soils. Surface complexation constants have been calculated for all minerals and soils. The pH dependence of sorption was modeled, fitting experimental data well, albeit the soil predictions were less accurate than for pure minerals. The ability of the average set of calculated surface complexation constants to predict B adsorption versus pH has been evaluated on four additional arid-zone soils. The model was able to predict B adsorption behavior on all four soils, giving reasonably well-fitted curves throughout the data. These results indicate the utility of such an average set for calculated sets to model the behavior of real soils. The research was extended in the next paper, where 22 soils from Midwestern USA, 5 alfisols, 14 mollisols, and 3 vertisols have been examined.[29] Some samples were taken from soil horizon A (topsoil) and some from horizon B (subsoil); in specific cases, both horizons were investigated. Three surface reactions were defined: protonation, dissociation, and B complexation and the corresponding equilibrium constants were calculated. Six different soils were chosen to compare the modeling of adsorption isotherms for both horizons. The results of calculations generally fitted well, but in some cases, there was a difference between calculated and experimental data, particularly for horizon B of Dennis soil—fine, mixed, and thermic Aquic Argiudoll. The authors explained this to be due to the equations having been determined on a set of soils primarily from California, developed under different pedological conditions.

Goldberg et al[30] studied boron sorption on soils as a function of equilibrium solution B concentration (0−250 mg/dm^3), solution pH (3−12), and electrical conductivity (0.03 or 0.78 S/m). Two arid-zone soils from the San Joaquin Valley of California have been investigated; all the physicochemical parameters (pH, surface area, organic and inorganic carbon content, amount of iron and aluminum) were very similar, except for the clay content (35.4% and 48.9%). Boron adsorption on both soils increased with increasing pH, reached a maximum near pH 9, and decreased with a further increase in the pH. Boron adsorption as a function of solution pH was independent of solution salinity from pH 3 to 9. Above pH 9, B adsorption was increased from the solution with higher electrical conductivity. Boron adsorption for both soils as a function of solution B concentration obeyed the Langmuir equation. The B adsorption maxima obtained with the Langmuir equation for both electric conductivities were not statistically significantly different at the 95% level of confidence. The constant capacitance model was able to describe B adsorption as a function of solution pH and B concentration.

Steiner et al. examined boron adsorption in lowland soils from Paraná State, Brazil[31]; samples of various origins (Gleissolo Háplico (GH)—basalt/alluvial sediments, Plintossolo Háplico (PH)—shale and Cambisolo Háplico (CH)—Furnas sandstone), composition, and texture were used; the effect of liming with calcium carbonate has been examined as well. The experimental data were fitted with the Langmuir isotherm:

$$Q_{eq} = \frac{K_L C_{eq} Q_{max}}{1 + K_L C_{eq}}$$

where Q_{eq} and Q_{max} are the equilibrium and maximum sorption capacities, respectively (milligrams per kilogram), K_L is the constant related to the bonding energy of boron to the soil (cubic decimeters per milligram), and C_{eq} is the equilibrium concentration of boron in solution (milligrams per cubic decimeter). Nonlinear solver has been used avoiding both the introduction of changes in the error distribution and the acquisition of influenced parameters (K_L, Q_{max}). In all cases, high determination coefficients were obtained ($R^2 \geq 0.98$), together with a p value <0.05. Soil samples GX, CX, and FX with liming showed a type H isotherm, according to Giles, Smith, and Huitson, indicating a high adsorption affinity. Q_{max} values varied significantly. For not-limed GX, CX, and FX it was 3.0, 6.5, and 13.9 mg/kg, respectively, while for limed samples, it was 22.2, 35.7, and 14.7, respectively.

Ranjbar and Jalali investigated release kinetics from calcareous soils.[32] Four models were tested:

- Parabolic diffusion model $q = a + b\sqrt{t}$,
- Power function equation $\ln q = \ln a + b \ln t$,
- Elovich equation $q = a + b \ln t$,
- First-order equation $\ln(q^\star - q) = a - bt$,

where q is the cumulative amount of B released at time t, $q^\star$ is the maximum B released, and a and b are constants; the latter describes B rate of release. The regression of linear forms of the above equations has been performed, with the results clearly showing that the best model is the first-order equation ($R^2 > 0.99$), followed by the Elovich equation ($R^2 = 0.97-0.99$), which fails for longer process times.

Diana et al. investigated the effect of mineral and organic fertilization on boron adsorption by soils.[33] Several isotherms have been studied, including Freundlich, Temkin, Langmuir, and Headie–Hofstee. The B adsorption decreased with increasing organic-mineral fertilizer (10% N), but was greater when mineral fertilizer (21% N ammonium nitrate) or distiller's residue (3.6% N) had been used. The lower soil-to-solution ratio $\left(\text{soil mass-to-solution volume } \frac{ms}{vs}\right)$ provided more contact between the soil and solution, giving rise to more favorable adsorption. The equilibrium sorption data are satisfactorily fitted in the order Freundlich > Temkin > Langmuir > Headie–Hofstee, for $\frac{ms}{vs} = 1$, with the Freundlich isotherm being the one of choice.

Lemarchand et al. published a paper on boron isotopic fractionation related to boron sorption on humic acid and the structure of surface complexes formed.[34] The ^{11}B MAS NMR and modeling revealed that this boron forms tetrahedrally coordinated five- or six-membered ring chelates, most likely 1,2-diol and 1,3-diol complexes at an alkaline pH ($8 < pH < 11$) and dicarboxylic complexes at near-neutral conditions ($6 < pH < 9$). The partition coefficient of boron between humic acids and solution, K_d, was determined as a function of pH. It has been found that for humic matter K_d reaches 40 at pH 9–10, while for clays it is usually between 0.5 and 3.[27] The adsorption obeys the Langmuir model, but the surface complexation model fit with three (Figure 6.15) and four complexes (diol complexes plus carboxylate complexes) gives an even better result.

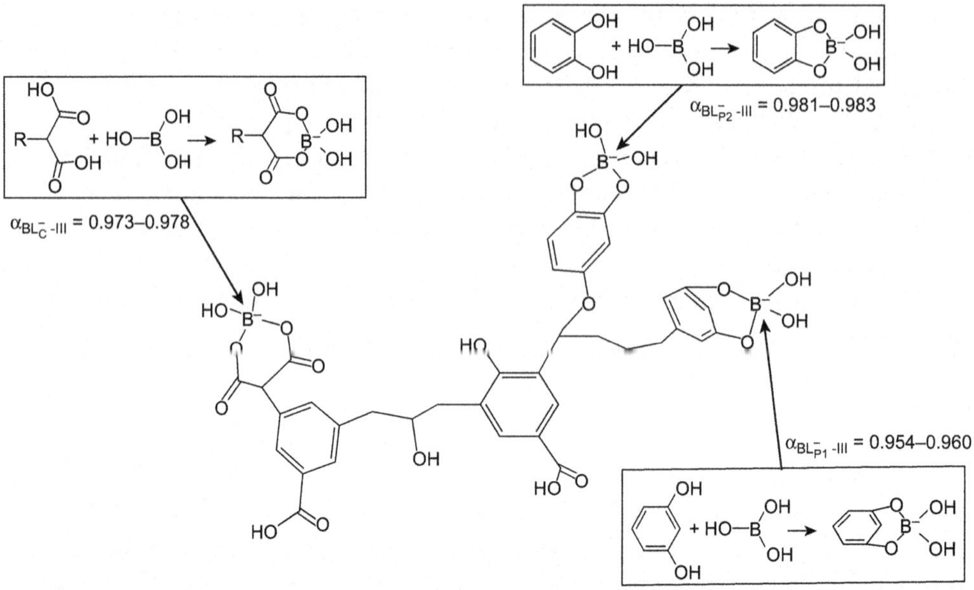

Figure 6.15 Schematic structures of boron surface complexes formed within the framework of the three-site and four-site models. It is assumed that the structures of boron surface complexes are the same in the two models with, in the case of the four-site model, an additional BL_C^- complex formed on weaker carboxylic functions.[34]

Tossell performed ab initio calculations, ^{11}B NMR measurements, and isotopic fractionations of surface complexes on humic acids.[35] Equilibrium geometries for $B(OH)_2L^-$ complexes have been calculated for L = oxalic acid, ethylene glycol, malonic acid, succinic acid, catechol, 1,8-dihydroxynaphthalene, acetic acid, glycolic acid, 4-ring carbonic acid, and corner-sharing carbonic acid; these ligands cover the vast majority of functionalities present in humic matter. The use of free $B(OH)_2L^-$ anions as models for the surface complexation, together with quantum mechanical calculations of their properties, leads to ^{11}B NMR shifts that agree well with the experimental data of Lemarchand and coworkers. There is one significant discrepancy between the measured ^{11}B value chemical shifts of the complexes and those assumed for $B(OH)_4^-$. This difference is due to the use of an inaccurate value for the equilibrium isotopic ^{11}B, ^{10}B fractionation constant for the $B(OH)_3$, $B(OH)_4^-$ pair. The use of corrected parameters results in a much better fit to the data.

REFERENCES

1. Reid RJ, Hayes JE, Post A, Stangoulis JCR, Graham RD. A critical analysis of the causes of boron toxicity in plants. *Plant Cell Environ* 2004;**27**(11):1405–14.
2. Camacho-Cristóbal JJ, Rexach J, González-Fontes A. Boron in plants: deficiency and toxicity. *J Integr Plant Biol* 2008;**50**(10):1247–55.

3. Nable RO, Bañuelos GS, Paull JG. Boron toxicity. *Plant Soil* 1997;**193**(2):181–98.
4. Cervilla LM, Blasco B, Rios JJ, Rosales MA, Sánchez-Rodríguez E, Rubio-Wilhelmi MM, et al. Parameters symptomatic for boron toxicity in leaves of tomato plants. *J Botany* 2012;**2012**(1–2):1–17.
5. Maas E. Salt tolerance of plants. In: Christie B, editor. *CRC handbook of plant science in agriculture*, vol. 2. Boca Raton, FL: CRC Press; 1987. pp. 57–75.
6. Ontario Ministry of Environment, Standards Development Branch. *Canadian water quality guidelines for the protection of aquatic life*. Boron; 2009. http://ceqg-rcqe.ccme.ca/download/en/221/.
7. Goldberg S, Glaubig R. Boron adsorption on aluminum and iron oxide minerals. *Soil Sci Soc Am J* 1985;**49**:1374–9.
8. Seki Y, Seyhan S, Yurdakoc M. Removal of boron from aqueous solution by adsorption on Al_2O_3 based materials using full factorial design. *J Hazard Mater* 2006;**138**(1):60–6.
9. Peak D, Luther GW, Sparks DL. ATR–FTIR spectroscopic studies of boric acid adsorption on hydrous ferric oxide. *Geochim Cosmochim Acta* 2003;**67**(14):2551–60.
10. Demetriou A, Pashalidis I. Adsorption of boron on iron-oxide in aqueous solutions. *Desalin Water Treat* 2012;**37**(1–3):315–20.
11. Goli E, Rahnemaie R, Hiemstra T, Malakouti MJ. The interaction of boron with goethite: experiments and CD–MUSIC modeling. *Chemosphere* 2011;**82**(10):1475–81.
12. Hiemstra T, van Riemsdijk W, Bolt G. Multisite proton adsorption modeling at the solid/solution interface of (hydr)oxides: a new approach. *J Colloid Interface Sci* 1989;**133**(1):91–104.
13. Hiemstra T, van Riemsdijk W. A surface structural approach to ion adsorption: the charge distribution (CD) model. *J Colloid Interface Sci* 1996;**179**(2):488–508.
14. Singh S, Mattigod S. Modeling boron adsorption on kaolinite. *Clays Clay Miner* 1992;**40**(2):192–205.
15. Hayes K, Leckie J. Modeling ionic strength effects on cation adsorption at hydrous oxide/solution interfaces. *J Colloid Interface Sci* 1987;**115**(2):564–72.
16. Goldberg S. Inconsistency in the triple layer model description of ionic strength dependent boron adsorption. *J Colloid Interface Sci* 2005;**285**(2):509–17.
17. Keren R, Sparks DL. Effect of pH and ionic strength on boron adsorption by pyrophyllite. *Soil Sci Soc Am J* 1994;**58**(4):1095–100.
18. Keren R, Grossl PR, Sparks DL. Equilibrium and kinetics of borate adsorption–desorption on pyrophyllite in aqueous suspensions. *Soil Sci Soc Am J* 1994;**58**(4):1116–22.
19. Kim Y, Kirkpatrick RJ. ^{11}B NMR investigation of boron interaction with mineral surfaces: results for boehmite, silica gel and illite. *Geochim Cosmochim Acta* 2006;**70**(13):3231–8.
20. Karahan S, Yurdakoç M, Seki Y, Yurdakoç K. Removal of boron from aqueous solution by clays and modified clays. *J Colloid Interface Sci* 2006;**293**(1):36–42.
21. Kehal M, Reinert L, Duclaux L. Characterization and boron adsorption capacity of vermiculite modified by thermal shock or H_2O_2 reaction and/or sonication. *Appl Clay Sci* 2010;**48**(4):561–8.
22. Tabelin CB, Igarashi T, Arima T, Sato D, Tatsuhara T, Tamoto S. Characterization and evaluation of arsenic and boron adsorption onto natural geologic materials, and their application in the disposal of excavated altered rock. *Geoderma* 2014;**213**:163–72.
23. Sasaki K, Qiu X, Hosomomi Y, Moriyama S, Hirajima T. Effect of natural dolomite calcination temperature on sorption of borate onto calcined products. *Microporous Mesoporous Mater* 2013;**171**:1–8.
24. Kentjono L, Liu J, Chang W, Irawan C. Removal of boron and iodine from optoelectronic wastewater using Mg–Al (NO_3) layered double hydroxide. *Desalination* 2010;**262**(1–3):280–3.
25. Speight JG. *Environmental technology handbook*. New York: Taylor & Francis; 2000.
26. Goldberg S. Reactions of boron with soil. *Plant Soil* 1997;**193**(2):35–48.
27. Keren R, Mezuman U. Boron adsorption by clay minerals using a phenomenological equation. *Clays Clay Miner* 1981;**29**(3):198–204.
28. Goldberg S. Reanalysis of boron adsorption on soils and soil minerals using the constant capacitance model. *Soil Sci Soc Am J* 1999;**63**(4):823–9.
29. Goldberg S, Suarez DL, Basta NT, Lesch SM. Predicting boron adsorption isotherms by midwestern soils using the constant capacitance model. *Soil Sci Soc Am J* 2004;**68**(3):795.
30. Goldberg S, Suarez DL, Shouse PJ. Influence of soil solution salinity on boron adsorption by soils. *Soil Sci* 2008;**173**(6):368–74.

31. Steiner F, Lana MDC, Zoz T, Fey R, Frandoloso JF. Boron adsorption in lowland soils from Paraná State. *Braz Semin., Ciênc. Agrár.* 2012;**33**(4):1391–402.
32. Ranjbar F, Jalali M. Release kinetics and distribution of boron in different fractions in some calcareous soils. *Environ Earth Sci* 2013;**70**(3):1169–77.
33. Diana G, Beni C, Marconi S. Comparison of adsorption isotherm equations to describe boron behavior in soils affected by organic and mineral fertilization. *Commun Soil Sci Plant Anal* 2010; **41**(9):1112–28.
34. Lemarchand E, Schott J, Gaillardet J. Boron isotopic fractionation related to boron sorption on humic acid and the structure of surface complexes formed. *Geochim Cosmochim Acta* 2005;**69**(14):3519–33.
35. Tossell J. Boric acid adsorption on humic acids: ab initio calculation of structures, stabilities, ^{11}B NMR and ^{11}B, ^{10}B isotopic fractionations of surface complexes. *Geochim Cosmochim Acta* 2006;**70**(20): 5089–103.

CHAPTER 7

Iminobis-Alkylene Diol Function as Alternative Boron-Chelating Group: Its Incorporation into Various Polymer Topologies for Removal of Trace Boron Via Direct Sorption and Polymer-Enhanced Ultrafiltration

Hasan Zerze[1], Levent Yilmaz[1], H. Onder Ozbelge[1], Niyazi Bicak[2]
[1]Chemical Engineering Department, Middle East Technical University, Ankara, Turkey
[2]Department of Chemistry, Istanbul Technical University, Istanbul, Turkey

7.1 INTRODUCTION

Low-molecular weight compounds or polymers with multihydroxy functions are known to bind boric acid or borates to some extent. However, increasing the number of hydroxyl groups is not enough by itself to attain better boron chelation and to remove trace boron. Interaction of hydroxyl functional polymers such as poly(vinyl alcohol) (PVA) with boric acid results in rapid gelation, but only a few percent of hydroxyl groups involve in boron chelation. In another words, most hydroxyl groups, up to 80%, remain untouched. Therefore, to attain high boron chelation with polymers, the boron-chelating ligands must be specially designed.

7.2 DESIGN CRITERIA FOR BORON-CHELATING POLYMERS

In view of practical uses, (1) boron selectivity, (2) regenerability, (3) high capacity, and (4) efficiency, against trace boron are the most important measures to quantify the performance of boron-removing materials. It is known that, aliphatic multihydroxy compounds are boron-chelating agents. Since these groups are unreactive to metal ions in water, multihydroxy compounds specifically act on boric acid. Therefore, the reaction can be considered as specific, at least selective for chelation with boric acid or borates.

Regeneration and recovery of boron-chelated polymers is generally achieved by acid leaching. To recover the chelated polymer as unchanged after each removal cycle, it should be hydrolysis proof. This is especially essential for long-term and repeated use of

Boron Separation Processes
ISBN 978-0-444-63454-2, http://dx.doi.org/10.1016/B978-0-444-63454-2.00007-1

the boron sorbent polymer. Presence of hydrolyzable linkages such as ester or amide groups is not preferable to attain high hydrolytic stabilities and long-term uses.

7.2.1 Alternative Boron-Chelating Ligands

Having more than four hydroxy groups, sugar derivatives such as *N*-methyl glucamine, sorbitol, and mannitol are known to form four-coordinated boron complex in water. Polymer-bound sugar derivatives behave similarly and form strong chelates with boric acid. For this reason sugar functional polymers have been extensively studied and used for boron removal.[1–3] However, those groups are susceptible to biological attack and may decompose during acid–base treatments in regenerating cycles. This phenomenon has prompted the search for new and alternate boron-chelating groups.

In this respect the iminobis-propanediol function (Figure 7.1) has found great attention. Having two vicinal hydroxyl groups around the nitrogen atom, this group is considered to form either four coordinated borate anion or neutral boron ester via four or three hydroxyl groups, respectively.

In both cases, Imninobis-propanediol (IBP) alone can bind boric acid without the necessity of additional hydroxyl groups. This means that, one IBP group is able to trap one boric acid. This implies that, soluble polymers bearing IBP groups will not undergo cross-linking during chelation with boric acid. Hydoxy groups of commercial PVA are also able to coordinate boric acid. However, in aqueous solution, PVA experiences rapid gelations by treating with dilute boric acid solution, as reported many times. Such a fast cross-linking results in more than half of hydroxy groups unoccupied. The overall result would then be low boron sorption capacity of PVA.[4] Figure 7.2 compares boron-binding characteristics of IBP functional polymer with that of PVA.

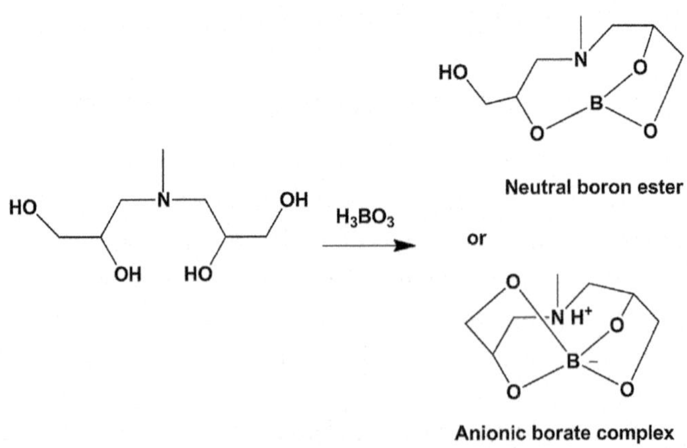

Figure 7.1 Schematic view of boron ester complexes formed by reaction of Imninobis-propanediol (IBP) group with boric acid.

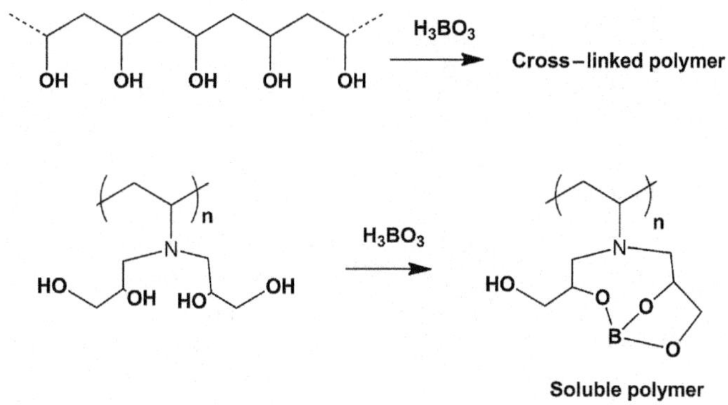

Figure 7.2 Comparison of the boron complexation of iminobis-propanediol (IBP) functional polymer with poly(vinyl alcohol).

Therefore, IBP functional polymer has many advantages over PVA and other well-known boron-specific resins with sugar functions.

These can be summarized as follows:

- Since each IBP function can hold one boric acid molecule, the polymer with IBP function does not undergo cross-linking in boron binding.
- Complexation of IBP with boron does not change hydrophilicity of the polymer and this causes nearly full occupation of the chelating sites, so that boron-binding capacities of the polymers become considerably high. The capacities attained by IBP and related groups might be as high as 3–4 mmol per gram, which are not attainable by sugar functional resin materials.
- IBP itself does not contain hydrolysable connections. This makes the polymers with IBP functions hydrolytically stable. Due to the hydrolytic stability, the boron-loaded polymer can be freed from the boron by acid leaching without structural change. The polymer can be recovered and recycled by this way many times. Due to these peculiarities, IBP group is superior to sorbitol, mannitol, and other sugar molecules.

7.2.2 Synthesis Strategies for IBP and Related Boron-Chelating Ligands

Three common strategies have been utilized for creating IBP and related functionalities on linear and resinous polymers.

1. Action of glycidol on polymer-bound amino groups (route a in Figure 7.3),
2. Reaction of epoxide functional polymers with amines followed by reaction of secondary amino groups with glycidol (route b in Figure 7.3).
3. Reaction of amino alcohols with glycidol and subsequent incorporation of the resulting amine to epoxy functional polymers (route c in Figure 7.3).

The transformation yields are generally greater than 90%. The reaction of the amino group can be achieved even in water. Some representative examples for generation of IBP and similar ligands on linear and resinous polymers are given below.

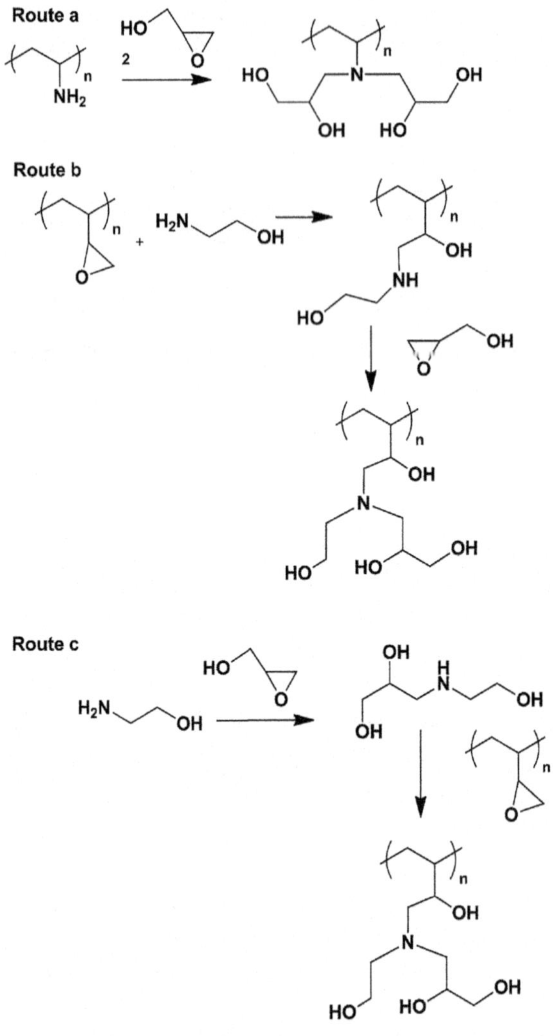

Figure 7.3 Synthesis strategies for functionalization of polymers with IBP and related boron-chelating ligands.

Example 1: IBP Functional Hairy Polymer via Amine Function[5]

Poly(vinyl amine) brushes generated on polystyrene-divinyl benzene (PS-DVB) resin beads by surface-initiated polymerization of *N*-vinyl formamide (NVF) (grafting degree: 1350%) and subsequent acid hydrolysis were used for creating IBP surface functions. Amine content of the polymer was determined (by acid titration) as 13.4 mmol per gram. A sample of this product (~2 g) swelled in 15 mL of distilled water overnight. Ten grams glycidol (0.135 mol) was added to the mixture at 0 °C and this solution was stirred for

6 h. The solution was then stirred at room temperature for 18 h. Finally, the reaction mixture was heated to 70 °C, maintained at that temperature for 30 min, and filtered. Then, the beads were washed with water (5 × 100 mL), methanol (2 × 50 mL), and diethyl ether (30 mL) and dried at 40 °C for 24 h under vacuum. Iminobis-propanediol (IBP) function density of the resulting resin was determined as 3.8 mmol per gram.

Example 2: Generation of IBP-like Ligands on Epoxy Functional Microbeads[6]

Epoxy functional terpolymer beads obtained by suspension polymerization of the monomer mixture constituted with 0.4 mol glycidyl methacrylate (GMA), 0.5 mol methyl methacrylate (MMA), and 0.1 mol DVB were used as starting polymer. Vacuum-dried bead product was sieved and 210–420 μm of the size fraction was used in further reaction. Epoxy content of this fraction was determined 3.4 ± 0.08 mmol/g by pyridine-HCl method.[7] A sample of this product (10.0 g) was soaked in a mixture of 20 mL ethylenediamine (0.3 mol) and 40 mL *N*-methyl pyrrolidinone at 0 °C. The mixture was shaken for 24 h at room temperature. The bead product obtained by filtration was left to stand in 250 mL water to remove amine residues. The filtered product was washed several times with water (5 × 200 mL) and dried under vacuum at room temperature for 48 h.

The dry product weighed 12.3 g. Amine content of the resulting product was determined as 5.52 mmol/g by acid titration. Then, 10.2 g of the above product was dispersed in 15 ml of 2-methyl pyrrolidone (Figure 7.4). While stirring mechanically

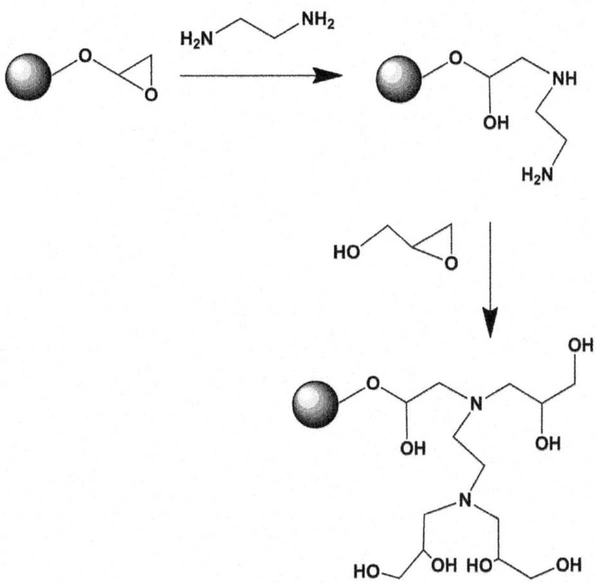

Figure 7.4 Generation of IBP-like ligands on epoxy functional polymer beads.

6 mL (0.0904 mol) of glycidol (3.1 equivalent of the amine content) was added dropwise to the mixture. The mixture was stirred for 6 h at room temperature and 1 h at 50 °C. The modified bead product was isolated by filtration. After washing with water (5 × 200 mL) the product was dried at 50 °C for 24 h under vacuum. The dried product weighed 16.2 g.

Example 3. Generation of 2-hydroxyethylamino, 2,3-propanediol Ligands on Epoxy Functional Microbeads[8]

This was carried out by the reaction sequences depicted in Figure 7.5. Thus, to a 250 mL two-necked canonical flask equipped with a reflux condenser and a dropping funnel, there was added ethanolamine (107 g, 1.5 mol). Glycidol (32.8 mL, 0.5 mol) was added dropwise to the flask while stirring at room temperature. The reaction was exothermic. The flask was then placed in an oil bath at 120 °C and excess ethanolamine was distilled off at 70–73 °C/0.5 mm to give 2-hydroxyethylamino propylene glycol as a viscous liquid (see Figure 7.5). The colorless liquid is pure enough for the next reaction as demonstrated by thin-layer chromatography (with basic alumina using dimethoxy ethane as solvent). The yield was 52.5 g (76.6%).

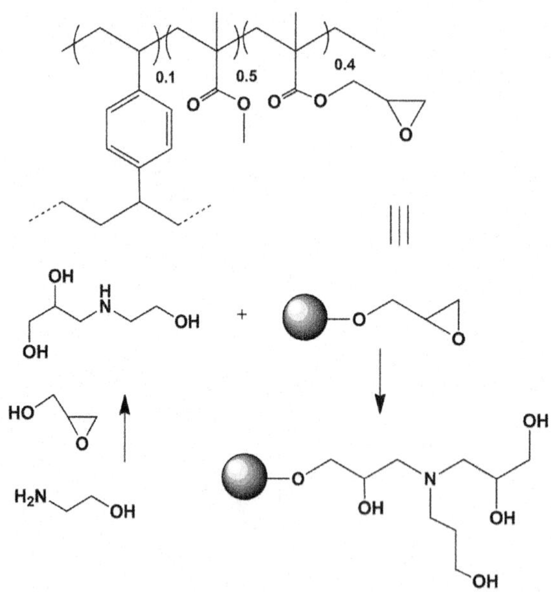

Figure 7.5 Generation of hydroxyethylimino propanediol (HEP) ligands on epoxy functional polymer beads.[8]

Five grams of the beaded polymer poly(glycidyl methacrylate) (PGMA) was added portionwise to the stirred solution of 7.3 g 2-hydroxyethylamino propylene glycol at room temperature. The reaction content was shaken for 24 h at room temperature and heated to 70 °C for 15 min. The bead product was transferred to a beaker with 250 mL water and left in contact for overnight. The filtered product was washed several times with water (5 × 200 mL) and dried under vacuum at 45 °C for 48 h. The dry product weighed 6.62 g. Acid titration of this product indicated 1.82 mmol of amine content per gram.

7.2.3 Other Boron-Chelating Groups

There appear to be few interesting reports dealing with unusual new boron-chelating groups. Among those, dextran has been found to be of interest. Recently, Kabay and Tuncer reported dextran-immobilized PGMA-based microbeads.[9] They reported two methods for immobilizing dextran onto the polymer microspheres. In the first route, azide function was generated on the microspheres by reacting epoxy group with sodium azide and then propiolic acid functional dextran was clicked onto the microsphere surface. In the second route, dextran was directly attached to cross-linked PGMA microbeads. They reported 1.8−0.5 mg boron sorption per gram resin in 100 ppm boron solutions.

In another paper, Doganay et al. reported synthesis of 4-vinyl-1,3-dioxalan-2-on copolymers with vinyl acetate.[10] Water-soluble hydroxy functional polymers obtained on hydrolysis (see Figure 7.6) of this copolymer have been employed for boron chelation in polymer-enhanced ultrafiltration (PEUF) system.

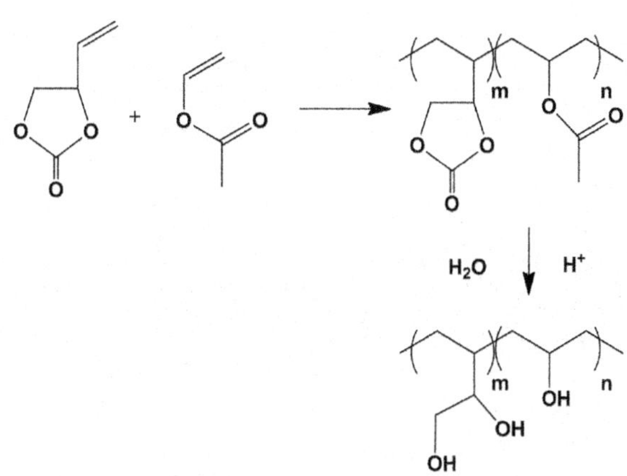

Figure 7.6 Synthesis of 4-vinyl-1,3-dioxalan-2-one-vinyl acetate copolymers and their hydrolysis for boron extraction.[10]

7.3 CARRIER POLYMERS

Chemical structure and topology of polymers used as carrier for boron-chelating ligands is of importance in their practical use for long-term boron uptake. Structurally, these polymers must be constituted with an inert backbone. Inertness of the backbone means unreactivity of the polymer including oxidative and hydrolytic stabilities in boron-chelation conditions. The polymers must carry a key function such as chloromethyl, chlorosulfonyl, or epoxy group or must be easily functionalized for tethering boron-chelating ligands. Considering these requirements, styrene and methacrylate-based polymers are mostly preferred trunk polymers as carriers for boron-chelating ligands. Polystyrene is relatively inert to most nucleophilic, electrophilic, and oxidation reactions in ordinary reactions. In contrast to ordinary ester functions, the ester group of methacrylate monomers including GMA and their polymers are known to have considerable stability against acid and base hydrolyses.[11]

These polymers have been widely used as carriers for boron-chelating ligands. Some natural polymers such as chitosan have also been used as a starting polymer, although it is not stable enough to hydrolysis.

Linear, gel, resinous, and brushed bead polymers are well-known topologies that can be employed as carriers for boron-chelating ligands (Figure 7.7). Those topologies have their own advantages and disadvantages as carrier boron-chelating ligands.

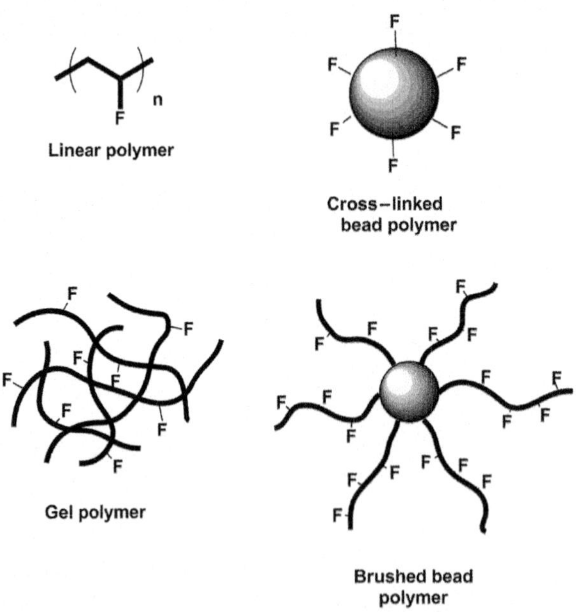

Figure 7.7 Linear, gel, cross-linked bead and brushed bead polymers can be employed as functional group carriers.

For instance, water-soluble linear polymers carrying boron-chelating functions has found applications for removal of trace boron in PEUF systems, as discussed below.

7.4 LINEAR BORON-CHELATING POLYMERS AND THEIR USE IN POLYMER-ENHANCED ULTRAFILTRATION: WHAT IS THE IDEA BEHIND IT?

In recent years, water-soluble polymers have been successfully used in combination with ultrafiltration (UF) systems for the separation of various hazardous compounds from water.[12–24] This hybrid separation scheme is usually referred to as polymer-enhanced (or polymer assisted) ultrafiltration (abbreviated as PEUF or PAUF).[15–17,20,21,23,24] In this section, some introductory information about PEUF is provided.

A clear understanding of this method necessitates differentiation from other pressure-driven membrane processes, namely, reverse osmosis (RO), nanofiltration (NF), UF, and microfiltration. These membrane processes are based on the establishment of a pressure difference between the two sides of the membrane and the separation is achieved mainly by differences in the particle sizes. General specifications of these pressure-driven membrane processes are represented in Table 7.1. It is illustrated in Table 7.1 that the size of the target molecule clearly dictates the type of the membrane process to be employed. Moreover, the required transmembrane pressure difference increases as membrane pore size decreases. This simply illustrates the fact that the separation of smaller molecules requires higher pressure and operating energy. An effort to find a solution to this issue led the researchers to study the complexation of target molecules with much larger molecules in order to enable separation using a membrane with larger pores. In this way, permeate flux can be increased along with much lower energy requirement for the separation. It is simply this idea that the PEUF is based on. As depicted in Figure 7.8, in principle the PEUF setup is similar to that of the other membrane processes. However, the use of water-soluble functional polymer is one important distinction of the PEUF technique.

Trace removal of boric acid or borate ion can be performed via NF at pressure differences of several tens of bars (see Table 7.1). However, in boron removal by PEUF,

Table 7.1 General Specifications of Four Different Pressure-Driven Membrane Processes

Specifications	Microfiltration	Ultrafiltration	Nanofiltration	Reverse Osmosis
Pore size	0.05–4 μm	0.005–0.05 μm	Dense	Dense
Pressure difference	Less than 2 bar	1–10 bar	5–35 bar	15–150 bar
Target species	Microorganisms and colloids	Macromolecules	Separation of ions with multiatoms or multivalency	Separation of ions with single atom or single valency

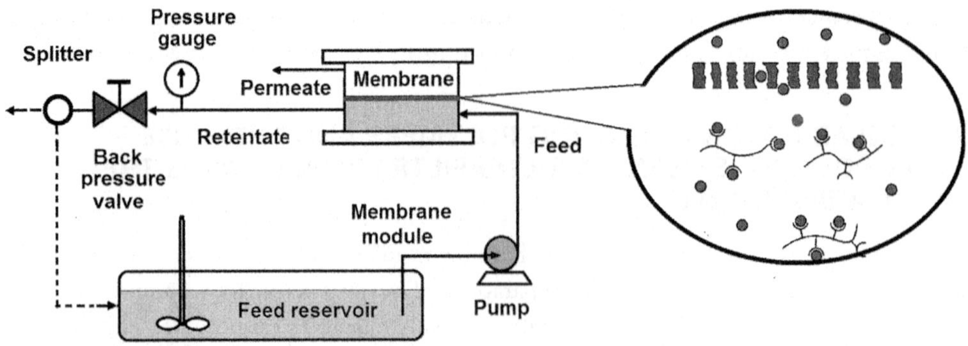

Figure 7.8 Representative setup for polymer-enhanced ultrafiltration system.

the separation is realized via binding of the boron by a boron-chelating water-soluble polymer in the homogeneous phase followed by the removal of the macromolecular complexes via the UF of the solution.

Polymer-boron chelation enables the use of UF membranes with much larger pore sizes, when compared to the RO or NF processes. In that sense, the main target of the PEUF studies has been to substantially lower the operational pressure drops, ΔP.

In general, performance of a membrane process is quantified by retention and flux, which can also be evaluated similarly in order to describe the performance of a PEUF process. Retention represents the degree of the removal of the target species and is expressed by the relation

$$R = 1 - \frac{C_P}{C_F}$$

where C_P is the concentration of the target compound in the permeating solution and C_F is the concentration in the feed stream. On the other hand, the flux is defined by the permeate flow rate per unit membrane surface area and it is used as a measure of the amount of water processed through unit membrane area. The amount of functional polymer used in PEUF is usually quantified by the term loading, which is defined by the boron-to-polymer mass ratio.

7.5 SYNTHESIS OF WATER-SOLUBLE BORON-BINDING FUNCTIONAL POLYMERS

In this section, the functional polymers used in PEUF studies are introduced and their syntheses are mentioned briefly. The studies have shown that the success of the PEUF mostly depends on the design of the water-soluble target-specific functional polymers.[10,25–28] In spite of the abundance of successful PEUF studies for a variety of hazardous compounds,[12–23] boron removal in such a scheme has been scarcely studied due to the challenges in designing suitable boron-binding polymers. The main aim has been to synthesize stable and reusable polymers that provide high retentions along with optimum permeate fluxes. Therefore, a pertinent polymer should preferably possess high solubility and high boron-binding affinity accompanied by the ease of the decomplexation in response to a change in one of the physicochemical conditions such as pH and temperature. Accordingly, several functional polymers have been synthesized and exploited for boron removal via PEUF. Mainly, solution polymerization technique has been used for the synthesis of the linear starting polymers. The chemical structures of the relevant polymers are demonstrated in Figure 7.9 with the corresponding abbreviations. As known from the synthesis of boron-binding polymeric adsorbents, ring-opening reactions of epoxides or lactones result in versatile routes to synthesize various functional polymers that constitute hydroxyl groups.

This reaction has been typically realized with compounds that possess amine or imine groups. In one of the early studies, poly(ethylene imine) (PEI) was functionalized using three different ligands, namely, polyethylene oxide, glycidol, and 3,4-dihydroxybutene-acetoketal.[25] The functionalization of PEI using glycidol is illustrated in Figure 7.10(a). The reactions could be carried out in mild temperatures due to vigorous reaction between imine and epoxide groups. Due to clear molecular weight difference between the synthesized functional polymers and the side products, diafiltration could be used for the purification of the resulting polymers monool-PEI and diol-PEI. Similarly, glucoheptonic lactone was used in order to derivatize PEI.[23] The reaction is represented in Figure 7.10(b). The resultant polymer has functional branches with crowded hydroxyl groups. However, this polymer has limited chemical stability due to hydrolysis of amide linkages involved.[23]

In another study, poly(vinyl amino-*N*,*N*′-bis-propane diol) (GPVA) was prepared by the functionalization of poly(vinyl amine) using glycidol (see Figure 7.10(c)).[26,27] The polymer was reported to exhibit high boron retentions. Since the opening of oxyrane ring results in the formation of *trans*-diol, it was speculated that GPVA complexation with boric acid must give neutral boron ester and the complexation would be reversible, when compared to the highly irreversible boron binding by a previously reported adsorbent possessing *cis*-diols instead.[29] In addition to its suitable structure favoring reversible boron-binding ability, GPVA was reported to be a highly water-soluble polymer and its copolymer with diallyl dimethyl ammonium chloride (DADMAC) was also synthesized, facilitating enhanced solubility with no reduction in the boron-binding efficiencies.[27]

Alternatively, readily reacting epoxide rings existing within the branches of a polymer can be treated with compounds that constitute amine functional such as alkanolamines and amino sugars. This clearly results in a wide range of chemical structures. Such a reaction can be illustrated as in Figure 7.10(d), which demonstrates the synthesis strategy in the study of Doganay et al.[10] In that study, PGMA was dissolved in tetrahydrofuran and

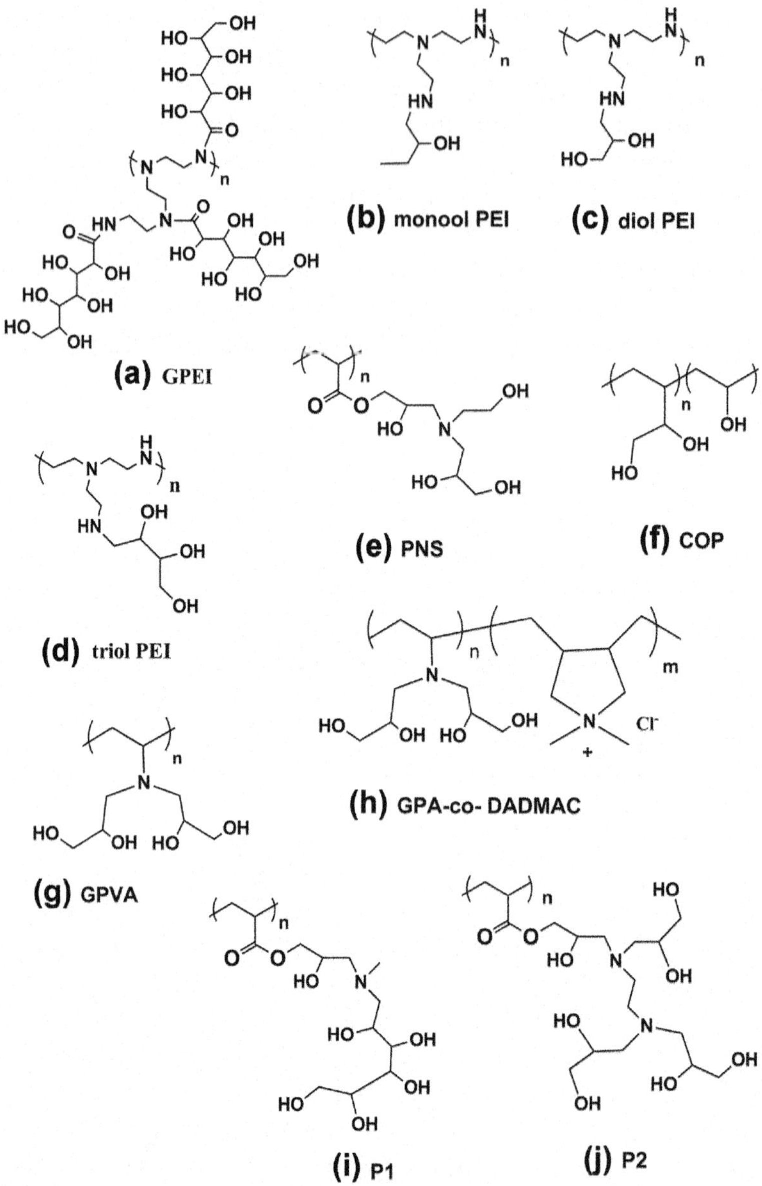

Figure 7.9 Boron-binding water-soluble functional polymers synthesized for use in PEUF experiments. Given short names are from original studies for (a) GPEI: glucoheptonic lactone derivative of poly(ethylene imine),[23] (b) Monool-PEI,[25] (c) Diol-PEI,[25] (d) Triol-PEI,[25] (e) PNS,[10] (f) COP: poly(vinylethane diol-co-vinyl alcohol),[10] (g) GPVA: poly(vinyl amino-*N*,*N*′-bis-propane diol),[26] (h) GPVA-co-DADMAC: poly(vinyl amino-*N*,*N*′-bis-propane diol-co-diallyl dimethyl ammonium chloride),[27] (i) P1: *N*-methyl-D-glucamine-grafted poly(glycidyl methacrylate),[28] (j) P2: iminodipropyleneglycol-grafted poly(glycidyl methacrylate).[28]

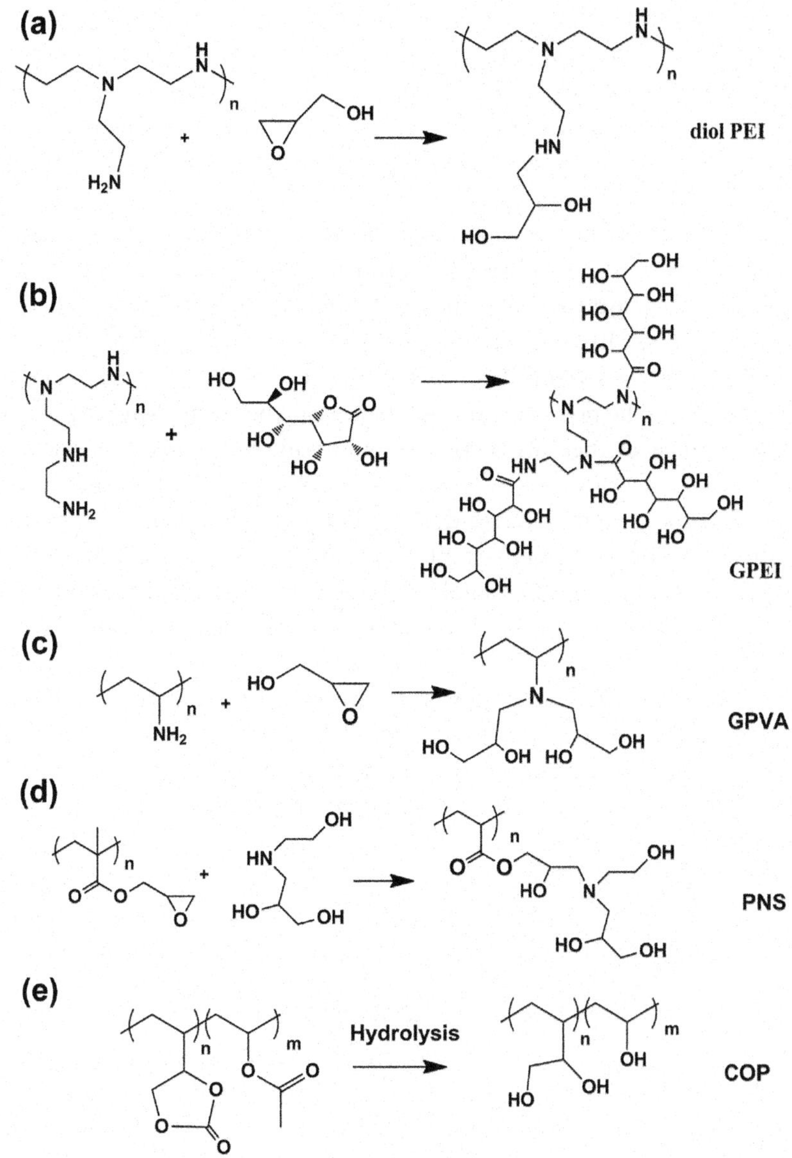

Figure 7.10 (a) Functionalization of poly(ethylene imine) (PEI) using glycidol.[25] (b) Functionalization of polyethylene imine (PEI) using glucoheptonic lactone.[23] (c) Synthesis of poly(vinyl amino-bis-propane diol) (GPVA).[26] (d) Functionalization of poly(glycidyl methacrylate) with hydroxyethylamino glycerol.[10] (e) Synthesis of poly(vinyl-ethanediol-*co*-vinyl alcohol).[10] GPEI, glucoheptonic lactone derivative of PEI; COP, poly(vinyl-ethanediol-co-vinyl alcohol).

reacted with 1-(2-hydroxyethylamino) glycerol in order to obtain PNS (hydroxy-ethylaminoglycerol functioned poly(glycidylmethacrylate)). The polymer was reported as chemically stable against extreme pHs.[10]

Another shortcut technique for the synthesis of polymers bearing hydroxyl groups, is the hydrolysis of polymeric acetates as in the synthesis of PVA from poly(vinyl acetate).

Doganay et al. copolymerized an equimolar mixture of 4-vinyl-1,3-dioxalan-2-one and vinyl acetate with 2,2′azobis-isobutyronitrile (AIBN) in bulk.[10] This copolymer was then hydrolyzed in order to obtain poly(vinyl-ethanediol-co-vinyl alcohol) (COP) as shown in Figure 7.10(e). In that study, the aim was to obtain a functional polymer superior to PVA. This polymer has vicinal hydroxyl groups that are directly attached to a carbon backbone. Therefore, the possibility of backbone—backbone cross-linking should still be considered at some level since any boron complexation might directly affect the freedom of backbone bond rotations. However, COP was reported to be chemically resistant to acid hydrolysis, which would indicate the reusability of such polymers.

Reaction of an amino sugar (*N*-methyl glucamine) with PGMA has recently been applied for the synthesis of a water-soluble functional polymer.[28] The resulting polymer (P1) was marked by its high hydroxyl population. However, complexation was weakly pH dependent.

It is clear that the reaction routes for the syntheses of linear boron-binding polymers are analogous to the ones used in the syntheses of cross-linked or tethered chelating agents. In spite of the similarities in the reaction types, synthesis conditions of water-soluble polymers differ significantly from polymeric adsorbents. First, the production of a water-soluble polymer usually suffers from tedious purification steps due to the necessity of finding a good nonsolvent after each reaction step,[26] which is not always trivial for linear polymers. Alternatively, membrane techniques can be successfully utilized for the purification and fractionation of the water-soluble macromolecules.[25,26]

7.5.1 PEUF Performances of Functional Polymers

Boron removal performances of the aforementioned functional polymers are discussed in the following paragraphs. Specifically, the polymers are evaluated according to their affinity to boron, solubility, reusability, chemical stability, and binding selectivity. In Table 7.2, some PEUF results and study parameters are given for a valid comparison among those polymers.

The studies have shown that higher boron affinities can be obtained by the increasing —OH functionality of the polymer. In order to illustrate this quantitatively, boron complexation constants for three different polymers are listed in Table 7.3. It is clearly seen that GPVA-co-DADMAC outcompetes the boron-binding performances of the other two polymers, most probably due to having denser population of hydroxyl groups.

Table 7.2 Evaluation of Functional Polymers Used in PEUF by Their Boron Removal Performances and the Study Parameters

Polymer	Boron Concentration Range (ppm)	Polymer Concentration (g/L)	pH Range	% Boron Retention	Comments or Issues Reported
Poly(vinyl alcohol)[24]	10	0.01–5	7.0–10	9–28	Insufficient solubility
PEI[25]	50–5000	10	7.5–10.5	55–65	Low selectivity due to ionic attraction
Monool-PEI[25]	50–5000	10	7.0–9.5	35–50	Low selectivity due to ionic attraction
Diol-PEI[25]	50–5000	10	6.5–9.5	40–78	Boron retentions seem promising but more dilute solutions need studying
Triol-PEI[25]	100–5000	10	7.5–9.0	30–90	Highly pH dependent high boron retentions seem promising
GPEI[23]	10	80	3.8–5.7	NA	Low chemical stability
PNS[10]	10	2	7.0–10	25–55	Single-step removal does not seem adequate
COP[10]	10	2	7.0–10	31–52	Significant decrease in pH during complexation and low pH dependency of complexation
GPVA[26]	10	10	7.0–10	52–96	High boron retentions were obtained even in the presence of other ions
GPVA-co-DADMAC[27]	10	10	7.0–10	55–92	Introduction of charged comonomers provided a better solubility
P1[28]	10	0.01	7–10	80–90	Low pH dependency of complexation
P2[28]	10	0.01	7–9	49–54	Low pH dependency of complexation

PEUF, polymer-enhanced ultrafiltration; PEI, poly(ethylene imine); COP, poly(vinyl-ethanediol-co-vinyl alcohol); GPVA, poly(vinyl amino-*N*,*N*′-bis-propane diol); GPVA-co-DADMAC, poly(vinyl amino-*N*,*N*′-bis-propane diol-co-diallyl dimethyl ammonium chloride); P1, *N*-methyl-D-glucamine-grafted poly(glycidyl methacrylate); P2, iminodipropyleneglycol-grafted poly(glycidyl methacrylate).
NA refers to "Not Available".

Table 7.3 Boron Complexation Constants of Three Different Ligands Expressed by $K_C = [BL]/[B][L]$

Polymer	pH	Temperature (°C)	K_C (L/mol)	References
Diol-PEI	9.87	20	21	25
Triol-PEI	9.30	20	111	25
GPVA-co-DADMAC	9.0	25	186	27

PEI, poly(ethylene imine); GPVA-co-DADMAC, poly(vinyl amino-*N,N′*-bis-propane diol-co-diallyl dimethyl ammonium chloride).

Therefore, comparative evaluation of these studies has shown that the boron affinities can be enhanced by the synthesis of polymers with relatively dense hydroxyl groups.

For an efficient UF process, even a molecular-level precipitation should be strictly avoided since it results in a scaling on the membrane surface and hence leads to significant reductions in permeate fluxes during operation. Therefore, some researchers have targeted to solve this issue by the design and synthesis of suitable branched polymers with highly dense hydroxyl groups. In this way, they obtained high polymer solubility and eliminated the risk of intramolecular cross-linking induced by boron binding. In that sense, hyperbranched glucoheptonic lactone derivative of PEI (GPEI)[23] and GPVA[26] are among such boron chelators that are composed of bulky functional branches.

It was suggested in another study that high polymer solubilities could be attained by the copolymerization using a suitable charged co-monomer.[27] In that study, GPVA-co-DADMAC was reported to have higher solubility than GPVA due to its permanent charges on DADMAC repeat units providing intramolecular ionic repulsion. It was reported that even small percentages (2–10%) of DADMAC within the copolymer, imparted sufficient solubility so that polymer precipitation was not observed.

Reusability of a polymer mostly depends on two things: easy decomplexation ability and chemical stability. A suitable polymer for a PEUF application, should provide a controlled switching between complexation and decomplexation by adjusting, for instance, pH or temperature. This necessity arises from the fact that the recovery of the polymer can be realized by reversing the complexation reaction. Although the effect of temperature on boron removal has not been investigated via PEUF yet, it is well known that boron binding by a hydroxyl group is exothermic.[30] Therefore, boron retention is predicted to increase in a PEUF system, when the system temperature decreases. However, pH adjustment has been regarded as a more practical process than the control of temperature and hence, the reversibility of the binding in PEUF studies has usually referred to the sensitivity of the complexation to the change in pH. In some studies, polymers with pH-sensitive boron binding were synthesized and successfully used in PEUF experiments. There have been two efficient boron chelators reported with highly pH-sensitive boron binding, namely, triol-PEI[25] and GPVA.[26] As demonstrated in Table 7.2, with the former polymer, boron retentions around 90% obtained at pH 9.0

was able to be reduced to 30% levels by changing the pH down to 7.5. It was reported for the latter polymer that boron retention of 96% could be achieved at pH 9.0 while shifting pH to 7.0 lowered the retention to 50% levels. In addition, the recovery and reusability of GPVA and GPVA-co-DADMAC were investigated.[31] It was found that 97% of polymers were decomplexed at pH equal to 4.0 and they were repeatedly used with no decrease in their boron removal performances.[31]

Chemical stability is another important concern in the design of reusable functional polymers. Smith et al. (1999) studied multiple complexation/decomplexation cycles using GPEI and reported undesired polymer loss in the feed solution due to acid-catalyzed hydrolysis in the presence of 0.01 M HCl.[23] Although extreme pH conditions may not be realistic in a potential PEUF application, the risk of such hydrolysis reactions most probably results in short polymer lifetimes and thus high operating costs. Similarly, other process conditions such as temperature and the chemical and biological content of water being processed should be taken into consideration for a proper polymer choice.

Testing the boron removal performances of polymers in multicomponent waters, seems a prerequisite for their use in real water treatment. It was shown that the presence of competing ions such as chloride and sulfate did not affect appreciably the high boron retentions obtained by GPVA and GPVA-co-DADMAC owing to the selective binding reaction.[31] In the same study, boron was separated with superior retentions from a geothermal water sample containing a rich ionic content.[31]

It is demonstrated in this section that the polymer chemistry has reached a level that fulfills the need for the proper design of functional polymers. However, a real PEUF implementation will necessitate, as a next step, some detailed studies performed in much larger process capacities in order to predict the technological challenges in a real application. The process should be clearly described for a concrete application and the proper pilot-scale equipment should be constructed. In this way, a much more valid comparison can be made against the other separation techniques, which are already being applied in the industry. This would probably provide significant knowledge on some real problems difficult to foresee otherwise and hence lead to beneficial feedbacks to the researchers for the design of more suitable functional polymers and/or separation schemes.

7.6 BORON-CHELATING GEL POLYMERS

Logically, gel polymers can also be employed as carriers for the boron-chelating ligands (Figure 7.11).

In the literature, however, there appears only one report on boron-chelating polymer gels. Two gels derived from *N*-methyl-D-glucamine and GMA have been described in this report.[32] In the first, a multihydroxy functional monomer derived by reaction of

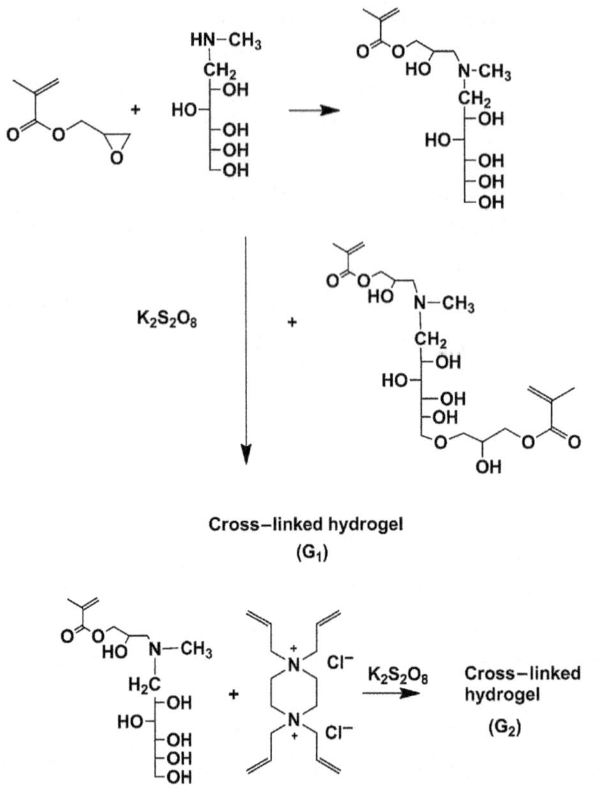

Figure 7.11 Synthesis pathway of the boron-chelating hydrogels from glycidyl methacrylate and *N*-methyl-D-glucamine.[32]

GMA with *N*-methyl-D-glucamine has been copolymerized by redox initiation to give boron-chelating hydrogel.

In the second route, GMA mixed with 3/2 equivalent *N*-methyl-D-glucamine has been directly polymerized to give a gel polymer. Both gels have been employed as reusable boron-specific sorbents. Capacities of the gels were around 2.2 mmol per gram. No capacity loss has been shown after 5 times of recycling. Common disadvantage of the gels was diffusion limitation during boric acid sorption.

7.7 RESIN BEADS WITH BORON-CHELATING LIGANDS

In general, cross-linked polymer beads are preferred as the carrier polymer for boron-binding functional groups. These materials are obtained by suspension polymerization methodology. Many receipts are now available in open literature for manufacturing bead polymers.[33]

Porosities of the resulting microbeads can be controlled by cross-linker ratio and by volume fraction of the porogen employed. These can be classified in three categories, which are gel-type resins, macroporous resins, and macroreticular resins. Gel-type resins contain a low percentage of cross-linker and flexible pores so that the pores are completely closed in dry state. In macroporous beads the pores are semiflexible that means some percent of pores have constant volume and remain open in a dry state or in solvent. The residual portion of the pores open in the presence of swelling solvents, but are closed in the dry state, due to moderate cross-link densities (10—30%). Macroreticular microbeads on the other hand are constituted with high percent cross-linker. As a result pores of these materials do not swell in solvents due to their constant pore volumes. Macroreticular resin beads, however, are very brittle and may easily be disintegrated mechanically. For this reason they have not found practical use. Very high swelling ratio of gel-type resins is disadvantageous in practical applications. For this reason, today macroporous beads are mostly the preferred resin materials used as carriers for chelating groups including boron-chelating ligands.[34] Macroporous resin beads containing both rigid and flexible pores are resistant to mechanical disintegration. This is one of the apparent advantages of macroporous resin beads.

The second prerequisite of the carrier polymer is the chemical inertness of its backbone. In other words, except for the functional group, the rest of the carrier polymer must be resistant to hydrolysis and oxidation reactions. In this respect, styrene and methacrylate-based polymers are of prime importance.

Not having hydrolysable and reactive groups, polystyrene can be considered as an inert polymer in mild conditions. The ester function of methacrylate polymers shows considerable hydrolytic stability against acid and base hydrolyses. Other acrylic polymers do not exhibit such a hydrolytic stability. Therefore, styrene and methacrylate ester polymers are most common carrier polymers that are utilized as linear or resinous carriers.

The third requirement is a hydrolysis-proof link between the functional group and the carrier polymer. In accord with those criteria, Bicak group inserted IBP groups to GMA-MMA-DVB terpolymer beads with 40% GMA (mol/mol), by reacting first with ethylenediamine.

Followed reaction of amino groups with glycidol yields a boron-chelating group consisting seven hydroxy groups around two nitrogen atoms.[6]

For 3.44 mmol/g amine content of the glycidol-modified resin, its boron-loading capacity, 3.0 mmol/g implies that each nitrogen atom acts as individual boron-chelating center. The ratio of the capacity to the amine content gives an average boron-chelation efficiency of 0.885 (Table 7.4, first row). Removal of the chelated boron by leaching with 4 M HCl solution regenerates the resin that is useful for reuse in boron binding. In a similar paper, cross-linked GMA beads have been modified to generate a boron-chelating center constituted by two hydroxyethyl groups and one dihydroxypropyl group around the amine nitrogen. Although boron-

Table 7.4 Boron Uptake Characteristics of the Resin Beads with Covalently Attached IBP and Related Ligands

Carrier Polymer	Boron-Chelating Ligand	Boron Uptake Capacity (mmol/g)	Efficiency	Regeneration	References
GMA-MMA-DVB	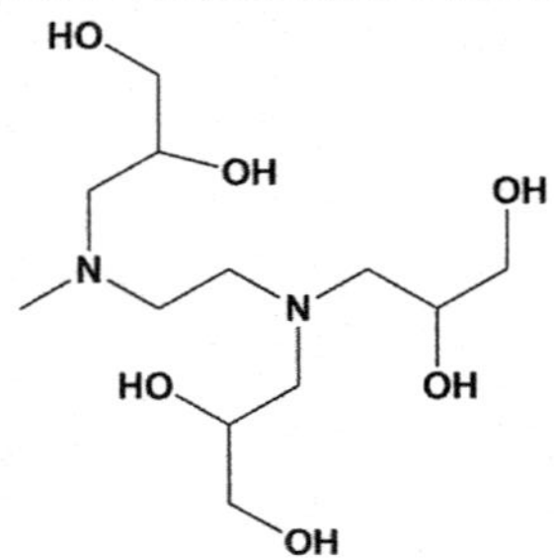	3.0 (pH: 6.8)	0.885	4 M HCl	6
GMA-MMA-DVB	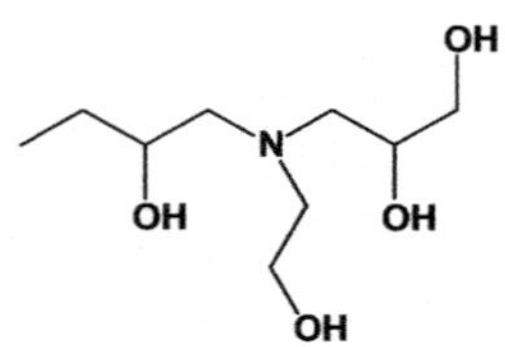	1.6 (pH: 7.4–8.0)	0.879	4 M HCl	8
GMA-MMA-ethylene glycol dimethacrylate (EGDMA)	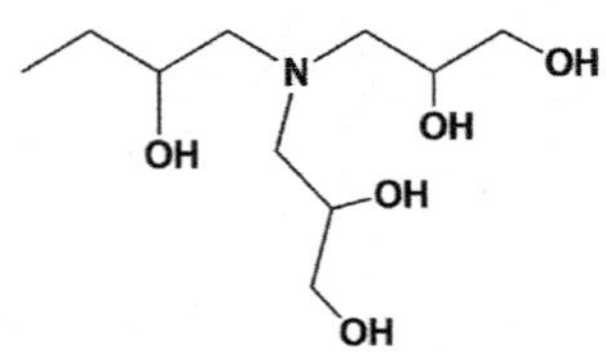 **Amino bis(propane-*cis*-2,3 diol)**	~1.73 (pH: 4.1–8.0)	~1.0	2 M HCl (53.5%)	29

Cross-linked chitosan beads	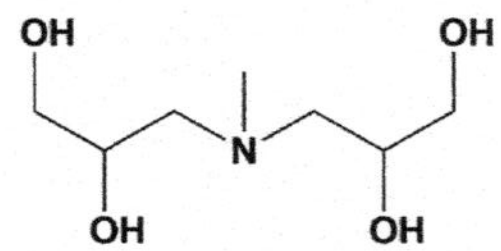	2.2 (pH: 4–8.0)	0.956	4 M HCl (91.3%)	35
Cross-liked polyethyleneimine (PEI) resin	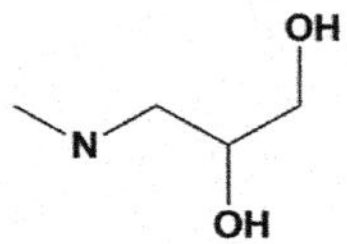	1.23 (pH: 6–8)	N.A	1.0 M HCl	36

NA refers to "Not Available".

loading capacity of the resulting resin is somewhat low (Table 7.4, second row), its efficiency and regeneration behavior was similar. Interestingly, *cis*-propane diol groups generated on PGMA beads by reacting with diallyl amine followed by *cis*-dihydroxylation showed a boron-loading capacity of 1.73 mmol/g and a 100% chelating efficiency. However, boron recovery with 2 M HCl was almost ineffective due to strong binding of boric acid with two *cis*-diol units (Table 7.4, third row). Although freely rotating hydoxy groups of IBP groups generated by glycidol reaction are expected to form also *cis*-diol-boron complex, strong chelation in the latter case contradicts this possibility.

IBP function has also been incorporated into post-cross-linked chitosan beads. The resulting material has been demonstrated to show a boron-loading capacity of 2.2 mmol/g and similar regeneration behavior (Table 7.4, fourth row). Mishra et al. described glycidol modified PEI resin, in which each amine group bears one propane diol group, instead of two.[37] This material was demonstrated to have a reasonable boron chelation ability (1.23 mmol/g) and similar regeneration behavior.

Those examples show that polymer-bound amine groups bearing one or two propane diol groups act as boron chelators.

7.8 IBP FUNCTIONAL SURFACE BRUSHES TETHERED TO CROSS-LINKED POLYMER MICROSPHERES

Recent developments in controlled/living polymerization techniques have led graft copolymerization from solid surfaces. There appear many reports dealing with grafting from solid surfaces using atom transfer radical polymerization (ATRP), reversible addition-fragmentation chain transfer polymerization, and stable free radical polymerization techniques.[37] Among these techniques ATRP has been found especially fruitful process for grafting from solid surfaces.[38] The use of brushes generated on the planar or bead surface as ligand or catalyst carrier has been considered as extremely beneficial. In such topologies ligands or functional groups on the brushes are expected to react fast due to flexibilities of the graft chains. In another words, the reactions of the functional groups on the brushes must take place according to nearly homogenous kinetics. Combination of this behavior with insolubility of the carrier polymer makes it possible to carry out the solid-phase reactions faster, provided that the solvent used is a good solvent for the free polymers of the brushes.

The use of surface brushes as functional group carriers, in general seems to be straightforward due to easy access of those structures by surface initiated-ATRP. The most crucial step of surface grafting with ATRP is generating active alkyl halides as initiating sites on the solid surfaces. One common way of anchoring alkyl halides is reaction of hydroxyl functional surfaces with bromo *iso*-butyryl bromide. An ester connecting group at this initiating site is not desirable due to the hydrolysis susceptibility of this group, especially when the surface brushes are considered as the functional group carrier. To avoid this drawback the author's group incorporated bromoacetyl groups onto PS-DVB microspheres.[39]

The polymer brushes generated by ATRP initiation from these groups must be linked to the surfaces with hydrolysis-proof linkages. Hydrolysis resistance of the connecting units allows processing of the polymers in harsh conditions (Table 7.5).

This approach has been employed in the design of new boron-chelating polymers constituted with a microspherical core and IBP functional brushes. The first example of this approach was reported in 2008.[40] In that report 2-hydroxyethylamino 2,3-propanediol was directly anchored to PGMA graft chains on cross-linked PS-DVB microspheres. Multihydroxy alkylamine functions so generated on the graft chains were demonstrated to bind boric acid selectively; similar to the ordinary bead polymers with the same chelating units tethered directly to the particle surfaces. However, the time span for the boron sorption of this polymer was less than 2 min, which is considerably shorter than for the ordinary bead polymers (ca. 30 min). Obviously, fast boron uptake of the polymer is due to partial mobility of dangling graft chains. Later on, it was noticed that initial boron uptake capacity of this material reduces from 3.28 mmol down to 1−1.2 mmol per gram, upon standing. The reason for such an unexpected capacity loss is currently ambiguous.

Recently, the Kabay and Tuncel group reported synthesis of monodisperse porous particles bearing dextran brushes[9] either via alkyne-azide click chemistry or by direct coupling with epoxy groups of PGMA resin beads. Maximum boron adsorptions of the dextran brushes formed by "grafting onto approach" were found to be 1.8 and 0.5 mgB/g resin for the particles prepared by click chemistry and direct attachment, respectively. Both resins were demonstrated to be regenerable. Although capacities of the resins are somewhat low, selective boron binding of the hydroxyl groups of dextran units is significant.

Table 7.5 Structural and Boron-Binding Characteristics of Some Hairy Grafts Tethered to Solid Microspheres

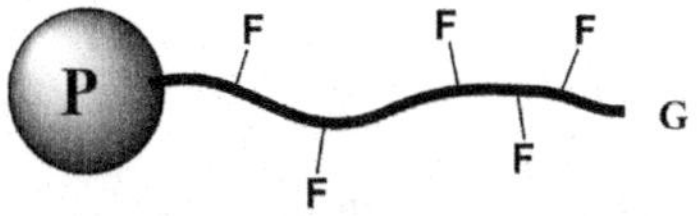

P	G	F	Boron Uptake Capacity	Efficiency	Eq. Time	References
Polystyrene-divinyl benzene (PS-DVB)	**PGMA**	**HEP[a]**	**3.20 mmol/g**	**0.95−1.0**	**<2 min**	**40**
P(GMA-EGDMA)	G + F: Dextran		0.5−1.8 mgB/g	NA	NA	9
PS-DVB	Poly(vinyl amine)	IBP	2.9−4.0 mmol/g	0.76−0.98	<1 min	5

PGMA, poly(glycidyl methacrylate).
NA refers to "Not Available".
[a]HEP denotes 2-hydroxyethyliminopropanediol.

The last significant development in this direction is the design and use of PS-DVB resin beads with dense IBP functional hairy grafts.[5] In this strategy, NVF monomer has been polymerized together with DADMAC from dithiocarbamate surface groups on PS-DVB microspheres. The grafting process carried out by photoiniferter technique using dithiocarbamate groups yields 13.5−15.1 times of mass increases.

Amino groups formed by hydrolysis of the resulting formamido groups were then reacted with glycidol to give boron-chelating IBP functions. Topological view of the resulting resins is depicted in Figure 7.12. Roles of the components involved in such structure can be summarized as follows:

- PS-DVB microsphere is inert carrier that provides easy separation by simple filtration.
- The linker is hydrolysis-proof that allows processing of the graft chains in acid or base medium
- Flexibility of the graft chains provides quasi-homogenous reaction kinetics for the boron chelation with IBP groups.
- IBP group selectively reacts with boric acid or borates and forms one-to-one complex within water.
- DADMAC segments having quaternary amine groups impart hydrophilicity and act as chain expander for the graft chains.

Boron sorption capacities of the resulting resins are proportional to IBP group densities (3.8−4.2 mmol/g) estimated by acid titration (Table 7.6), but almost equal for the resin with 20% DADMAC.

Careful assignment of the boron sorption efficiencies have shown them to be proportional to the DADMAC content, so that the resin sample with 20% of DADMAC shows nearly 100% boron-chelating efficiency in nonbuffered solutions. Also, the highest capacity ever reported (4.0 mmol/g), has been attained by using this resin

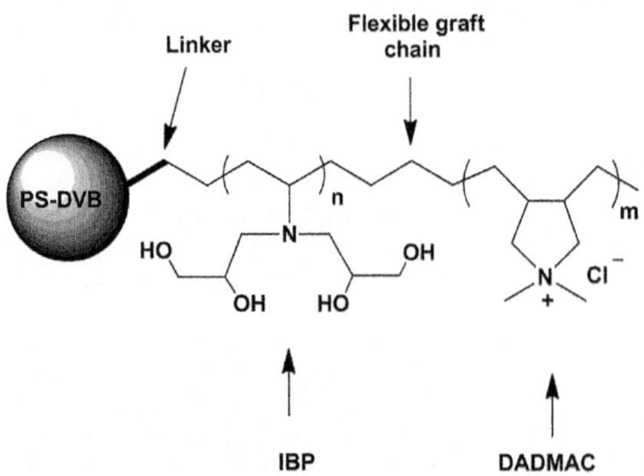

Figure 7.12 Schematic view of IBP functional polymer brushes tethered to polystyrene-divinyl benzene (PS-DVB) microspheres.[5] DADMAC, diallyl dimethyl ammonium chloride.

Table 7.6 Comparison of Boron-Binding Characteristics of the Resins with IBP Functional Hairy Grafts Possessing Various Percent of DADMAC

DADMAC Content[a] (%)	IBP Content[b]	Boron Uptake (mmol/g)	Efficiency[c]
0.0	3.8	2.9 ± 0.1	0.763
10.0	4.2	3.7 ± 0.1	0.881
20.0	4.1	4.0 ± 0.1	0.976

DADMAC, diallyl dimethyl ammonium chloride.
[a]In moles of DADMAC per mole total grafted monomer.
[b]Estimated by acid titration.
[c]Efficiency was defined as moles of boron uptake per mole of IBP function.

material. This can be ascribed to the chain-expanding effect of DADMAC segments of the surface brushes. In another words, mutual electrostatic repulsion of the quaternary groups makes the graft chains strained, so that all the IBP functions become accessible for the chelation with boron.

It is important to note that equilibrations of the boron loading are established in less than a minute for all the resin samples. This is considerably short time in comparison to those for the ordinary resin beads (ca. 30—35 min). Not having hydrolysable chemical bonds, these resins can be regenerated easily using 10% HCl solution without capacity loss.

New topology attained by special design in this report is very attractive and promising due to its high and fast boron uptakes. Such topology might be of interest for large-scale boron removal processes. This can be realized by an improved procedure for surface initiated polymerization of NVF, instead of the photoiniferter technique employed.

7.9 BORON BINDING SELECTIVITY OF IBP AND RELATED FUNCTIONS: EFFECT OF FOREIGN IONS

Possessing tertiary amino group, IBP function can be considered as electron-donating ligand, which tends to form coordinative bonds with transition metal ions to give donor—acceptor complexes or so-called Werner complexes. Ethylenediamine and its alkylated derivatives are well-known ligands forming chelate complexes with various metal ions. Taking this into consideration IBP can be considered also as an electron donating ligand forming metal complexes with accompanying foreign metal ions during boron sorption.

In principle, this phenomenon is expected to reduce the boron-binding capacity of IBP functional polymers. The literature directed to foreign effect on boron chelation of IBP function is scarce. Simplest estimation of this effect is conducting the boron sorption experiments in the presence of the foreign ions. It was demonstrated that, boron-binding capacity of IBP functional groups (3 mmol/g) generated on GMA-MMA-DVB terpolymer beads[6] remains almost constant in the presence of Ca(II), Mg(II), Fe(III) ions in 0.2 M concentrations (Table 7.7).

Table 7.7 Effect of Foreign Ions on Boron Binding of IBP-like Groups Tethered to Various Supports

Boron-Chelating Polymer	H_3BO_3	Foreign Ion	Foreign Ion Sorption (mmol/g)	Boron Capacity (mmol/g)	References
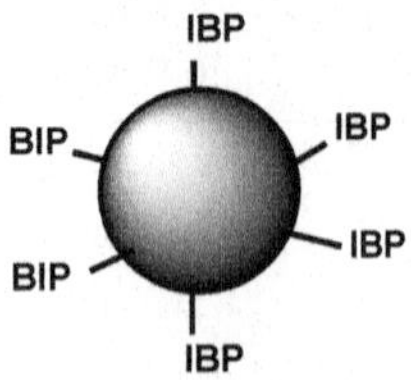	0.41 M	–	–	3.0	6
		0.14 M Ca(II)	0.3	3.0	
		0.14 M Mg(II)	0.2	3.2	
		0.14 M Fe(III)	0.8	nd	
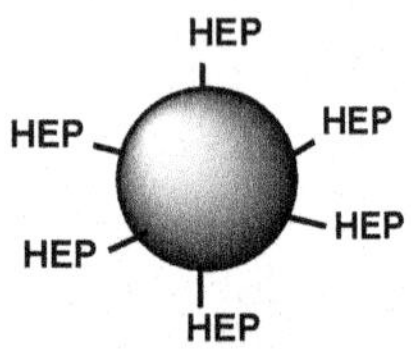	0.41 M	–	–	1.60	8
		0.1 M Ca(II)	0.13	1.60	
		0.1 M Mg(II)	0.1	1.62	
		0.1 M Fe(III)	0.34	1.55	
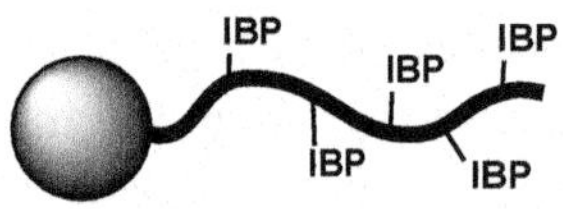	0.166 M	–	–	3.70	5
		0.1 M Ca(II)		3.65	
		0.1 M Mg(II)		3.80	
		0.1 M Na(I)		3.20	

nd refers to "not determined".

By contrast, a slight capacity increment (from 3 to 3.1 mmol/g) has been observed for the case of Mg(II). Concentrations of these ions slightly decrease in the presence of boric acid, indicating 0.2–0.8/mmol of metal ion sorption in addition to boron sorption. In fact these metal ions form insoluble hydroxides by the action of tertiary amino group of IBP function, rather than coordination, due to oxophilic nature these ions. The precipitated hydroxides of iron, calcium, and magnesium are known as natural inorganic sorbents for boron as reported many times in the literature. Therefore, presence of those common ions cannot be considered as interfering entities.

Similar behavior has been reported for the case of 2-hyroxyethyiminopropanediol (HEP) functional resin beads.[8] In this case boron-binding capacity of the resin showed a slight decrease (from 1.62 to 1.55 mmol/g) in the presence of 0.2 M Fe(III). Foreign ion sorption of this resin has been found to be 0.1–0.34 mmol/g.

Moreover, boron-loading capacity of the resin with IBP functional surface brushes has been demonstrated to be unchanged in the presence of Ca(II) or Mg(II) ions in 0.1 M concentrations.[5] Similarly, slight depletions in their concentrations revealed 0.15 and 0.46 mmol of virtual Ca(II) and Mg(II) sorption, respectively.

However, a significant decrease in the boron-binding capacity has been detected for the case of NaCl, so that in 0.1 M NaCl solutions, the original capacity of the resin material decreases from 3.7 to 3.2 mmol/g. Such a capacity loss (13.5%) can be ascribed to a typical "salt effect." Since salinity of sea water is about 0.1 M, boron-binding capacity of this resin material would be 13.5% less, in comparison to the original capacity.

As a result, presence of Ca(II), Mg(II), and Fe(III) ions does not bring negative interfering effect on boron-chelating abilities of IBP and HEP functions practically. Despite a capacity loss in salt water, boron chelating of IBP function is still high enough and can be considered as selective.

7.10 CONCLUDING REMARKS

In this report, iminobis-alkylene diol functional group was highlighted as an efficient alternative boron-chelating unit that can be fabricated in various topologies. The presented synthesis routes were successful in tuning the properties of the final product. The sorbents possessing IBP function showed remarkably high boron-binding capacities. In addition, it was demonstrated that capability of these structures for easy boron desorption could be attained by the synthesis of *cis*-isomer. This easy decomplexation ability of IBP, along with the chemically inert nature of the synthesized products, enables their use in numerous cycles of sorption–desorption without any degradation. Furthermore, the polymers presented with IBP functionality showed high selectivity to boron binding regardless of the topology.

REFERENCES

1. Kabay N, Güler E, Bryjak M. Boron in seawater and methods for its separation - a review. *Desalination* 2010;**261**:212–7.
2. Hilal N, Kim GJ, Somerfield C. Boron removal from saline water: a comprehensive review. *Desalination* 2011;**273**:23–35.
3. Wolska J, Bryjak M. Methods for boron removal from aqueous solutions – a review. *Desalination* 2013;**310**:18–24.
4. Sanderson BR. *Removal of boron from solution: an examination of the co-removal of boron with polyvinyl alcohol.* Technical Report. TR-77-16. London: Borax Technical, Ltd.; 1977.
5. Ince A, Karagoz B, Bicak N. Solid tethered imino-bis-propanediol and quaternary amine functional copolymer brushes for rapid extraction of trace boron. *Desalination* 2013;**310**:60–6.
6. Senkal BF, Bicak N. Polymer supported iminodipropylene glycol functions for removal of boron. *React Funct Polym* 2003;**55**:27–33.
7. Sidney S. *Quantitative organic analysis.* 3rd ed. New York: Wiley; 1967.
8. Gazi M, Bicak N. Selective boron extraction by polymer supported 2-hydroxylethylamino propylene glycol functions. *React Funct Polym* 2007;**67**:936–42.
9. Samatya S, Orhan E, Kabay N, Tuncel A. Comparative boron removal performance of monodisperse-porous particles with molecular brushes via "click chemistry" and direct coupling. *Colloids Surf A* 2010;**372**:102–6.
10. Doğanay CO, Ozbelge HO, Bıçak N, Aydoğan N, Yılmaz L. Use of specially tailored chelating polymers for boron removal from aqueous solutions by polymer enhanced ultrafiltration. *Separ Sci Technol* 2011;**46**:581–91.

11. Semen J, Lando JB. The acid hydrolysis of isotactic and syndiotactic poly (methyl methacrylate). *Macromolecules* 1969;**2**:570–5.
12. Jiao AY, Li ZS, Bao LC. Poly (acrylic acid-co-maleic acid) for the enhanced treatment of Cu(II)-loaded aqueous solution and its reuse by ultrafiltration-electrolytic process. *Desalination* 2013;**322**:29–36.
13. Palencia M, Rivas BL, Pereira E. Metal ion recovery by polymer-enhanced ultrafiltration using poly (vinyl sulfonic acid): fouling description and membrane-metal ion interaction. *J Membr Sci* 2009;**345**:191–200.
14. Zamariotto D, Lakard B, Fievet P, Rouge NF. Retention of Cu(II)- and Ni(II)- polyaminocarboxylate complexes by ultrafiltration assisted with polyamines. *Desalination* 2010;**258**:87–92.
15. Canizares P, Perez A, Camarillo R, Villajos MT. Improvement and modelling of a batch polyelectrolyte enhanced ultrafiltration process for the recovery of copper. *Desalination* 2005;**184**: 357–66.
16. Canizares P, Perez A, Camarillo R, Llanos J, Lopez ML. Selective separation of Pb from hard water by semi-continuous polymer-enhanced ultrafiltration process (PEUF). *Desalination* 2007;**206**:602–13.
17. Aroua MK, Zuki FM, Sulaiman NM. Removal of chromium ions from aqueous solutions by polymer-enhanced ultrafiltration. *J Hazard Mater* 2007;**147**(3):752–8.
18. Chikhi M, Meniai AH, Meterfi S, Khelfaoui A, Nedjar Z. Experimental and theoretical heavy metal complexation prior to elimination by ultrafiltration. *Desalination* 2008;**229**:342–7.
19. Rivas BL, Aguirra MDC. Water-soluble polymers: optimization of arsenate species retention by ultrafiltration. *J Appl Polym Sci* 2009;**112**:2327–33.
20. Uludag Y, Özbelge HÖ, Yilmaz L. Removal of mercury from aqueous solutions via polymer-enhanced ultrafiltration. *J Membr Sci* 1997;**129**:93–9.
21. Muslehiddinoglu J, Uludag Y, Özbelge HÖ, Yilmaz L. Effect of operating parameters on selective separation of heavy metals from binary mixtures via polymer enhanced ultrafiltration. *J Membr Sci* 1998;**140**:251.
22. Volchek K, Krentsel E, Zhilin Y, Shtereva G, Dytnersky Y. Polymer binding/ultrafiltration as a method for concentration and separation of metals. *J Membr Sci* 1993;**79**(2–3):253–72.
23. Smith BM, Todd P, Bowman CN. Hyperbranched chelating polymers for the polymer-assisted ultrafiltration of boric acid. *Separ Sci Technol* 1999;**34**(10):1925–45.
24. Dilek Ç, Özbelge HÖ, Bıçak N, Yılmaz L. Removal of boron from aqueous solutions by continuous polymer-enhanced ultrafiltration with polyvinyl alcohol. *Separ Sci Technol* 2002;**37**(6):1257–71.
25. Smith BF, Robison TW, Carlson BJ, Labouriau A, Khalsa GRK, Schroeder NC, et al. Boric acid recovery using polymer filtration: studies with alkyl monool, diol, and triol containing polyethylenimines. *J Appl Polym Sci* 2005;**97**:1590–604.
26. Zerze H, Karagoz B, Ozbelge HO, Bicak N, Aydogan N, Yilmaz L. Imino-bis-propane diol functional polymer for efficient boron removal from aqueous solutions via continuous PEUF process. *Desalination* 2013;**310**:158–68.
27. Zerze H, Ozbelge HO, Bicak N, Aydogan N, Yilmaz L. Novel boron specific copolymers with quaternary amine segments for efficient boron removal via PEUF. *Desalination* 2013;**310**:169–79.
28. Yurum A, Taralp A, Bicak N, Ozbelge HO, Yilmaz L. High performance ligands for the removal of aqueous boron species by continuous polymer enhanced ultrafiltration. *Desalination* 2013;**320**:33–9.
29. Bicak N, Gazi M, Senkal BF. Polymer supported amino bis-(*cis*-propane 2, 3 diol) functions for removal of trace boron from water. *React Funct Polym* 2005;**65**:143–8.
30. Sinton SW. Complexation chemistry of sodium borate with poly (vinyl alcohol) and small diols: a boron-11 NMR study. *Macromolecules* 1987;**20**(10):2430–41.
31. Zerze H. *Boron removal from aqueous solutions via polymer enhanced ultrafiltration using novel imino-bis-propane diol functional polymers* [Ph.D. thesis]. Ankara: Middle East Technical University; 2012.
32. Bicak N, Ozbelge HO, Yilmaz L, Senkal BF. Cross-linked polymer gels for boron extraction derived from *N*-glucidol-*N*-methyl-2-hydroxypropyl methacrylate. *Macromol Chem Phys* 2000;**201**:577–84.
33. Dowding PJ, Vincent B. Suspension polymerization to form polymer beads. *Colloids Surf A* 2000;**161**:259–69.

34. Sherrington DC. Preparation, structure and morphology of polymer supports. *Chem Commun* 1998; **21**:2275–86.
35. Gazi M, Shahmohammadi S. Removal of trace boron from aqueous solution using iminobis-(propylene glycol) modified chitosan beads. *React Funct Polym* 2012;**72**:680–6.
36. Mishra H, Yu C, Chen DP, Goddard WA, Dalleska NF, Hoffmann MR, et al. Branched polymeric media: boron-chelating resins from hyperbranched polyethylenimine. *Environ Sci Technol* 2012;**46**:8998–9004.
37. Braunecker WA, Matyjaszewski K. Controlled/living radical polymerization: features, developments and perspectives. *Prog Polym Sci* 2007;**32**:93–146.
38. Ejaz M, Ohno K, Tsujii Y, Fukuda T. Controlled grafting of a well-defined glycopolymer on a solid surface by surface-initiated atom transfer radical polymerization. *Macromolecules* 2000;**33**:2870–4.
39. Karagoz B, Bayramoglu G, Altintas B, Bicak N, Arica MY. Poly(glycidylmethacrylate)-polystyrene copolymer grafted nano-composite P(S-DVB) microspheres from surface initiated-atom transfer radical polymerization: immobilization of lipase and application in esters synthesis. *Ind Eng Chem Res* 2010;**49**:9655–65.
40. Gazi M, Galli G, Bicak N. The rapid boron uptake by multi-hydroxyl functional hairy polymers. *Sep Purif Technol* 2008;**62**:484–8.

CHAPTER 8

Boron Removal Using Membranes

Viatcheslav (Slava) Freger, Hilla Shemer, Abraham (Avi) Sagiv, Raphael (Rafi) Semiat
Chemical Engineering Department, Technion—Israel Institute of Technology, Haifa, Israel

8.1 INTRODUCTION

Boron is widely distributed in the environment, occurring naturally or coming from anthropogenic contamination, mainly in the form of boric acid or borate salts. It can be found in seawater (5 mg/L as boron), drinking and brackish waters, or in treated wastewater (effluent). The amount of boron in fresh water depends on the geochemical nature of the drainage area, proximity to marine coastal regions, and pollution by industrial and municipal effluents.[1]

A low concentration of boron is required for certain metabolic activities in humans and plants. However, at slightly higher concentrations it can be quite toxic to some ornamental plants. Therefore, boron is regulated in drinking water and treated wastewaters.

For drinking water, the European Union defined an upper limit of 1 mg/L boron and the World Health Organization (WHO) recommended limits of 0.3 mg/L. In 2011, the Drinking Water Quality Committee of WHO revised the boron guideline value to 2.4 mg/L based on human health perspective.[2] Some utilities may set seawater desalination plant product water limits as low as 0.3 mg/L to reflect agriculture-related issues. It is realized that the permitted level of boron varies between countries and regions.

Occurrence of boron in natural and wastewaters dictates the need to remove it from these aqueous media. This chapter focuses on removal of boron by desalination processes. Both thermal (evaporation) and membrane techniques are used in desalination. The evaporation methods include multistage flash evaporation, multieffect distillation, and vapor compression (thermal or mechanical), while the relevant membrane techniques are reverse osmosis (RO) and electrodialysis (ED). Since the thermal techniques are based on water evaporation, they are not sensitive to the dissolved solids in the feed water hence, produce clean water without boron.

A single-stage RO desalination unit can decrease the boron concentration in seawater (4–5 mg/L) to about 0.9–1.8 mg/L in the permeate, depending on the recovery ratio. In some cases even a two-stage RO process operated at normal conditions is insufficient for reducing boron concentrations to meet the requirements for drinking water quality.[1–8] This is due to the small size of the boric acid molecule that allows it to diffuse through the membranes as a nonionic species, similar to carbonic acid or water.[6] Borate

Boron Separation Processes
ISBN 978-0-444-63454-2, http://dx.doi.org/10.1016/B978-0-444-63454-2.00008-3

ions are much better rejected by the membrane due to its charge and size, but their fraction is negligible below pH 8.

There are several methods used in seawater desalination to manage the boron concentration in the product water[9–11]:

1. Use of improved membranes with higher boron rejection. Common RO membranes reject boron to a level of about 60–96% while the rejection of seawater salt is above 99.6%.[1,6,7] All major manufacturers of RO membranes have invested significant efforts in developing new tight grades of desalination membranes having boron rejection as high as 93–96%. Table 8.1 displays the performance of representative commercial membranes of both regular and high-rejection types.
2. Increasing the dissociation of boric acid by increasing the pH with caustic soda (or other alkaline agents) prior to seawater reverse osmosis (SWRO). Then re-acidify the treated water to adjust it to the desired pH.
3. Utilizing a second-pass in reverse osmosis systems, potentially coupled with pH adjustment.
4. Passing the feed water through an ion exchanger (IX) that removes the boron.
5. Passing the desalinated water through a series of brackish water reverse osmosis (BWRO) membrane at high pH treatment.
6. Adding an ED stage after RO treatment.

Table 8.1 Performance of some Commercial 8″ RO Membrane Elements for Seawater Desalination. Information Obtained from the Manufacturers' Web Sites[a]

Manufacturer	Product Type	Nominal Area, ft^2, (m^2)	Permeate Flow, gallon per day (gpd), (m^3/day)	Salt Rejection, %	B Rejection, %
Dow-Filmtec[b]	SW30HR-380	380 (35.3)	6000 (22.7)	99.70	n/a
	SW30XHR-440i	440 (41)	6600 (25.0)	99.82	93
	SW30HR LE-400i	400 (37.2)	7500 (28.4)	99.75	91
	SW30HR-320	320 (29.7)	6000 (22.7)	99.75	91
	SW30XLE-400i	400 (37.2)	9000 (34.1)	99.70	88
Nitto-Hydranautics[c]	SWC4B MAX	440 (40.8)	7200 (27.3)	99.8	95
	SWC4B	400 (37.2)	6500 (24.6)	99.8	95
	SWC4+		6500 (24.6)	99.8	93
	SWC3+		7000 (26.5)	99.83	92
	SWC5		9000 (34.1)	99.83	92
Toray[b]	TM820K-440	440 (41)	6400 (24.2)	99.86	96
	TM820A-370	370 (34)	5500 (21)	99.80	94–96
	TM820-370	370 (34)	6000 (23)	99.75	91–93
	SU-820	315 (29)	5100 (19)	99.75	91–93
	TM820H-370	370 (34)	6000 (23)	99.8	91–93
	SU-820 BCM	315 (29)	6000 (23)	99.83	91–93

[a]Typical or stabilized values.
[b]Testing conditions: feed 32,000 mg/L NaCl, 5 mg/L B, pressure 800 psi (5.5 MPa), 77 °F (25 °C), pH 8, 8% recovery.
[c]Testing conditions: feed 32,000 mg/L TDS, 5 mg/L B, pressure = 800 psi, 25 °C, pH 7, 10% recovery

7. Adding an adsorption stage, complexing molecules, or ion exchange stage followed by an ultrafiltration (UF)/microfiltration (MF) membrane to remove adsorbent and boron complexes from the solution.

Most of the current solutions specified above cost between 4 and 8 US ¢/m^3. This calls for new and improved methods that would rely either on improved techniques or integration of existing methods.

8.2 BORON REJECTION BY RO MEMBRANES

The removal of boron by RO is affected by many factors, such as, temperature, pressure, pH, feed flow rate, feed salinity or ionic strength, an initial boron concentration, and recovery ratio. The effect of these parameters is independent of the membrane type.

Boron occurs predominantly in the form of boric acid in the aquatic environment. Boric acid is a weak acid that in water dissociates according to

$$H_3BO_3 \leftrightarrow H^+ + H_2BO_3^- \; pK_a \; 9.14 \tag{8.1}$$

$$H_2BO_3^- \leftrightarrow H^+ + HBO_3^{-2} \; pK_2 \; 12.74 \tag{8.2}$$

$$HBO_3^{-2} \leftrightarrow H^+ + BO_3^{-3} \; pK_3 \; 13.8 \tag{8.3}$$

The concentration is expressed as total boron (*tB*), which includes all species in terms of boron weight in unit water volume:

$$tB = [H_3BO_3] + [H_2BO_3^-] + [HBO_3^{-2}] + [BO_3^{-3}] \; mg/L \tag{8.4}$$

The speciation of borate, according to dissociation reactions in Eqns. (8.1)–(8.3) is presented in Figure 8.1 as function of the pH. In the pH range of interest for drinking

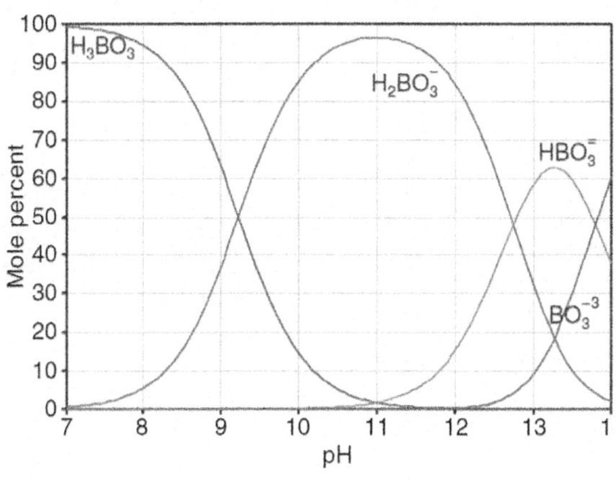

Figure 8.1 Speciation diagram of boric acid.[12] Dissociation of neutral boric acid is low at low pH. The size of the boric acid molecule is small in comparison with the borate ion that due to its negative charge attracts positive hydronium ions. Therefore, mostly borates can be rejected by an RO membrane, so only at high pH, above 10, the rejection of the boron species approaches 100%.

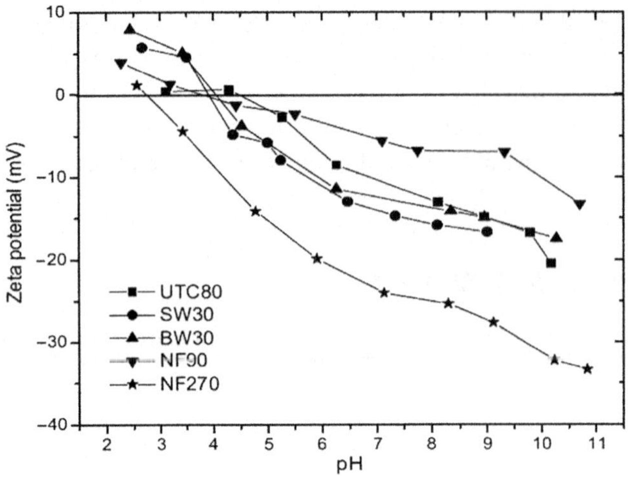

Figure 8.2 Zeta-potential of selected membranes, deduced from the tangential streaming potential measured at 25 °C in a background electrolyte solution of 10 mM KCl; pH was adjusted using HCl and KOH.[17]

water systems the borate ions concentration present in Eqns. (8.2) and (8.3) is low and most of the boron is present as a nonionic species, boric acid. The ionized borate species are highly rejected by the membrane due to their charge (see next section), whereas boric acid, a small uncharged molecule, readily diffuses through the membrane. Therefore, higher rejection of boron by RO membranes is observed only at pH ≥ 8, i.e., the pH at which boric acid transforms to borate ion.[1–8,13–15]

The pH affects not only the boron speciation but also the surface characters of the membrane and, in particular, membrane charge and zeta potential, as shown in Figures 8.2 and 8.3 and Table 8.2. Zeta potential (Figure 8.2) indicates that the membranes

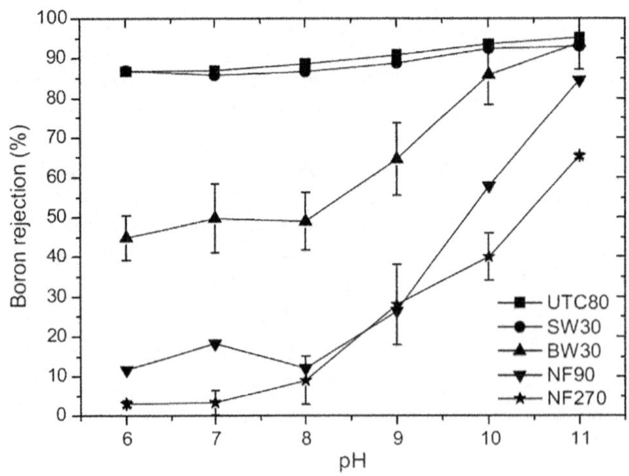

Figure 8.3 Boron rejection by different commercial membranes as a function of solution pH. Feed solution contains 10 mM NaCl, 1 mM $CaCl_2$, 1 mM $NaHCO_3$, and 0.43 mM$B(OH)_3$ (4.6 mg/L as B). Feed temperature −20 °C, permeate flux 40 L/m^2 h, cross-flow velocity 30.4 cm/s. Four repetitions were made on each point.[16]

Table 8.2 Boron Rejection Utilizing the Boron Rejection Capabilities of Hydranautics SWC Membrane[18]

Source	Feed pH	Temp. °C	Recovery	Flux GFD[a]	Boron Rejection %
Mediterranean	6.9	24.0	49	8.1	89
Pacific	7.7	22.2	50	7.8	85
Pacific	8.4	31.4	50	8.0	85
Gulf of Mexico	6.6	34.3	61	8.8	78
Mediterranean	6.6	21.0	40	7.2	90

[a]GFD comes for Gallons per square Feet per Day. Each GFD equal to 0.57 LMH (Liter per square Meter per Hour)

become increasingly negatively charged as pH increases.[16,19] A larger negative charge leads to a higher rejection due to an enhanced electrostatic interaction between the negative charges in membrane and the borate anions. This is seen in Figure 8.3 showing an increased boron rejection at elevated pH. The best results are obtained at pH 10.5. Therefore, in practice, when it is economically justifiable, increasing pH is an established method to increase boron removal by RO membranes.[20,21]

The efficiency of boron rejection by various nanofiltration (NF) and RO membranes and feed water qualities was evaluated in Refs. 22–27. Table 8.2 displays results on boron rejection of various feed water sources and operating conditions using Hydranautics SWC membrane. As seen, the rejection achieved was between 78% and 90%.

The treated water pH may also affect the lifetime of the membrane. RO membranes deteriorate with the operation time and their wear is accelerated at high pH.[8] Therefore, the changes of RO membrane lifetime needs to be taken in account when considering the parameters affecting boron permeability.

Temperature and ionic strength show a negligible effect on boron speciation with pH according to Eqn. (8.1) (Figure 8.4). Yet, they can affect the RO membrane performance. For example, boron removal decreases with increased salinity due to the effect of screening of the membrane charge and surface potential at high salinity.[28] Increased temperature also results in lower boron rejection. Although the total flux through RO membranes increases with temperature, the solute exclusion becomes weaker and the boron permeation increases more significantly than the water flux (cf. Eqn. (8.5) in the next section).

8.3 REJECTION MECHANISM AND MEMBRANE DEVELOPMENT FOR IMPROVED BORON REJECTION

The rejection of a solute is in general determined by its permeability ω_s (sometimes designated as B), which may be expressed as follows[29]:

$$\omega_s = \frac{DK}{\delta}, \tag{8.5}$$

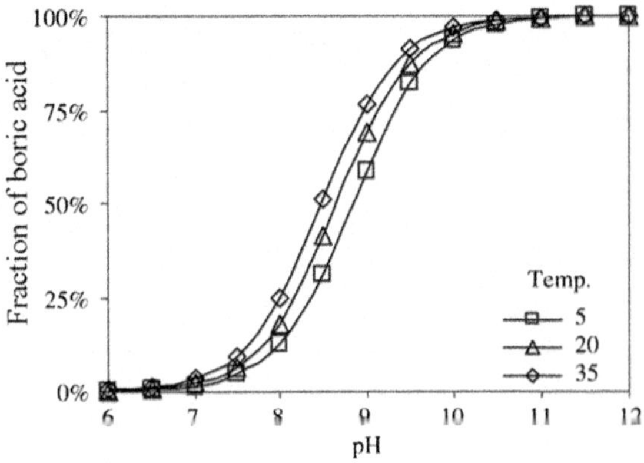

Figure 8.4 Effect of temperature on boric acid dissociation as function of pH.[22]

where, D and K are the diffusion and partitioning coefficients of the solute in the membrane and δ is the membrane (active layer) thickness. For very thin composite membranes used today the convection is small and a reasonable approximation in most cases is the solution–diffusion model, which predicts that the rejection is simply related to ω_s and total volume flux J_V as follows:

$$R = \frac{J_V}{J_V + \omega_s} \tag{8.6}$$

Eqn. (8.5) explicitly shows that a low permeability is required for high rejection. For charged solutes, e.g., the borate ion, the major factor responsible for their low permeability is the small partitioning coefficient K rather than D. The small K, i.e., strong exclusion of the solute from the membrane may be related to the solvation energy as follows

$$K = K_0 \exp\left(-\frac{\Delta W}{kT}\right) \tag{8.7}$$

where, K_0 is the factor that incorporates the entropic effects of steric and Donnan exclusion and ΔW is the difference in solvation energy per molecule between the membrane and water phases. For charged solutes (ions) the K_0 factor may be small due to the fixed charge of the membrane (Donnan exclusion) and the exponential factor is small since the solvation energy W of ions is large in a hydrophobic membrane.[30–32] For this reason hydrophobicity, i.e., a low dielectric constant of the membrane, and, to a lesser extent, fixed charges are important for achieving a high salt rejection.

Unfortunately, these factors become inactive for uncharged boric acid, since its solvation energy W in the membrane loses the large electrostatic component and sharply drops and the Donnan exclusion is switched off. In absence of these mechanisms the small W and the remaining weak steric effects included in K_0 and D (Eqn. (8.4) and (8.6))

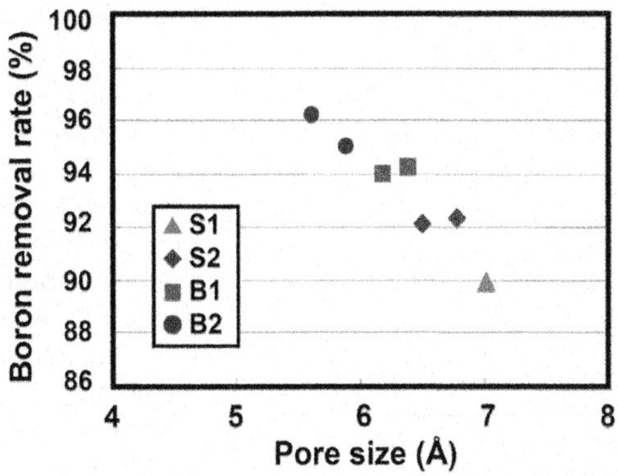

Figure 8.5 The correlation between the boron rejection and average pore size of the polyamide layer in RO membranes probed by positron annihilation lifetime spectroscopy.[34]

are unable to ensure a low solute permeability ω_s. Moreover, the boric acid in water shows an unusual salting-in behavior, i.e., its solubility increases upon addition of salts.[33] This suggests that the fixed charges within the membrane, acting to exclude salts, may actually promote uptake of the neutral boric by the membrane and thus reduce its rejection.

It is seen that the same factors that are beneficial for high salt rejection, which has been the main criterion for optimization of desalination membranes, become inefficient or even detrimental for boric acid. This is further aggravated by the fact that boric acid, rich in OH groups, tends to strongly associate with water thereby it is hard to decouple its permeation from that of water. This explains the formidable challenge of developing membranes that can efficiently pass water yet reject both salt and boron to a high degree around neutral pH.

The main approach taken by most manufacturers in developing "high B rejection" grades of RO membranes is to "tighten" the membranes and increase boron rejection at the expense of reduced permeability. Eqns. (8.4) and (8.5) suggest that "tightening" may be achieved by increasing the thickness δ, but it is also possible to vary the density, pore size, or chemical structure and thus affect D and K. For instance, Kurihara[34] recently demonstrated that the average size of the free volume cavities in polyamide layer probed by positron annihilation lifetime spectroscopy correlates well with boron rejection (Figure 8.5). The pore size, which is essentially the size of the free volume cavities present in the polyamide, affects water permeability as well, yet the effect on boric acid, the larger solute, is stronger. Apparently, manufacturers extensively explore this approach. Figure 8.6, adapted from Ref. 35 indicates that the relation between B rejection and permeability falls on a single trend for "conventional" membranes, as would be in the case when just the thickness is varied. However, a new series of improved membranes shows a different trend, with a significantly higher B rejection for the same flux. This

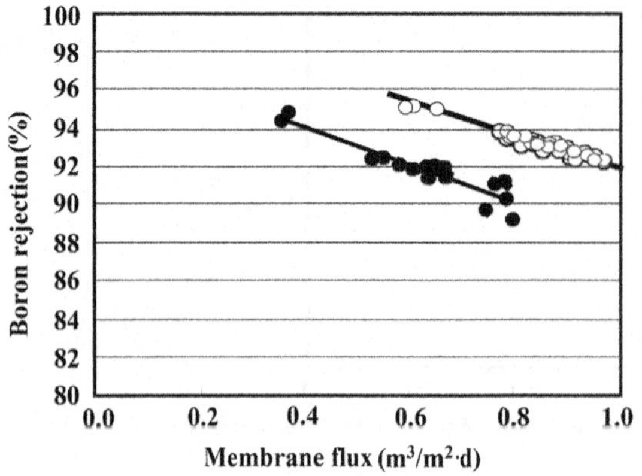

Figure 8.6 The correlation between B rejection and flux in standard testing conditions (feed 32,000 mg/L NaCl, 5 mg/L B, pressure 55 bar, pH 6.5, temperature 25 °C). The membranes produced using an unspecified modified approach (open symbols) show a relation to the flux that is distinctly improved compared to "conventional" RO polyamide membranes (solid symbols). In Table 8.1 Toray's membranes TM820A-370 and TM820H-370 represent such modified and conventional types, respectively.[35]

indicates that there is much room for improving B rejection through improving the membrane structure, in which the permeability is not necessarily sacrificed.

A conceptually different but related approach for improving boron rejection and keeping a high salt rejection was pointed out in Refs. 33,36. It was demonstrated that solubility of boric acid drops faster than solubility of water as the organic medium becomes less polar, i.e., more hydrophobic. This indicates that "tightening" may be also achieved by making the membrane more hydrophobic, again at the expense of water flux. It was proposed to use an in situ surface grafting with mildly hydrophobic monomers for this purpose, which can be done both as part of manufacturing or as on-site tuning of performance through appropriate chemical treatment.[33,36] Modification with a glycidyl methacrylate, which may both graft-polymerize on the surface and form extra bonds through its epoxy groups reacting with amine residues in polyamide, was the most efficient and boron in the permeate could be reduced by about a third. Curiously, this treatment also significantly improved the salt rejection and the overall performance of the modified element was far better than for another commercial BWRO membrane of similar permeability.[33] It is still unclear whether the improvement came from the rejection properties of the extra layer of graft polymer on top of polyamide or from an enhanced performance of polyamide layer itself due to defect plugging, pore size reduction, or a minor change of chemical composition. The available data[33,37] strongly suggest that the main effect comes from improved performance of the polyamide layer. This explains why the treatment was efficient for more open BWRO membranes and less efficient for tight SWRO membranes, which may be already very tight and have much fewer defects to plug.[33]

A few other authors reported modification of commercial membranes for better rejection of boron. A hypochlorite treatment of thin-film composite RO membranes, which is known to increase the salt rejection, probably through increased hydrophobicity, was reported.[38] Their results, however, showed good rejection only at pH 11. Another attempt to improve RO membranes was claimed by Wei and colleagues,[39] yet their results seem more like a development of an adsorptive membrane.

8.4 RO SYSTEMS CONFIGURATIONS FOR BORON REDUCTION

The main stages of the desalination process are described in detail elsewhere.[40] In order to reduce the concentration of boron in RO permeate and comply with the regulatory requirements, several configurations have been developed. These configurations were studied and reviewed extensively[41–47] along with reports on field experience.[12,15,48–50]

Conventional single-pass SWRO unit (Figure 8.7) operates at a recovery rate of 40–50%, permeate flux of 7–9 gallons per square feet per day, and pH ranging from 6.0 to 8.2. Under these conditions the permeate salinity is within potable limits (<500 mg/L total dissolved solids); however, boron content of 0.4–1.0 mg/L cannot be reliably achieved especially at high water temperatures.

In a two-pass system (Figure 8.8) the first unit operates at a recovery rate of 40–50% and the permeate is processed by the second unit at a recovery of 85–90%. The use of a two-pass RO system without pH adjustment was investigated by Farhat and colleagues.[86] Addition of sodium hydroxide prior to the second unit increases the boron rejection to 80–95%. There is a sodium hydroxide addition point prior to the second pass, which increases the feed pH to 9.5–10.5 and thus increases the boron rejection.[85] Although addition of sodium hydroxide significantly increases the boron rejection, it may also enhance the precipitation rate of calcium carbonate and, at pH $\geq$ 10.5, precipitation of magnesium hydroxide as well. One solution to prevent calcium carbonate precipitation is to acidify the feed water by removing the CO_2, prior to dosing of the NaOH. The whole process is, however, costly resulting in a significant increase in the production cost. Furthermore, in this method the possibility of magnesium hydroxide precipitation is not addressed hence this technique is not in use.

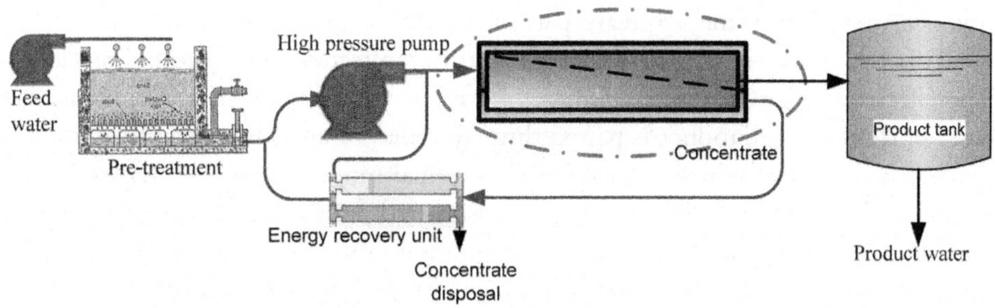

Figure 8.7 Schematic presentation of SWRO desalination system. The RO membranes—the heart of the desalination process—are encircled.

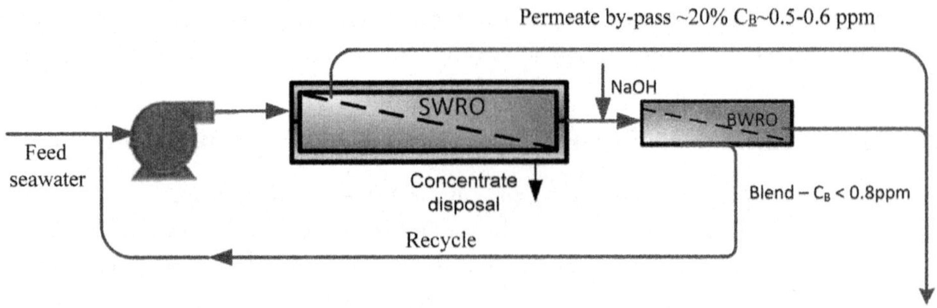

Figure 8.8 Two-pass RO system design.

An approach of minimizing the usage of sodium hydroxide injected in a single pass RO process of Pacific Ocean seawater was shown in Ref. 51. A mechanistic predictive model of boron rejection by RO membranes and the influence of fouling were simulated in Ref. 52.

A hybrid process of RO combined with the use of an ion exchange resin (IER) to remove most of the boron in the form of borate is also used. Resin regeneration may be performed by using the concentrate leaving the plant and the overall costs may be comparable with the RO-only alternative processes. As a result, the two main membranes-based options considered in the industry for boron removal are:

1. Partial treatment of the desalinated water by a set of BWRO units with pH change between units.
2. Partial treatment of the product by parallel systems of ion exchange and a secondary RO membrane.

The permeate produced at the upstream portion of the membrane desalination unit (close to the feed seawater entrance) has a lower boron content than the permeate produced at the downstream portion of the membrane module (close to the concentrate exit). This is due to the fact that the latter is produced from the concentrate containing up to twice the salt concentration and almost twice the boron concentration compared to the feed seawater. Based on this consideration, there are two alternative processes. In both alternatives a small portion of the product (usually around 25%) containing 0.5–0.6 mg/L boron is removed from the upstream part of the RO train, as shown in Figures 8.9 and 8.10. This side stream is not treated further and is transferred directly to the product line for post-treatment (i.e., re-mineralization). In the first method, displayed in Figure 8.9, a second portion of the product is passed through a bed of an IER, exiting it with about 0.1 mg/L boron, after which it is added to the final product line. The remaining third portion is fed to a second RO stage at low pressure and pH elevated to about 10, reaching around 90% recovery. The total boron concentration in the final product depends on the ratio between the three streams and is maintained usually at 0.3–0.4 mg/L boron.

In the second alternative, the side stream with relatively low boron concentration, taken from the upstream part of the membrane train, is used in the same way as in the

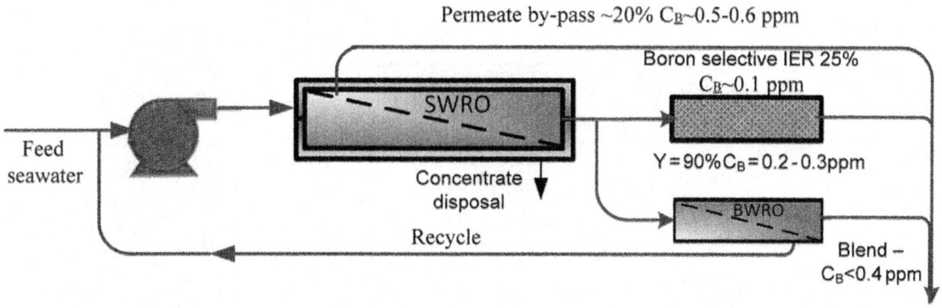

Figure 8.9 Boron removal with ion exchange resin in parallel with BWRO unit.

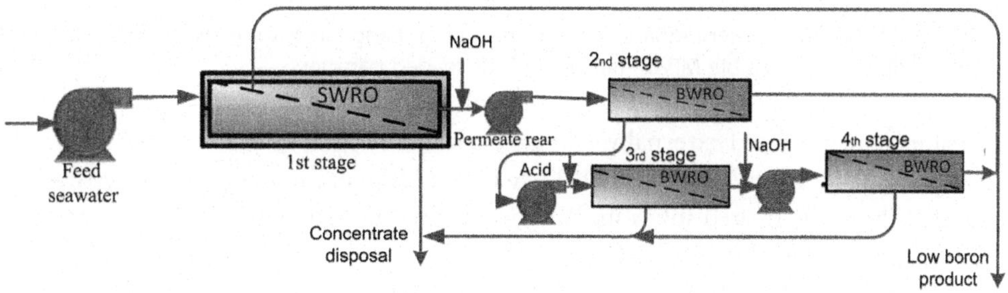

Figure 8.10 Boron removal with three BWRO units at different pH levels.

first technique, as seen in Figure 8.10. Then, the rest of the product, leaving the concentrate side of the membrane modules, is fed to a second-stage BWRO at pH elevated above 10.5 enabling a very high borate rejection. Anti-scalants are used occasionally in order to prevent precipitation of calcium and magnesium salts on the membrane. High recovery is gained at this stage. The concentrate is acidified and enters a third stage consisting of a smaller BWRO unit. The boron content of the permeate leaving the third stage is high since its feed contains about 10 times the boron content of the product leaving the first stage, due to the high recovery of the second stage. Therefore, this stream is further treated in a fourth stage at elevated pH using BWRO unit. The product of this final stage is of good quality in terms of boron content. The industrial application of this four-stage process was reported in Refs. 12,53.

8.5 OTHER POSSIBLE RO/UF/MF TECHNIQUES FOR BORON REMOVAL

Methods of boron removal by different adsorbents have been developed. Examples of such processes are shown schematically in Figures 8.11 and 8.12. The different methods may be divided into two categories: ion exchange and adsorption.

The IX may be used as fixed or fluidized beds. In the latter loaded beds are frequently replaced by regenerated beds. The spent beds are removed from the bottom of a fluidized

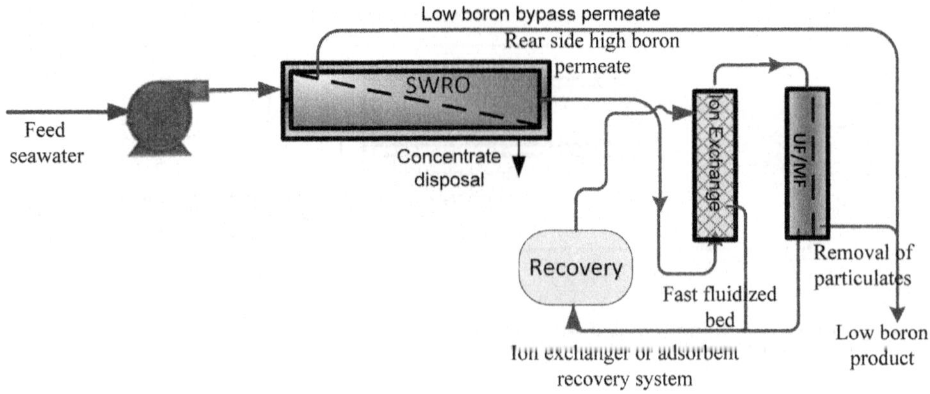

Figure 8.11 Schematic presentation of continuous ion exchange in fixed or fluidized bed, with IX recovery, followed by possible MF/UF filtration of the loaded particles.

bed column, regenerated externally, and fed back at the top of the fluidized column. This technique is demonstrated in Figure 8.11, where a UF membrane is used to separate the spent beds from the treated solution. A hybrid technique where the RO permeate is fed to an ion exchange floating bed followed by UF or MF membrane separation was reported in Refs. 54,55. In this case BWRO membranes were used for removal of high boron content from geothermal water on a pilot scale. In a similar study[56] the BWRO membranes were replaced with SWRO membranes for the same application.

Adsorption of boron may be achieved using molecules, small particles, or even nano-scale particles that exhibit high surface area and hence high adsorption capacity. The adsorption is based on a continuous slurry dosage of the adsorbent with the feed water as shown in Figure 8.12. The adsorbent slurry is then filtered using MF or UF membranes that accumulate the adsorbent. Backwash of the membranes is followed by regeneration of the adsorbent by removing the boron. This option may allow using a small amount of adsorbent/ion exchange particles, as it can be recovered and reused.

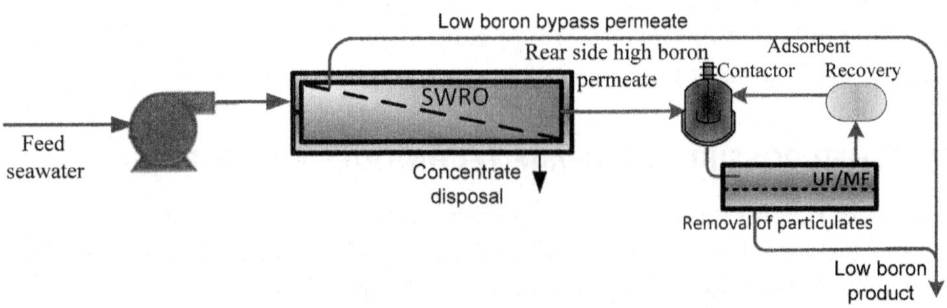

Figure 8.12 Schematic presentation of continuous adsorption/ion exchange in mixed reactor, with adsorbent/ion exchange recovery using MF/UF membranes.

The hybrid processes of adsorption—membrane filtration for boron removal were reviewed in Ref. 57. The suggested processes include iron/RO,[58] activated alumina/RO,[59] ion exchange/RO,[60–64] grounded chelating resin/MF,[65,66] and activated carbon/MF.[67]

New directions of hybrid or stand-alone processes to enhance boron rejection by membranes include:

- Coupling NF/RO membrane separation with complexation between compounds containing multiple hydroxyl groups (polyols) and boric acid. For example, complexation of boric acid with mannitol (a polyol compound) improved the boron rejection by both RO and NF membranes.[68] Similarly, the use of polyvinyl alcohols as chelating agent with UF membranes was shown to increase the boron removal.[69]
- Development of novel hollow fiber membranes that can adsorb boron.[70,71]
- Development of a boron sorption membrane.[26]
- Development of monodisperse porous particles with dextran-based molecular brushes as a sorbent for boron combined with membrane separation.[72]

8.6 BORON REMOVAL BY ELECTRODIALYSIS

Electrodialysis is used for desalination of brackish water. The technique is based on a stack of flat-sheet anion and cation exchange membranes (AEM and CEM, respectively) that form two parallel sets of alternating channels, as can be seen in Figure 8.13. Feed

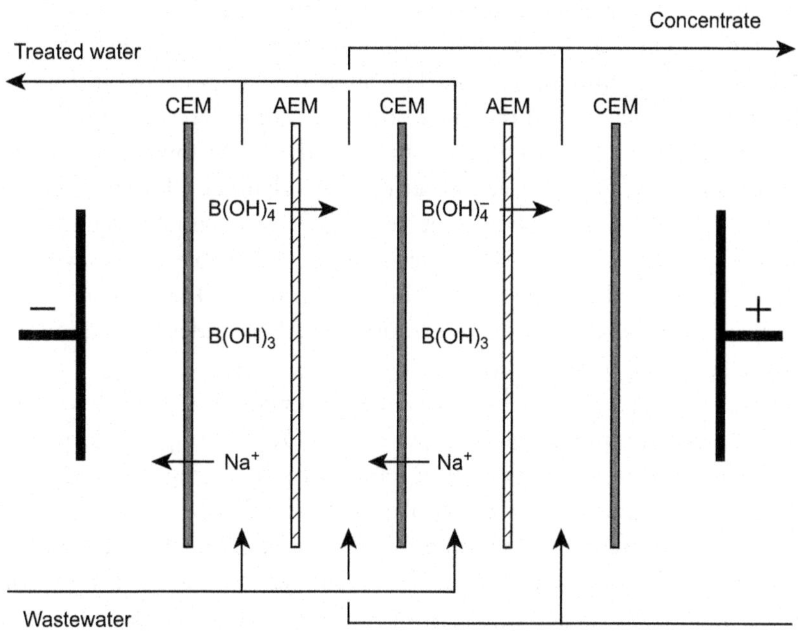

Figure 8.13 The concept of electrodialysis desalination.[73]

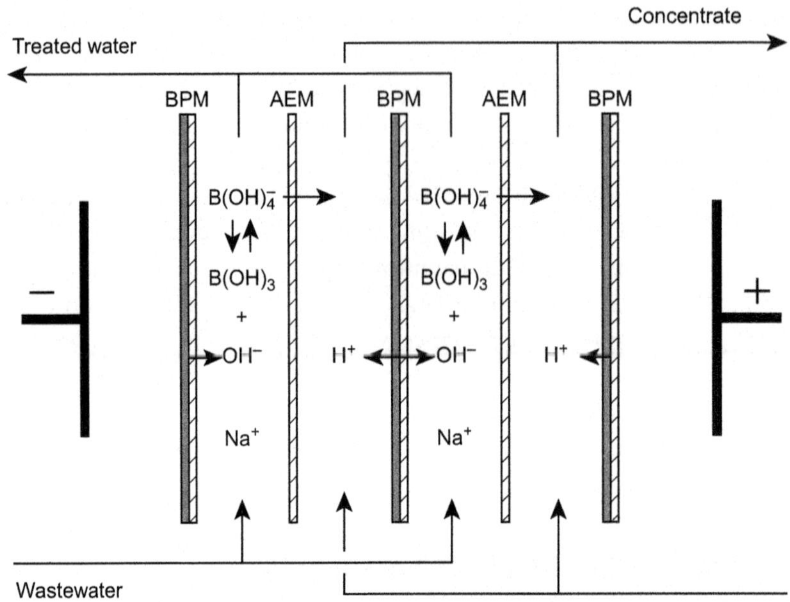

Figure 8.14 Boron removal with bipolar membranes.[73]

water is pumped between AEM and CEM through the feed (diluate) channels, when an electrical field is applied across the membranes by means of two electrodes located on the two sides of the membranes stack. The applied voltage between the membranes causes the ions to move toward the electrodes (i.e., anions are moving through AEM toward the anode and cations are moving through the CEM toward the cathode). As a result, the ions accumulate in a second set of brine channels. The feed water is desalinated along the feed channel while the second set of channels accumulates the salt coming from both sides to form the concentrate. It is possible to remove boron from water when the pH of the feed stream is high, as in the RO membranes. Figure 8.13 demonstrates the passage of borate ions through the AEM under high pH while sodium ions are moving to the opposite side.

A few research groups[32,67,74–77] have demonstrated the ability to use electrodeionization to remove boron from RO permeates. Other groups[78–81] investigated and reviewed the application of ED for boron removal from varying water sources.

Replacement of the CEM with bipolar (BP) membrane was suggested in Refs. 73,82. In this improved method, a high pH is generated in the channel between the BP membrane and the AEM, enabling a better passage of the generated borate ions through the AEM without passage of anions through the BP membrane, as shown in Figure 8.14.

8.7 THE COST OF BORON REMOVAL

It is difficult to give a single answer to the question of how much it costs to remove boron by membrane processes. The total cost depends on the content of boron in the feed water, the required level in the product water and the cost of the different techniques, namely, required investment in the special equipment and the amount of consumables needed. It should also be noted that the cost varies among regions, depending on the local conditions such as manpower, automation, and energy costs. For example, various scenarios for boron removal from seawater using single-pass SWRO, a second BWRO pass, and SWRO + adsorbent were analyzed in Ref. 35. It was concluded that the cost of the three options were within 1 US cent/m^3 for boron concentration above 0.5 mg/L. On the other hand, the difference between the single RO stage and the other techniques for boron removal below 0.5 mg/L was as high as 5–8 US cent/m^3, depending on the feed water quality.

The cost of boron removal by a second BWRO pass and with boron-selective IERs was compared in Ref. 83. It was estimated that the boron removal by IER process was within the same range of cost as the treatment using a second pass, i.e., it adds an average of 7–9 US cents/m^3 of product water to the partial streams undergoing an additional treatment for enhanced boron removal. However, the product salinity in the RO-only process is much lower.

The cost of boron removal from the permeate of a pilot plant desalination system using IXs and RO membranes was calculated in Ref. 84. The cost of removing the boron below the detection limit was between 0.2 and 3.2 EU cents/m^3, while reducing it below 0.5 mg/L cost between 0.1 and 1.2 EU cents/m^3. The rather wide range of estimates is a result of variation in the initial boron content and the type of membranes and resins used. In comparison to the RO systems, the cost of using ED to reduce the boron concentration from 75 to 0.8 mg/L was estimated at 22 US cent/m^3.[81]

At the current stage of knowledge, the boron removal process cost is still expensive and calls for further research efforts. Cost reduction may be potentially achieved through better RO membranes, cheaper ion exchange resins, and better utilization of the chemicals used.

REFERENCES

1. Loizou E, Kanari PN, Kyriacou G, Aletrari M. Boron determination in the Multi element national water monitoring program: the absence of legal limits. *J Verbr Leb* 2010;**5**:459–63.
2. WHO/HSE/WSH/09.01/2. *Boron in drinking water, background document for development of WHO guidelines for drinking water quality*; 2009.
3. Yasumoto M, Takako A, Shouichi K, Masaki I, Minoru K, Mutuo K, et al. The behavior of inorganic constituents and disinfection by products in reverse osmosis water desalination processes. *Water Sci Technol* 1996;**34**:141–8.
4. Prats D, Chillon-Arias MF, Rodriguez-Pastor M. Analysis of the influence of pH and pressure on the elimination of boron in reverse osmosis. *Desalination* 2000;**128**:269–73.

5. Kawasaki M, Hirose M, Ohara T, Kimura S. Simulation of boron reduction system for reverse osmosis seawater desalination unit. *Membrane* 1999;**24**:296–303.
6. Rodriguez-Pastor M, Ruiz AF, Chillon MF, Rico DP. Influence of pH in the elimination of boron by means of reverse osmosis. *Desalination* 2001;**140**:145–52.
7. Nadav N. Boron removal from seawater reverse osmosis permeate utilizing selective ion exchange resin. *Desalination* 1999;**124**:131–5.
8. Fukunaga K, Matsukata M, Ueyama K, Kimura S. Reduction of boron concentration in water produced by a reverse osmosis seawater desalination unit. *Membrane* 1997;**22**:211–6.
9. Sagiv A, Semiat R. Analysis of parameters affecting boron permeation through reverse osmosis membranes. *J Membr Sci* 2004;**243**:79–87.
10. Glueckstern P, Priel M. Optimization of boron removal in old and new SWRO systems. *Desalination* 2003;**156**:219–28.
11. Redondo J, Busch M, De-Witte JP. Boron removal from seawater using FILMTEC^{TM} high rejection SWRO membranes. *Desalination* 2003;**156**:229–38.
12. Faigon M, Hefer D. Boron rejection in SWRO at high pH conditions versus cascade design. *Desalination* 2008;**223**:10–6.
13. Shigecki S, Mitsuharu F, Kashu O. Seawater desalination apparatus, Pat Number JP10128325 1998.
14. Torataro M. Methods of removing boron in reverse osmosis seawater desalination, Pat Number JP10225682 1998.
15. Magara Y, Tabata A, Kohki M, Kawasaki M, Hirose M. Development of boron reduction system for sea water desalination. *Desalination* 1998;**118**:25–34.
16. Tu KL, Chivas AR, Nghiem LD. Enhanced boron rejection by NF/RO membranes by complexation with polyols: measurement and mechanisms. *Desalination* 2013;**310**:115–21.
17. Tu KL, Nghiem LD, Chivas AR. Boron removal by reverse osmosis membranes in seawater desalination applications. *Sep Purif Technol* 2010;**75**:87–101.
18. Hydranautics Technical Application Bulletin. Boron removal by hydranautics RO membranes. *TAB* 2005;**113**.
19. Oo MH, Ong SL. Implication of zeta potential at different salinities on boron removal by RO membranes. *J Membr Sci* 2010;**352**:1–6.
20. Koseoglu H, Kabay N, Yüksel M, Kitis M. The removal of boron from model solutions and seawater using reverse osmosis membranes. *Desalination* 2008;**223**:126–33.
21. Koseoglu H, Harman BI, Yigit NO, Guler E, Kabay N, Kitis M. The effects of operating conditions on boron removal from geothermal waters by membrane processes. *Desalination* 2010;**258**:72–8.
22. Huehmer RP, Wang F, Lozier J, Henthorne L. Enhancing boron rejection in seawater reverse osmosis facilities. *Water Sci Technol* 2008;**8**:519–25.
23. Sarp S, Lee S, Ren X, Lee E, Chon K, Choi SH, et al. Boron removal from seawater using NF and RO membranes, and effects of boron on HEK 293 human embryonic kidney cell with respect to toxicities. *Desalination* 2008;**223**:23–30.
24. Cengeloglu Y, Arslan G, Tor A, Kocak I, Dursun N. Removal of boron from water by using reverse osmosis. *Sep Purif Technol* 2008;**64**:141–6.
25. Henmi M, Fusaoka Y, Tomioka H, Kurihara M. High performance RO membranes for desalination and wastewater reclamation and their operation results. *Water Sci Technol* 2010;**62**:2134–40.
26. Güler E, Kabay N, Yüksel M, Yavuz E, Yüksel Ü. A comparative study for boron removal from seawater by two types of polyamide thin film composite SWRO membranes. *Desalination* 2011;**273**:81–4.
27. Dominguez-Tagle C, Romero-Ternero VJ, Delgado-Torres AM. Boron removal efficiency in small seawater reverse osmosis systems. *Desalination* 2011;**265**:43–8.
28. Oo MH, Song LF. Effect of pH and ionic strength on boron removal by RO membranes. *Desalination* 2009;**246**:605–12.
29. Bason S, Oren Y, Freger V. Ion transport in the polyamide layer of RO membranes: composite membranes and free-standing films. *J Membr Sci* 2011;**367**:119–26.
30. Bason S, Kaufman Y, Freger V. Analysis of ion transport in nanofiltration using phenomenological coefficients and structural characteristics. *J Phys Chem B* 2010;**114**:3510–7.

31. Bason S, Freger V. Phenomenological analysis of transport of mono- and divalent ions in nanofiltration. *J Membr Sci* 2010;**360**:389–96.
32. Oren Y, Linder C, Daltrophe N, Mirsky Y, Skorka J, Kedem O. Boron removal from desalinated seawater and brackish water by improved electrodialysis. *Desalination* 2006;**199**:52–4.
33. Bernstein R, Belfer S, Freger V. Toward improved boron removal in RO by membrane modification: feasibility and challenge. *Environ Sci Technol* 2011;**45**:3613–20.
34. Kurihara M. *The pursuit of ultimate membrane technology and the progress of future essential membrane technology, presented at the meeting "Advances in materials and processes for polymeric membrane mediated water Purification,"*; February 24–27, 2013. Asilomar, CA.
35. Taniguchi M, Fusaoka Y, Nishikawa T, Kurihara M. Boron removal in RO seawater desalination. *Desalination* 2004;**167**:419–26.
36. Bernstein R, Belfer S, Freger V. Improving performance of spiral wound RO elements by in-situ concentration polarization-enhanced radical graft polymerization. *J Membr Sci* 2012;**405–406**: 79–84.
37. Ben-David A, Bernstein R, Belfer S, Dosoretz C, Freger V. Facile surface modification of nanofiltration membranes to target the removal of endocrine-disrupting compounds. *J Membr Sci* 2010;**357**:152–9.
38. Zhai X, Meng J, Li R, Ni L, Zhang Y. Hypochlorite treatment on thin film composite RO membrane to improve boron removal performance. *Desalination* 2011;**274**:136–43.
39. Wei YT, Zheng YM, Chen JP. Functionalization of regenerated cellulose membrane via surface initiated atom transfer radical polymerization for boron removal from aqueous solution. *Langmuir* 2011;**27**:6018–25.
40. Semiat R. Desalination – present and Future. *Water Int* 2000;**25**:54–65.
41. Hung PVX, Cho SH, Moon SH. Prediction of boron transport through seawater reverse osmosis membranes using solution-diffusion model. *Desalination* 2009;**247**:33–44.
42. Kabay N, Güler E, Bryjak M. Boron in seawater and methods for its separation - a review. *Desalination* 2010;**261**(3):212–7.
43. Criscuoli A, Rossi E, Cofone F, Drioli E. Boron removal by membrane contactors: the water that purifies water. *Clean Tech Environ Policy* 2010;**12**:53–61.
44. Hilal N, Kim GJ, Somerfield C. Boron removal from saline water: a comprehensive review. *Desalination* 2011;**273**:23–35.
45. Yavuz E, Arar T, Yüksel M, Yüksel T, Kabay N. Removal of boron from geothermal water by RO system-II-effect of pH. *Desalination* 2013;**310**:135–9.
46. Yavuz E, Arar T, Yüksel T, Yüksel M, Kabay N. Removal of boron from geothermal water by RO System-III-Utilization of SWRO system. *Desalination* 2013;**310**:140–4.
47. Rahmawati K, Ghaffour N, Aubry C, Amy GL. Boron removal efficiency from Red Sea water using different SWRO/BWRO membranes. *J Membr Sci* 2012;**423–424**:522–9.
48. Nadav N, Priel M, Glueckstern P. Boron removal from the permeate of a large SWRO plant in Eilat. *Desalination* 2005;**185**:121–9.
49. Glueckstern P, Priel M. Boron removal in brackish water desalination systems. *Desalination* 2007;**205**: 178–84.
50. Kim j, Wilf M, Park JS, Brown J. *Boron rejection by reverse osmosis membranes: national reconnaissance and mechanism study*; 2009. Desalination and water purification research and Development program report No. 127, USA Department of Interior, Bureau of Reclamation.
51. Hasson D, Shemer H, Brook I, Zaslavschi I, Semiat R, Wilf M, et al. Scaling propensity of seawater in RO boron removal processes. *J Membr Sci* 2011;**384**:198–204.
52. Park PK, Lee S, Cho JS, Kim JH. Full-scale simulation of seawater reverse osmosis desalination processes for boron removal: effect of membrane fouling. *Water Res* 2012;**46**:3796–804.
53. Gorenflo A, Brusilovsky M, Faigon M, Liberman B. High pH operation in seawater reverse osmosis permeate: first results from the world's largest SWRO plant in Ashkelon. *Desalination* 2007;**203**: 82–90.
54. Kabay N, Köseoğlu P, Yapici D, Yüksel Ü, Yüksel M. Coupling ion exchange with ultrafiltration for boron removal from geothermal water-investigation of process parameters and recycle tests. *Desalination* 2013;**316**:17–22.

55. Kabay N, Köseoğlu P, Yavuz E, Yüksel Ü, Yüksel M. An innovative integrated system for boron removal from geothermal water using RO process and ion exchange-ultrafiltration hybrid method. *Desalination* 2013;**316**:1–7.
56. Yavuz E, Güler E, Sert G, Arar T, Yüksel M, Yüksel T, et al. Removal of boron from geothermal water by RO system-I - Effect of membrane configuration and applied pressure. *Desalination* 2013;**310**: 130–4.
57. Kabay N, Bryjak M, Schlosser S, Kitis M, Avlonitis S, Matejka Z, et al. Adsorption-membrane filtration (AMF) hybrid process for boron removal from seawater: an overview. *Desalination* 2008;**223**:38–48.
58. Qin JJ, Oo MH, Wai MN, Cao YM. Enhancement of boron removal in treatment of spent rinse from a plating operation using RO. *Desalination* 2005;**172**:151–6.
59. Bouguerra W, Mnif A, Hamrouni B, Dhahbi M. Boron removal by adsorption onto activated alumina and by reverse osmosis. *Desalination* 2008;**223**:31–7.
60. Kabay N, Yilmaz-Ipek I, Soroko I, Makowski M, Kirmizisakal O, Yag S, et al. Removal of boron from Balcova geothermal water by ion exchange-microfiltration hybrid process. *Desalination* 2009;**241**:167–73.
61. Melnyk L, Goncharuk V, Butnyk I, Tsapiuk E. Boron removal from natural and wastewaters using combined sorption/membrane process. *Desalination* 2005;**185**:147–57.
62. Melnyk L, Goncharuk V, Butnyk I, Tsapiuk E. Development of the sorption-membrane "green" technology for boron removal from natural and wastewaters. *Desalination* 2007;**205**:206–13.
63. Kabay N, Yilmaz I, Bryjak M, Yüksel M. Removal of boron from aqueous solutions by a hybrid ion exchange-membrane process. *Desalination* 2006;**198**:158–65.
64. Babak YV, Goncharuk VV, Mel'nik LA, Badekha VP. Removal of boron compounds in pressure-driven desalination of the Black Sea water. *J Water Chem Technol* 2012;**34**:288–93.
65. Bryjak M, Wolska J, Kabay N. Removal of boron from seawater by adsorption-membrane hybrid process: implementation and challenges. *Desalination* 2008;**223**:57–62.
66. Blahušiak M, Schlosser Š. Simulation of the adsorption-microfiltration process for boron removal from RO permeate. *Desalination* 2009;**241**:156–66.
67. Kang JS, Eusebio RC, Kim HS. Boron removal by activated carbon and microfiltration for pre-treatment of seawater desalination. *Water Sci Technol* 2011;**11**:560–7.
68. Geffen N, Semiat R, Eisen M, Balazs Y, Katz I, Dosoretz CG. Boron removal from water by complexation to polyol compounds. *J Membr Sci* 2006;**286**:45–51.
69. Doğanay CO, Özbelge HO, Biçak N, Aydoğan N, Yilmaz L. Use of specifically tailored chelating polymers for boron removal from aqueous solutions by polymer enhanced ultrafiltration. *Separ. Sci Technol* 2011;**46**:581–91.
70. Ni L, Meng J, Yuan J, Zhang Y. Surface glycosylation of PVDF microporous membrane towards a novel affinity membrane for boron removal. In: *Proceedings of the 3rd international conference on environmental technology and knowledge transfer*; 2010. pp. 631–4.
71. Meng J, Yuan J, Kang Y, Zhang Y, Du Q. Surface glycosylation of polysulfone membrane towards a novel complexing membrane for boron removal. *J Colloid Interf Sci* 2012;**368**:197–207.
72. Samatya S, Orhan E, Kabay N, Tuncel A. Comparative boron removal performance of monodisperse-porous particles with molecular brushes via "click chemistry" and direct coupling. *Colloid. Surf a.* 2010;**372**:102–6.
73. Nagasawa H, Iizuka A, Yamasaki A, Yanagisawa Y. Utilization of bipolar membrane electrodialysis for the removal of boron from aqueous solution. *Ind Eng Chem Res* 2011;**50**:6325–30.
74. Melnik L, Vysotskaja O, Kornilovich B. Boron behavior during desalination of sea and underground water by electrodialysis. *Desalination* 1999;**124**:125–30.
75. Ayyildiz HF, Kara H. Boron removal by ion exchange membranes. *Desalination* 2005;**180**:99–108.
76. Banasiak LJ, Schäfer AI. Removal of boron, fluoride and nitrate by electrodialysis in the presence of organic matter. *J Membr Sci* 2009;**334**:101–9.
77. Arar T, Yüksel T, Kabay N, Yüksel M. Application of electrodeionization (EDI) for removal of boron and silica from reverse osmosis (RO) permeate of geothermal water. *Desalination* 2013;**310**: 25–33.

78. Dydo P, Turek M. Boron transport and removal using ion-exchange membranes: a critical review. *Desalination* 2013;**310**:2–8.
79. Turek M, Dydo P, Trojanowska J, Bandura B. Electrodialytic treatment of boron-containing wastewater. *Desalination* 2007;**205**:185–91.
80. Turek M, Bandura B, Dydo P. Electrodialytic boron removal from SWRO permeate. *Desalination* 2008;**223**:17–22.
81. Kijański M, Bandura-Zalska B, Dydo P, Turek M. The concept of a system for electrodialytic boron removal into alkaline concentrate. *Desalination* 2013;**310**:75–80.
82. Nagasawa H, Iizuka A, Yamasaki A, Yanagisawa Y. Boron removal from aqueous solution by bipolar electrodialysis. In: *AIChE annual meeting, conference proceedings*; 2008.
83. Busch M, Mickols WE, Prabhakaran S, Lomax I, Tonner J. *Boron removal at the lowest cost, IDA*; 2005. World Congress on Desalination and Water Reuse (SP05-024).
84. Chillón Arias MF, Valero i Bru L, Prats Rico D, Varó Galvañ P. Approximate cost of the elimination of boron in desalinated water by reverse osmosis and ion exchange resins. *Desalination* 2011;**273**:421–7.
85. Xu J, Gao X, Chen G, Zou L, Gao C. High performance boron removal from seawater by two-pass SWRO system with different membranes. *Water Sci Technol* 2010;**10**:327–36.
86. Farhat A, Ahmad F, Hilal N, Arafat HA. Boron removal in new generation reverse osmosis (RO) membranes using two-pass RO without pH adjustment. *Desalination* 2013;**310**:50–9.

CHAPTER 9

Boron Removal From Seawater Using Reverse Osmosis Integrated Processes

Nalan Kabay[1], Marek Bryjak[2]

[1]Ege University, Chemical Engineering Department, Faculty of Engineering, Izmir, Turkey
[2]Wrocław University of Technology, Faculty of Chemistry, Department of Polymer and Carbon Materials, Wrocław, Poland

9.1 INTRODUCTION

Due to increasing demand for potable and irrigation water, connected with reduction of fresh water reservoirs, suppliers have to turn to alternative resources. Desalination of seawater by the reverse osmosis (RO) process has become the target for production of potable water. However, the use of seawater results in passing of some trace contaminants to the final product water. One of the most common elements that exists in the seawater reverse osmosis (SWRO) permeate is boron. Typical boron concentration in seawater can be as high as 7 mg/L in the Arabian Gulf and usually is about 4–5 mg/L.[1] With the conventional RO membranes at natural seawater pH, it is not easy to reduce the boron concentration in the permeate to below 1 mg/L.

A major limiting factor for the presence of boron in desalinated water is the possible damage to some plants and consequent reduction of crop yields. Most crops are sensitive to boron concentrations greater than 1.0 mg/L in water. Although boron is vital as a trace element for plant growth, as it is required for their metabolic activities, it could be detrimental at high concentrations. Among the more sensitive plants are citrus trees, which show leaf damage at boron levels more than 0.3 mg/L in irrigation water. Excess boron reduces fruit yield or induces premature ripening in other species. The boron levels of 0.3–0.5 mg/L for citrus and grapes, and 0.5–0.75 mg/L for corn are the highest limit for irrigation waters.[1]

The problem of high boron concentration was observed after processing an SWRO plant in Eilat, Israel, in 1997. Farmers using product water obtained from a desalination plant for irrigation noticed poisoning of crops and partly discolored leaves. Later, boron was identified as the toxic element responsible for these effects. Since then, several post-RO treatment methods have been developed for boron removal in desalination plants.[2]

The World Health Organization for many years recommended the drinking water limits for boron as low as 0.5 mg B/L. However, since 2011 the Drinking-Water Quality Committee decided to revise the Boron Guideline Value to 2.4 mg/L and that value is

Boron Separation Processes
ISBN 978-0-444-63454-2, http://dx.doi.org/10.1016/B978-0-444-63454-2.00009-5

incorporated into the Guidelines. The document notes "although the new guideline value is based on a human health perspective, some utilities may set seawater desalination plants product water limits as low as 0.5 mg/L to reflect agricultural-related issues. These issues include boron's herbicidal effect on some plant species, which is a particular concern in areas of low rainfall." Hence, the limit value of boron as 0.3–0.5 mg/L in the permeate of the RO process is still effective if the product water is going to be used for irrigation.[3]

9.2 BORON CHEMISTRY

Boric acid is a very weak acid with a pKa value of 9.2. At a lower pH than 7, boron is present in its nondissociated form (boric acid) and at a pH greater than 10.5, it is present in the dissociated borate form. The exact percentage of boric acid and borate in any aqueous system is basically dependent on pH. The borate monovalent anion $B(OH)_4^-$ dominates at higher pH while nonionized boric acid $B(OH)_3$ is prevalent at lower pH. Between pH 6 and 11 and at high concentration (>0.025 mol/L), highly water-soluble polyborate ions such as $B_3O_3(OH)_4^-$, $B_4O_5(OH)_4^-$, and $B_5O_6(OH)_4^-$ are formed.[4] There are two alternative routes to the formation of borate trimers. Due to the ring type of structure of trimeric hydroxyborate $B_3O_3(OH)_4^-$, and its high stability, the first option results in $B_3O_3(OH)_4^-$ structure.[5] The second option leads to $B_3(OH)_{10}^-$ structure. At high boron concentration, the formation of tetraborates and pentaborates can appear.[6]

At seawater pH (8.2), boron mostly presents in boric acid (H_3BO_3) form. The rejection of nonionized boric acid by RO is low due to its smaller size and lack of charge. The ionic form borate ($B(OH)_4^-$), however, which exists at higher pH is well hydrated and is mostly rejected by RO membrane due to its large radius and presence of its negative charge.[7]

9.3 REMOVAL OF BORON FROM SEAWATER BY SEAWATER REVERSE OSMOSIS PROCESS

Typical RO membranes have high rejection of the charged form of boron, dominating at pH > pKa, whereas rejection is low for the uncharged species. Under standard test conditions (32 g/L NaCl, 8% recovery, 55 bar feed pressure), high-rejection SWRO membranes display boron rejection between 88% and 91%, but brackish water RO (BWRO) membranes can reject boron from seawater between 30% and 80%.[8] Typical characteristics of SWRO permeate samples obtained from the mini-pilot scale desalination tests conducted at different operating conditions are summarized in Table 9.1.

Table 9.1 Characteristics of SWRO Permeates

	Permeate 1	Permeate 2
Temperature (°C)	19.2	20.0
Conductivity (μS/cm)	980	1950
pH	6.6	6.8
TDS (mg/L)	492	975
Salinity (‰)	0.3	0.8
Boron concentration (mg/L)	1.65	1.95

9.3.1 Effect of Operation Parameters of SWRO System for Boron Removal from Seawater

Boron rejection depends on several process variables such as pH, temperature, and salt concentration.[7,9–11] A shift to pH 10 elevates rejection of SWRO membranes to about 99% and BWRO membranes to 93% while at pH 11 SWRO membranes can reject 99.5% while BWRO 99% of boron.[12] Boron removal by RO membranes therefore requires elevated pH values. In a single pass, using high pH is problematic due to the precipitation of scaling layers. Increased pH is therefore used primarily in double-pass operation at the second RO pass. Alternatively, boron-selective resins can be used instead of a second RO stage.[8]

Quality of SWRO permeate is highly dependent on seawater temperature. As temperature decreases, lower conductivity but higher mineral rejection is obtained in permeate. In terms of boron removal, higher values of rejection are obtained at lower temperatures as summarized in Table 9.2.[11]

Pressure is the other important parameter affecting the performance of SWRO processes. Lower conductivity of permeates were obtained at higher pressures. According to the published data, increasing operation pressure facilitated boron removal from seawater as depicted in Table 9.3.[11]

Figure 9.1 shows the average removal percentages of chemical species from seawater by SWRO process at 55 bar. All ions were rejected by the SWRO membranes at a

Table 9.2 Effect of Temperature on Quality of Product Water Obtained at 55 bar by SWRO Process (Concentrations in Milligrams Per Liter)

T (°C)	(Na^+)	(Mg^{2+})	(Ca^{2+})	(K^+)	(HCO_3^-)	(Cl^-)	(SO_4^{2-})	(B)
9.8	68.4	3.1	0.7	4.8	0.9	128.6	9.6	0.7
13.2	79.4	4.0	0.7	6.3	2.1	131.7	11.4	1.1
14.2	80.3	3.4	0.8	6.3	0.6	132.1	12.2	1.2
15.4	86.0	3.2	0.8	5.8	1.8	177.6	15.4	0.9

Adapted from Ref. 11.

Table 9.3 Effect of Pressure on Quality of Product Water Obtained at 14 °C by SWRO Process (Concentrations in Milligrams Per Liter)

P (bar)	(Na^+)	(Mg^{2+})	(Ca^{2+})	(K^+)	(HCO_3^-)	(Cl^-)	(SO_4^{2-})	(B)
55	80.3	3.4	0.8	6.3	0.6	132.1	12.2	1.2
60	71.6	3.2	0.7	4.6	0.7	116.3	23.2	0.7
62	65.4	3.0	0.6	4.2	0.1	105.6	10.6	0.7

Adapted from Ref. 11.

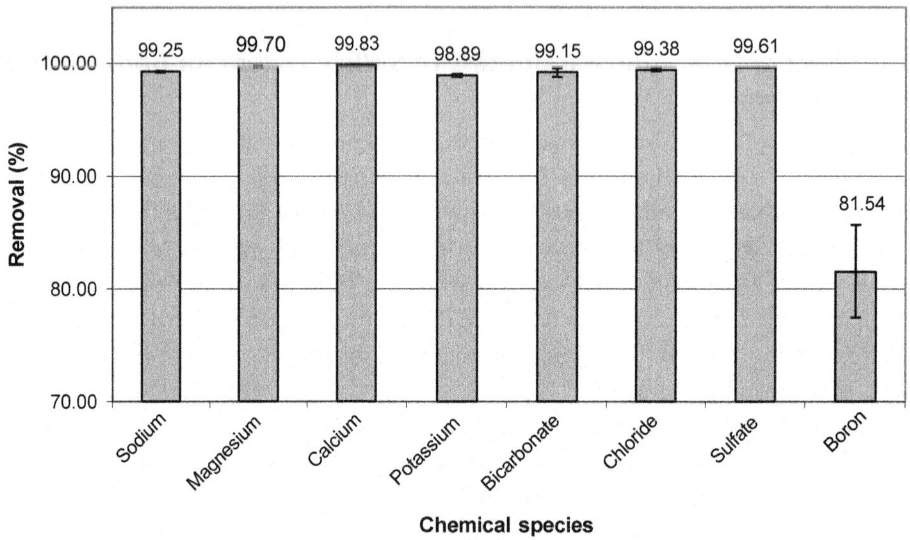

Figure 9.1 Removal efficiencies for various ions and boron in seawater. *(Adapted from Ref. 11).*

percentage of more than 99%; however, the separation efficiency for boron was the smallest among all seawater mineral species.

9.3.2 Effect of Membrane-Type SWRO System for Boron Removal from Seawater

The membrane type has a critical effect on boron removal from seawater in the desalination process. According to the published literature, the old-type membranes (SW-8040) have low rejection efficiency of boron (75%) especially after a long time of membrane service.[12] It was reported that the new types of membranes such as FilmTec SW 380HR, SW 30HR, and Toray UTC-80-AB showed much better removal efficiency for boron.[12] The concentration of boron in permeate reaches a level lower than 1 mg/L by using these membranes in seawater desalination.

According to the obtained laboratory tests, boron rejections by Toray and FilmTec membranes at pH 8.2, at various operating conditions, ranged between 85% and 92%.

On the other hand, boron removal was higher than 98% by the same membranes at pH 10.5 and the resulting boron concentration in the permeate was 0.1 mg/L.[12]

In a published study, the performance of a small-scale SWRO unit that consists of two types of FilmTec polyamide membranes (high-rejection FILMTEC SW30XHR-2540 and FILMTEC SW30-2540 membranes) was checked for desalination and boron rejection.[13] The effects of feed temperature, operating pressure, and pH on the quality of product water and boron rejection were monitored. For both types of membranes, it was observed that permeate flux and flow recovery increased with increasing temperature. It was estimated that membranes productivity increased by 2% per each degree of temperature increase. The new-generation SW30XHR-2540 high-rejection membrane had comparatively higher flux and water recovery values than conventional SW30-2540 membrane. Besides these, the SW30XHR-2540 high-rejection membrane exhibited the higher values of salt and boron rejections. A high dependence of boron rejection on pH was observed for both membranes. Although pH of the natural seawater did not exceed 8.3, it was possible to see the enhancement of boron rejection at natural pH when compared to the rejection obtained using acidified feed at pH 7.0. At natural pH values, boron rejection was as high as 89% for the high-rejection membrane.[13]

9.4 REMOVAL OF BORON FROM SEAWATER BY INTEGRATED PROCESSES

To date, most systems use multiple stages or combinations of different processes for the efficient removal of boron. The main options for boron removal are as follows[8]:

- Single-pass SWRO with high boron rejection membranes
- SWRO followed by BWRO
- SWRO followed by boron selective ion exchange resin (BSR)
- SWRO followed by a hybrid process of BSR and BWRO

9.5 REMOVAL OF BORON FROM SEAWATER BY ION EXCHANGE

Using boron-selective chelating ion exchanger is a good alternative to achieve high boron removal level efficiently and economically.[14] BSRs are primarily classified as macroporous cross-linked polystyrenic resins functionalized with *N*-methyl-D-glucamine (1-amino-1-deoxy-D-glucitol, NMDG) group as shown in Figure 9.2.[15]

Boron-selective chelating fiber (Smopex DS-248v) based on viscose fiber grafted with glycidyl methacrylate was also used for boron sorption from seawater. Its structure is given in Figure 9.3.[16]

In contrast to standard ion exchange processes, the NMDG moieties of BSRs capture boron via a covalent chemical reaction and an internal coordination complexation. Over

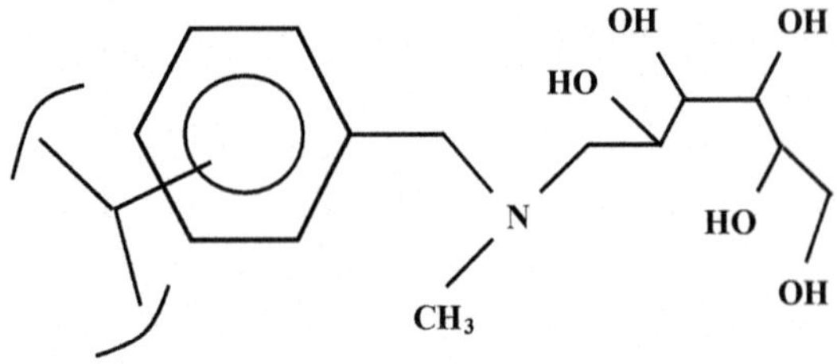

Figure 9.2 Structure of the boron selective ion exchange resin. *(Adapted from Ref. 15).*

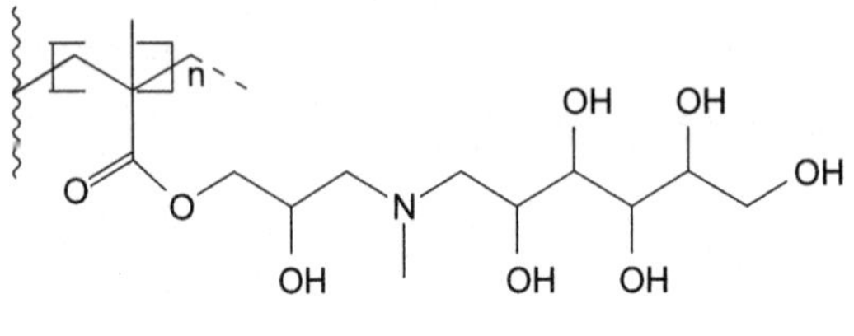

Figure 9.3 Structure of the boron selective ion exchange fiber. *(Adapted from Ref. 16).*

a wide range of pH, boric acid "adds" across one of the *cis*-diol pairs of the functional group to form the relatively stable *cis*-diol borate ester complex as given in Figure 9.4.[15]

In a boron removal process, once the BSR has achieved its maximum loading of boron, the bond between polymer functional groups and borate is subsequently hydrolyzed and the boron is eluted from the resin by an acid solution. The boron liberating hydrolysis is facilitated at pH < 1.0. Therefore, relatively high concentrations of acid are required to complete hydrolysis and to elute the boric acid from BSRs.[15]

Some studies on application of batch and column mode tests were carried out with BSRs containing NMDG groups for boron removal from model seawater and natural SWRO permeate spiked with H_3BO_3.[17,18]

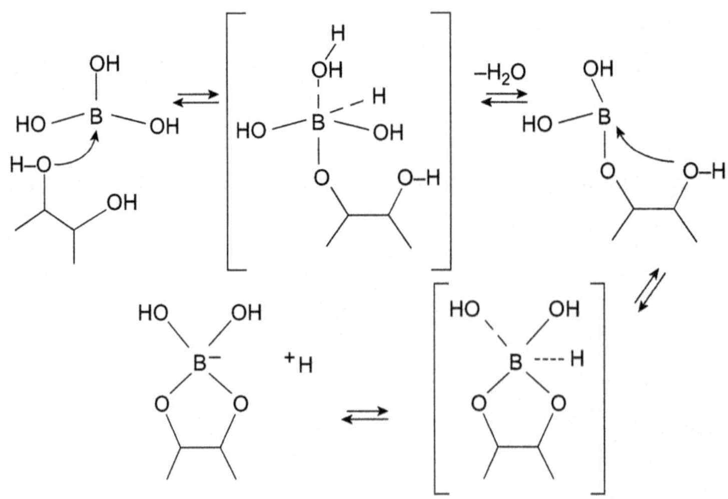

Figure 9.4 Capturing boric acid by *cis*-diol groups of NMDG group. *(Adapted from Ref. 15).*

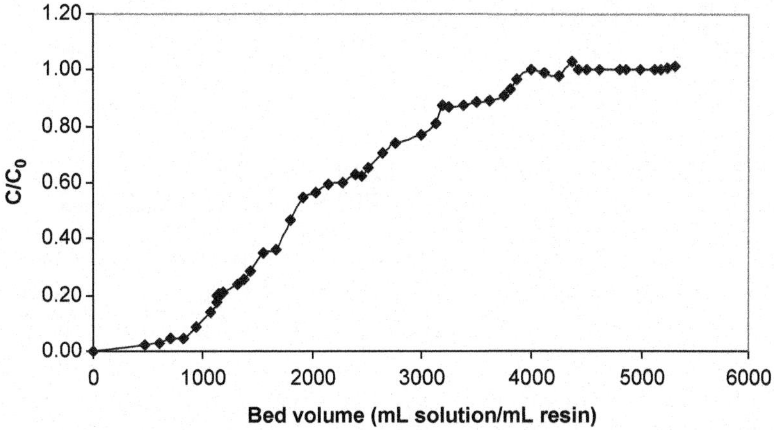

Figure 9.5 Breakthrough curve of boron. *(Adapted from Ref. 3).*

As shown in Figure 9.5, it was possible to remove boron from SWRO permeate containing boron with a concentration of 1.7 mg/L by using BSR. Breakthrough capacity of the resin was calculated as 1.80 mg B/mL-resin by assuming that breakthrough point (0.29 mg B/L) was reached at 1120 bed volume of SWRO permeate.

As shown in Figure 9.6, elution of boron from the resin was completed with 20 bed volumes of H_2SO_4 (5%). Elution efficiency was 98.9%.

Despite *N*-methyl-D-glutamine ligand being commonly used for the preparation of boron-selective resin, some other compounds show an affinity to borates. Matrices of glycidyl methacrylate copolymers modified with ethylene diamine and glycidol,[19] hydroxyl-ethylamine propylene glycol,[20] or diallylamine[21] formed excellent sorbents for

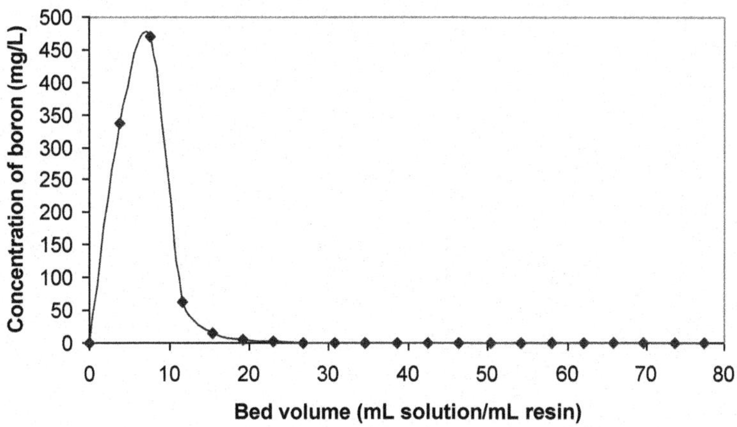

Figure 9.6 Elution curve of boron. *(Adapted from Ref. 3).*

boron removal. Poly(vinyl chloride) particles grafted with poly(glycidyl methacrylate) and modified with ethylenediamine and glycidol formed brushlike sorbents characterized by fast kinetics.[22]

9.6 REMOVAL OF BORON FROM SEAWATER BY SORPTION–MEMBRANE FILTRATION HYBRID PROCESS

Recent development in "materials science" offer new types of fine sorbents and binding agents. They are prepared to establish an outstanding selectivity toward some particular species that appear in trace quantities. After sorption, the fine particles are separated by means of filtration membranes. The combination of the properties of binding agents with membrane filtration reveals many advantages compared to the systems used conventionally. The main benefit is the high efficiency and lower costs of the hybrid process. At first, the sorbents may be prepared as very fine particles, which causes an increase in surface area and results in improvement of the process rate. Second, synergetic interactions may intensify the sorption process when sorbent is deposited on the membrane surface.[23] Recent knowledge about membrane processes enables the high productivity of the process to be reached despite the very small size of sorbent particles.[24,25]

The typical sorption–membrane filtration hybrid system includes two steps (Figure 9.7):

Step 1: Binding of boron (B) on a specific sorbent (S) is followed by separation of complex (BS) from water (W) by means of microporous membrane. The process is

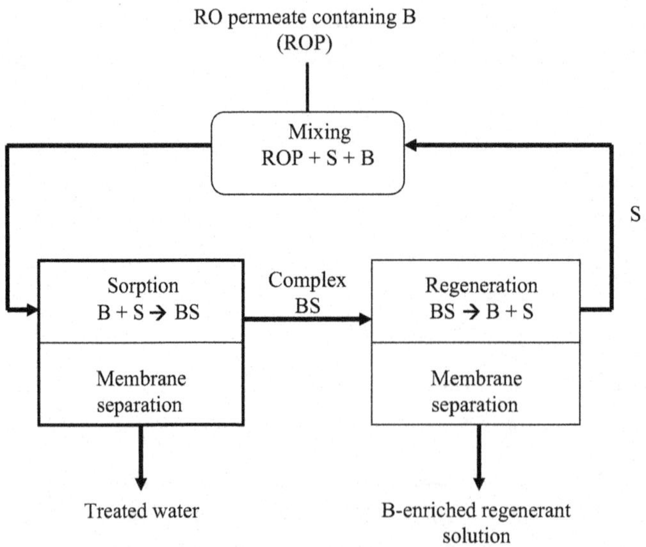

Figure 9.7 Flow sheet of the sorption–membrane filtration hybrid system (S: sorbent). *(Adapted from Ref. 24).*

controlled by balancing of fluxes. Pure water (W) is the main product while (BS) passes to regeneration.

Step 2: Splitting of the complex (BS) onto the free sorbent (S) and boron (B). It allows reuse of the sorbent and produces concentrated brine.

Kabay et al[24] investigated the efficiency of sorption–membrane filtration hybrid process for boron removal from model solution. The same authors[25] evaluated the sorption–submerged membrane hybrid process using BSRs bearing NMDG groups. The suitability and performance of the hybrid process was investigated for boron removal from seawater and SWRO permeate.[26–28] Additionally, fine BSRs were used in a hybrid system for boron removal from geothermal water.[29]

An integrated system coupling SWRO process with sorption–membrane filtration hybrid method was studied by Guler et al[30] BSRs with an average particle size of 20 μm were employed for removal of boron from SWRO permeate. A submerged hollow fiber type ZW-1 ultrafiltration (UF) membrane module was used for the sorption–membrane filtration hybrid system. The flow sheet of the systems is shown in Figure 9.8. Removal of boron was performed on fine BSR and boron-loaded resin was separated by submerged UF membranes later on.

It was reported that sorption–membrane filtration hybrid process can be considered to be good alternative for polishing SWRO permeate. The process variables such as ion exchange resin concentration in suspension, rate of replacements for fresh and saturated resins, and flow rate of permeate stream should be optimized for a particular system.[30]

As seen in Figure 9.9, an increase in the amount of resin in the suspension tank from 0.5 g/L to 1 g/L reduced boron concentration in the permeate dramatically. It was also

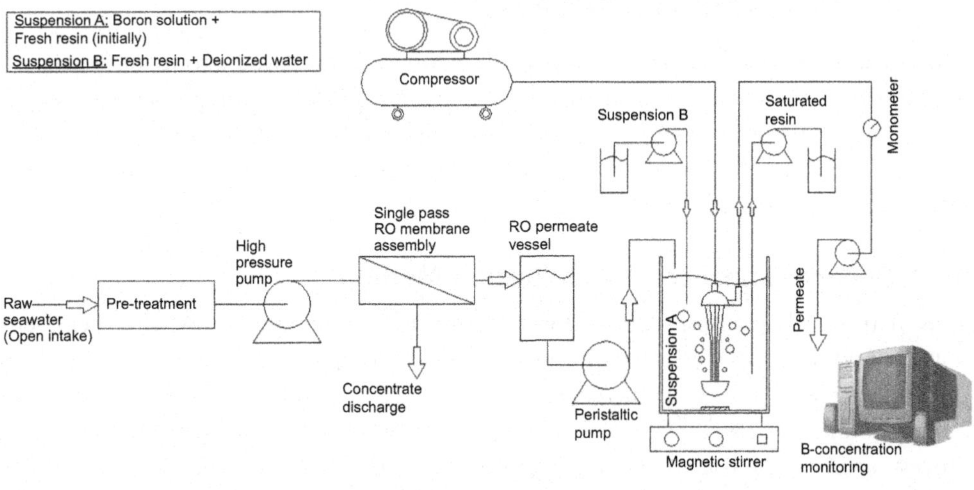

Figure 9.8 Integrated RO process with the sorption–membrane filtration hybrid system. *(Adapted from Ref. 30).*

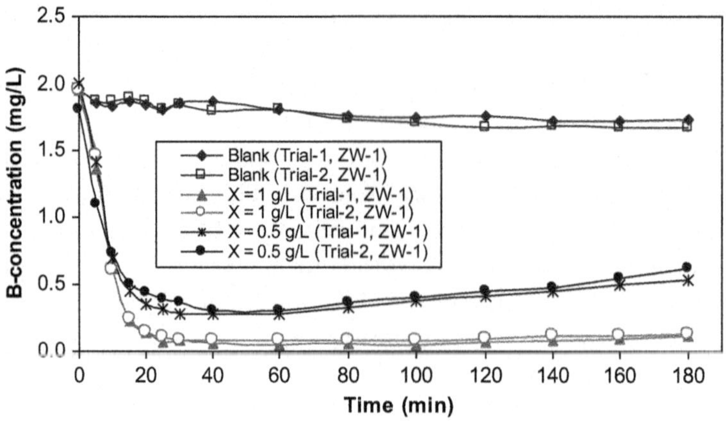

Figure 9.9 Effect of resin concentration in hybrid system on boron removal from seawater RO permeate. *(Adapted from Ref. 30).*

possible to observe a decline of boron concentration in this system for the blank tests. This is because of the dilution effect of fresh resin solution in deionized water. It was possible to eliminate boron from RO permeate with a resin concentration of 1 g/L.[30]

The more extensive studies on the use of submerged membrane modules were conducted by Blahušiak et al[31,32] and Onderková et al.[33] They implied that the power input and the consumption of chemicals in the hybrid process were much lower than in fixed-bed ion exchange processes.

9.7 OTHER MEMBRANE-BASED HYBRID PROCESSES FOR REMOVAL OF BORON FROM RO PERMEATE

The sorption–membrane hybrid processes belongs to the wide group of separation methods that combine two phenomena: (i) sorption on fine coupling agents and (ii) membrane separation of these complexes. The general classification of these methods is reported elsewhere.[34] The most popular membrane-based hybrid methods of boron separation are described below.

9.7.1 Polymer-Enhanced Ultrafiltration Method

This method is based on coupling boron with water-soluble polymers with a concentration of up to 1% wt. and separation of the formed complexes by UF membranes. This method is called polymer-enhanced ultrafiltration (PEUF) and is used for the removal of trace amounts of ions.[35,36] The benefit of the method is the fact that the separation process occurs in the homogeneous medium. It is a highly efficient and fast process. The drawbacks are the requirement of using high-molecular weight water-soluble polymers, membrane fouling, and difficulties in its regeneration. The polymers used for boron separation via PEUF are poly(vinyl alcohol),[37] some derivatives of poly(amidoamine) and poly(ethyleneimine),[38] poly(glycidyl methacrylate) modified with glycerol compounds,[39,40] as well as alkyl monool-, diol-, or triol-containing polyethylenimines.[40,41] Sometimes, in micellar-enhanced UF, where micelles are used for boron removal, their structure is reinforced by water-soluble polymers.[42]

9.7.2 Polyol-Enhanced Filtration

This approach is used for the case when boron can pass through the membrane. Coupling boron with low-molecular weight polyols allows formation of a larger substance that cannot penetrate through the membranes. Geffen et al[43] used mannitol to improve boron rejection on nanofiltration membranes. Dydo et al[44] used D-mannitol, D-gluconic acid, and *N*-methyl-D-glucamine for tests conducted by RO membranes. Both authors noted a significant increase in boron rejection after addition of polyols.

9.8 OTHER MEMBRANE-BASED SEPARATION METHODS FOR BORON REMOVAL

9.8.1 Electrodialysis

Many papers dealing with the use of electrodialysis (ED) for boron separation are available in the literature. To limit the number of references, the reader is directed to the newly published review of Dydo and Turek.[45] In that paper, they discussed the efficiency of boron removal for feed with pH lower or higher than 9.0. The authors noted that for the lower value of pH and boron content typical of seawater, the removal efficiency was not higher than 40%. The higher values of boron removal, usually 75–95%, were obtained for pH of feed above 9. The authors note the poor feasibility of borate removal by ED method. According to them, it is the result of limited current efficiency and competition of anions transport when pH of feed is raised.

9.8.2 Membrane Distillation

Membrane distillation (MD) is a newly emerging process that involves transport of water vapor through pores of a hydrophobic membrane. Vapor, formed on the entrance of pores, is transferred to the opposite side where it is then condensed. When solution of nonvolatile components is used, only water vapor can pass the membrane. The advantages of MD are as follows: (i) lower operating temperature than for conventional distillation, (ii) lower operating pressure than for RO, (iii) almost 100% rejection of nonvolatile solutes, and (iv) no effect of osmotic pressure. There are few reports on the use of MD for boron separation. Hou et al[46] studied the effects of the use of Polyvinylidene fluoride (PVDF) membranes for the removal of boron. They found that in all cases boron rejection was above 99.8% independent of the feed pH value and temperature. An increase in feed temperature caused an exponential increase in permeate flux. A similar conclusion was drawn by Macedonio and Drioli.[47] They did not investigate the removal of boron, but they focused on integration of MD with SWRO. According to them, such an integration system was able to produce permeate with water recovery as large as 80% and still had an acceptable boron rejection. The provided data on the cost of production of 1 m^3 of fresh water were 0.39 USD for RO system and 0.73 USD for RO-MD combined system. The evaluation was carried out when the units were equipped with a Pelton turbine. The fact of the limited interest for the implementation of MD for desalination was explained by the too-high energy consumption, membrane scaling, and the high cost of membrane replacement and maintenance. To reduce the energy cost, some preliminary studies on the use of solar energy were also conducted.[48] It was noted there was 99.998% desalination of seawater with a reduction of boron, arsenic, and fluoride concentrations to the level far below standards for drinking water. The permeate flux was 5 L/m^2 day and it did not change during the tests.

9.8.3 Liquid Membranes

Stability of the liquid membrane is the main obstacle decelerating the commercial use of emulsion or supported liquid membranes. It seems that the problem can be bypassed when ionic liquids are applied as a liquid phase. Some preliminary studies[49] showed that supported liquid membranes impregnated with ionic liquids could extract boron from water and they are stable for a long time. However, it is difficult to predict now if this method will be developed in the future and be used for removal of boron from seawater in industrial scale.

9.9 COST OF BORON REMOVAL FOR SWRO DESALINATION

The cost of SWRO desalination has decreased dramatically. One of the key reasons for this phenomenon is the reduction of unit costs for membrane elements combined with the increase in membrane element productivity (flux) and rejection performance. Another significant cost reduction factor is the decrease in energy costs due to the development of more efficient devices for energy recovery. In addition, improved pretreatment technologies and comprehensive operational experience allowed increasing membrane lifetime and thereby reducing costs for membrane replacement and cleaning.[50]

Energy is the largest cost component in the operation of a desalination plant and offers the greatest potential for further cost reduction. The share of energy on overall cost varies with the operation parameters of the desalination plant and its location.[51]

The cost of boron removal from seawater depends on few stages of the process. Busch estimated that the reduction of boron concentration to level of 0.3–0.5 mg/L by FilmTech membranes should cost about 7–10 US cents/m^3.[3,8] It is important to note that these membranes can improve water quality, so at least 2–3 US cents may be attributed to the reduction of NaCl content in the final product. An overall estimation would be therefore between 5–7 US cents/m^3, which is between 10% and 14% of the total cost of desalinated seawater.

Several concepts were developed to achieve an efficient boron removal at reasonable low costs. Among them the following designs can be listed[52]:

1. Single SWRO pass with natural seawater feed pH as well as lower or higher feed pH.
2. Double passes with increased pH, especially in the second pass. This includes options with high and low recovery in the second pass.
3. Double passes with BSRs, with options treating a part of the first pass permeate, which does not feed the second pass.
4. Three passes with low- and high-pH stages in the second pass.

Boron removal by the ion exchange process is in about the same range of cost as the second-pass treatment, i.e., it adds in average 0.07–0.09 \$/m^3 of product water.[8] The system combining a second pass with a BSR has been proposed for seawater containing 4.5–6.0 mg/L boron and with a request for boron concentration of less than 0.4 mg/L in the final product.[52] This combination was projected and offered for large desalination plants with unit costs in the range of 0.50–0.55 \$/m^3 of product water.[52]

9.10 COMPARATIVE ANALYSIS OF PROCESSES USED FOR BORON REMOVAL FROM SEAWATER

Table 9.3 shows the SWOT (strengths, weaknesses, opportunities, and threats) chart of the processes used for boron removal. The first two columns describe the internal power of the methods while the last two are focused on the external problems.[53] It is seen that it is not easy to select the best method for boron removal. Each of them has its own benefits and drawbacks. The search for efficient technology of boron removal is mostly related to local conditions and to experiences of decision makers. One positive feature can be noted: the membrane separations are well-recognized and they are implemented to the new desalination plants.

9.11 CONCLUSIONS

Boron concentration in seawater is generally in the range of 4–5 mg/L while the tolerable limits of plants to boron in irrigation water change according to the type of the plant. The growth of some plants can be badly affected when the concentration of boron in irrigation water is above 1 mg/L. The most widely used desalination method in order to produce potable water from seawater is RO. In seawater, boron is usually present as the undissociated boric acid form. On the other hand, typical SWRO membranes have a high rejection of the charged form, dominating at pH values above the pKa, whereas

Table 9.4 Comparison of Desalination Processes Used for Boron Removal

Method	Strengths	Weaknesses	Opportunities	Threats
Ion exchange	Well-recognized process Industry implementation	The use of boron-selective resins Regeneration	Experiences—column systems are used in industry	Cost Waste management
Second and higher RO stage	Modular system Adaptable to the particular request	Increase energy needs Relatively high cost of membrane replacement	Better membranes are offered, most useful technique High level of competition among membrane manufacturers	Selective membranes Waste management
pH increase of feed and/or first-stage RO permeate	Efficiency	Chemicals consumption Potential membrane damage due to alkaline conditions Handling, storage, and dosing of chemicals	Easy to implement Relatively low cost of chemicals used	Permitting and disposal of high-pH solution
Electrodialysis of first-stage RO permeate	Reduced energy requirement Regenerant free	Temperature control B-selective membranes Low conductivity of dialysate	Easy to combine with RO Well recognized	Power consumption Waste management
Hybrid system—coupling agent and NF	Homogenous Fast kinetics	Process control problems Separation of complexes Regeneration and reuse of coupling agent	The use of waste materials	Recycling of boron-containing residue
Hybrid system—suspended sorbent and MF	Fast removal Easy to separate sorbent Low pressure	Unknown process	Possibility to use scraps from sorbent producer Waste organic matter can be used	Recycling of boron-containing residue

Abbreviations: MF, membrane filtration; NF, nanofiltration.
Adapted from Ref. 53.

rejection is low for the uncharged species, which dominates at lower pH. Thus, several posttreatment methods have been developed for boron removal in RO permeate. Most systems are based on multiple stages or combinations of different processes for lowering the level of boron in product water. Among them, hybrid systems integrating RO process with sorption–membrane filtration (or UF) process seemed to be the most promising one for elimination of boron from seawater (Table 9.4).

REFERENCES

1. Redondo J, Busch M, De Witte J. Boron removal from seawater using FILMTEC™ high rejection SWRO membranes. *Desalination* 2003;**156**:229–38.
2. Fritzmann C, Löwenberg J, Wintgens T, Melin T. State-of-the-art of reverse osmosis desalination. *Desalination* 2006;**216**:1–76.
3. Kabay N, Guler E, Bryjak M. Boron in seawater and its separation—a review. *Desalination* 2010;**261**:212–7.
4. WHO. *Environmental health criteria 204 for boron*. Geneva (Switzerland): International Programme on Chemical Safety; 1998, ISBN 92 4 1572043.
5. Edwards JO. Detection of anionic complexes by pH measurement: I. Polymeric borates. *J Am Chem Soc* 1953;**75**:6151–4.
6. Ross VF, Edwards JO. The structural chemistry of borates. In: Muetterties EL, editor. *Chemistry of boron and its compounds*. New York (NY): Wiley; 1967.
7. Busch M, Mickols WE, Jons S, De Witte J, Redondo J. *Boron removal in seawater desalination*. BAH03-039. Bahrain: IDA World Congress; 2003.
8. Busch M, Mickols ME, Prabhakaran S, Lomax I, Conner J. *Boron removal at the lowest cost*. Singapore: IDA World Congress; 2005.
9. Koseoglu H, Kabay N, Yuksel M, Kitis M. The removal of boron from model solutions and seawater using reverse osmosis membranes. *Desalination* 2008;**223**:126–33.
10. Koseoglu H, Kabay N, Yuksel M, Sarp S, Arar O, Kitis M. Boron removal from seawater using high rejection SWRO membranes—impact of pH, feed concentration, pressure and cross-flow velocity. *Desalination* 2008;**227**:253–63.
11. Guler E, Piekacz J, Ozakdag D, Kujawski W, Arda M, Yuksel M, et al. Influence of chosen process parameters on the efficiency of seawater desalination: SWRO pilot plant results at Urla Bay seashore. *Desalin Water Treat* 2009;**5**:167–71.
12. Glueckstern P, Priel M. Optimization of boron in old and new SWRO systems. *Desalination* 2003;**156**:219–28.
13. Guler E, Kabay N, Yuksel M, Yavuz E, Yuksel U. A comparative study for boron removal from seawater by two types of polyamide thin film composite SWRO membranes. *Desalination* 2011; **273**:81–4.
14. Kunin R, Preuss AF. Characterization of a boron-specific ion exchange resin. *Ind Eng Chem Prod Res Dev* 1964:304–6.
15. Marston C, Busch M, Prabhakaran S. *A boron selective resin for seawater desalination. Paper for presentation at European desalination society conference on desalination and the environment*; 2005. Santa Margherita Ligure(Italy).
16. Parschova H, Mistova E, Matejka Z, Jelinek L, Kabay N, Kauppinen P. Comparison of several polymeric sorbents for selective boron removal from reverse osmosis permeate. *React Funct Polym* 2007;**67**:1622–7.
17. Kabay N, Sarp S, Yuksel M, Arar O, Bryjak M. Removal of boron from seawater by ion exchange resins. *React Funct Polym* 2007;**67**(12):1643–50.
18. Kabay N, Sarp S, Yuksel M, Kitis M, Koseoglu H, Arar O, et al. Removal of boron from SWRO permeate by boron selective ion exchange resins containing containing N-methyl glucamine groups. *Desalination* 2008;**223**:49–56.

19. Senkal BF, Bicak N. Polymer supported iminodipropylene glycol functions for removal of boron. *React Funct Polym* 2003;**55**:27–33.
20. Bicak N, Ozbelge O, Yilmaz L, Senkal BF. Cross-linked polymer gels for boron extraction derived from N-glucidol-N-methyl-2-hydroxypropyl methacrylate. *Macromol Chem Phys* 2000;**201**:577–84.
21. Bicak N, Gazi M, Senkal BF. Polymer supported amino bis-(cis-propane 2,3 diol) functions for removal of trace boron from water. *React Funct Polym* 2005;**65**:143–8.
22. Yavuz E, Gursel Y, Senkal BF. Modification of poly(glycidyl methacrylate) grafted onto cross-linked PVC with iminopropylene glycol group and use for removing boron from water. *Desalination* 2013;**310**:145–50.
23. Koltuniewicz A, Witek A, Bezak K. Efficiency of membrane-sorption integrated processes. *J Membr Sci* 2004;**239**:129–41.
24. Kabay N, Yılmaz I, Bryjak M, Yuksel M. Removal of Boron from aqueous solutions by ion exchange-membrane hybrid process. *Desalination* 2006;**198**:74–81.
25. Yılmaz I, Kabay N, Bryjak M, Yuksel M, Wolska J, Koltuniewicz A. A submerged-ion exchange hybrid process for boron removal. *Desalination* 2006;**198**:310–5.
26. Kabay N, Bryjak M, Schlosser S, Kitis M, Avlonitis S, Matejka Z, et al. Adsorption-membrane filtration (AMF) hybrid process for boron removal from seawater—an overview. *Desalination* 2008;**223**:38–48.
27. Bryjak M, Wolska J, Kabay N. Removal of boron from seawater by adsorption-membrane hybrid process: implementation and challenges. *Desalination* 2008;**223**:57–62.
28. Bryjak M, Wolska J, Soroko I, Kabay N. Adsorption-membrane filtration process in boron removal from first stage seawater RO permeate. *Desalination* 2009;**241**:127–32.
29. Kabay N, Yilmaz-Ipek I, Soroko I, Makowski M, Kirmizisakal O, Yag S, et al. Removal of boron from Balcova geothermal water by ion exchange–microfiltration hybrid process. *Desalination* 2009;**241**(1–3):167–73.
30. Guler E, Kabay N, Yuksel M, Yiğit NÖ, Kitiş M, Bryjak M. Integrated solution for boron removal from seawater using RO process and sorption-membrane filtration hybrid method. *J Membr Sci* 2011;**375**:249–57.
31. Blahušiak M, Onderková B, Schlosser S, Annus J. Microfiltration of microparticulate boron adsorbent suspensions in submerged hollow fibre and capillary modules. *Desalination* 2009;**241**:138–47.
32. Blahušiak M, Schlosser S. Simulation of the adsorption-microfiltration process for boron removal from RO permeate. *Desalination* 2009;**241**:156–66.
33. Onderková B, Šchlosser S, Blahušiak M, Búgel M. Microfiltration of suspensions of microparticulate boron adsorbent through a ceramic membrane. *Desalination* 2009;**241**:148–55.
34. Kabay N, Bryjak M. In: Hoek EMV, Tarabara VV, editors. *Hybrid processes combining sorption and membrane filtration, encyclopedia of membrane science and technology.* Wiley 2013; 2013.
35. Rivas BL, Pereira ED, Palencia M, Sanchez J. Water-soluble functional polymers in conjunction with membranes to remove pollutant ions from aqueous solutions. *Prog Polym Sci* 2011;**36**:294–322.
36. Rivas BL, Pereira ED, Villoslada IM. Water-soluble polymer-metal ion interactions. *Prog Polym Sci* 2003;**28**:173–208.
37. Dilek C, Ozbelge HO, Bicak N, Yilmaz L. Removal of boron from aqueous solutions by continuous polymer-enhanced ultrafiltration with polyvinyl alcohol. *Sep Sci Technol* 2002;**37**(6):1257–71.
38. Smith BM, Todd P, Bowman CN. Hyper branched chelating polymers for the polymer-assisted ultrafiltration of boric acid. *Sep Sci Technol* 1999;**34**(10):1925–45.
39. Doganay CO, Ozbelge HO, Yilmaz L, Bicak N. Removal and recovery of metal ions via functional polymer based PEUF. *Desalination* 2006;**200**:286–7.
40. Doganay CO, Ozbelge HO, Bicak N, Aydogan N, Yilmaz L. Use of specifically tailored chelating polymers for boron removal from aqueous solutions by polymer enhanced ultrafiltration. *Sep Sci Technol* 2011;**46**:581–91.
41. Smith BF, Robinson TW, Carlson BJ, Labouriau A, Khalsa GRK, Schroeder NC, et al. Boric acid recovery using polymer filtration: studies with alkyl monool, diol, and triol containing polyethylenimines. *J Appl Polym Sci* 2005;**97**:1590–604.

42. Bryjak M, Duraj I, Pozniak G. Colloid-enhanced ultrafiltration in removal of traces amounts of borates from water. *Environ Geochem Health* 2010;**32**:275–7.
43. Geffen N, Semiat R, Eisen MS, Balazs Y, Katz I, Dosoretz CG. Boron removal from water by complexation to polyol compounds. *J Membr Sci* 2006;**286**:45–51.
44. Dydo P, Nems I, Turek M. Boron removal and its concentration by reverse osmosis in the presence of polyol compounds. *Sep Purif Technol* 2012;**89**:171–80.
45. Dydo P, Turek M. Boron transport and removal using ion-exchange membranes: a critical review. *Desalination* 2013;**310**:2–8.
46. Hou D, Wang J, Sun X, Luan Z, Zhao C, Ren X. Boron removal from aqueous solution by direct contact membrane distillation. *J Hazard Mater* 2010;**177**:613–9.
47. Macedoni F, Drioli E. Pressure-driven membrane operations and membrane distillation technology integration for water purification. *Desalination* 2008;**223**:396–409.
48. Zwijnenberg HJ, Koops GH, Wessling M. Solar driven membrane pervaporation for desalination processes. *J Membr Sci* 2005;**250**:235–46.
49. Fortuny A, Coll MT, Sastre AM. Use of methyltrioctyl/decylammonium bis 2,4,4-(trimethylpentyl) phosphinate ionic liquid (ALiCY IL) on the boron extraction in chloride media. *Sep Purif Technol* 2012;**97**:137–41.
50. Semiat R, Chapman M, Price P, Hasson D. Desalination project costs. In: *Proceedings of international conference on desalination costing, Limassol, Cyprus*. Muscat (Oman): The Middle East Desalination Research Center; 2004.
51. Wilf M. *Fundamentals of RO-NF technology*. International Conference on Desalination Costing, Limassol; 2004.
52. Busch M. *Boron removal in sea water desalination*. Bahrain: International Desalination Association World Congress on Desalination and Water Reuse; March 2002.
53. MEDRC Research Project (Project No 04-AS-004). *Study of the adsorption-membrane filtration (AMF) hybrid process for removal of boron from seawater*; 2007.

CHAPTER 10

Boron Removal From Water by Sorption—Membrane Filtration Hybrid Process

Marek Bryjak[1], Nalan Kabay[2]

[1]Wrocław University of Technology, Faculty of Chemistry, Department of Polymer and Carbon Materials, Wrocław, Poland
[2]Ege University, Chemical Engineering Department, Faculty of Engineering, Izmir, Turkey

10.1 INTRODUCTION

There is no simple technology for boron removal from aqueous solutions. Some conventional and advanced treatment methods, such as chemical precipitation, ion exchange, adsorption, liquid—liquid extraction, electrocoagulation, reverse osmosis (RO), and electrodialysis (ED), are critically reviewed by Xu and Jiang.[1] The authors pointed at chemical precipitation, electrocoagulation, and adsorption as the most promising techniques for boron removal from water. The first two can be used for solutions with high boron concentration while the last one can effectively treat diluted solutions of boron. When the treatment of crude solution is not problematic from a technical point of view, handling diluted solution has still to be improved. So far the following methods have been tested[2]:

- Single-pass RO system equipped with high boron-rejecting membranes,
- Double-pass RO with raised pH at the second stage,
- Sorption on boron-selective ion exchange resins,
- ED.

It should be emphasized here that the use of sorption on boron-selective ion exchange resins is still one of the most viable and economically justified approaches.

Membrane-enhanced hybrid process can be considered as an alternative process for boron removal from diluted solutions. It combines sorption and membrane separation and merges advantages of both methods. In the hybrid process, solutes are adsorbed by fine coupling agents and complexes are separated by membranes. The main benefit of hybrid process over any conventional sorption process is the higher efficiency and lower costs of the first one as compared to classical fixed-bed approach. In the hybrid systems coupling agents with small diameter can be used. Consequently, it is possible to reduce the amount of sorbent required and decrease the cost of the process significantly.

Boron Separation Processes
ISBN 978-0-444-63454-2, http://dx.doi.org/10.1016/B978-0-444-63454-2.00010-1

By using new types of binding agents discovered recently, new technologies have been developed for the removal of trace compounds from aqueous solutions. The new binding agents comprise organic and inorganic components, chelating and ion exchange resins, molecularly imprinted polymers, and/or coordinating agents. They can appear in various forms as solid particles, micelles, colloids, water-soluble polymers, or even low-molecular weight components. The general principle for their use in the membrane-enhanced separations is their affinity to the target molecules and formation of large associates that are easily removed during membrane filtration. In the second step, the separated complexes are decomposed to binding agent and the target compound. Binding agent is recycled to the next sorption step. That approach allows removal or recovery of some substances that appear in the solution even at a very low concentration. Below is schematic description of typical membrane-enhanced process divided into five stages[3]:

1. Entrapment of target molecules by coupling agent
2. Membrane separation of complex target molecule: coupling agent
3. Desorption of target molecule from the complex
4. Membrane separation and regeneration of coupling agent
5. Regulation/activation of coupling agent for use in the first stage.

Depending on the kind of coupling agent, the hybrid systems can be divided into several groups. There is not any consensus on the clear nomenclature of the processes and some ambiguity can be found.[4] It seems that the process is based on the formation of complexes of small target molecules with larger structure of binding agent and separation of such complexes through membrane filtration. The type of membrane employed depends on the size of separated complexes. Hence, the mechanism of complexation with the binding agents seems to have secondary importance to the nature of the process. The target molecules can employ ionic forces, hydrophobic interactions, or chelating or other forms of affinity to bind to the coupling agent while the membrane-enhanced process will not change its nature. The elimination of the complexes controls its efficiency for the most part. For simplicity, taking into account the complexing agent size, membrane-enhanced processes can be divided as follows:

Membrane-Enhanced Process	Size of Binding Agent
Molecule-enhanced membrane separation (MEMS)	Less than nanometer
Polymer-enhanced ultrafiltration (PEUF)	Several nanometers
Micellar-enhanced ultrafiltration (MEUF)	Several nanometers
Colloid-enhanced ultrafiltration (CEUF)	Dozens of nanometers
Suspension-enhanced microfiltration (SEMF)	Several micrometers

Considering the case of water deboronation, the above processes can be useful for removal of boron from aqueous solutions. They are described in the following sections.

10.2 MOLECULE-ENHANCED MEMBRANE SEPARATION

Some organic molecules carrying vicinal hydroxyl groups can be used as coupling agents for removal of boric acid and borates. Geffen et al.[5] used mannitol to improve boron rejection on nanofiltration membranes. They noted that addition of this polyol increased boron rejection up to 90%.

Dydo et al.[6] analyzed the effect of 1,2 diol-containing polyols on boron removal with RO membranes. They evaluated the effect of presence of the following polyols: D-mannitol, sodium D-gluconate, and *N*-methyl D-glucamine and concluded that borate complexes were rejected by membranes more effectively than monoborate. In the last paper Dydo[7] provided the mathematical modeling of the transport process for boric acid, borates, and boron complexes. On its base he was able to point at the profits of the use of using polydiols. He wrote that "adding alcohol greatly decreased the pH required to achieve a given boron rejection. A rejection of 90% was achieved at a retentate pH of 9.5 in a plain boric acid system, while the same rejection was achieved at pH 8.5 in the presence of D-mannitol or sodium D-gluconate and at pH 6.5 in the presence of *N*-methyl D-glucamine. Complex formation was found to be the cause of the above pH effect." Complexation of boron with D-mannitol was used[8] for reduction of boron concentration in diluate during ED also. No transport of boron—polyol complexes through ion exchange membranes was observed during the process.

10.3 POLYMER-ENHANCED ULTRAFILTRATION

This process has various names—some authors called it liquid-phase polymer-based retention (LPR),[9] while some others used the term polyelectrolyte-enhanced ultrafiltration (UF). The process is based on the complexation of borate by water-soluble polymers followed by separation of formed complexes on UF membranes. The advantage of the process is the need for lower transmembrane pressure than in ordinary nanofiltration and large permeate flux. The running process in the homogenous phase is the next advantage. The great disadvantage is the need to use water-soluble polymers with large molecular weight and sorption of these molecules on membrane surfaces. This leads to membrane fouling and malfunctioning of the process. Some problems can be created during the regeneration step.

Polymers, selective for boron removal, have been synthesized by attachment of sugarlike ligands to the polymeric chains.[10] The following polymers were used in the polymer-enhanced ultrafiltration (PEUF) process: poly(vinyl alcohol)[11]; glucoheptanamide derivatives of poly(amidoamine) and poly(ethyleneimine)[12]; poly(glycidyl methacrylate) and poly(*N*,*N*′-diallyl morpholinium bromide) modified with hydroxyethylaminoglycerol[13]; hydroxyethylaminoglycerol functionalized to poly(glycidylmethacrylate) and poly(4-vinyl-1,3-dioxalan-2-one-co-vinyl acetate)[14]; alkyl monool, diol, or triol containing polyethylenimines[10]; or diallyl dimethyl ammonium chloride copolymerized with poly(vinyl amino-*N*,*N*′-bis-propane diol).[15] The studies tested the effect of such process parameters as the ratio of boron to polymer, pH, molecular weight of water-soluble polymer, regeneration of polymer–boron complexes, and so on.

PEUF process can be run in batch, semicontinuous, and continuous modes of operations.[3] The bottleneck step in PEUF is polymer regeneration. There are two sets of methods for regeneration of water-soluble polymers and recycling them in the hybrid process: splitting the complex by altering solution pH or by electroregeneration. In the first case, the regeneration is simple and not so expensive. However, low concentration of boron in brine and high water consumption are its main drawbacks. The electrochemical regeneration does not produce large volumes of effluents but its cost seems to be a strong barrier. According to the authors' knowledge, this process may not be applied for regeneration of boron-loaded water-soluble polymers.

10.4 MICELLAR-ENHANCED ULTRAFILTRATION AND COLLOID-ENHANCED ULTRAFILTRATION

In micellar-enhanced ultrafiltration, surfactants are added to solution to the level equal to or higher than critical micelle concentration. In such conditions, surfactant molecules assemble and organize aggregates—micelles that are able to bind borate and make complexes. To the best knowledge of the authors, such structures were not used for removal of boron so far. However, when micelles formed by sugarlike surfactants are reinforced by water-soluble polymers, the obtained aggregates—colloids—could be employed for removal of boron.[16] As the surfactant, 1,4-sorbitol oleate, was applied to obtain micelles and chitosan as the reinforcing polymer. It was noted that presence of chitosan significantly improved separation of boron from aqueous solutions. Besides, positively charged filtration membrane was not so prone for fouling with processed colloids.

10.5 SUSPENSION-ENHANCED MICROFILTRATION OR ULTRAFILTRATION

The interesting point of view on the use of suspension-enhanced microfiltration (SEMF)/UF in water treatment was given lately by Abdulgader et al.[17] These authors mention that the growing interest in the use of hybrid processes with powdered ion exchange resins results from the following facts: enhancement of filtration performances of the system; possibility to prepare high-quality water; significant decrease in discharge waste streams, and reduction of cost. The suspension of resins used can either work as the coupling agent and selectively reduce the concentration of pollutant in water, or decrease the membrane fouling.[18] In the case of boron removal, the author pointed out that the use of boron-selective resins (BSR) in the fixed-bed system for RO permeate significantly reduces the capital and operating costs for water preparation.[18] When the conventional RO systems are considered, the reduction of boron concentration is achieved in the second RO pass, where water pH is raised by addition of caustic soda. The use of antiscalants in such conditions is mandatory. In consequence, the footprint of such installation increases costs for chemicals significantly. Jacob[18] has compared the effects of the use of double-pass RO system vs. hybrid RO—ion exchange system. He pointed at the following benefits coming from the use of hybrid alternatives: saving on the membrane area for the first RO stage, reduction of wastes, and an increase in bypass volume.

SEMF/UF processes have been considered as an interesting alternative to the system discussed above. The major advantage of the SEMF/UF process over a conventional sorption processes is the improvement of boron uptake and faster kinetics. To see the difference, the reader is asked to look at Figure 10.1. It shows the kinetics of boron sorption on the same (BSR-1, Dow Chem) sorbent. The only variable is the particle diameter.

It is obvious that for all evaluated particles, the same sorption equilibrium should be reached. However, it is reached after minutes for the 20 μm particles and after 48 h for the 300 μm particles. The particle diffusion of borate within the sorbent grains is the critical process. Hence, the smaller the particle, the faster the process. Additionally, the process pressure is controlled by the pressure drop on the membrane in the case of SEMU/UF hybrid. When the fixed-bed system is used, the pressure drop increases dramatically. Hence, this allows us to point to two benefits coming from the use of the

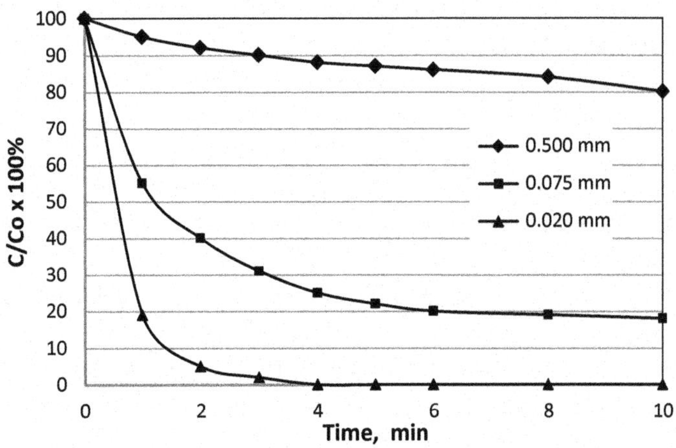

Figure 10.1 Boron sorption kinetics on BSR1 (Dow Chem). Boron concentration 2 mg/L. Sorbent: solution ratio 1 g: 1 L. *(Adopted from Ref. 20).*

SEMF/UF hybrid: increase in the process speed and decrease in the energy consumption. A third benefit was shown by Blahusiak et al.[19] It was pointed out that the consumption of chemicals for the hybrid system was smaller in comparison to sorbent in a classical fixed-bed column.

The efficiency of SEMF/UF hybrid system in the removal of boron from aqueous solutions was reported in Refs. 2,3,20–23. The authors compared the effectiveness of the use of two commercial resins bearing *N*-methyl-D-glucamine (NMDG) ligands: Diaion CRB 02 (Mitsubishi) and BSR1 (Dow Chem). The membrane filtration module consisted of microfiltration polypropylene capillary membranes with a pore diameter 0.4 μm working in the submerged mode. Checking the process parameters, the authors determined the impact of resin particle size, suspension rate delivery, resin concentration in suspension, and permeate flux. It was observed that there is significant reduction of boron concentration in permeate in the first 15 min of the process. The observation leads to the conclusion that sorbent particle size largely affected the economics of the SEMF hybrid system. Unfortunately, commercially available boron-selective ion exchange resins prepared for use in fixed-bed systems have particle size in a range of 300–500 μm. They are too large to be used in SEMF/UF systems. Attempts to ground sorbents to smaller beads resulted in large size heterogeneity and irregular shape of particles and it is not difficult to predict that such particles have negative effects on process stability and reliability. Therefore, to obtain monodisperse microspheres with predicted dimensions, it is necessary to establish a new procedure for the synthesis of fine, round particles.

Preparation of monodisperse polymer microspheres, whose dimensions can be controlled during their synthesis, becomes critical for developing of SEMF/UF hybrids. Among several methods, a few can be considered as most promising: membrane emulsification followed by polymerization of monomers[24,25] and "seed polymerization."[26] The first leads to preparation of NMDG-modified microspheres of poly(styrene-vinyl benzyl chloride) copolymer, with diameter of 30 μm and very narrow size distribution. The second method offers brushlike particles, based on 6 μm core structure of poly(glycidyl methacrylate) polymer, covered by dextran molecules. The first sorbent was tested in a laboratory SEMF/UF system and showed boron uptake comparable to commercial resins and shape stability after several sorption–desorption cycles.[24] The second sorbent showed an extraordinary good sorption for boron. According to the authors' suggestion it was the result of brushlike structure of dextran chains grafted to the surface of particles. Similar outcome can be found in the paper of Gazi et al.[27] and Yavuz et al.[28] describing the performances of brushlike sorbents. These findings lead to point that core–shell structures of polymer sorbents have great potential for use in SEMF/UF systems. They have large capacity for boron sorption and show fast sorption kinetics.

The synthesis of fine monodispersed particles was described by Samatya et al.[29–31] In the first paper porous poly(glycidyl methacrylate-co-ethylene methacrylate) particles carrying diol functionality were reported to have a potential as BSRs.[29] These resins were used for the removal of boron from geothermal water and its RO permeate in the second paper.[30] Finally, the third paper described a new sorbent with a high boron adsorption capacity based on porous poly(vinylbenzyl chloride-co-divinylbenzene) beads of 8.5 μm diameter. They were synthesized by a modified seeded polymerization technique and were derivatized by direct reaction with NMDG.[31]

The SEMF/UF hybrid process combines two separate steps: sorption of boron on suspended sorbent particles and desorption of boron followed by sorbent regeneration (Figure 10.2). While the first step is pretty well recognized now[3,20,23,32] the regeneration still needs some additional studies on its mechanism and optimization of operational cost.[17] The sorbent–boron complex can be treated by diluted solutions of either hydrochloric acid or sulfuric acid.[33] After acid treatment, the sorbent should be regenerated by sodium hydroxide solution. Such alternative exposition to various pH results with shrinking and swelling of sorbent particles, and consequently deterioration of their structure can appear. It is dangerous for ground particles that keep the frozen stress in their structures. For that reason some studies on the stability of sorbent particles were performed.[34] The authors repeated 10 cycles of sorption–washing–elution–washing–regeneration–washing batch processes for ground BSR1 sorbent with average particle diameter of 20 μm. A decrease in boron removal after the fifth cycle was noted (see Figure 10.3). The observed phenomenon was explained by the loss of sorbent in the batch processes, by not fully stripping off boron from sorbents as well as by scaling of sorbent sample by species presented in feed.

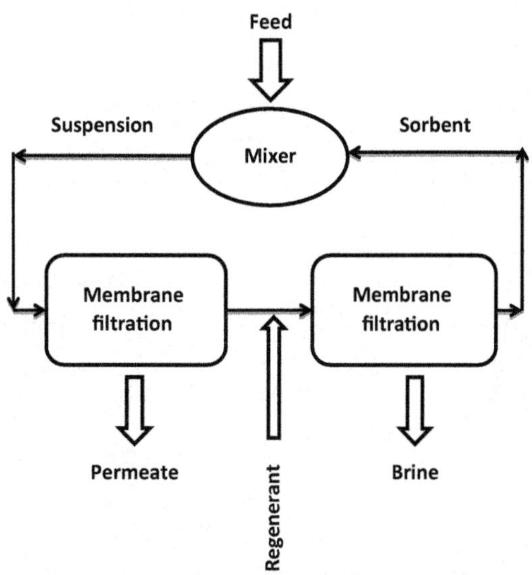

Figure 10.2 Block scheme of SEMF/UF system.[20]

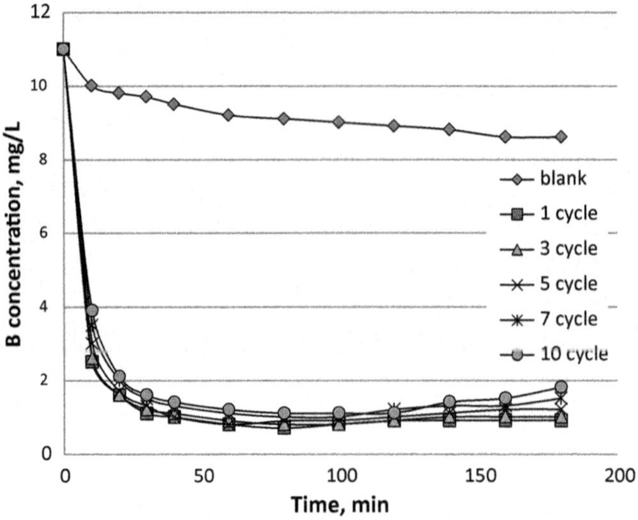

Figure 10.3 Boron concentration in permeate after cycling use of sorbent. *(Adopted from Ref. 34).*

However, when fresh sorbent was added to the hybrid system after each cycle in amounts counterbalancing the loss of boron capacity, the cyclic use of BSR1 did not affect the efficiency of the SEUF process (Figure 10.4).

As was mentioned above, the application of fine, round-chapped monodisperse particles of boron-selective sorbent seems to be critical for long-lasting separation.

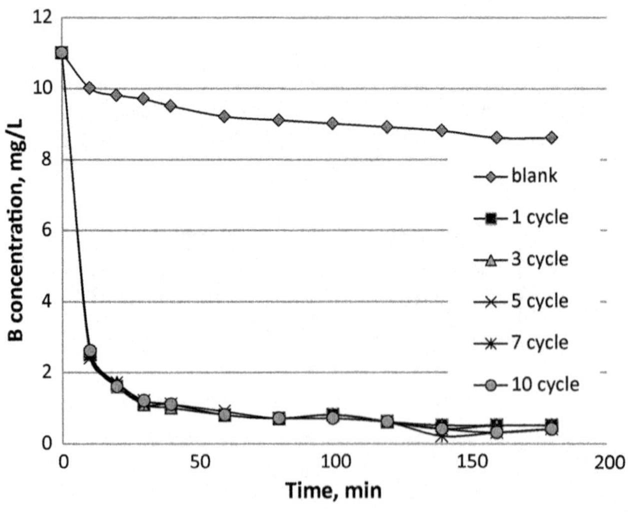

Figure 10.4 Boron concentration in permeate after cycling use of supplemented sorbent. *(Adopted from Ref. 34).*

It should be profitable to use continuous sorbent regeneration step with membrane separator that protects the particles to be washed out of the system.

To model the SEMF/UF system one should consider the boron balance (Figure 10.5).

The hybrid system is fed with a boron stream (J_f) that contains boron at the C_f(B) level. At the same time, suspension of the sorbent is delivered to the system (J_s(in)). The stream contains X(in) sorbents that is loaded with boron at q(in) level. Two streams leave the system: boron loaded suspension (J_s(out), X(out) q(out)) and water (J_p, C_p(B)). With the following assumptions,

1. $J_f = J_p$
2. $J_s(\text{in}) = J_s(\text{out})$
3. $X(\text{in}) = X(\text{out})$
4. Sorbent residence time is long enough to allow boron uptake to reach the maximal value

one is able to balance the system

$$J_f C_f(\text{B}) + J_s(\text{in})X(\text{in})q(\text{in}) = J_p C_p(\text{B}) + J_s(\text{out})\, X(\text{out})q(\text{out})$$

The simple rearrangement of the above balance leads to the following equation

$$C_p(B) = C_f(B) - \frac{J_s}{J_f} X(\text{out})q(\text{out})$$

that correlates boron concentration in permeate to ratio of suspension flow to feed flow and to the amount of sorbent in the system.[23]

Recently, studies on the use of submerged membrane systems have shown some advantages over conventional membrane filtration approach.[2,19,35] The particles of sorbent were not adsorbed on the surface of porous membranes and the whole system worked in the stable mode. This phenomenon can be explained for two reasons: at first the filtration process took place below the critical flux limit thus there was no chance of

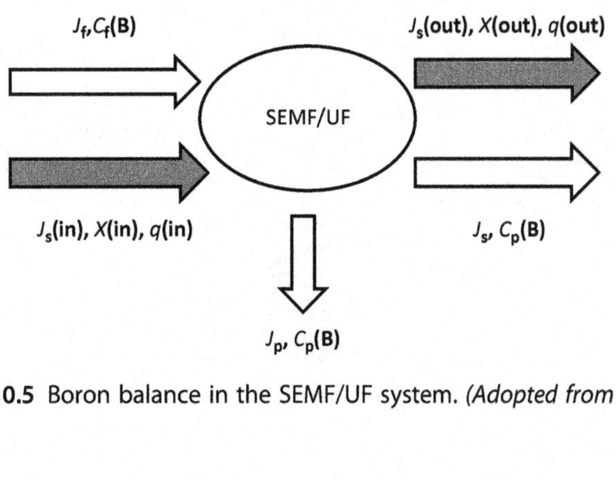

Figure 10.5 Boron balance in the SEMF/UF system. *(Adopted from Ref. 23).*

membrane fouling, second, the bubbles of air streams injected at the bottom caused gentle mixing of the whole system and vibration of filtration membranes. The conditions did not allow particles to be adsorbed. It seems that the use of submerged membrane systems, very popular for wastewater treatment plants, can be suitable for the construction of SEMF/UF hybrids.

The fact that SEMF/UF hybrids are not so widely implemented today is commented on by Abdulgader et al.[17] The reason is the cost of regeneration of the sorbent. The cost is generated by the frequent need to regenerate sorbents and the cost of chemicals used. The authors suggested the need to find a way to reduce regeneration frequency and use less expensive chemicals. The last step can be done by integration of hybrid with other processes. It is possible to perform electrodialytic regeneration of acid used for splitting sorbent–boron complexes[36] or by applying Donnan dialysis with RO-concentrated brine as the driving solution.[37] In the first application, ED concentrate, with a high content of hydrochloric acid, could be used directly for column regeneration, hence the consumption of fresh acid will be reduced. The second approach used the Donnan dialysis idea for removal of boron from the sorbent suspension. Highly concentrated NaCl brine and anion exchange membranes were used for sorbent regeneration. The scheme of separation is shown in Figure 10.6.

It was suggested to use brine from the first stage of seawater RO desalination as the driving solution and porous anion exchange membranes as a separator. In such an approach, sorbent regeneration was slow as it was based on the diffusion process. Its economy, however, can be improved by utilization of seawater RO brine.

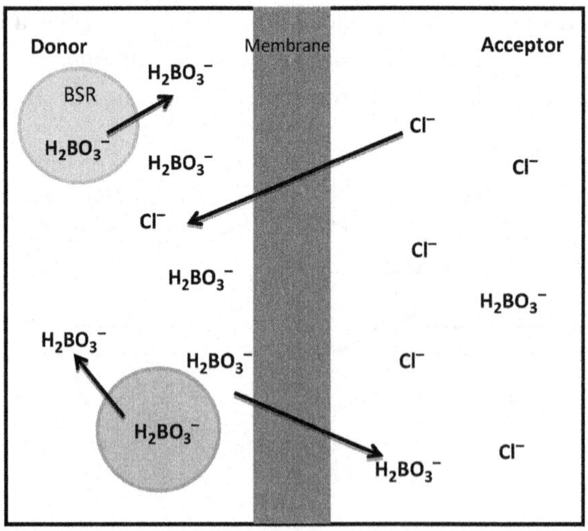

Figure 10.6 Scheme of sorbent regeneration by donnan dialysis. *(Adopted from Ref. 37).*

10.6 CONCLUSIONS

The membrane-based hybrid processes become powerful separation tools for boron removal from water. They allow removal and/or recovery of boron that can exist in water even at trace amounts. It is evident that hybrid processes seem to be more profitable with lower consumption of chemicals and energy. However, the laboratory data should be verified in the pilot-scale process that runs for a long time.

REFERENCES

1. Xu Y, Jiang JQ. Technologies for Boron removal. *Ind Eng Chem Res* 2008;**47**:16–24.
2. Guler E, Kabay N, Yuksel M, Yigit NO, Kitis M, Bryjak M. Integrated solution for boron removal from seawater using RO process and sorption-membrane filtration hybrid method. *J Memb Sci* 2011;**375**:249–57.
3. Kabay N, Yilmaz I, Bryjak M, Yuksel M. Removal of boron from aqueous solutions by a hybrid ion exchange-membrane process. *Desalination* 2006;**198**:158–65.
4. Kabay N, Bryjak M. Hybrid processes combining sorption and membrane filtration. In: Hoek E, Tarabara VV, editors. *Encyclopedia of membrane science and technology*. Wiley; 2013.
5. Geffen N, Semiat R, Eisen MS, Balazs Y, Katz I, Dosoretz CG. Boron removal from water by complexation to polyol compounds. *J Membr Sci* 2006;**286**:45–51.
6. Dydo P, Nemś I, Turek M. Boron removal and its concentration by reverse osmosis in the presence of polyol compounds. *Sep Pur Technol* 2012;**89**:171–80.
7. Dydo P. Transport model for boric acid, monoborate and borate complexes across thin-film composite reverse osmosis membrane. *Desalination* 2013;**311**:69–79.
8. Dydo P. The influence of D-mannitol on the effectiveness of boric acid transport during electrodialytic desalination of aqueous solution. *J Membr Sci* 2013;**429**:130–8.
9. Sanchez J, Rivas B, Nazar E, Bryjak M, Kabay N. Boron removal by liquid-phase polymer-based retention technique using poly(glycidyl methacrylate *N*-methyl –D-glucamine). *J Appl Polym Sci* 2013;**129**:1541–5.
10. Smith BF, Robinson TW, Carlson BJ, Labouriau A, Khalsa GRK, Schroeder NC, et al. Boric acid recovery using polymer filtration: studies with alkyl monool, diol, and triol containing polyethylenimines. *J Appl Polym Sci* 2005;**97**:1590–604.
11. Dilek C, Ozbelge HO, Bicak N, Yilmaz L. Removal of boron from aqueous solutions by continuous polymer-enhanced ultrafiltration with polyvinyl alcohol. *Sep Sci Technol* 2002;**37**(6):1257–71.
12. Smith BM, Todd P, Bowman CN. Hyper branched chelating polymers for the polymer-assisted ultrafiltration of boric acid. *Sep Sci Technol* 1999;**34**(10):1925–45.
13. Doganay CO, Ozbelge HO, Yilmaz L, Bicak N. Removal and recovery of metal ions via functional polymer based PEUF. *Desalination* 2006;**200**:286–7.
14. Doganay O, Ozbelge HO, Bicak N, Aydogan N, Yilmaz L. Use of specifically tailored chelating polymers for boron removal from aqueous solutions by polymer enhanced ultrafiltration. *Sep Sci Technol* 2011;**46**:581–91.
15. Zerze H, Ozbelge HO, Bicak N, Aydogan N, Yilmaz L. Novel boron specific copolymers with quaternary amine segments for efficient boron removal via PEUF. *Desalination* 2013;**310**:169–79.
16. Bryjak M, Duraj I, Pozniak G. Colloid-enhanced ultrafiltration in removal of trace amounts of borates from water. *Environ Geochem Health* 2010;**32**:275–7.
17. Abdulgader HA, Kochkodan V, Hilal N. Hybrid ion exchange-pressure driven membrane processes in water treatment: a review. *Sep Purif Techol* 2013;**116**:253–64.
18. Jacob C. Seawater desalination: boron removal by ion exchange technology. *Desalination* 2007;**205**:47–52.
19. Blahusiak M, Onderkova B, Schlosser S, Annus J. Microfiltration of microparticulate boron adsorbent suspensions in submerged hollow fibre and capillary modules. *Desalination* 2009;**241**:138–47.

20. Bryjak M, Wolska J, Soroko I, Kabay N. Adsorption—membrane filtration process in boron removal from first stage seawater RO permeate. *Desalination* 2009;**241**:127—32.
21. Kabay N, Bryjak M, Schlosser S, Kitis M, Avlonitis S, Matejka Z, et al. Adsorption—membrane filtration (AMF) hybrid process for boron removal from seawater: an overview. *Desalination* 2008;**223**:38—48.
22. Yilmaz I, Kabay N, Bryjak M, Yuksel M, Wolska J, Kołtuniewicz A. A submerged membrane-ion-exchange hybrid process for boron removal. *Desalination* 2006;**198**:310—5.
23. Bryjak M, Wolska J, Kabay N. Removal of boron from seawater by adsorption membrane hybrid process: implementation and challenges. *Desalination* 2008;**223**:57—62.
24. Wolska J, Bryjak M. Preparation of polymeric microspheres for removal of boron by means of sorption—membrane filtration hybrid. *Desalination* 2011;**283**:193—7.
25. Wolska J, Bryjak M, Kabay N. Polymeric microspheres with *N*-methyl-D-glucamine ligands for boron removal from water solution by adsorption-membrane filtration process. *Environ Geochem Health* 2010;**32**:349—52.
26. Samatya S, Orhan E, Kabay N, Tuncel A. Comparative boron removal performance of monodisperse-porous particles with molecular brushes via "click chemistry" and direct coupling. *Colloids Surf A Physicochem Eng Asp* 2010;**372**:102—6.
27. Gazi M, Galli G, Bicak N. The rapid boron uptake by multi-hydroxyl functional hairy polymers. *Sep Purif Technol* 2008;**62**:484—8.
28. Yavuz E, Gursel Y, Senkal BF. Modification of poly(glycidyl methacrylate) grafted onto cross-linked PVC with iminopropylene glycol group and use for removing boron from water. *Desalination* 2013;**310**:145—50.
29. Samatya S, Kabay N, Tuncel A. A hydrophilic matrix for boron isolation: monodisperse porous poly(glycidyl methacrylate-co-ethylene dimethacrylate) particles carrying diol functionality. *React Func Polym* 2010;**70**:555—62.
30. Samatya S, Tuncel A, Kabay N. Boron removal from geothermal water by a novel monodisperse porous poly(GMA-co-EDM) resin containing *N*-methyl-D-glucamine functional group. *Solv Ext Ion Exch* 2012;**30**(4):341—9.
31. Samatya S, Kabay N, Tuncel A. Monodisperse-porous *N*-methyl-D-glucamine functionalized poly(vinylbenzyl chloride-co-divinylbenzene) beads as boron selective sorbent. *J Appl Polym Sci* 2012;**126**(4):1475—83.
32. Yilmaz-Ipek I, Holdich R, Kabay N, Bryjak M, Yuksel M. Kinetic behaviour of boron selective resins for boron removal using seeded microfiltration system. *React Funct Polym* 2007;**67**:1628—34.
33. Marston C, Busch M, Prabhakaran S. A boron selective resin for seawater desalination. In: *Proceedings of european desalination society conference on desalination and the environment, Santa Margherita Ligure, Italy, 22—26 may 2005*; 2005.
34. Kabay N, Koseoglu P, Yapıcı D, Yuksel U, Yuksel M. Coupling ion exchange with ultrafiltration for boron removal from geothermal water-investigation of process parameters and recycle tests. *Desalination* 2013;**316**:17—22.
35. Blahusiak M, Schlosser S. Simulation of the adsorption-microfiltration process for boron removal from RO permeate. *Desalination* 2009;**241**:156—66.
36. Turek M, Dydo P, Trojanowska J. Electrodialytic utilization of boron IE column post-regeneration. *Desalination* 2008;**223**:113—8.
37. Bryjak M, Pozniak G, Kabay N. Donnan dialysis of borate anions through anion exchange membranes: a new method for regeneration of boron selective resins. *Reac Funct Polym* 2007;**67**:1635—42.

CHAPTER 11

Boron Removal Using Ion Exchange Membranes

Piotr Dydo, Marian Turek
Silesian University of Technology, Gliwice, Poland

11.1 INTRODUCTION

The low boron content limit for drinking water of 2.4 mg/L has led to the necessity for a reduction of the boron content in drinking waters; irrigation waters for some plants require even lower concentrations.[1–3] Conventional methods for water treatment, however, do not significantly remove boron. The methods proven to be efficient in the reduction of the boron content in waters are either adsorption or membrane based.[4] Among the membrane-based methods, reverse osmosis (RO) is most commonly reported[2–4]; however, some achievements in boron removal using ion exchange membranes (IEMs) have also been discussed. These achievements include the reports on boron removal in electrodialysis (ED), Donnan dialysis (DD), and electrodeionization (EDI) systems. IEMs are either cation exchange membranes (CEMs) or anion exchange membranes (AEMs). IEMs are usually made of synthetic ion exchangers and produced in the form of a thin foil membrane. In an ED system CEMs and AEMs are placed alternately and exposed to an external electric field as shown in Figure 11.1.

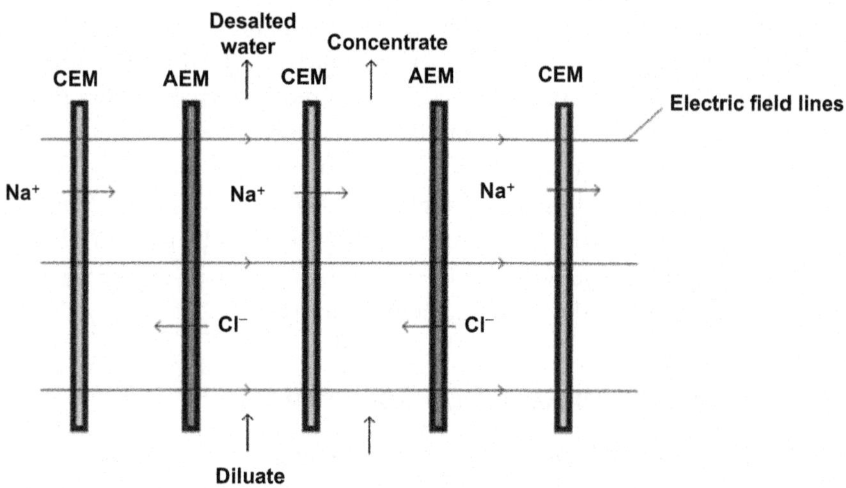

Figure 11.1 The scheme of ED.

Boron Separation Processes
ISBN 978-0-444-63454-2, http://dx.doi.org/10.1016/B978-0-444-63454-2.00011-3

Since electric field lines are positioned perpendicularly to the membrane, both the cations and anions start moving across the membranes, however, in the opposite directions. They move from the so-called diluate compartment into the concentrate compartment. Anions migrate across AEMs while cations migrate across CEMs. The rates of cation transport across AEM and anion transport across CEM are so small, so these are considered to be negligible in the majority of cases. Diluate and concentrate compartments are fed alternately therefore diluate solution becomes ion-depleted while concentrate solution becomes ion-enriched with the progress of ED. As a result ionic salts are effectively removed from the diluate. The main driving force for ED desalination is electric potential gradient.

Unlike ED, in a DD process there is no external electric field. Also, membranes of one type only (CEMs or AEMs) are applied. In the course of DD, e.g., borate is transported from the so-called feed solution in the receiving solution across AEM. However, to meet the feed solution electroneutrality requirement, an equivalent amount of another anion needs to be transported from the receiving solution in the feeding solution. The latter anion is referred to as the driver anion. The driver ion could have been chloride, sulfate, or even hydroxyl. The whole process proceeds until a certain ratio between the concentration of the borate and driver ion in both receiving and feed solutions is achieved. This ratio is defined by a Donnan equilibrium. Due to the equilibrium nature of the process, a large excess of driver over the removed ion is usually required.

In an EDI system a mixed-bed (MB) ion exchange (IE) resin is usually placed between the CEM and AEM in a diluate compartment of an ED unit (there are several different EDI designs, in an "all filled" design concentrate compartment is also resin filled).[5] Water, preliminarily deionized, usually by RO or ED, to salinity below 100 mg/L, enters the enhanced transfer zone of an EDI unit. In this zone ions are transported across the diluate compartment by electromigration enhanced by the presence of the resin increasing the compartment conductivity. In the electroregeneration zone due to low water salinity ED is operated over limiting current density that leads to the water dissociation (water splitting) and the resins become electrochemically regenerated to the H^+ and OH^- forms. Under these conditions, the resin acts as a continuously regenerated MB IE column, exchanging H^+ and OH^- ions with ions present in the solution. Also, dissociation of weak electrolytes is enhanced in the electroregeneration zone thus EDI removes more effectively, than classical ion exchange, such species as silica, carbon dioxide, ammonia, and boric acid.

11.2 BORON SPECIES IN AQUEOUS SOLUTION

In nature, boron exists mainly in the form of borates or boric acid. Boric acid is a weak electrolyte, with the reported pK_a values falling in the range of approximately 8.68–9.25[6,7] depending on temperature and salinity. These pK_a values only consider the ionic equilibrium between boric acid (H_3BO_3) and monoborate ($B(OH)_4^-$). A high value for pK_a indicates that at low pH (<9), boric acid is present, while at high pH (>9) and in the diluted solutions, the monoborate anion dominates as shown in Figure 11.2.

However, at higher concentrations (>0.025 mol/L of B), triborate ($B_3O_4(OH)_3^-$) and tetraborate ($B_4O_5(OH)_4^-$) ion formation must also be considered[2,3,8] as shown in Figure 11.3.

The two species, boric acid and monoborate, differ significantly in their character; most importantly, boric acid molecules are electrically neutral, while monoborate exhibits a distinct electric charge. Polyborate ions also present a distinct electric charge. This difference thus influences the extent to which the species interfere with charged membranes. Therefore, the effectiveness of boron transport and removal across IEMs must be discussed with respect to the type of species that dominates in the feedwater.

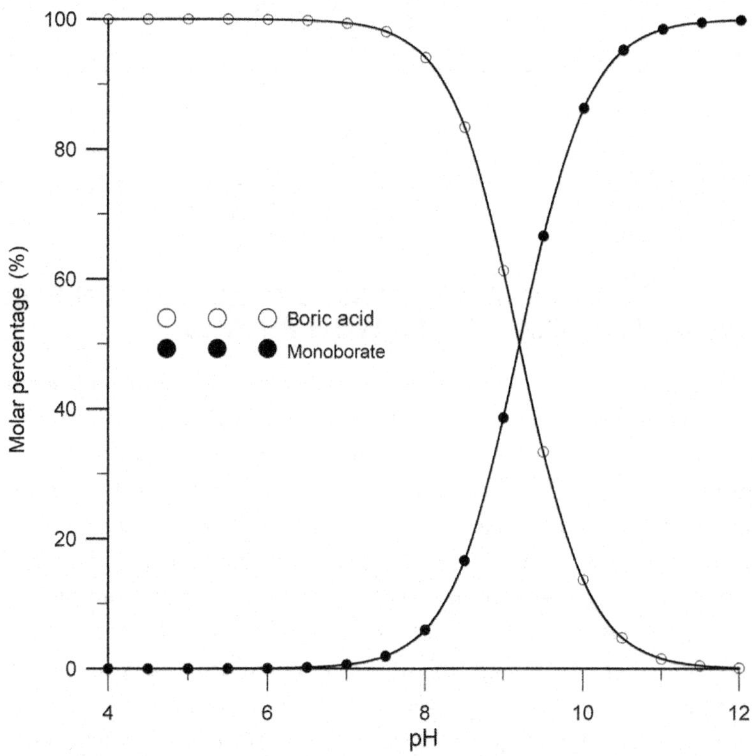

Figure 11.2 Molar fraction of boron species vs pH in a solutions with boron content less than 0.025 mol/L calculated for pKa = 9.2.

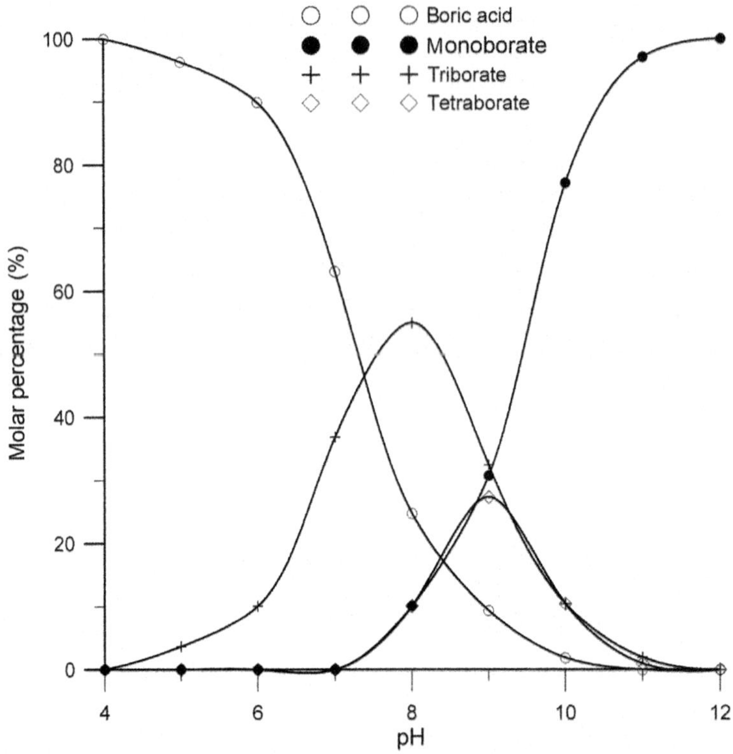

Figure 11.3 Molar fraction of boron species vs pH in a solution with boron content of 0.2 mol/L. *(Based on Ref. 8).*

11.3 REPORTS ON BORIC ACID TRANSPORT ACROSS ION EXCHANGE MEMBRANES FROM WATERS WITH PH <9.0

Reports on boric acid transport in ED are the most widely found in the existing literature.[2,9–21] However, the effectiveness of boric acid removal or the boron transport electric current efficiency in feedwaters with pH <9 in ED is poor when compared with the removal of ionic species.[9–14] According to the data presented by Melnik et al.[9] the electric current efficiency of boric acid transport in ED of seawater and underground water does not exceed 1.2%. In addition, boron fluxes remained low (<15 $\mu g/m^2$ s of B). The maximum degree of boron removal reported in this work was 87%. Turek et al.[10–12] reported that it is only possible to remove less than 12.4% of boron from the waters of low salinity at pH <9 into the neutral concentrate. However, when the pH of the concentrate was brought up to above 11.2 the effectiveness of boron removal increased up to 85%.[12] The electric current efficiency for boron removal did not exceed 15%. Also the electric current efficiency for ionic species was low. A significant fraction of the electric current was used in the above process, which resulted in reduced electric current efficiency for boron removal. Kabay et al.[13] reported that at a pH of 9.0, only 20% of boron can be removed, while at higher pH, the effectiveness of boron removal can be much higher. In addition, Banasiak and Schafer[14] reported that at pH <9, a maximum of 40% of boron can be removed from water of moderate salinity (5 g/L of NaCl). A summary of the experiments reported in Refs 9–14 is presented in Table 11.1.

All of the above reports indicate that boric acid fluxes in ED systems are relatively low in a wide diluate salinity range. The low boric acid fluxes can be understood when considering that the only likely mechanism for boric acid transport across IEMs is diffusion. Because diffusion is a relatively slow process, the rate of boric acid transport and the effectiveness of its removal remain low. In contrast, ions are transported via migration in an external electric field with diffusion usually acting in the opposite direction. Because the potential gradient within IEMs is high, and membrane resistivity is usually low, the observed ion fluxes are high. These ion fluxes are also high in comparison to boric acid fluxes.

In fact, boric acid fluxes across IEMs are so small that the possibility of boric acid separation from ionic species using ED has been strongly considered.[10,11,15–18] Melnik et al.[15,16] reported that by applying ED, it is possible to separate mineral acids from boric acid. That group examined the effectiveness of ED in the removal of acidity from boric acid mixtures using HCl or H_2SO_4. These solutions simulated wastes produced on the regeneration of the boron-selective sorbents. The researchers observed that even when 99% of the strong acid was removed from the diluate, a drop in the boron content of the diluate reached only 12% and the majority of boron remained in the diluate. Turek et al. examined removal of acidity from HCl and boric acid mixture obtained in the course of regeneration of the boron-selective sorbent. The IE postregeneration lye containing 17.57 g/L of HCl and 2.33 g/L of boron was deacidified to 0.41 g/L of HCl while a drop in the boron content of the diluate reached only 5.6%.[19] This finding led the authors[15,16,19] to the conclusion that ED should be effective in separating "the acidulous solutions." Similarly, Turek et al.[10,11] demonstrated the possibility of salt separation from boric acid using ED. The examined wastewater was of low salinity but contained approximately 75 mg/L of B. The researchers reported that at a diluate pH of approximately 3.2, only 0.8% of boron could be removed despite a 90% or more reduction in salinity. Under similar conditions at the diluate pH of approximately 7, only 5% of boron was removed from the diluate. Bandura-Zalska et al. analyzed the effect of the diluate pH, membrane pair voltage drop, and desalination degree on the same wastewater.[17] Those researchers observed that the boron content in the concentrate increased mainly with the pH of the diluate and with the desalination degree. However, only a slight, though negative, impact of the voltage drop on the boron content in the concentrate was observed. Despite the high concentration of boron in the feedwater (75 mg/L), the possibility of a high (approximately 99%) reduction in salinity was reported. The produced concentrate contained less than 1 mg/L of B. Dydo discussed the effect of

Table 11.1 The Summary of Reports on Boric Acid Removal by ED (at diluate pHs lower than 9.2)

Report By	Initial Diluate Salinity, Boron Content, and pH	Boron Flux, μg/m^2 s	Current Efficiency for Boron, %	Other
Melnik et al.[9]	3.0–15.0 g/L NaCl, 1.5–4.0 mg/L B, pH 5.5	<15	<1.2	Up to 87% of boron removed. Membranes applied: MK-40 and MA-40 (Russia); MK-100 and MA-100 (Ukraine), ionics
Turek et al.[10]	1.2–2.3 g/L TDS, 62.5 and 76.5 mg/L B, pH 3.2 and 8.3	<115	<1	Up to 12.4% of boron removed; electric current efficiency for Cl– and SO_4^{2-} equaled to 52% only. Membranes applied: AMX–CMX and ACS–CMS (Neosepta)
Turek et al.[11]	0.6–2.3 g/L TDS, 62.5 and 76.5 mg/L B, pH 3.2 and 6.95	<115	n/a	Up to 5.1% of boron removed membranes applied: AMX–CMX (Neosepta)
Turek et al.[12]	0.4–0.6 g/L TDS, 1.3 and 2.25 mg/L B, pH <9	<45	<12	Up to 85% of boron removed. The concentrate pHs were 11.2 and 12 membranes applied: AMX–CMX (Neosepta)
Kabay et al.[13]	25–100 mg/L B, pH 8.5 and 9.0	n/a	n/a	Up to 20% of boron removed. Membranes applied: AMX–CMX (Neosepta)
Banasiak et al.[14]	5 g/L NaCl, 10 mg/L B, pH 3–9	<0.25	n/a	Up to 40% of boron removed membranes applied: AMX–CMX (Neosepta)

n/a, data not available; TDS, total dissolved solids.

process parameters on the rate of boric acid transport during the ED desalination of aqueous solutions containing selected salts.[18] The boron flux was reported to depend upon several factors, such as the type of membrane, the boron concentration in the diluate, the type of ion that migrates across selected membranes, and the electric current density. Most importantly, the boron fluxes across AEMs were always observed to be much higher than those across CEMs. This information indicates that the majority of the boric acid is transported across AEMs. Thus, boric acid transport across AEMs should be of special interest in any research work. It was also noted that regardless of the membrane type, the flux of boron across AEM in the presence of sulfate was considerably higher than that in the presence of nitrate and chloride.[18] Thus, to reduce boron flux, it would be wise to reduce the sulfate content, if possible. Similar to the AEMs, the flux of boron across CEMs decreased with cation type in the following series: $K^+ > Na^+ > Ca^{2+} > Mg^{2+}$. This cation series confirms the findings of Goli et al.[20] Those researchers examined the effect of salt and acids on the diffusion of boric acid across CEMs and observed that boric acid is transported across CEMs solely by the diffusion of boric acid molecules, without any significant contribution of negatively charged borates. The existing differences in the membrane diffusivities of boric acid with respect to the cations were explained as being the result of a change in the mean viscosity of the confined solution in the membranes.[20] In addition, the mechanism of boric acid transport during electrodialytic desalination was proposed.[21] This result confirms that boric acid is transported across both AEMs and CEMs via diffusion during ED. However, a certain degree of a "convective" drag transport mechanism also appears to be present. This drag mechanism manifests itself as an increase in boron flux with increasing ion flux at a fixed concentration gradient. This mechanism can be understood as a type of coupling between boric acid molecules and ions themselves or boric acid. Such a coupling is analogous to the solvent–solute coupling reported in RO.[6] The effect of the membrane transport coefficient is discussed in Ref. 21.

Exceptionally high boron fluxes and electric current efficiencies of up to 220% were reported by Turek et al.[12,22] when boron was transported from neutral wastewater into an alkaline concentrate (pH >11) during ED. The electric current efficiency was determined by applying Faraday's laws of electrolysis. Electric current efficiency higher than 100% was definitely a result of something other than borate migration-driven transport of boric acid from the diluate, which was not explained by Faraday's equations. The examined wastewater contained 75 mg/L of B and was preliminarily desalinated. As a result of this process, up to a 40% reduction in the diluate boron content was reported with a maximum boron flux of approximately 860 $\mu g/m^2$ s.[22] High electric current efficiencies were first explained by Turek et al. as resulting from DD.[22] Due to the hydroxyl diffusion from the concentrate to the diluate, borate ions were thought to be produced in the AEM membrane adjacent to the boundary layer. Hydroxyl was also thought to act as a carrier and drive borate to the concentrate until a Donnan equilibrium was achieved.[22] In light of recent reports on boric acid diffusion across AEMs, simple boric acid diffusion could also have explained this effect. Because the diluate was preliminarily demineralized, the ion flux and its related electric current density were low even at a high membrane pair voltage drop. However, the diffusive transport of neutral boric acid species does not contribute to the electric current at all. Therefore, when the molar flux of monovalent ions was much lower than that of boric acid, the predicted electric current efficiency of boron transport was higher than 100%. This result occurs because more moles of B were transported than should have been transported based solely on the electric current. In the latter mechanism, the only role of the electric field was to maintain the difference in pH between the ED compartments.

The above described findings concerning boron transport into alkaline ED concentrate led to the development of a concept of a two-stage electrodialytic process for boron removal.[23] Boron-containing water is first desalinated at low pH to remove calcium and magnesium cations; the process is carried out at low pH to avoid borate transport. Then, in the subsequent step boric acid is transferred from ion-depleted diluate into the alkaline concentrate across IEMs. As a result the purified water that contains less than 1 mg/L of boron and a relatively concentrated borate solution are obtained. The latter solution can be used as raw material for the manufacturing of boron compounds for commercial purposes. The effectiveness of boron transport into the alkaline concentrate during the course of ED was examined in a batch mode. The effect of the concentrated boron content and pH as well as voltage drop was investigated. The obtained results were evaluated statistically and the mathematical model for the boron flux was revealed. The obtained results were applied to propose a system for ED boron removal and its concentration. An asymptotic decrease in the diluate boric acid content and an asymptotic increase in boric acid concentration in the concentrate with time were observed. The observed asymptotes were approximately at equal concentrations of boric acid in the diluate and in the concentrate. This was observed despite the varying concentrate total boron content and pH values as well as the varying membrane pair voltage drop. Such a behavior suggested that the driving force for boron transport was the difference in boric acid concentration between the diluate and the concentrate. It was then concluded that a simple, one-dimensional diffusion of boric acid across IEMs controlled the rate of the boron transport from the diluate in the alkaline concentrate. The minute boron flux values obtained from the batch experiments were then analyzed using the multiple regression method. It was found that a boric acid mass transfer coefficient, with its value $(50 \pm 2)\cdot 10^{-6}$ m/s, was the only statistically significant estimate of j_B. Based on the results concerning boron flux in the experiments carried out in batch mode a cascade of electrodialysers was suggested and verified using an ED unit with an effective membrane length equal to 42 cm. Two envisioned cascade stages were tested: the first of 5000 mg/L of boron in the concentrate and the second of 1000 mg/L of boron. The measured results were compared with those calculated based on the mathematical model that comprised the boric acid mass transport equation and mass balances. It was found that the developed model can be used for ED cascade design, as it enables one to predict diluate boron content with satisfying accuracy. The calculated and measured diluate boron contents did not differ by more than 5% in both cascade stages. The values of electric current efficiency equal to 790% and 940% in the first and second stages, respectively, confirmed that the mechanism of electrodialytic boron removal into alkaline concentrate was boric acid diffusion through the AEM. The role of the electric field in that process is to maintain the high pH value of ED concentrate to assure high boric acid dissociation rate in this solution and high boric acid concentration gradient between the diluate and the concentrate compartments being the driving force. It was found that a cascade of three electrodialysers with gradually decreasing concentrate boron content in the subsequent stages was required to reduce boron content in the diluate below 1 mg/L. It was assumed that the ratio between the initial diluate boron content and boron concentration in the concentrate should be equal for every cascade stage. This ratio was equal to 66.7. The feed boron content in the first stage of the cascade was 75 mg/L; the concentrate boron content was 5000 mg/L and the concentrate pH was 12.

11.4 REPORTS ON BORATE TRANSPORT ACROSS ION EXCHANGE MEMBRANES FROM WATERS WITH PH >9

Most of the reports on the effectiveness of borate transport across IEMs concerns ED. It is widely reported that as the pH of boron-containing water is increased to just above a pH of 9, a significant increase in the rate of boron transport in ED is observed.[9–16,24] The reported borate fluxes lie in the approximate range of 80–420 $\mu g/m^2$ s. Melnik et al.[9] reported that for heterogeneous membranes, an increase in the boron transport rate across an AEM membrane at pH >9 was accompanied by a decrease in the rate of boron transport across the CEM. It is clear that under these conditions (pH >9), borates are the boron species that dominates. As a negatively charged species, borates are transported across AEMs rather than across CEMs.

Moreover, the following increases in the boron removal degree have been reported in the literature: from 12.4% at an initial diluate pH of 8.31 to 17.8% at a pH of 10.5[10]; from 0.8% at an initial diluate pH of 3.2 to 97% at a pH of around 10[11]; from less than 20% at a diluate pH of 9.0 to approximately 80% at a pH of 10.5[13]; from approximately 20% at a diluate pH <7 to 60% at a pH of 12[14]; from 33% at a diluate pH of 3.6 to approximately 75% at a pH of 12.8.[15] This suggests that with an increase in pH, the rise in borate content creates the conditions for more efficient transport and removal of boron by ED. The summary of the above-mentioned experiment conditions is presented in Table 11.2.

Yazicigil and Oztekin,[24] however, arrived at a contradictory conclusion. They reported that in a 0.1 mol/L boron-containing solution a maximum rate of boron transport in ED was observed at a diluate pH near 9.0 with AHA (Tokuyama Soda)

Table 11.2 The Comparison of Reports on Boron Removal by ED at Different Diluate pHs

Report By	Initial Diluate Salinity, pH, and Boron Content	Boron Flux, μg/m^2 s	Current Efficiency for Boron, %	Other
Turek et al.[10]	1.2 g/L TDS, pH 8.31, 63.5 mg/L B	115	3.66	Up to 12.4% of boron removed, electric current efficiency of Cl− and SO_4^{2-} equaled to just 52%
	1.2 g/L TDS, pH 10.5, 59 mg/L B	130	3.76	Up to 18% of boron removed, current efficiency of Cl− and SO_4^{2-} equaled to just 43%. Membranes applied: CMS−ACS (Neosepta)
Turek et al.[11]	1.8 g/L TDS, pH 3.2, 76.5 mg/L B	<115	n/a	Up to 0.8% of boron removed
	360 mg/L TDS, pH 9.25−10, 59.6 mg/L B	416	Up to 25%	Up to 97% of boron removed membranes applied: AMX−CMX (Neosepta)
Kabay et al.[13]	100 mg/L B, pH 9.0	n/a	n/a	Up to 20% of boron removed
	100 mg/L B, pH 10.5	n/a	n/a	Up to 80% of boron removed membranes applied: AMX−CMX (Neosepta)
Banasiak et al.[14]	5 g/L NaCl, 10 mg/L H_3BO_3, pH 7	<28	n/a	Up to 20% of boron removed
	5 g/L NaCl, 10 mg/L H_3BO_3, pH 12	n/a	n/a	Up to 60% of boron removed membranes applied: AMX−CMX (Neosepta)
Melnyk et al.[15]	0.2 mol/L HCl, 162 mg/L B, pH 3.6	n/a	n/a	Up to 33% of boron removed
	0.2 mol/L HCl, 162 mg/L B, pH 12.8	n/a	n/a	Up to 75% of boron removed membranes applied: MK-40 and MA-40 (Russia)

n/a, data not available; TDS, total dissolved solids.

membrane. At higher pH, there was a dramatic drop in boron flux. Their finding was justified by the presence of polyborate ions at high boron concentrations. It appears that the transport rate of some of these polyborate ions is much higher than that of other ions, and that the polyborate dominates at a pH near 9.5. As shown in Figure 11.3 at a pH of around 9 there is a maximum fraction of tetraborate. Thus, the polyborate with a high transport rate could be $B_4O_5(OH)_4^{2-}$. The findings of Yazicigil and Ozetkin were similar to those reported by Ayyildiz and Kara,[25] related to the borate removal by DD presented in the following section.

A consensus has been reached that an increase in diluate boron concentration causes an increase in the boron (borate) flux in ED[12,13,24] and that boron (borate) is transported faster in the presence of chloride ions than in the presence of sulfates or nitrates. This difference was stated[13,24] to be the result of the difference in ion charge and hydrated anion radius; however, no detailed explanation of how it affects the flux of boron was provided.

Turek et al.[10–12] analyzed the effect of salinity on the electric current efficiency of boron (borate) transport in ED. Those researchers observed that the electric current efficiencies were low as long as the salinity of the alkaline diluate remained high. According to the authors,[10–12] during the initial part of the experiments, no (or almost no) boron was transported from the diluate. Then, when more than 90% of the Cl^- was removed from the diluate, a dramatic increase in the boron (borate) electric current efficiency was observed. This increase was accompanied by an increase in the boron flux. Such an effect can be identified as the result of low borate mobility (diffusivity) as compared with that of the other ions, such as Cl^- present in the waters; as long as ions of higher mobility than borates are present in the diluate, the boron (borate) electric current efficiency and flux remain low in ED. Even in an ion-depleted diluate, the electric current efficiency of boron (borate) transport does not exceed 30%. The remaining percentage of the electric current was most likely utilized for hydroxyl ion transport. A considerable drop in the pH of the diluate was reported afterward with a reduction in the flux of boron because all of the borate present was converted into boric acid.

The above-discussed literature reports indicate that boron can be removed effectively only from alkaline diluate, by ED. However, at high pHs a serious risk of membrane scaling, e.g., with insoluble compounds of calcium or magnesium, occurs. Therefore, the ED diluate needs to be desalinated prior to boron removal, e.g., by ED as shown in Section 3, from waters with pH <9. From so desalinated waters, borate can be removed at high rate and the highest possible, though limited, electric current efficiency.

Recently, tentative but promising results of borate removal by ED with bipolar membranes (BPMs) were presented by Nagasawa et al.[26,27] BPM is a membrane that contains anion exchanger on one side and cation exchanger on the other side. In such an arrangement solution in membrane pores does not contain ions, just water molecules. In an external electric field water dissociation within the BPM takes place and hydroxyl and hydrogen ions are produced. These ions are transported toward the anode and cathode, respectively. As a result the solution adjacent to the anion exchange part of BPM becomes alkaline, which according to Nagasawa et al. results in borate formation in boron-containing feedwater. Borate is then transported across AEM to the concentrate solution, which is adjacent to the cation exchange part of the BPM. Within that cation exchange part hydrogen ions are liberated and transported in the concentrate, which results in borate neutralization. Therefore, the feed solution becomes boron depleted and the concentrate solution contains boric acid. Nagasawa et al. reported[27] that with the use of ED with BPM it is possible to remove more than 90% of boron from water that initially contained 100 mg/L of boron and NaCl at 0–20 mmol/L concentration. This was observed over a wide range of feedwater pHs (2.3–12). Electric current efficiencies of boron transport approached 25% in the case of water that did not contain NaCl, but it quickly diminished with NaCl content to less than 5% in solutions that contain more than 10 mmol/L NaCl. Moreover, the electric current efficiency diminished with initial pH of the feedwater from around 20% at pH 2 to less than 3% at pH 12. Such a behavior was explained by the authors to be the result of low mobility of borate as compared to that of hydroxyl. In most cases, the electric current efficiencies reported for ED with BPMs were similar to those reported for simple ED. We concluded that ED with BPMs will be as energy efficient as simple ED in borate removal. Nagasawa et al.[27] also demonstrated a rapid increase in the treated solution pH to the value of above 11 during the course of ED with BPMs. To that extent ED with BPMs reassembled simple ED at elevated pHs. At so high a pH value severe precipitation of calcium and magnesium in the form of insoluble compounds inside the ED module (scaling) is suspected. Therefore, as in the simple ED, the treated solution needs to be desalinated prior to borate removal or the applicability of ED should be limited to the waters of low scaling potential even at high pHs.

11.5 REPORTS ON BORATE TRANSPORT ACROSS ION EXCHANGE MEMBRANES BY DONNAN DIALYSIS

As an alternative to the reports on borate removal by ED, tentative information concerning the possibility of borate removal in DD has also been presented. Ayyldiz and Kara examined the effectiveness of borate transport across AEMs in DD process.[25] They observed that the boron flux in DD depends upon the type of the membrane, the concentration of boron in the feed solution, the pH of the feed and the receiving solution, the presence of accompanying ions in the feed solution, and the type of the carrier anion in the receiving solutions. The effect of the pH of the feed solution was found to be complex. At high boron concentration (of 0.1 mol/L), the maximum boron flux was observed near a pH of 9.5, while for a dilute solution (0.001 mol/L), the maximum boron flux was observed for the maximum examined feedwater pH of 11.5.

Such behavior was explained by the formation of polyborate ions at high boron concentrations and their absence in diluted solutions. The maximum rate of boron transport at a pH of approximately 9.5 for the concentrated boron solution was also observed in ED as mentioned in the previous section. It could be attributed to tetraborate ion formation as explained previously. Moreover, the boron-accompanying anions, the chlorides, bicarbonates, and sulfates, were observed to affect the rate of boron transport in DD. The maximum boron flux was observed in the presence of bicarbonate. However, the highest boron transport rates were reported when using sodium chloride in the receiving solution. Neosepta AHA and AMH membranes produced similar fluxes of boron, while the superiority of AFN was clearly observed. The above difference was attributed to be the result of the highest water content in the AFN membrane than in the remaining membranes. The maximum boron flux reported in Ref. 25 was approximately 3500 $\mu g/m^2$ s. Kir et al.[28] reported that plasma modification of the existing AEMs may result in a significant enhancement of the rate of boron transport in a DD process. The maximum flux of boron reported in Ref. 28 was approximately 410 $\mu g/m^2$ s.

Bryjak et al.[29] proposed a DD-based method for the regeneration of finely pulverized boron-selective resin (BSR) DOWEX XUS 43594.00. In this process, a BSR slurry with boron adsorbed on it is fed into the DD feed compartment. The authors assumed that some minute amounts of borate are always present in the feedwater at equilibrium with the BSR. These amounts were subject to transport into the receiving solution due to DD, which should ultimately result in the complete removal of boron from the BSR. The net effect of the process would be a regenerated resin in the chloride form and a boron (borate)-rich receiving solution. As stated by Bryjak et al.,[29] considerably high boron fluxes were observed during their experiments (up to 1500 $\mu g/m^2$ s), although the process kinetics were governed by boron desorption from BSR. This method may constitute an interesting approach for the regeneration of BSR used in an adsorption—membrane filtration process for boron removal.[30]

11.6 BORON REMOVAL BY EDI

Boron is not well removed by conventional anion exchange resin processes due to its poor ionization and low selectivity.[31] However, a combination of IE with ED, in the form of EDI, can dramatically increase the boron removal efficiency.

Wen et al.[32] investigated the effects of electric current, feed conductivity, flow rate of diluate and concentrate, and feed pH on boron removal by EDI. They revealed that less than 1 $\mu g/L$ boron in product water can be obtained under optimum conditions. Waters with so low a boron content can be utilized for ultra-large-scale integrated circuits production. In Ref. 33 a water purifying system capable of efficiently producing treated water containing boron at a low concentration was described. Water to be treated was first fed to the RO membrane apparatus and the permeate obtained was passed through the boron absorption column. The minute amounts of boron from the eluate were then removed by EDI. The product water contained boron at around 2 ppb.

RO units typically remove only a small portion of the boron from the feedwater; however, it is widely applied to remove ionic contaminants. When waters for power and semiconductor plants were treated in RO-EDI hybrid systems boron concentration was reduced from 22–280 ppb to 14–170 ppb (24.1–39.3% reduction) only by RO. In the consequent EDI step, boron concentration was reduced further to 0.45–2.75 ppb, which corresponds to 96.1–98.9% reduction in boron content.[34]

Darbouret and Kano[35] studied silica and boron removal through various steps in a water purification chain. An optimized system configuration was proposed that combines RO and EDI technologies in the pretreatment phase, and results in the efficient removal of boron. The system contributed significantly to eliminating trace levels of boron, from 240 ng/L in EDI feedwater to 12 ng/L in final ultrapure water.

A hybrid process coupling RO with EDI to remove boron and silica from geothermal water was also investigated.[36] The effects of applied voltage, feed flow rate, Na_2SO_4 concentration in the electrode compartments, membrane type, and IE resin bed configuration on the removal of boron and silica have been examined. The concentration of boron remaining in the permeate was 5.9 mg/L. It was decreased to 0.4 mg/L with a layered bed configuration of EDI system operated at 40 V. In an alternative, MB EDI system, boron concentration in the product water was as low as 1.60 mg/L at 40 V.

11.7 THE REPORTED COSTS OF BORON REMOVAL WITH ION EXCHANGE MEMBRANES

Reports on the costs of boron removal with IEMs are limited. According to Turek et al.[10,11] costs of ED boron removal from waters with high boron content (ca. 70 mg/L) ranges from \$0.96 to \$1.27/m^3 and the cost of removal of 1 g of B should be less than \$0.02. The costs of boron removal from the same feed into the alkaline concentrate are even lower: \$0.614/m^3 and <\$0.01/g B.[22] In the three stages of the ED system for boron removal from 75 mg/L to 0.8 mg/L, with its simultaneous concentration to 5000 mg/L,[23] the energy consumption was equal to \$0.094/m^3 while the total unit costs of boron removal were estimated at \$0.22/m^3.

The estimated costs of electricity for ED boron removal with BPMs from feedwater of 100 mg/L B content are \$0.13/m^3 and \$0.013/g B.[26] In contrast, the cost of ED boron removal from waters with lower boron content (2.25 mg/L) equals to \$0.23/m^3 and the cost of removal of 1 g of B equals to approximately \$0.10.[12] High unit costs of boron removal at lower boron contents are definitely the results of high share of investment costs in the total costs. No reports on energy requirements or costs of boron removal with DD are available; however, since the fluxes of boron in DD are similar to those in ED, as discussed in Section 5, it can be assumed that the costs of boron removal with DD should be similar.

The reported costs for boron removal by ED are similar to the costs of boron removal with IE or RO. According to Nadav et al.[37] the costs of boron removal by IE from RO permeate of 1.6 mg/L B content, ranges from \$0.0085 to 0.0107/m^3 (ca. \$0.04/g B). Jacob[38] reported the costs of RO boron removal from feedwater of similar boron content as \$0.05/m^3 (\$0.04/g B) while the cost of IE boron removal was \$0.09/m^3 (\$0.08/g B).

The similar costs of boron removal with IEMs and RO or IE suggest that IEMs offer a promising alternative to the other methods, especially at high boron concentrations. However, the necessity for initial diluate demineralization prior to boron (borate) removal at high pH needs to be taken into account. This necessity is the result of possible scale formation at high pHs and constitute the fundamental limitation to the applicability of IEMs in borate removal. The costs of such a demineralization should depend on the salinity. Turek et al.[11] reported a cost of \$0.19/m^3 in the case of desalination of feed-waters with low salinity (0.6 g/L TDS) and 75 mg/L boron content. It constituted only about 20% of the whole costs for boron removal (\$1.27/m^3). Similar cost of desalination (\$0.30/m^3) was reported for waters with 1.8 g/L TDS and 75 mg/L content.[17] At higher salinity, however, the share of the demineralization cost could have been higher and needs to be taken into account when assessing the ED economic feasibility of boron removal from waters.

11.8 CONCLUSIONS

The survey of reports presented here on boron transport and removal using IEMs revealed that ED is the most commonly discussed method for boron removal. The reported rates of boric acid transport in ED systems are low (<150 μg/(m^2 s)) as compared to borate fluxes in ED and DD systems (up to 3500 μg/(m^2 s)), and the possibility of boric acid separation from acids and salts was proven. In this case, most of the boric acid was retained in the diluate, while the majority of the ionic species was transported throughout the IEMs. This resembles simple electrodialytic desalination systems. According to Hilal et al.,[3] in 2008, approximately 2,220,000 m^3 of water was desalted every day using ED. Therefore, the economic feasibility of salt removal from waters by ED seems to be proven. The costs of production of desalinated boric acid solution should be similar to the costs of ED desalination of water with similar ionic composition. These costs can be assessed for each water type by analyzing the ED desalination economic reports.

In contrast, the economic and technical feasibility of borate removal using ED can be questioned. Although borate fluxes across IEMs are higher than those of boric acid, strongly limited electric current efficiencies for borate transport have been reported. Moreover, borates are effectively transported across IEMs only at high pH, at which there is a serious risk of scaling with basic compounds such as those of calcium and magnesium. At a very high pH (>13) AEMs could be also slightly attacked by caustic media. These findings imply that diluate demineralization is necessary before borate removal by ED. In addition, even in the preliminarily desalinated solutions, the borate electric current efficiency in ED remains low because hydroxyl anions migrate faster than borate, which results in borate conversion into neutral boric acid molecules that remain in the diluate.

The tentative reports on borate transport in DD systems do not provide much information about the effectiveness of boron removal nor its final concentration in the product water. The economic and technical applicability of DD for boron removal cannot be assessed outright. It can only be noted that the reported fluxes of boron are similar to those of borate in ED systems. Thus, a similar membrane area will be required for DD and ED and maintenance costs should be similar. As in the case of ED, a deep demineralization prior to borate removal will be necessary. It is clear that the applicability of borate removal by DD will be limited by the same factors as ED.

EDI is a method that effectively removes boron but it is applicable for treatment of waters with low salinity. Therefore, EDI should always be preceded by desalination by RO or ED.

REFERENCES

1. WHO. *Guidelines for drinking-water quality.* 4th ed. Geneva: World Health Organization; 2011.
2. Kabay N, Guler E, Bryjak M. Boron in seawater and methods for its separation - a review. *Desalination* 2010;**261**:212–7.
3. Hilal N, Kim GJ, Somerfield C. Boron removal from saline water: a comprehensive review. *Desalination* 2011;**273**:23–35.
4. WHO. *Boron in drinking water. Background document for development of WHO guidelines for drinking-water quality.* Geneva: World Health Organization; 2009.
5. Tanaka Y. *Ion exchange membranes, fundamentals and applications.* 1st ed. Elsevier Science; 2007.
6. Hyung H, Kim JH. A mechanistic study on boron rejection by seawater reverse osmosis membranes. *J Membr Sci* 2006;**286**:269–78.
7. Lide DR. *CRC handbook of chemistry and physics.* 85th ed. Boca Raton, FL: CRC Press; 2004.
8. Garrett DE. *Borates. Handbook of deposits, properties, processing, and use.* Academic Press; 1998.
9. Melnik L, Vysotskaja O, Kornilovich B. Boron behavior during desalination of sea and underground water by electrodialysis. *Desalination* 1999;**124**:125–30.
10. Turek M, Dydo P, Ciba J, Trojanowska J, Kluczka J, Palka-Kupczak B. Electrodialytic treatment of boron-containing wastewater with univalent permselective membranes. *Desalination* 2005; **185**:139–45.
11. Turek M, Dydo P, Trojanowska J, Bandura B. Electrodialytic treatment of boron-containing wastewater. *Desalination* 2007;**205**:185–91.
12. Turek M, Bandura B, Dydo P. Electrodialytic boron removal from SWRO permeate. *Desalination* 2008;**223**:17–22.
13. Kabay N, Arar O, Acara F, Ghazal A, Yuksel U, Yuksel M. Removal of boron from water by electrodialysis: effect of feed characteristics and interfering ions. *Desalination* 2008;**223**:63–72.
14. Banasiak LJ, Schafer AI. Removal of boron, fluoride and nitrate by electrodialysis in the presence of organic matter. *J Membr Sci* 2009;**334**:101–9.

15. Melnik L, Goncharuk V, Butnyk I, Tsapiuk E. Boron removal from natural and wastewaters using combined sorption/membrane process. *Desalination* 2005;**185**:147–57.
16. Melnik L, Goncharuk V, Butnyk I, Tsapiuk E. Development of the sorption-membrane "green" technology for boron removal from natural and wastewaters. *Desalination* 2007;**205**:206–13.
17. Bandura-Zalska B, Dydo P, Turek M. Desalination of boron-containing wastewater at no boron transport. *Desalination* 2009;**241**:133–7.
18. Dydo P. The effect of process parameters on boric acid transport during the electrodialytic desalination of aqueous solutions containing selected salts. *Desalination* 2013;**310**:43–9.
19. Turek M, Dydo P, Trojanowska J. Electrodialytic utilization of IE column post regeneration lyes. *Desalination* 2008;**223**:113–8.
20. Goli E, Hiemstra T, Van Riemsdijk WH, Rahnemaie R, Malakouti MJ. Diffusion of neutral and ionic species in charged membranes: boric acid, arsenite, and water. *Anal Chem* 2010;**82**:8438–45.
21. Dydo P. The mechanism of boric acid transport during an electrodialytic desalination process. *J Membr Sci* 2012;**407-408**:202–10.
22. Turek M, Bandura B, Dydo P. The influence of concentrate alkalinity on electrodialytic boron transport. *Desalination* 2008;**223**:119–25.
23. Kijański M, Bandura-Zalska B, Dydo P, Turek M. The concept of a system for electrodialytic boron removal into alkaline concentrate. *Desalination* 2013;**310**:75–80.
24. Yazicigil Z, Oztekin Y. Boron removal by electrodialysis with anion-exchange membranes. *Desalination* 2006;**190**:71–8.
25. Ayyildiz HF, Kara H. Boron removal by ion exchange membranes. *Desalination* 2005;**180**:99–108.
26. Nagasawa H, Iizuka A, Yamasaki A, Yanagisawa Y. Boron removal from aqueous solution by bipolar membrane electrodialysis. In: *AIChE annual meeting, AIChE, 2008, Philadelphia, PA*; 2008.
27. Nagasawa H, Iizuka A, Yamasaki A, Yanagisawa Y. Utilization of bipolar membrane electrodialysis for the removal of boron from aqueous solutions. *Ind Eng Chem Res* 2011;**50**:6325–30.
28. Kir E, Gurler B, Gulec A. Boron removal from aqueous solution by using plasma-modified and unmodified anion-exchange membranes. *Desalination* 2011;**267**:114–7.
29. Bryjak M, Poźniak G, Kabay N. Donnan dialysis of borate anions through anion exchange membranes: a new method for regeneration of boron selective resins. *React Funct Polym* 2007;**67**:1635–42.
30. Kabay N, Bryjak M, Schlosser S, Kitis M, Avlonitis S, Matejka Z, et al. Adsorption-membrane filtration (AMF) hybrid process for boron removal from seawater: an overview. *Desalination* 2008;**223**:38–48.
31. Yagi Y, Hayashi F, Uchitomi Y. Evaluation of boron behavior in ultrapure water manufacturing system. In: *Proceedings of the semiconductor pure water and chemical conference*; 1994. pp. 54–62. San Jose, CA, March 8–10.
32. Wen R, Deng S, Zhang Y. The removal of silicon and boron from ultra-pure water by electrodeionization. *Desalination* 2005;**181**:153–9.
33. Osawa M, Yamada S. *Water purifying system.* United States Patent 7699968.
34. Hernon B, Zanapalidou H, Prato T., Zhang L. *Removal of Weakly-Ionized species by EDI, GE water & process technologies*. TP1075EN.doc Mar-10.
35. Darbouret D, Kano I. Ultrapure water blank for boron trace analysis. *J Anal At Spectrom* 2000;**15**:1395–9.
36. Arar O, Yuksel U, Kabay N, Yuksel M. Application of electrodeionization (EDI) for removal of boron and silica from reverse osmosis (RO) permeate of geothermal water. *Desalination* 2013;**310**:25–33.
37. Nadav N, Priel M, Glueckstern P. Boron removal from the permeate of a large SWRO plant in Eilat. *Desalination* 2005;**185**:121–9.
38. Jacob C. Seawater desalination: boron removal by ion exchange technology. *Desalination* 2007; **205**:47–52.

CHAPTER 12

Boron Removal From Geothermal Water Using Membrane Processes

Nalan Kabay
Ege University, Chemical Engineering Department, Faculty of Engineering, Izmir, Turkey

12.1 INTRODUCTION

Geothermal energy for electricity generation has been used commercially since 1913. Recently, the utilization of geothermal energy has increased very rapidly. In 2000, geothermal resources have been found in more than 80 countries and the usage of geothermal energy was identified in 58 countries in the world.[1] Geothermal energy is the energy contained as heat in the Earth's interior. The origin of this heat is linked with the internal structure of our planet and the physical processes occurring there. The heat moves from the Earth's interior toward the surface where it dissipates. Its existence is known because the temperature of rocks increases with depth, with a geothermal gradient, averages 30 °C/km of depth.[2] Generally, the heat is transferred from depth to subsurface regions first by conduction and then by convection, with geothermal fluids acting as the carrier in this case. These fluids are essentially rainwater that has penetrated into the Earth's crust from the recharge areas, has been heated on contact with the hot rocks, and has accumulated in aquifers, occasionally at high pressures and temperatures (up to above 300 °C). These aquifers (reservoirs) are the essential parts of most geothermal fields.[2]

The utilization of geothermal resources can be investigated in two categories such as electricity generation and direct use. Direct utilization refers to the immediate use of heat energy rather than to its conversion to some other form, such as electrical energy.[3] Besides electricity generation, geothermal resources have found a use in geothermal heat pumps, balneology, cooling and snow melting, industrial uses, agriculture, aquaculture, greenhouse heating, and space heating.[4]

In Turkey about 95% of the 170 geothermal fields are suitable for direct use applications. Among the remaining fields, Denizli–Kizildere (200–242 °C), Aydin-Germencik (232 °C), Çanakkale-Tuzla (174 °C), Aydın-Salavatlı (171 °C), Kütahya-Simav (162 °C), Manisa-Salihli (150 °C) and Izmir-Seferihisar (153 °C) are high-enthalpy fields, which are suitable for electrical energy production.[3] Turkey is among five leading countries in its geothermal direct use applications. Direct uses of geothermal resources includes, heat pumps, industrial processes like dry ice production

Boron Separation Processes
ISBN 978-0-444-63454-2, http://dx.doi.org/10.1016/B978-0-444-63454-2.00012-5

Table 12.1 Chemical Composition of Geothermal Water, Izmir, Turkey

Cations	Concentration (mg/L)	Anions	Concentration (mg/L)
Na^+	366	Cl^-	188
K^+	26.3	SO_4^{2-}	109
Ca^{2+}	26.2	F^-	4.45
Mg^{2+}	3.70	HCO_3^-	622
pH		8.6	
EC (μS/cm)		1770	
TDS (mg/L)		885	
Salinity (‰)		0.7	
Turbidity (Nephelometric turbidity unit (NTU))		0.15	
B (mg/L)		10.5	

EC, electrocoagulation, TDS, total dissolved solids.

or whitening material, district heating, health spas, greenhouse heating, and some other uses such as drying food and agriculture products.[3,5] In geothermal waters, solute concentrations vary greatly and these differences are due to variations in temperature, gas content, heat source, rock type, permeability, age of the hydrothermal system, and fluid source or mixing (for example, with sea water). They contain various kinds of species including boron and arsenic. Also the following species are common:

- Anions: Cl^-, HCO_3^-, SO_4^{2-}, F^-, Br^-, I^-
- Cations: Na^+, K^+, Li^+, Ca^{2+}, Mg^{2+}, Rb^+, Cs^+, Mn^{2+}, Fe^{2+}
- Other species: SiO_2, NH_3, B, As, noble gases.
- Gases such as CO_2, H_2S, NH_3, N_2, H_2, and CH_4 are generally found in geothermal steams and are invariably present in geothermal discharges from both neutral features and wells.[2]

Chemical composition of geothermal water in Izmir region, Turkey, is given in Table 12.1.

12.2 BORON IN GEOTHERMAL WATER AND ITS REMOVAL

Boron can exist in natural waters as a result of natural or anthropogenic sources.[6] One of the reasons of natural existence of boron in water is considered to be due to the geothermal discharges, leaching from a large variety of rocks, or the mixing of groundwater with oil field water or fossil brines.[6] Since the boron concentration is generally high, geothermal waters may cause environmental problems in groundwaters and surface waters.[6] The Kızıldere geothermal field is one of the major geothermal energy sources of Turkey. However, due to the high boron concentration of geothermal water, a suitable method should be considered for boron removal from waste brine if it is going to be discharged to the Menderes River. Another approach is to reinject the geothermal water back to the underground.[7]

Geothermal waters are sometimes used for irrigation in agricultural areas. Their boron contents accumulate in the soil and this may cause a change in the characteristics of it. Also, these waters could mix with underground waters by passing through the soil and constitute complexes with Pb, Cu, Cd, and Ni ions. The toxicities of these complexes are reported to be higher than those of the heavy metals alone.[8] Removal of boron from geothermal water then becomes an important concern because of these reasons.

There is no easy method for removing boron from geothermal water. In coprecipitation, the removal efficiency is about 90% using $Al_2(SO_4)_3$ and $Ca(OH)_2$. This method is not effective due to the sludge production at the end of the process.[9] Electrocoagulation was used for the removal of boron from thermal waters and the efficiency of boron removal from geothermal water was up to 95% at optimal conditions.[10] Sorption is the most widely used method for removal of boron from water and wastewater. Among the inorganic adsorbents tested, activated carbon, activated alumina, hydrous cerium oxide, and hydrous lanthanium oxide exhibited a high sorption performance for boron.[11] Also, some clays and nonylammonium chloride-modified clays as adsorbents for boron elimination were tested by Karahan et al.[12] In addition, fly ashes,[13,14] some natural adsorbents,[15,16] and other inorganic adsorbents[9,17–21] were used for removal of boron from water.

Yoshizuka et al. published a review on the elimination of boron from geothermal water.[22] The use of boron-selective chelating ion exchange resins was reported as the most efficient method for boron removal.[7,22–37]

Commercially available boron-selective chelating ion exchange resins are generally prepared from macroporous poly(styrene-co-divinylbenzene) by functionalization with *N*-methyl-D-glucamine (NMDG). The NMG groups bind boron through a covalent attachment and an internal coordination complex formation.[36] These resins were employed for boron removal from geothermal water.[25–27,37] The same resins were also tested in the geothermal field.[28]

According to the recent literature, the synthesis of novel boron-selective resins with high capacity, high selectivity, and high sorption rate is of importance for removal of boron from water. Wolska et al. prepared boron-selective small particles by reacting NMG with styrene-vinylbenzylchloride and divinylbenzene by membrane emulsification method followed by polymerization. The resulted chelating resins showed good performance for removal of boron from water.[38,39] Samatya et al. reported on boron-removal performances of monodisperse porous particles with dextran-based molecular brushes attached to the particles via click chemistry and direct coupling. They implied that these particles were efficient in boron removal from aqueous solution.[40] Elsewhere, monodisperse porous poly(glycidyl methacrylate-co-ethylene methacrylate) particles carrying diol functionality were reported to have a potential as boron-selective resins.[41] These resins were used for removal of boron from geothermal water and its reverse osmosis (RO) permeate also.[42] Samatya et al. also produced a new sorbent with high boron-adsorption capacity, using monodisperse-porous poly(vinylbenzyl chloride-co-divinylbenzene), beads 8.5 μm in size synthesized by a new "modified seeded polymerization" technique. By using their chloromethyl functionality, the beads were derivatized by a simple, direct reaction with a boron-selective ligand, *N*-methyl-D-glucamine (NMDG).[43] The resulted chelating resins were applied for boron removal from geothermal water and its RO permeate.[44]

Yavuz et al. reported that polymer-supported core—shell type iminodipropylene glycol functions are very effective in chelation with boric acid and could be used for boron removal at trace levels. For this, they performed graft copolymerization of glycidyl methacrylate (GMA) onto dehydrochlorinated poly(vinyl chloride) by Atom transfer radical polymerization (ATRP) method.[45] Most recently, Santander et al. published a paper on the synthesis of a novel boron-selective resin prepared from *N*-(4-vinylbenzyl)-*N*-methyl-D-glucamine as monomer and *N,N*-methylene-*bis*-acrylamide as cross-linking agent.[46]

12.2.1 Removal of Boron from Geothermal Water by Reverse Osmosis Processes

Cengeloglu et al.[8] performed the removal of boron from water by using the RO technique with SWHR, BW-30 (FilmTec), and AG (GE Osmonics) membranes. The effect of pH and concentration of the feedwater and operating pressure on the boron rejection was investigated. The rejection efficiency of used membranes was compared. Finally, under optimal conditions, RO with SWHR membrane was applied to the natural (ground) water samples containing 24.8 and 9.4 mg/L of boron. According to the obtained results RO could be efficiently used (with >95% rejection) for removal of boron from groundwaters.[8]

Koseoglu et al.[47] emphasized the importance of desalination of the geothermal waters and boron rejection from geothermal waters. Boron removal from geothermal water was studied to determine the optimum operating conditions and permeate flux, employing lab-scale tests. Totally, eight different seawater reverse osmosis (SWRO), brackish water reverse osmosis (BWRO), and nanofiltration membranes were tested under operating pressure ranging from 100 to 750 psi. Effect of increasing pH from 8 to 10.5 on boron rejection and permeate flux was also evaluated. Silica, sulfate, and conductivity rejection performances of used membranes were investigated. It was found that the tested geothermal water may be effectively desalinated by BWRO membranes and used for irrigation purposes.[47]

A comparative study for the removal of boron and silica from geothermal water by cross-flow flat sheet RO method was performed by Öner et al.[48] In order to investigate the effect of membrane type at 15 bar, four different commercially available RO membranes (AG, AD, AK, and BW-30) were used. As shown in Figure 12.1, boron

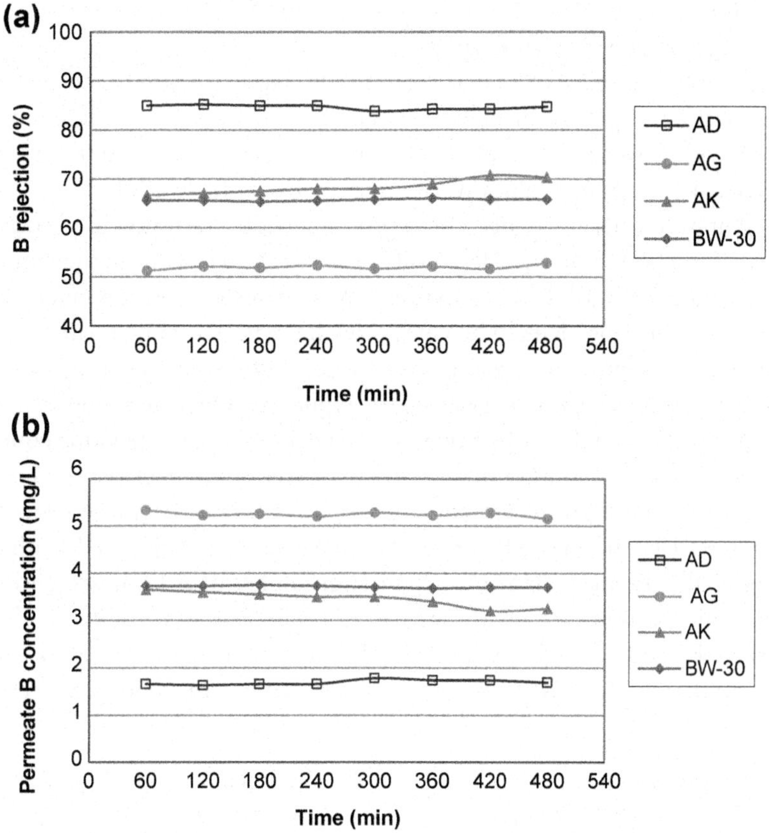

Figure 12.1 Effect of membrane type on (a) boron removal from geothermal water at 15 bar and (b) permeate boron concentration. *(Adapted from Ref. 48).*

rejection efficiencies of used membranes at 15 bar were found to be in the order of AG < BW-30 < AK < AD.[48] According to the obtained results,

- Boron rejection and permeate flux depend strongly on membrane type and operating pressure.
- For all membranes, pressure increases affected both permeate quality and quantity positively.
- At all operating conditions salt and silica were rejected effectively (higher than 96.5%) by all membranes.
- SWRO (AD) membrane provided higher boron rejections with a large difference. By using this membrane, a permeate that was most suitable for EU drinking water standard and irrigation water standard was obtained in terms of boron concentration. But when the concentrations of anions and cations were taken into consideration, this water should be even mineralized before going to be used as irrigation or potable water.

Increase in flow rate and thus the cross-flow velocity did not affect permeate flux, salt, boron, and silica rejections significantly.[48]

To investigate the effect of pH on boron, salt, silica rejections, and permeate flux, geothermal water was passed through AG membrane at 15 bar. Four different pH values were tested: ~8.5 (natural pH of the geothermal water), 9.5, 10.0, and 10.5. The obtained data at various pH values were summarized in Table 12.2.

The significant impact of pH on boron rejection was shown in Figures 12.2. The positive impact of elevated pH levels on boron rejection was due to the shift toward the formation of charged (ionized) boron species. The acid dissociation constant (pKa) of boric acid is 9.2. Therefore, the charged boron species are expected to be formed at pH values >9.2. For many compounds, it is known that charged species are rejected to a greater extent by RO membranes through electrostatic repulsion. This is also supported by finding higher boron rejections at pH of 10.5 than those at 8.5 at constant operating conditions. By increasing geothermal water pH from 8.5 to 10.5, B rejection increased by about 47%. At pH 9.5, average permeate B concentration was found as 3.0 mg/L with an average rejection of 70.6%. At pH 10.0, permeate B concentration decreased to 1.55 mg/L with an average rejection of 85.0%. Permeate boron concentrations of less than 1.0 mg/L (~0.72 mg/L) were easily achieved at pH 10.5 with uppermost B rejection value of 92.9%.[48]

Tomaszewska and Bodzek studied the desalination of geothermal waters using a hybrid ultrafiltration (UF)-RO process by performing some pilot-scale tests.[49] They

Table 12.2 Average Measured and Calculated Data for RO Tests for Removal of Boron and Silica from Geothermal Water as a Function of pH

pH	8.5	9.5	10.0	10.5
J_V (L/m^{2}h)	7.1	5.9	6.1	8.1
EC_f (μS/cm)	1854	2109	2359	2890
EC_p (μS/cm)	62.3	107	136	194
TDS_f (mg/L)	926	1054	1180	1443
TDS_p (mg/L)	31.1	53.5	68.0	96.7
R_{salt} (%)	96.6	94.9	94.2	93.3
B_f (mg/L)	10.7	10.3	10.3	10.2
B_p (mg/L)	5.5	3.0	1.6	0.7
R_B (%)	49.2	70.6	85.0	92.9
Si_f (mg/L)	71.1	67.1	69.5	68.3
Si_p (mg/L)	2.15	2.42	1.93	1.20
R_{Si} (%)	97.0	96.4	97.4	98.3

J_v: Average permeate flux, L/m^{2}h.
EC_f, EC_p: Average electric conductivities of feed and permeate, μS/cm.
TDS_f, TDS_p: Total dissolved solids of feed and permeate, mg/L.
R_{salt}, R_B, R_{Si}: Average rejections of salt, B, and Si, %.
B_P, Si_P: Average B and Si concentrations of permeate, mg/L.
B_f, Si_f: Average B and Si concentrations of feed, mg/L.
Adapted from Ref. 48.

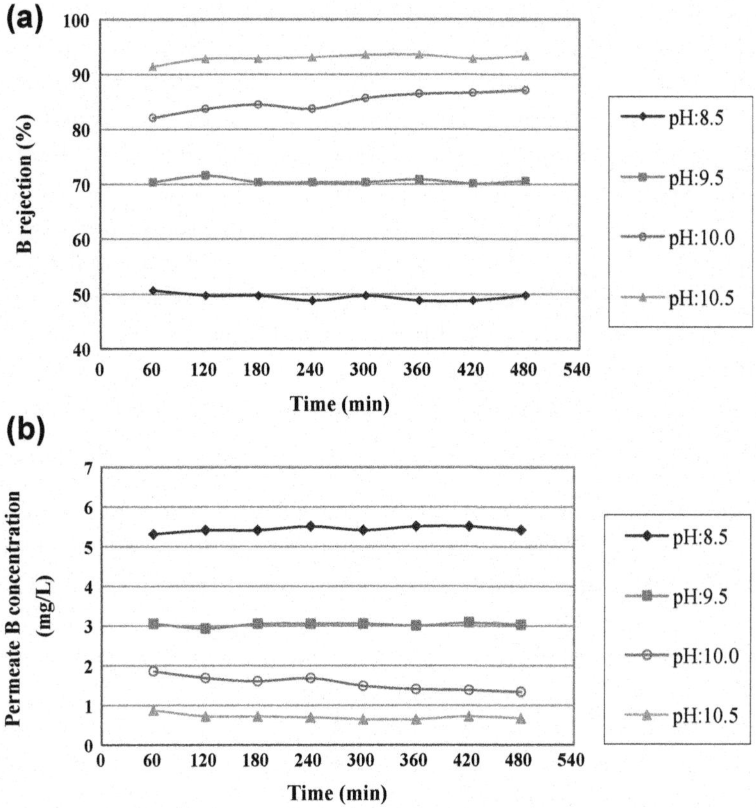

Figure 12.2 Effect of feed pH on (a) boron removal from geothermal water and (b) permeate boron concentration. *(Adapted from Ref. 48).*

have reported that high-quality water may be obtained even after the first RO stage. However, the addition of a second step of RO process with pH adjustment was needed for boron removal. They have obtained high boron-rejection ratios of 96% and 97% with a feedwater pH of 10 and 11 and up to 9.5 mg B/L. On the other hand, in waters with higher salinity and a very high boron content, the boron rejection was only 66% due to the fact that dissociated boron was only 88.4% at pH of 10 where boron concentration is 96.7 mg/L while the dissociated boron was 95.5% at the same pH in a geothermal water with lower concentration of boron (9.5 mg B/L). According to the long-term test results, the authors suggested that geothermal water should be acidified first (pH 5) for desalination of geothermal water with high level of hardness prior to first stage of RO. During first stage of RO conducted at pH 5.0, high rejections for conductivity (97%), hardness (99%), silica (92%), fluoride (92%), and arsenic (84%) were obtained. Boron rejection was 97% at pH 10 after second stage of RO process.[49]

Yavuz et al. used a mini pilot-scale RO system containing two spiral wound FilmTec BW30-2540 membranes installed at a geothermal area. They investigated the effects of operating conditions on desalination performance of the system and boron removal from geothermal water with a boron concentration of 10–11 mg B/L. When the operating pressure was increased, a high permeate recovery, permeate flux, and salt rejection were obtained with both single- and double-membrane configurations. The maximum boron rejection was 47% at 12 bar of applied pressure. Boron rejection increased to 49% at 15 bar of applied pressure. The lowest average boron concentration in the product water was 4.7 mg/L, which is still higher than the permissible level of boron for irrigation.[50] As also obtained by Tomaszewska and Bodzek, boron rejection increased to 95% at pH 10.5 when the operating pressure was 12 bar.[51] Some characteristics and boron concentrations of product water obtained from geothermal water at various pH values were summarized in Table 12.3.

As reported before, in geothermal water at pH around 7.0–8.0, boron is mostly present as boric acid $B(OH)_3$, which is in the molecular form since pH of the geothermal water is below the pKa value of $B(OH)_3$, which is 9.2. The percent distributions of $B(OH)_3$ and $B(OH)_4^-$ depend on the solution pH.[51] It is not easy for an RO membrane to reject the uncharged and small $B(OH)_3$ molecules easily. The dissociated form, $B(OH)_4^-$ ion, however, will be highly rejected by RO membrane due to its fully hydrated form and thus a larger radius and its negative charge.[52]

According to Yavuz et al., it was possible to decrease the concentration of boron in geothermal water to the permissible level for irrigation water when high-rejection SWRO membranes were employed in SWRO system.[52] Also, pH adjustment was not needed to decrease the boron concentration in the RO permeate. An average boron rejection of 88% and a salt rejection of 99% were obtained at 20 bar of operating pressure by a high-rejection XUS SW30XHR-2540 membrane while the respective average boron rejection value of SW30-2540 membrane was 84% at 20 bar. Tables 12.4 and 12.5 summarize the product water characteristics obtained by FilmTec™ SW30-2540 and FilmTec™ XUS SW30XHR-2540 SWRO membranes.[53]

Table 12.3 Effect of Operating pH on Characteristics of RO Permeate of Geothermal Water

		Operating pH					
Parameter	**Unit**	**8.0**	**8.5**	**9.0**	**9.5**	**10.0**	**10.5**
pH	–	5.98	6.40	7.49	8.90	9.66	10.3
Conductivity	μS/cm	19.2	15.5	15.8	19.0	31.4	74.4
Salinity	‰	BDL	BDL	BDL	BDL	BDL	BDL
TDS	mg/L	9.44	8.00	7.78	9.44	15.7	37.1
Si	mg/L	0.40	0.39	0.47	0.38	0.26	0.25
B	mg/L	5.26	4.63	4.08	2.66	1.14	0.55

Adapted from Ref. 51.

Table 12.4 Characteristics of RO Permeate of Geothermal Water Treated with FilmTec™[52] SW30-2540 Membrane

		Operating Pressure (bar)				
Parameter	**Unit**	**15**	**20**	**25**	**30**	**40**
pH	–	5.34	5.43	5.28	5.03	5.24
Conductivity	μS/cm	33.09	28.92	18.73	14.38	18.42
Salinity	‰	BDL[a]	BDL	BDL	BDL	BDL
TDS	mg/L	16.33	14.44	9.33	7.11	9.22
Si	mg/L	0.52	0.56	0.34	0.22	0.39
B	mg/L	2.56	1.81	1.50	1.79	1.77

[a]BDL: Below detection limit.
Adapted from Ref. 51.

Table 12.5 Characteristics of RO Permeate of Geothermal Water Treated with FilmTec™ XUS SW30XHR-2540 Membrane

		Operating Pressure (bar)				
Parameter	**Unit**	**15**	**20**	**25**	**30**	**40**
pH	–	5.39	5.40	5.18	5.21	5.10
Conductivity	μS/cm	17.50	13.83	14.36	13.89	13.30
Salinity	‰	BDL	BDL	BDL	BDL	BDL
TDS	mg/L	9.00	7.00	7.11	7.11	7.00
Si	mg/L	0.24	0.31	0.17	0.19	0.14
B	mg/L	2.37	1.28	1.85	1.84	1.61

Adapted from Ref. 52.

12.2.2 Removal of Boron from Geothermal Water by Sorption–Membrane Filtration Hybrid Method

The main advantage of the sorption–membrane filtration integrated system when compared to fixed-bed systems is the possibility of using fine particles having high surface area and higher efficiency and better kinetics. In fixed-bed columns, the pumping cost is inversely proportional with particle size (d_p) according to pressure drop relation (Eqn (12.1))[53]:

$$\Delta P = u_o \cdot \mu \cdot h \cdot \frac{72 \cdot \tau \cdot (1-\varepsilon)^2}{\varepsilon^3 \cdot d_p^2} \tag{12.1}$$

where d_p is particle diameter (meters), ε is porosity, τ is tortuosity (s), μ is viscosity (pascal second), h is the height of the layer (meters), and u_o (meters per second) is superficial velocity referred to the cross-section of the sorbent layer.

In sorption—membrane filtration system, the pressure drop is independent on particle size as shown in Eqn (12.2)[53]:

$$\Delta P = J \cdot \mu \cdot R_m \tag{12.2}$$

where J is permeate flux (cubic meters per square meter second) and R_m is membrane flow resistance.

Boron-selective chelating ion exchange resins were tested also in sorption—membrane filtration hybrid method for boron removal from geothermal water. During this process, boron sorption was carried out by using boron-selective chelating ion exchange resins while the boron-loaded resin was filtered through submerged microfiltration or UF membranes.[49,54,55]

To specify the diffusion coefficient of boron inside the resin particles and mass transfer coefficient in liquid phase, nonequilibrium kinetic modeling was applied to compare the theoretical data with the experimental results obtained using different boron-selective ion exchange resins and seeded microfiltration system for boron removal from geothermal water.[56] The effective diffusion and mass transfer coefficients that give the best fitting of theoretical data calculated using nonequilibrium kinetic modeling and experimental results are summarized in Table 12.6.

A hybrid process coupling ion exchange with UF was applied for removal of boron from geothermal water. For this, boron-selective chelating ion exchange resins Dowex XUS-43594.00 with an average particle diameter of 20 μm and a submerged hollow fiber-type UF membrane module (ZW-1, GE) were used.[57] Process variables such as concentration of ion exchange resin in the suspension, replacement rates of fresh and saturated resins in the stirred vessel, and flow rate of permeate influenced the efficiency of the hybrid process for boron removal from geothermal water. It was possible to decrease the boron concentration in the geothermal water below 1.0 mg B/L by hybrid process when optimal conditions were employed as shown in Figure 12.3.[55]

According to the obtained results, the recycle performance of the hybrid system could be kept steady if there is no resin loss during the process and process can be run in a closed loop.[55]

Elsewhere, removal of boron from the geothermal water was investigated by an integrated process combining RO with sorption—UF hybrid method.[57] As experienced

Table 12.6 Diffusion and Mass Transfer Coefficients for Boron-Selective Ion Exchange Resins

Ion Exchange Resin	Effective Diffusion Coefficient, D_{eff} (cm^2/s)	Mass Transfer Coefficients, K (cm/s)
Diaion CRB 02	8.0×10^{-9}	2.2×10^{-1}
Dowex (XUS 43594.00)	5.0×10^{-9}	1.0×10^{-1}

Adapted from Ref. 56.

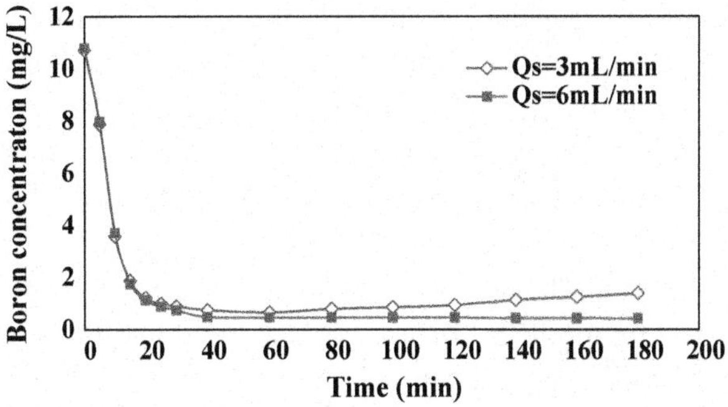

Figure 12.3 Effect of replacement speed of saturated and fresh resins on permeate boron concentration of geothermal water (resin concentration: 2 g/L, permeate flow rate: 5 mL/min). *(Adapted from Ref. 57).*

in the case of geothermal water, similar process parameters influenced the efficiency of the hybrid process for boron removal from RO permeate of geothermal water also.

Most recently, monodisperse porous poly (GMA-co-Ethylene dimethacrylate (EDM)) and poly(Vinylbenzylchloride (VBC)-co-Divinylbenzene (DVB)) particles with a particle size range of 5—20 μm were synthesized using modified seeded polymerization method. These particles were functionalized with NMDG to obtain boron-selective resins. The efficiency of these particles for boron removal from geothermal water was investigated by a hybrid process coupling sorption with UF using a submerged hollow fiber-type UF membrane module. It was possible to lower the concentration of boron in geothermal water from 11.0 mg/L to around 1.0 mg/L in 20 min when the concentration of boron-selective resins is 4 g/L, speeds of replacement of fresh and saturated resins 6 mL/min, and flow rates of feed and permeate 5 mL/min.[58]

12.2.3 Removal of Boron from Geothermal Water by Electromembrane Methods

In the literature, some research activities on boron removal using ion exchange membranes have also been reported. These activities include research on boron removal by electrodialysis (ED) and Donnan dialysis (DD) systems. Recently, a review of the existing literature reports on boron removal using ion exchange membranes was published by Dydo and Turek.[59] It was reported that, removal of boric acid by ED from a solution with $pH < 9$ is difficult compared with that of other ions that exist in the solution.[60–64] According to the published literature, the fluxes of boric acid in ED systems are quite low in a wide salinity range since diffusion is considered to be the only mechanism that is responsible for boric acid transport. It was reported that the removal of boric acid becomes difficult since the diffusion is a slow process.[65] In addition, boron flux was found to depend on several factors, such as the membrane type, the boron concentration, the electric current density, and the type of ion migrating through the membranes.[65] Turek et al.[61–63] and Kabay et al.[64] showed that boron can be removed effectively by ED only from alkaline solutions. On the other hand, the scaling problem on the membrane surface at high pH due to the formation of insoluble salts of calcium and magnesium may cause a serious problem. To eliminate this, removal of multivalent ions by RO before boron separation by ED is suggested as an alternative way.[63]

Elsewhere, Nagasawa et al. suggested use of bipolar membranes for efficient separation of boron by ED method.[66,67] Bryjak et al. used DD method for the regeneration of boron-selective resins.[68]

Arar et al. studied demineralization of RO permeate of geothermal water by the electrodeionization (EDI) method.[69–71] In an EDI cell with layered-bed configuration, the product water quality increased when the applied potential and the feed flow rate were increased.[69] The conductivity of RO permeate of geothermal water decreased from 15.0 μS/cm to 2.9 μS/cm at a 40 V of applied voltage. It was possible to obtain product water with an average conductivity of 1.0 μS/cm with a feed flow rate of 1 L/h.[69] For demineralization of geothermal water RO permeate, EDI system where the ion exchange resins were filled in the mixed-bed configuration into the central compartment was also investigated by Arar et al.[70] The thickness of the membrane was found to be also an important parameter for the quality of product water obtained. When thick membranes were applied in EDI operation, the product water quality was much better with a very low conductivity value.[70]

Arar et al. studied EDI method for removal of boron and silica from geothermal water in an integrated process including RO. As shown in Table 12.7, the concentration of boron that remained in the RO permeate was 5.90 mg/L. But it was possible to decrease the boron concentration from 5.90 mg/L to 0.30 mg/L and silica concentration from 0.3 mg/L to 0.1 mg/L with EDI system having mixed-bed configuration in the central compartment.[72]

Table 12.7 Quality of Product Water Obtained by RO and RO + EDI Integrated Methods

Characteristics	Product Water of RO[a] Concentration (mg/L)	Product Water of RO + EDI[b] Concentration (mg/L)
pH	6.71	6.27
EC (μS/cm)	15.0	0.90
TDS (mg/L)	7.50	0.40
B (mg/L)	5.90	0.30
SiO_2 (mg/L)	0.34	0.11

[a]RO: 12 bar.
[b]EDI: feed flow rate 1.0 L/h, membranes Selemion AME and CME, electrode solution flow rate: 0.6 L/h, applied voltage 40 V, electrode solution (Na_2SO_4 (aq) 500 μS/cm conductivity).
Adapted from Ref. 71.

12.3 CONCLUSIONS

Although geothermal waters could be considered as an alternative energy source to fossil fuels, their environmental impacts should be taken into consideration before they are used extensively. Therefore, geothermal fields should be planned carefully. The mineral contents of geothermal waters in many fields are generally very high. Thus, disposal of these waters can contaminate water resources. Their disposal should be carefully done to protect the groundwater resources.[72]

Geothermal waters may contaminate the irrigation water in agricultural areas. Their boron contents accumulate in the soil and this may cause a change in the characteristics of it. Another risk is that these waters could mix with underground waters by passing through the soil and form complexes with some heavy metal ions. The toxicities of these complexes are mentioned to be more serious than those of the heavy metals alone.[73] Thus, removal of boron from geothermal water is an important concern especially because this water is discharged into the receiving environments without any treatment.

The excess concentration of boron in irrigation water may affect plant growth negatively and cause boron poisoning. The tolerance of different plants to boron in irrigation water is depicted in Table 12.8.

Thus, removal of boron from geothermal water is an important concern especially if this water is discharged into the receiving environments without any treatment.

It is possible to obtain a high quality of water by desalination of geothermal waters using low-pressure RO membranes even after the first RO stage. However, it is not possible to remove boron at a high ratio with RO process if the pH of geothermal water is not adjusted to 10–11.

Table 12.8 The Tolerance of Different Crops in Response to Boron in Irrigation Water

< 1 mg B/L	<2 mg B/L	<4 mg B/L
Walnut	Sunflower	Asparagus
Plum	Potato	Palm
Pear	Cotton	Sugar beet
Apple	Tomato	Bean
Grape	Pea	Onion
Fig	Olive	Cabbage
Cherry	Barley	Lettuce
Peach	Wheat	Carrot
Apricot	Corn	
Orange		
Grapefruit		
Lemon		

Adapted from Ref. 10.

A hybrid process coupling ion exchange with membrane filtration could be a potential method for boron removal from geothermal water by employing optimal process variables such as concentration of ion exchange resin in the suspension, replacement rates of fresh and saturated resins, and flow rate of permeate. Such hybrid processes can be considered a suitable method for the secondary treatment of desalinated geothermal water coming from the RO process. By this method the obtained product water could be used for irrigation without any problem in terms of its salinity and boron content.

ACKNOWLEDGMENT

The author would like to thank her coworkers (Profs. M. Yuksel, U. Yuksel, M. Bryjak, A. Tuncel, M. Kitis, K. Yoshizuka, R. Holdich, A. Ozdural, B. Rivas; Drs. M. Arda, I. Yilmaz Ipek, O. Arar, S. Samatya, M. Badruk, E. Guler, S. Sarp, J. Wolska I. Soroko, M. Makowski, P. Santander, N.O. Yiğit, H. Koseoğlu, S. Nishihama, and T. Iwanaga; graduate students P. Koseoğlu, E. Yavuz, G. Sert, G. Oner, and D. Yapıcı) cited in the various references of this chapter and other students/coworkers for their great collaborations. The support of Izmir Geothermal Co., Izmir, Turkey, was always helpful for getting geothermal water samples and running membrane tests in the geothermal heating center. The financial support of BOREN (National Boron Research Institute, Turkey) is greatly acknowledged for the research project on boron removal from geothermal water (Project Number: 2008-G-0192). N. Kabay thanks MEDRC (04-AS-004), NATO (CLG 980131 and CLG 981422), TUBITAK (CAYDAG-104I096), Ege University Scientific Research Projects Commission (2004-BIL-004; 2004-MUH-026), NEDO-Japan, and 7FP-MC Actions Grant so-called CHILTURPOL2 (Grant Number: 2691537FP- MC-IRSES) for their financial supports to let them get experience on boron removal from water for the last 13 years.

REFERENCES

1. Fridleifsson IB. Geothermal energy for the benefit of the people. *Renew Sustain Energy Rev* 2001;**5**:299–312.
2. Barbier E. Geothermal energy technology and current status: an overview. *Renew Sustain Energy Rev* 2002;**6**:3–65.
3. Hepbasli A, Ozgener L. Development of geothermal energy utilization in Turkey: a review. *Renew Sustain Energy Rev* 2004;**8**:433–60.
4. Lund JW, Freeston DH, Boyd TL. Direct application of geothermal energy: worldwide review. *Geothermics* 2005;**34**:691–727.
5. Serpen U, Aksoy N, Ongur T, Korkmaz ED. Geothermal energy in Turkey: 2008 update. *Geothermics* 2009;**38**:227–37.
6. Gemici U, Tarcan G, Helvacı C, Somay AM. High arsenic and boron concentrations in groundwaters related to mining activity in the bigadic borate deposits (Western Turkey). *Appl Geochem* 2008;**23**:2462–76.
7. Recepoglu O, Beker U. A preliminary study of boron removal from Kizildere Turkey geothermal wastewater. *Geothermics* 1991;**20**:83–9.
8. Cengeloglu Y, Arslan G, Tor A, Kocak I, Dursun N. Removal of boron from water by using reverse osmosis. *Sep Purif Technol* 2008;**64**:141–6.
9. Yurdakoc M, Seki Y, Karahan S, Yurdakoc K. Kinetic and thermodynamic studies of boron removal by Siral 5, Siral 40, and Siral 80. *J Colloid Interface Sci* 2005;**286**(2):440–6.
10. Yilmaz AE, Boncukcuoglu R, Kocakerim MM, Yilmaz MT, Paluluoglu C. Boron removal from geothermal waters by electrocoagulation. *J Hazard Mater* 2008;**153**:146–51.

11. Ooi K, Kanoh H, Sonoda A, Hirotsu T. Screening of adsorbents for boron in brine. *J Ion Exch* 1996;**7**(3):166–72.
12. Karahan S, Yurdakoc M, Seki Y, Yurdakoc K. Removal of boron from aqueous solution by clays and modified clays. *J Colloid Interface Sci* 2006;**293**:36–42.
13. Polat H, Vengosh A, Pankratov I, Polat M. A new methodology for removal of boron from water by coal and fly ash. *Desalination* 2004;**164**:173–88.
14. Ozturk N, Kavak D. Adsorption of boron from aqueous solutions using fly ash: batch and column studies. *J Hazard Mater* 2005;**127**:81–8.
15. Inukai Y, Tanaka Y, Matsuda T, Mihara N, Yamada K, Nambu N, et al. Removal of boron (III) by *N*-methylglucamine-type cellulose derivatives with higher adsorption rate. *Anal Chim Acta* 2004;**511**:261–5.
16. Smith BM, Owens JL, Bowman CH, Todd P. Thermodynamics of borate ester formation by three readily grafted carbohydrates. *Carbohydr Res* 1998;**308**:173–9.
17. Ay AN, Karan-Zumreoglu B, Temel A. Boron removal by hydrotalcite-like, carbonate-free Mg–Al–NO_3-LDH and a rationale on the mechanism. *Microporous Mesoporous Mater* 2007; **98**(1–3):1–5.
18. Seyhan S, Seki Y, Yurdakoc M, Merdivan M. Application of iron-rich natural clays in Camlica, Turkey for boron sorption from water and its determination by fluorimetric-azomethine-H method. *J Hazard Mater* 2007;**146**:180–5.
19. Seki Y, Seyhan S, Yurdakoc M. Removal of boron from aqueous solution by adsorption on Al_2O_3 based materials using full factorial design. *J Hazard Mater* 2006;**138**(1):60–6.
20. Ozturk N, Kavak D. Boron removal from aqueous solutions by adsorption on waste sepiolite and activated waste sepiolite using full factorial design. *Adsorption* 2004;**10**:245–57.
21. Ozturk N, Kavak D. Boron removal from aqueous solutions by batch adsorption onto cerium oxide using full factorial design. *Desalination* 2008;**223**:106–12.
22. Yoshizuka K, Kabay N, Bryjak M. Arsenic and Boron in geothermal water and their elimination. In: Kabay N, Bundschuh J, Hendry B, Bryjak M, Yoshizuka K, Bhattacharya P, et al., editors. *"The global arsenic problem: challenges for safe water production", arsenic in the environment. Arsenic in the environment series,* **2**. CRC-Taylor & Francis, Inc; April 2010. pp. 103–20.
23. Kunin R, Preuss AF. Characterization of boron-specific ion exchange resins. *Industrial Eng Chem Research-Product Res Dev* 1964;**3**(4):304–6.
24. Okay O, Guclu H, Soner E, Balkas T. Boron pollution in Simav River, Turkey and various methods for boron removal. *Water Res* 1985;**19**:857–62.
25. Badruk M, Kabay N, Demircioglu M, Mordogan H, Ipekoglu U. Removal of boron from wastewater of geothermal power plant by selective ion exchange resins. I. Batch sorption-elution studies. *Sep Sci Technol* 1999;**34**(13):2553–69.
26. Badruk M, Kabay N, Demircioglu M, Mordogan H, Ipekoglu U. Removal of boron from wastewater of geothermal power plant by selective ion exchange resins. II. Column mode sorption-elution studies. *Sep Sci Technol* 1999;**34**(15):2981–95.
27. Kabay N, Yilmaz I, Yamac S, Samatya S, Yuksel M, Yuksel U, et al. Removal and recovery of boron from geothermal wastewater by selective ion exchange Resins. I. Laboratory tests. *React Funct Polym* 2004;**60**:163–70.
28. Kabay N, Yilmaz I, Yamac S, Yuksel M, Yuksel U, Yildirim N, et al. Removal and recovery of boron from geothermal wastewater by selective ion exchange resins-II. Field studies. *Desalination* 2004;**167**:427–38.
29. Kabay N, Sarp S, Yuksel M, Arar O, Bryjak M. Removal of Boron from seawater by boron selective ion exchange resins. *React Funct Polym* 2007;**67**:1643–50.
30. Kabay N, Sarp S, Yuksel M, Kitis M, Koseoglu H, Arar O, et al. Removal of boron from SWRO Permeate by boron selective ion exchange resins containing *N*-methyl glucamine groups. *Desalination* 2008;**223**:49–56.
31. Parshova H, Mistova E, Matejka Z, Jelinek L, Kabay N, Kauppinen P. Comparison of several polymeric sorbents for selective boron removal from reverse osmosis permeate. *React Funct Polym* 2007;**67**:1622–7.

32. Yilmaz I, Kabay N, Yuksel M, Holdich R, Bryjak M. Effect of ionic strength of solution on boron mass transfer by ion exchange separation. *Sep Sci Technol* 2007;**42**(5):1013–29.
33. Yilmaz-Ipek I, Kabay N, Yuksel M, Kirmizisakal O, Bryjak M. Removal of boron from Balçova-Izmir geothermal water by ion exchange process: batch and column processes. *Chem Eng Communucations* 2009;**196**:277–89.
34. Bicak N, Bulutçu N, Şenkal BF, Gazi M. Modification of crosslinked glycidyl methacrylate-based polymers for boron-specific column extraction. *React Funct Polym* 2001;**47**:175–84.
35. Marston C, Busch M, Prabhakaran S. A boron selective resin for seawater desalination. In: *Proceedings of european desalination society conference on desalination and the environment, Santa Margherita Ligure, Italy, 22–26 may 2005*; 2005.
36. Senkal BF, Bicak N. Polymer supported iminodipropylene glycol functions for removal of boron. *React Funct Polym* 2003;**55**:27–33.
37. Koseoglu P, Yoshizuka K, Nishihama S, Yuksel U, Kabay N. Removal of boron and arsenic from geothermal water in Kyushu island, Japan, by using selective ion exchange resins. *Solvent Extr Ion Exch* 2011;**29**(3):440–57.
38. Wolska J, Bryjak M, Kabay N. Polymeric microspheres with *N*-methyl-D-glucamine ligands for boron removal from water solution by adsorption-membrane filtration process. *Environ Geochem Health* 2010;**32**:349–52.
39. Wolska J, Bryjak M. Preparation of polymeric microspheres for removal of boron by means of sorption-membrane filtration hybrid. *Desalination* 2011;**283**:193–7.
40. Samatya S, Orhan E, Kabay N, Tuncel A. Comparative boron removal performance of monodisperse-porous particles with molecular brushes via "click chemistry" and direct coupling. *Colloids Surfaces A: Physicochem Eng Aspects* 2010;**372**:102–6.
41. Samatya S, Kabay N, Tuncel A. A hydrophilic matrix for boron isolation: monodisperse porous poly(glycidyl methacrylate-co-ethylene dimethacrylate) particles carrying diol functionality. *React Funct Polym* 2010;**70**:555–62.
42. Samatya S, Tuncel A, Kabay N. Boron removal from geothermal water by a novel monodisperse porous poly(GMA-co-EDM) resin containing *N*-methyl-D-glucamine functional group. *Solvent Extr Ion Exch* 2012;**30**(4):341–9.
43. Samatya S, Kabay N, Tuncel A. Monodisperse-porous *N*-methyl-D-glucamine functionalized poly(-vinylbenzyl chloride-*co*-divinylbenzene) beads as boron selective sorbent. *J Appl Polym Sci* 2012; **126**(4):1475–83.
44. Samatya S, Tuncel A, Kabay N. Boron removal from RO permeate of geothermal water by monodisperse poly(vinylbenzyl chloride-co-divinylbenzene) beads containing N-methyl-D-glucamine. *Desalination*. http://dx.doi.org/10.1016/j.desal.2014.01.029.
45. Yavuz E, Gürsel Y, Şenkal BF. Modification of poly(glycidyl methacrylate) grafted onto cross-linked PVC with iminopropylene glycol group and use for removing boron from water. *Desalination* 2013;**310**:145–50.
46. Santander P, Rivas BL, Urbano BF, Ipek IY, Özkula G, Arda M, et al. Removal of boron from geothermal water by a novel boron selective resin. *Desalination* 2013;**310**:102–8.
47. Koseoglu H, Harman BI, Yigit NO, Guler E, Kabay N, Kitis M. The effects of operating conditions on boron removal from geothermal waters by membrane processes. *Desalination* 2010;**258**:72–8.
48. Öner ŞG, Kabay N, Güler E, Kitiş M, Yüksel M. A comparative study for the removal of boron and silica from geothermal water by cross-flow flat sheet reverse osmosis method. *Desalination* 2011;**283**:10–5.
49. Tomaszewska B, Bodzek M. Desalination of geothermal waters using a hybrid UFRO process. Part I: boron removal in pilot-scale tests. *Desalination* 2013;**319**:99–106.
50. Yavuz E, Guler E, Sert G, Arar O, Yuksel M, Yuksel U, et al. Removal of boron from geothermal water by RO system-I-effect of membrane configuration and applied pressure. *Desalination* 2013;**310**:130–4.
51. Yavuz E, Arar O, Yuksel M, Yuksel U, Kabay N. Removal of boron from geothermal water by RO system-II-effect of pH. *Desalination* 2013;**310**:135–9.

52. Yavuz E, Arar O, Yuksel U, Yuksel M, Kabay N. Removal of boron from geothermal water by RO system-III-Utilization of SWRO system. *Desalination* 2013;**310**:140–4.
53. Koltuniewicz AB, Witek A, Bezak K. Efficiency of membrane-sorption integrated processes. *J Membr Sci* 2004;**239**:129–41.
54. Kabay N, Ipek IY, Soroko I, Makowski M, Kirmizisakal O, Yag S, et al. Removal of boron from Balcova geothermal water by ion exchange-microfiltration hybrid process. *Desalination* 2009; **241**(1–3):167–73.
55. Kabay N, Köseoğlu P, Yapıcı D, Yüksel Ü, Yüksel M. Coupling ion exchange with ultrafiltration for boron removal from geothermal water-investigation of process parameters and recycle tests. *Desalination* 2013;**316**:17–22.
56. Ipek IY, Kabay N, Ozdural AR. Non-equilibrium sorption modeling for Boron removal from geothermal water using sorption-microfiltration hybrid method. *Chem Eng Process Process Intensif* 2011;**50**:599–607.
57. Kabay N, Köseoğlu P, Yavuz E, Yüksel U, Yüksel M. An innovative integrated system for boron removal from geothermal water using RO process and ion exchange-ultrafiltration hybrid method. *Desalination* 2013;**316**:1–7.
58. Samatya S, Köseoğlu P, Kabay N, Tuncel A, Yüksel M. Application of sorption-ultrafiltration hybrid method for Boron removal from geothermal water by monodisperse nanoporous polymer beads containing N-methyl-d-glucamine groups, presented during PPM-2013 Conference, Sept. 2013, Çeşme, Izmir, Turkey.
59. Dydo P, Turek M. Boron transport and removal using ion-exchange membranes: a critical review. *Desalination* 2013;**310**:2–8.
60. Melnik L, Vysotskaja O, Kornilovich B. Boron behavior during desalination of sea and underground water by electrodialysis. *Desalination* 1999;**124**:125–30.
61. Turek M, Dydo P, Ciba J, Trojanowska J, Kluczka J, Palka-Kupczak B. Electrodialytic treatment of boron-containing wastewater with univalent permselective membranes. *Desalination* 2005; **185**:139–45.
62. Turek M, Dydo P, Trojanowska J, Bandura B. Electrodialytic treatment of boron containing wastewater. *Desalination* 2007;**205**:185–91.
63. Turek M, Bandura B, Dydo P. Electrodialytic boron removal from SWRO permeate. *Desalination* 2008;**223**:17–22.
64. Kabay N, Arar O, Acar F, Ghazal A, Yuksel U, Yuksel M. Removal of boron from water by electrodialysis: effect of feed characteristics and interfering ions. *Desalination* 2008;**223**:63–72.
65. Banasiak LJ, Schafer AI. Removal of boron, fluoride and nitrate by electrodialysis in the presence of organic matter. *J Membr Sci* 2009;**334**:101–9.
66. Nagasawa H, Iizuka A, Yamasaki A, Yanagisawa Y. Boron removal from aqueous solution by bipolar membrane electrodialysis. In: *AIChE annual meeting*. Philadelphia, PA: AIChE; 2008.
67. Nagasawa H, Iizuka A, Yamasaki A, Yanagisawa Y. Utilization of bipolar membrane electrodialysis for the removal of boron from aqueous solutions. *Industrial Eng Chem Res* 2011;**50**:6325–30.
68. Bryjak M, Poźniak G, Kabay N. Donnan dialysis of borate anions through anion exchange membranes: a new method for regeneration of boron selective resins. *React Funct Polym* 2007;**67**:1635–42.
69. Arar O, Yuksel U, Kabay N, Yuksel M. Demineralization of geothermal water reverse osmosis (RO) permeate by electrodeionization (EDI) with layered bed configuration. *Desalination* 2013;**317**:48–54.
70. Arar O, Yuksel U, Kabay N, Yuksel M. Demineralization of geothermal water reverse osmosis (RO) permeate by electrodeionization (EDI) with mixed bed configuration. *Desalination* 2014;**342**:23–8. http://dx.doi.org/10.1016/j.desal.2013.08.015.
71. Arar O, Yuksel U, Kabay N, Yuksel M. Application of electrodeionization (EDI) for removal of boron and silica from reverse osmosis (RO) permeate of geothermal water. *Desalination* 2013;**310**:25–33.
72. Aksoy N, Simşek C, Gunduz O. Groundwater contamination mechanism in a geothermal field: a case study of Balcova, Turkey. *J Contam Hydrol* 2009;**103**:13–28.
73. Cengeloglu Y, Tor A, Arslan G, Ersoz M, Gezgin S. Removal of boron from aqueous solution by using neutralized red mud. *J Hazard Mater* 2007;**142**:412–7.

CHAPTER 13

Basic Principles of Simulating Boron Removal in Reverse Osmosis Processes

Jae-Hong Kim[1], **Pyung-Kyu Park**[2]

[1]Department of Chemical and Environmental Engineering, Yale University, New Haven, CT, USA
[2]Department of Environmental Engineering, Yonsei University, Wonju-Shi, Gangwon-Do, South Korea

13.1 WATER PERMEATION, SOLUTE TRANSPORT, AND CONCENTRATION POLARIZATION

Transport of water and solute across an RO membrane is commonly expressed using the following irreversible thermodynamic model (often referred to as the Kedem-Katchalsky or Spiegler-Kedem model)[1,2]:

$$J_v = -p_h\left(\frac{dP}{dX} - \sigma\frac{d\pi}{dX}\right) \tag{13.1}$$

$$J_s = -p_s\frac{dC}{dX} + (1-\sigma)J_vC \tag{13.2}$$

where J_v = volumetric water flux [LT^{-1}]; J_s = gravimetric solute flux [$ML^{-2}T^{-1}$]; p_h = specific hydraulic permeability coefficient [$M^{-1}L^3T$]; p_s = local solute permeability coefficient [L^2T^{-1}]; P = hydraulic pressure [$ML^{-1}T^{-2}$]; π = osmotic pressure [$ML^{-1}T^{-2}$]; σ = reflection coefficient [dimensionless]; and C = superficial solute concentration [ML^{-3}] which is assumed to be in equilibrium with concentration of solute in the membrane phase.

Equation (13.1) implies that water permeation through an RO membrane is proportional to the difference between applied hydraulic pressure and osmotic pressure. The effect of the osmotic pressure is influenced by a reflection coefficient (σ), which represents the extent of solute-water coupling. The reflection coefficient approaches unity for an ideal RO membrane and zero for a porous membrane (i.e., no osmotic pressure). Equation (13.2) represents the solute transport through the membrane. The first term on the right hand side denotes the solute transport by diffusion, which is proportional to a concentration gradient. The second term represents the solute transport by convection, which is determined by the degree of coupling between solutes and water, a solvent flux, and solute concentrations. When there is little or no coupling between the solutes and the solvents (i.e., $\sigma \approx 1$), the solute transport by convection becomes negligible.

Boron Separation Processes
ISBN 978-0-444-63454-2, http://dx.doi.org/10.1016/B978-0-444-63454-2.00013-7

The application of the above equations should consider the fact that the concentration of solutes, both salts and boron, near the membrane surface is different from that in the bulk phase due to concentration polarization, which results from the accumulation of solutes rejected by the membrane. This phenomenon, termed as concentration polarization, is expressed as follows[9]:

$$\frac{C_m - C_p}{C_f - C_p} = \exp\left(\frac{J_v}{k}\right) \tag{13.3}$$

where C_f = feed concentration $[ML^{-3}]$; C_p = permeate concentration $[ML^{-3}]$; C_m = concentration at the membrane surface $[ML^{-3}]$ and k = mass transfer coefficient $[LT^{-1}]$. Combining Eqns (13.1) and (13.3), an apparent rejection (R_0) of the solutes by the membrane is expressed as follows:

$$\frac{R_0}{1 - R_0} = \frac{C_f - C_p}{C_p} = \frac{\sigma}{1-\sigma} \cdot \frac{1 - \exp(-J_v \cdot (1-\sigma)/P_s)}{\exp(J_v/k)} \tag{13.4}$$

where $R_0 = (C_f - C_p)/C_f$ = apparent rejection [dimensionless]; $P_s = p_s/\Delta X$ = overall solute permeability constant $[LT^{-1}]$ and ΔX = thickness of the separation layer [L]. These expressions apply to both salts and boron. Note that the degree of concentration polarization layer also depends on the types of membranes due to differences in rejection performances.

Unlike salt, modeling boron transport should further consider the fact that boron exists as boric acid (H_3BO_3) and deprotonated borate ion ($H_2BO_3^-$) with the corresponding first acid dissociation constant (pK_{a1}) of 9.14 at 25 °C in a low ionic strength solution.[3,4,5] Complexation of boric acid and borate ion with other metal ions is negligible.[6] At a natural pH range, a majority of boron exists as uncharged boric acid. However, the fraction of negatively-charged borate ion increases as pH increases and borate ion becomes a dominant species as pH increases beyond pK_{a1}:

$$H_3BO_3 \rightarrow H_2BO_3^- + H^+, \quad pK_{a1} = 9.14 \tag{13.5}$$

This acid-base speciation is of particular importance since the rejection of charged species by RO membranes is much higher than that of uncharged species due to negative charges on the active layer in typical thin-film composite membranes. For example, polyamide is often the material of choice for active layers and contains terminal carboxylic groups which are weak acid. Consequently, as pH increases, the charge repulsion between negatively charged borate ion and the negatively charged membrane surface plays a more important role on the overall rejection of boron. In other words, increased charge repulsion at higher pH resulted in decreased diffusive transport of boron through the membrane (decreased p_s in Eqn (13.2)) and reduced solute-solvent coupling (increased σ in Eqn (13.2)). This phenomenon is conceptually illustrated in Figure 13.1.

This pH-dependent speciation can be modeled by describing the overall transport of boron as the sum of individual and independent contributions from boric acid and borate ion. The overall permeability of boron species can be expressed as:

$$p_{s,boron} = \alpha_0 \times p_{s,H_3BO_3} + \alpha_1 \times p_{s,H_2BO_3^-} \tag{13.6}$$

where p_{s,H_3BO_3} = permeability constant of boric acid [LT^{-1}] and $p_{s,H_2BO_3^-}$ = permeability constant of borate ion [LT^{-1}]. Similarly, the reflection coefficient of boron at any pH can be expressed as:

$$\sigma_{boron} = \alpha_0 \times \sigma_{H_3BO_3} + \alpha_1 \times \sigma_{H_2BO_3^-} \tag{13.7}$$

where $\sigma_{H_3BO_3}$ = reflection coefficient of boric acid [dimensionless] and $\sigma_{H_2BO_3^-}$ = reflection coefficient of borate ion [dimensionless]. In the above equations, α_0 and α_1 represent the ionization fraction of boric acid and borate ion, respectively, and are defined as follows:

$$\alpha_0 = \frac{\{H^+\}}{\{H^+\} + K'_{a1}} = \frac{[H_3BO_3]}{C_B} \tag{13.8}$$

$$\alpha_1 = \frac{K_{a1}}{\{H^+\} + K'_{a1}} = \frac{[H_2BO_3^-]}{C_B} = 1 - \alpha_0 \tag{13.9}$$

$$K'_{a1} = \frac{[H_2BO_3^-]\{H^+\}}{[H_3BO_3]} \tag{13.10}$$

where $[H_3BO_3]$ = concentration of boric acid [ML^{-3}]; $[H_2BO_3^-]$ = concentration of borate ion [ML^{-3}]; $\{H^+\}$ = activity of proton [ML^{-3}]; K'_{a1} = apparent first acid dissociation constant for boric acid in sea water [ML^{-3}]; and C_B=total concentration of boron species [ML^{-3}]. Note that the ionic strength of brackish and sea waters cannot be neglected and activities should be used instead of concentration in these expressions.[7]

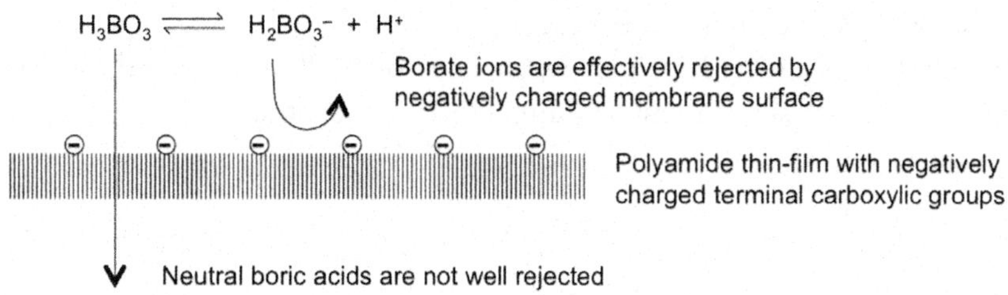

Figure 13.1 Conceptual illustration of the effect of boron speciation and rejection by RO membranes.

However, it is more convenient to use the terms that are readily measured by common instruments. Therefore, K'_{a1} is defined herein based on the concentrations of boron species and the activity of protons. The value of K'_{a1} depends on salinity and temperature according to the following empirical equation[8]:

$$-\log K'_{a1} = \frac{2291.9}{T} + 0.01756 \times T - 3.385 - 0.32051 \times S^{1/3} \quad (13.11)$$

where T=absolute temperature [K] and S= total salt concentration (salinity) in ppm [ML^{-3}]. At a representative sea water salinity of 34,000 ppm and 25 °C, for example, pK'_{a1} is estimated at 8.68, which is much lower than 9.14 in a dilute solution. Note that the salinity near the membrane surface further increases due to concentration polarization, discussed above.

13.2 SPIRAL WOUND ELEMENT SIMULATION

In a spiral wound element, one needs to consider the fact that salinity, flow rate, and pressure change from the inlet to the outlet as permeates are produced. A two-dimensional finite element analysis can be employed to take into account the nonhomogeneous condition within the element. Each leaf in the spiral wound element consists of two membrane sheets, a permeate flow channel between the membranes and feed flow channels and turbulent promoters on the opposite side of each membrane. The simulation practice presented herein (Figure 13.2) conceptually divides each sheet into $m \times n$ subelements (i.e., m in the direction parallel to the axis of the membrane module, x axis, and n in the orthogonal direction, y axis), while the number of elements can be adjusted. The dimensions of each subelement, Δx and Δy, are obtained as follows:

$$\Delta x = \frac{L}{m} \quad (13.12)$$

$$\Delta y = \frac{W}{n} \quad (13.13)$$

where L= length of a membrane sheet (along the axis of the membrane module) [m; L] and W= width of a membrane sheet [m; L].

Equations (13.1) and (13.2), i.e., water and solute (both salt and boron species) transport through each subelement, and Eqn (13.3), concentration polarization, then can be expressed as follows[10,11]:

$$J_v(i,j) = P_h\left[\left\{P_f(i,j) - P_p(i,j)\right\} - \sigma\left\{\pi_w(i,j) - \pi_p(i,j)\right\}\right] \quad (13.14)$$

$$J_s(i,j) = P_s\left\{C_w(i,j) - C_p(i,j)\right\} + (1-\sigma)J_v(i,j)\overline{C} \quad (13.15)$$

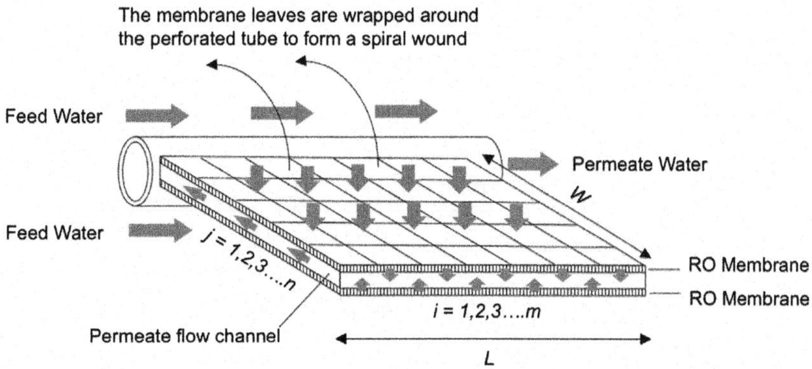

Figure 13.2 Finite element analysis of spiral wound element. A single membrane leaf attached to the center tube is illustrated.

$$C_w(i,j) - C_p(i,j) = \{C_f(i,j) - C_p(i,j)\}\exp\left(\frac{J_v(i,j)}{k(i,j)}\right) \tag{13.16}$$

where i = index number along x axis; j = index number along y axis; $P_h = p_h/\Delta X$ = overall water permeability coefficient; and $\overline{C}$ = average value of solute concentrations in the feed and permeate sides [ML^{-3}].

Equation (13.14) through (13.16) describe the permeation of water, salt, and boron only at local finite elements within the membrane leaf for which the model parameters are assumed to be uniform. Since the model parameters change depending on the location within the membrane leaf as a result of solute concentration and feed pressure drop, a mass balance approach should be used to predict the overall performance of the entire spiral wound module. First, pressure, feed flow rate, and solute concentration in the feed channel for all grid elements at the inlet of the spiral wound element (i = 1 and j = 1 to n) can be defined as follows:

$$\left.\begin{aligned} P_f(i,j) &= P_{f0} \\ Q_f(i,j) &= \frac{Q_{f0}}{n \cdot n_L} \\ C_f(i,j) &= C_{f0} \end{aligned}\right\} \quad \text{for } i = 1 \text{ and } j = 1 \text{ to } n \tag{13.17}$$

$$P_p = 1 \text{ atm} \quad \text{for } i = 1 \text{ to } m \text{ and } j = n \tag{13.18}$$

where P_{f0} = initial feed (operating) pressure [$ML^{-1}T^{-2}$]; Q_f = flow rate in a feed channel [L^3T^{-1}]; Q_{f0} = initial feed flow rate [L^3T^{-1}]; n_L = number of membrane leaves [dimensionless]; C_{f0} = initial feed solute concentration [ML^{-3}].

The feed flow rate and feed solute concentration for the next grid element in each leaf are determined as follows:

$$Q_f(i+1,j) = Q_f(i,j) - Q_p(i,j) \tag{13.19}$$

$$C_f(i+1,j) = \frac{Q_f(i,j)C_f(i,j) - Q_p(i,j)C_p(i,j)}{Q_f(i+1,j)} \tag{13.20}$$

$$Q_p(i,j) = 2\cdot J_v(i,j)\Delta x\Delta y \tag{13.21}$$

The hydraulic pressure drop in the feed channel along the x axis and in the permeate channels along the y axis can estimated based on Darcy's Law as follows[12]:

$$P_f(i,j) - P_f(i+1,j) = \left(k_{fb}\mu\Delta x^2 n_F\right)U_f^{n_F-1}\sum_1^i\left(\frac{2\cdot J_v(i,j)}{h_b}\right) \tag{13.22}$$

$$P_p(i,j+1) - P_p(i,j) = k_{fp}\mu\Delta y^2\sum_1^j\left(\frac{2\cdot J_v(i,j)}{h_p}\right) \tag{13.23}$$

where n_F = empirical fitting parameter [dimensionless]; k_{fb} = friction parameter for feed channel of a module [L^{-2}]; k_{fp} = friction parameter for permeate channel of a module [L^{-2}]; h_b = thickness of a feed channel [L]; and h_p = thickness of a permeate channel [L].

Once the water and solute permeations were predicted for each grid element by simultaneously solving Eqns (13.14)–(13.23), the overall permeate recovery and salt and boron rejection can be determined from the following permeate balance and total solute balance relationships:

$$\text{Overall permeate recovery} = \frac{Q_{pT}}{Q_{f0}} \tag{13.24}$$

$$\text{Overall rejection} = \left(1 - \frac{C_{pT}}{C_{f0}}\right) \tag{13.25}$$

where

$$Q_{pT} = n_L\sum_{i=1}^{m}\sum_{j=1}^{n} Q_p(i,j) \tag{13.26}$$

$$C_{pT} = \frac{n_L\sum_{i=1}^{m}\sum_{j=1}^{n} Q_p(i,j)C_p(i,j)}{Q_{pT}} \tag{13.27}$$

Q_{pT} = overall permeate flow rate [L^3T^{-1}] and C_{pT} = overall solute concentration in the product water [ML^{-3}].

13.3 MODEL PARAMETER ESTIMATION

This section describes some example empirical approaches to obtain parameters that are required to perform numerical simulations. The empirical approaches are instrumental in predicting the parameters that change as the filtration proceeds as well as depending on operating conditions. Temperature dependence of permeability coefficients for water and salts, P_h and P_s, can be estimated using the Arrhenius relationship[13,14]:

$$P_h = P_{h,0} \exp\left\{-\frac{E_{A,h}}{R}\left(\frac{1}{T}-\frac{1}{T_0}\right)\right\} \tag{13.28}$$

$$P_s = P_{s,0} \exp\left\{-\frac{E_{A,s}}{R}\left(\frac{1}{T}-\frac{1}{T_0}\right)\right\} \tag{13.29}$$

where $P_{h,0}$ = water permeability coefficient measured at T_0; $P_{s,0}$ = salt and boron permeability coefficient measured at T_0; T_0 = 298.15 K; $E_{A,h}$ = activation energy for transport of a water molecule through the membrane [J/(mol K)]; $E_{A,s}$ = activation energy for transport of salt and boron through membrane [J/(mol K)]; R = ideal gas constant = 8.3145 J/(mol K). The temperature dependence of the reflection coefficient for salt and boron, albeit relatively small, can be estimated by[19]:

$$\sigma_B = 0.997 - 4.98 \times 10^{-5} t \tag{13.30}$$

where t = relative temperature (°C). Equation (13.6) can be modified as follows to consider the effect of temperature on boron transport:[19]

$$P_{s,boron} = \alpha_0 P_{s,H_3BO_3,0} \exp\{a(T-T_0)\} + \alpha_1 P_{s,H_2BO_3^-} \exp\{b(T-T_0)\} \tag{13.31}$$

where $P_{s,H_3BO_3,0}$ = permeability constant of boric acid (H_3BO_3) estimated at T_0 = 298.15 K [LT^{-1}]; $P_{s,H_2BO_3^-,0}$ = permeability constant of borate ($H_2BO_3^-$) estimated at T_0 [L^2T^{-1}]; The temperature dependence of σ_{boron} is negligible. Coefficients a and b can be experimentally determined and α_0 and α_1 can be calculated using Eqns (13.8)–(13.10).

Osmotic pressure, π, in Pa can be estimated using the following empirical equations.[15]

$$\pi(C,t) = (0.6955 + 0.0025t) \times 10^8 \left(\frac{C}{\rho}\right) \tag{13.32}$$

$$\rho = 498.4M + \sqrt{248,400M^2 + 752.4MC} \quad \left(M = 1.0069 - 2.757 \times 10^{-4} t\right) \tag{13.33}$$

where ρ = density [kg/m^3; ML^{-3}]. Mass transfer coefficient, k, can be estimated using the empirical equation developed by Winograd et al., among others, that takes into account the effect of geometry of spacers and feed flow channels, fluid properties (dynamic viscosity) and solute properties (diffusivity).[16,17]

$$k = 0.753\left(\frac{K}{2-K}\right)^{0.5}\left(\frac{D}{h_b}\right)S_c^{-1/6}\left(\frac{Peh_b}{\Delta L}\right)^{0.5} \tag{13.34}$$

where K = efficiency of mixing net (i.e., spacer) [dimensionless]; D = diffusion coefficient [m^2/s; L^{2}T^{-1}]; S_c = Schmidt number = $\mu/\rho D$, [dimensionless]; Pe = Péclet number = $2h_bU_b/D$, [dimensionless]; ΔL = characteristic length of mixing net [m; L]; μ = dynamic viscosity [Pa s; ML^{-1}T^{-1}]. The solute diffusion coefficient and water viscosity can be estimated as follows:

$$D = 6.725 \times 10^{-6}\exp\left(0.1546 \times 10^{-3}C - \frac{2,513}{T}\right) \tag{13.35}$$

$$\mu = 1.234 \times 10^{-6}\exp\left(0.00212C + \frac{1,965}{T}\right) \tag{13.36}$$

13.4 PILOT- AND FULL-SCALE SIMULATION

Pilot- and full-scale RO process simulation for boron rejection involves simulating each spiral wound element in the system and setting up a mass balance. The simulation of each spiral wound element can be performed following the approach presented in Chapters 13.1–13.3. The set of equations can be grouped into a function that can be called in, for example, a numerical solver, several times during the mass balance. Setting up mass balances depends on how the entire system is configured, for example, single-pass versus double-pass and single-stage versus double-stage. In each pressure vessel, more than one (typically six to eight in a full-scale design) spiral wound elements are placed in series (Figure 13.3). Therefore, the simulation on a pressure vessel involves performing the calculations on spiral wound elements consecutively from the leading element (i.e., inlet) to the last subelement (i.e., outlet). The output from the preceding element is the inlet condition for the following element. The output from the last element of the vessel is the output from the vessel. More than one vessel are placed in parallel to constitute a train. For a train, therefore, outputs from pressure vessels that constitute the train are simply added. More than one train constitutes each pass and each stage. Additional mass balances are needed to either simply add the outputs from the trains or subsequent feeding of outputs from the train to the following trains.

A sample simulation of boron rejection in a hypothetical full-scale single-stage single-pass RO process is presented in Figure 13.4.[18] The process employs six or eight

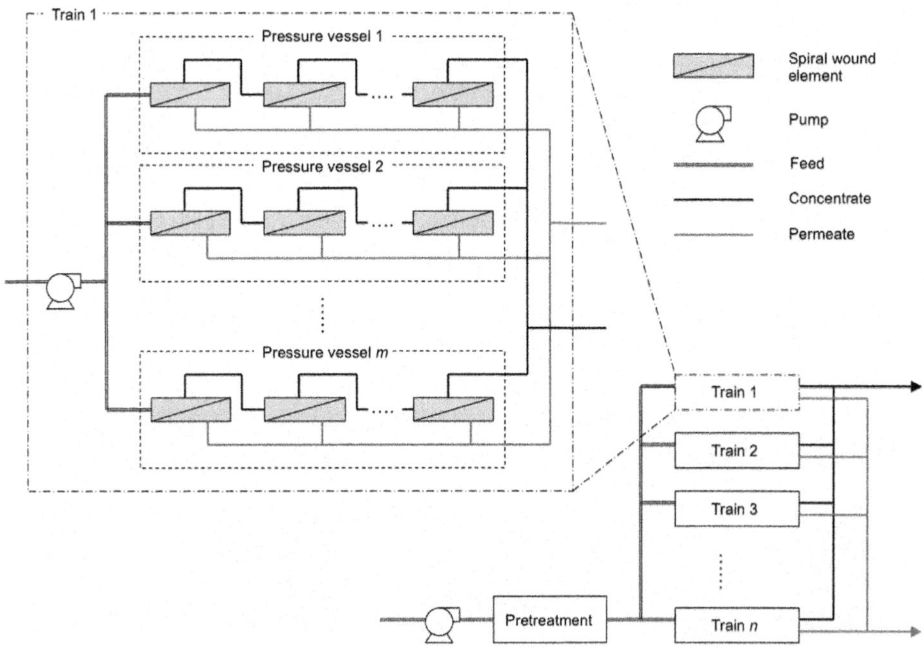

Figure 13.3 An example of a single-stage, single-pass, full-scale plant configuration.

commercial spiral-wound elements placed in series to treat 250 m^3/day of model seawater containing 32,000 mg/L of salt and 5 mg/L of boron. Individual spiral wound membrane elements along the pressure vessel are sequentially simulated (i.e. from the feed side to the concentrate side such that the output from the preceding element is the input for the subsequent element) as discussed above. The feed pressure (pressure at the inlet of the first element) is changed from 600 to 1000 psi and temperature from 15 to 35 °C The results from each pressure vessel are combined following the aforementioned mass balance approach to predict the overall permeate production and boron rejection in the combined flow of the entire process.

The simulation results suggest that the boron rejection in the full-scale plant increases as pH and pressure increase, with greater dependency at higher temperatures (Figure 13.4(a)). For the simulation with six elements in series, the highest boron rejection (98.7%) is observed at 15 °C, pH 12, and 1000 psi, and the lowest rejection (71.7%) at 35 °C, pH 6, and 600 psi (Figure 13.4(a)). For the simulation with eight elements, the highest boron rejection (98.4%) is observed at 15 °C, pH 12, and 1000 psi, and the lowest rejection (68.4%) at 35 °C, pH 6, and 600 psi. It is noted that the plant with eight elements in series consistently show lower boron rejection than the plant with six elements. This results since the former achieves higher recovery. At high recoveries, the boron concentration in the concentrate becomes high (particularly in the tailing

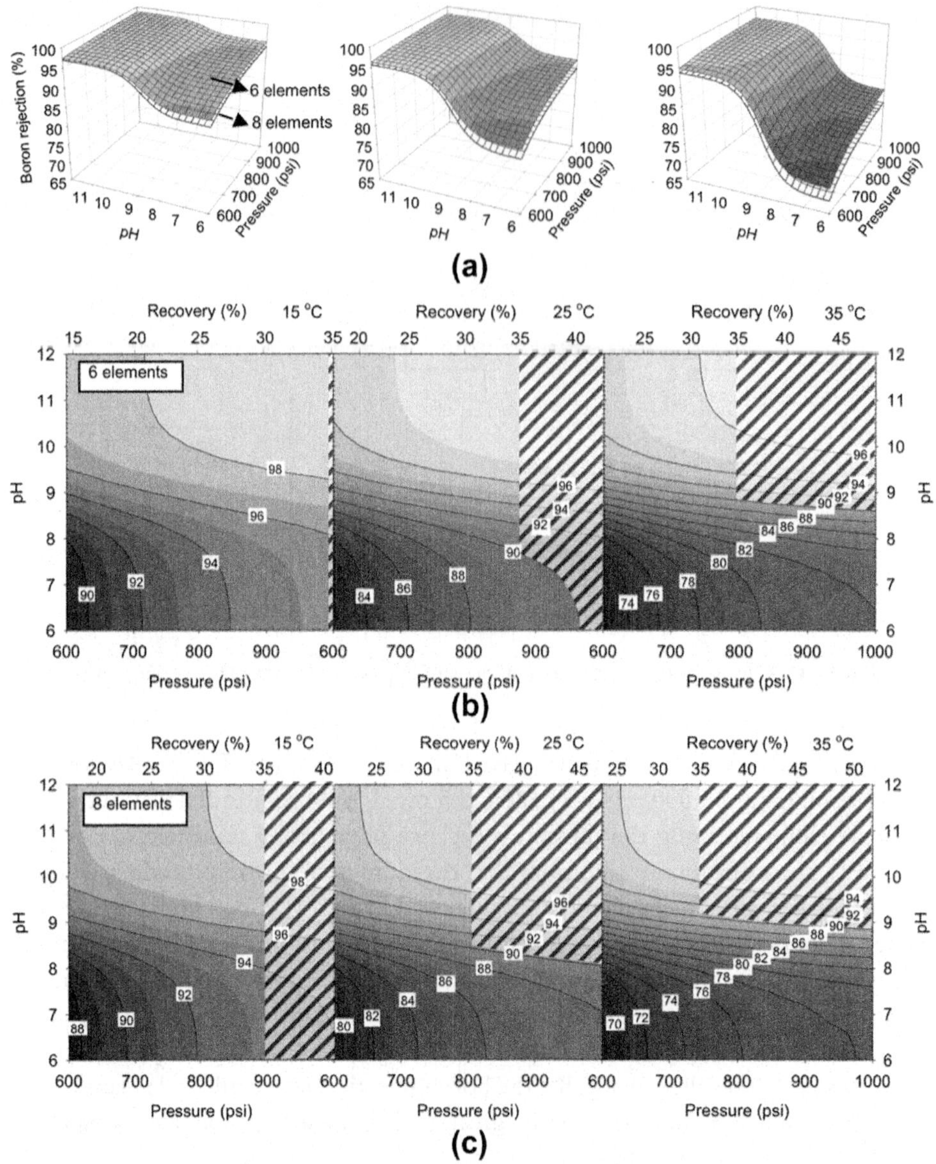

Figure 13.4 Model simulations of boron rejection by a hypothetical full-scale single-stage, single-pass RO process employing six and eight spiral-wound elements in series. 3-D plots in (a) shows boron rejection at varying feed water pHs and operating pressures at 15, 25, and 35 °C. Contour plots in (b) and (c) represent the combination of pH and pressure with water recovery at the three different temperatures for six and eight elements, respectively.

elements within the train), leading to greater boron permeation, as permeation is driven by the concentration gradient. At the same time, as concentrate salt concentration increases at higher recoveries, osmotic pressure increases, which results in decreased water permeation.

Figures 13.4(b) and (c) show example pH/pressure/recovery contour diagrams of boron rejection at three temperatures for the six- and eight-element systems. Vertically aligned contour lines indicate that boron rejection would change significantly with pressure but not with pH. When the contour lines are aligned horizontally, boron rejection is highly dependent on pH and less dependent on pressure. For example, contour lines in each plot become flatter near pH 9, suggesting greater dependency of boron rejection near this pH value. As temperature increases, the distance between contour lines becomes narrower, indicating that the boron rejection is more sensitive to changes in operating conditions.

In these contour diagrams, shaded regions represent combinations of pH and pressure that allow full-scale operation with at least 35% recovery and at least 90% boron rejection (values arbitrarily chosen). At 15 °C, boron rejection is greater than 90% with recovery greater than 35% at all operating pHs as long as the inlet pressure is higher than 996 psi and 896 psi for six and eight elements, respectively. However, at 25 °C, pH control might be required in addition to pressure control to achieve 90% boron rejection and 35% process recovery. For example, when the inlet pressure is less than 965 psi for the plant with six elements, the pH should be higher than the value defined by the 90% contour curve. For the plant with eight elements, in which boron rejection is further decreased, pH control needs to be considered for the entire range of operating pressures (up to 1000 psi). For example, at 804 psi (equivalent to 35% recovery), pH should be adjusted to higher than 8.4. At 35 °C, the shaded region becomes even smaller and additional system control is required to achieve the same process goals. For the six-element plant operating at a feed pressure of 976 psi (35% recovery), pH should be higher than 8.8 to achieve 90% boron rejection. For the plant with eight elements operating at the same recovery (and pressure of 745 psi), pH needs to be raised to at least 9.1. For these conditions, pH control using caustic becomes necessary. In all cases, boron rejection appears more challenging as system recovery increased (i.e., as the number of membrane elements increased).

13.5 SUMMARY

A mechanistic model presented in this chapter can be used to simulate boron rejection by pilot- and full-scale SWRO processes under varying operating conditions. The model represents a variation of existing models that simulate salt rejection, i.e., including boron speciation as a function of pH. As presented by sample simulation results, pH, pressure (or recovery), temperature, and the process configuration all affect the boron rejection, with the pH having the most significant impact. This fact is central to the strategy to control boron by raising pH in many SWRO processes.

REFERENCES

1. Kedem O, Katchalsky A. Physical interpretation of phenomenological coefficients of membrane permeability. *J Gen Physiol* 1961;**45**(1):143.
2. Spiegler KS, Kedem O. Thermodynamics of hyperfiltration (reverse osmosis): criteria for efficient membranes. *Desalination* 1966;**1**:311–26.
3. Magara Y, Tabata A, Kohki M, Kawasaki M, Hirose M. Development of boron reduction system for sea water desalination. *Desalination* 1998;**118**:25–33.
4. Prats D, Chillon-Arias MF, Rodriguez-Pastor M. Analysis of the influence of pH and pressure on the elimination of boron in reverse osmosis. *Desalination* 2000;**128**(3):269–73.
5. Stumm W, Morgan JJ. NetLibrary Inc. Aquatic chemistry: chemical equilibria and rates in natural waters.
6. Stumm. W, Morgan JJ. *Aquatic chemistry.* 3rd ed. New York: Wiley-Interscience; 1996.
7. Riley JP, Skirrow G. *Chemical oceanography.* 2nd ed., vol. 2. New York: Academic; 1975.
8. Gieskes JM. In: Goldberg ED, editor. *The sea*, vol. 5. Wiley-Interscience; 1974.
9. Cussler EL. *Diffusion*. Cambridge University Press; 1984.
10. Spiegler KS. Diffusion of gases across porous media. *Ind Eng Chem Fund* 1966;**5**(4):529–32.
11. Marinas BJ, Urama RI. Modeling concentration polarization in reverse osmosis spiral-wound elements. *J Environ Eng ASCE* 1996;**122**(4):292–8.
12. Senthilmurugan S, Ahluwalia A, Gupta SK. Modeling of a spiral-wound module and estimation of model parameters using numerical techniques. *Desalination* 2005;**173**(3):269–86.
13. Mehdizadeh H, Dickson JM, Eriksson PK. Temperature effects on the performance of thin-film composite, aromatic polyamide membranes. *Ind Eng Chem Res* 1989;**28**(6):814–24.
14. Sharma RR, Agrawal R, Chellam S. Temperature effects on sieving characteristics of thin-film composite nanofiltration membranes: pore size distributions and transport parameters. *J Membr Sci* 2003;**223**(1–2):69–87.
15. Taniguchi M, Kimura S. Estimation of transport parameters of RO membranes for seawater desalination. *AIChE J* 2000;**46**(10):1967–73.
16. Winograd Y, Solan A, Toren M. Mass-transfer in narrow channels in presence of turbulence promoters. *Desalination* 1973;**13**(2):171–86.
17. Benboudinar M, Hanbury WT, Avlonitis S. Numerical-simulation and optimization of spiral-wound modules. *Desalination* 1992;**86**(3):273–90.
18. Mane PP, Park PK, Hyung H, Brown JC, Kim JH. Modeling boron rejection in pilot- and full-scale reverse osmosis desalination processes. *J Membr Sci* 2009;**338**(1–2):119–27.
19. Hyung H, Kim JH. Mechanistic study on Boron rejection by sea water reverse osmosis membranes. *J Membr Sci* 2009:269–78.

CHAPTER 14

Single SWRO Pass Boron Removal at High pH: Prospects and Challenges

Oded Nir, Ori Lahav
Faculty of Civil and Environmental Engineering, Technion—Israel Institute of Technology, Haifa, Israel

14.1 INTRODUCTION AND PROSPECTS

The worldwide demand for freshwater is constantly increasing. Seawater desalination is probably the most sustainable solution for maintaining stable freshwater supply in water scarce regions, adjacent to a shore. In the recent decade, reverse osmosis (RO) has emerged as the most cost-effective technology for this purpose, and it is expected to remain as such in the foreseeable future.[1] With 69% of the global water supply allocated to irrigation, desalinated water is also increasingly considered for this purpose. Fresh water produced by seawater reverse osmosis (SWRO) can be applied for irrigation either directly, as done in Spain at a commercial scale, or indirectly through domestic wastewater reuse, as performed in Israel at a national scale.[2] This trend will, in the near future, most likely spread to other countries located in arid and semiarid regions.

Due to its detrimental effect on plants at relatively low concentrations, the removal of boron species (B) from desalinated water is a significant component in the process design of many SWRO desalination plants. At concentrations above ~1.0 mg B/L, boric acid is known to damage various agricultural crops, as well as plant species used for municipal gardening. For example, crops such as avocado and most citrus types are sensitive to B at the concentration range 0.5—0.75 mgB/L.[3] Therefore, although the World Health Organization had recently updated up the guidelines for B concentration in drinking water from 0.5 to 2.4 mg/L (for strictly human health-related reasons), it is very likely (and also apparent from the results of recent international bids) that the requirement for low B concentration (0.3—0.8 mgB/L) in desalinated waters will remain unchanged.

Boron appears in natural fresh waters as a weak acid, with a thermodynamic pK value of 9.23, related to the equilibrium reaction shown in Eqn (14.1). At pH values lower than the pK, the protonated, neutral, boric acid species ($B(OH)_3$) dominates, while above it the negatively charged borate ion ($B(OH)_4^-$) prevails. In seawater and seawater desalination brines, the apparent pK′ value is in the range of ~8.8—8.4 (at 250 °C). The boric acid species are poorly rejected by RO membranes as compared to the borate species which is almost fully rejected (typical rejection > 99%), similarly to other ions. As a result, B rejection is highly pH depended in the pH range of 7.5—10.5, where both

Boron Separation Processes
ISBN 978-0-444-63454-2, http://dx.doi.org/10.1016/B978-0-444-63454-2.00014-9

B species appear at significant concentrations. At pH close to the pK′, a small diversion in pH significantly affects B rejection.

$$B(OH)_3 + H_2O \leftrightarrow B(OH)_4^- + H^+ \quad pK = 9.23 \tag{14.1}$$

The average total concentration of the boron weak-acid system in seawater is ~5 mg-B/l. Given the common operational pH of SWRO, which is pH~8.2 (natural seawater pH) or lower, the practical B rejection attained upon the application of standard SWRO membranes is only ~65–85%, corresponding to ~0.75–1.8 mgB/L in the permeate.[4] Thus, either ion-exchange-based posttreatment (PT) or the operation of a second (brackish) RO pass is typically implemented in order to meet stricter B regulations.[3]

Application of a second pass includes dosage of a strong base to permeate of the first SWRO pass, in order to elevate pH to 9.5–10.5, prior to its introduction into the brackish water RO (BWRO) membrane. The pH elevation diverts the boric acid species toward the borate ion, whose rejection by RO membranes is much more efficient. Ion exchange (IX) technology utilizes a resin with a high affinity toward B, which adsorbs $B(OH)_4^-$ at basic to neutral conditions. Strong acid is required for the regeneration of the resin and a strong base is required for neutralization of the brine solution thereafter. Several process configurations were developed that make use of these technologies, including combinations of the two.[3,4]

Published attempts at operational cost approximations for B removal from RO permeates at the posttreatment stage,[3–5] resulted in a roughly similar cost range, i.e., between $\$0.04/m^3$ and $\$0.1/m^3$ for either the IX- or BWRO-based methods. While energy consumption and membrane replacement are the major cost factors for the operation of BWRO B removal, consumption of chemicals and resin renewal are the most significant cost items associated with the IX approach.

The current use of posttreatment methods to remove B after the first SWRO pass has several shortcomings, especially when viewed in light of recent trends in membrane development and process design, evidently moving toward higher flux and higher recovery values. Figure 14.1 shows the advance in membrane technology achieved by Dow's Filmtec brand. A continuous improvement in both element flow and NaCl rejection over the years can be observed, which is mostly due to the advance in the industrial process and the elimination of membrane defects.[6]

Although it seems that the classic polyamide membrane technology has come close to its full potential in tackling the tradeoff between water permeability and selectivity, recent developments further extend the ability to improve selectivity without impeding flux, and vice versa. Integration of nanoparticles into the membrane polymer is a promising development in this regard.[7] Membranes produced by this technology, known as nanocomposite membranes, are already commercial and show the highest flux and NaCl rejection in their category.

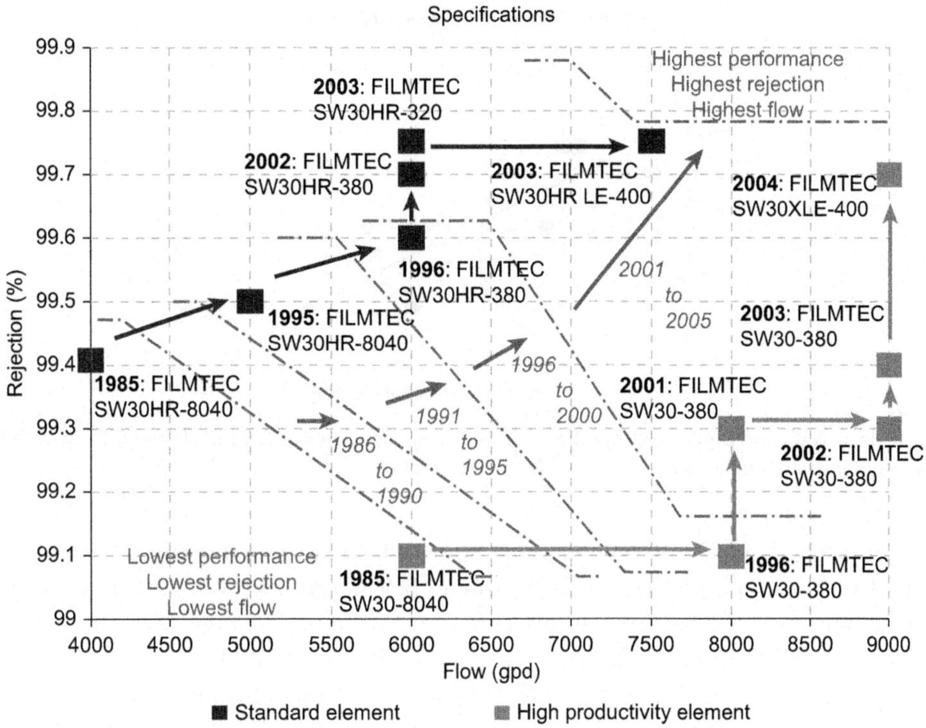

Figure 14.1 Membrane element flow-rate and salt rejection: historic trends. *(Ref. [6]).*

In contrast, B rejection depends strongly on water permeability as shown in Figure 14.2, which compares B and NaCl rejection of last generation commercial membranes from four different brands, characterized by different water flow at standard test conditions. Breaking the tradeoff between water permeability and B rejection will require a paradigm shift in membrane materials and production.[8] Cutting edge membrane technologies, e.g., carbon nanotubes,[7] aquaporine,[9] and graphene[10] membranes, may promote this paradigm shift in the more distant future, while polymer-based membranes will probably continue to be the workhorse for SWRO in the foreseeable future.

Higher flux membranes have the potential to reduce the energy consumption and the required membrane area for SWRO, thus reducing the total cost of the product water.[11] Elimelech and Phillip[8] pointed out that the reduction in energy in the first pass is expected to be marginal, since SWRO desalination plants already operate close to the practical thermodynamic limit imposed by the osmotic pressure of the concentrate. It was therefore stated by these authors that the main potential for cost reduction lies in the pretreatment and posttreatment steps.

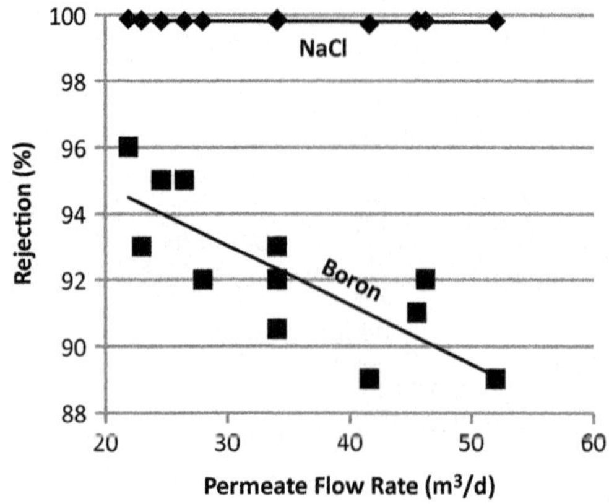

Figure 14.2 NaCl and boron rejection trends by latest-generation membrane elements from different brands. Membrane active area is 37.4 m^2 for all elements.

Nevertheless, this outlook does not render the high-flux membrane technology redundant. First, considering the large scale of SWRO plants, even a small decrease in energy consumption may result in substantial cost savings. Secondly, there are potential benefits for other steps in the process. Zhu et al.[12] studied the effect of high-flux membranes on the overall cost of SWRO plant at various operational conditions and concluded that the optimal recovery ratio had shifted to a higher value. This conclusion was backed up by cost-analysis of different process scenarios carried out by membrane manufacturers.[13] Operating SWRO at higher recovery values may reduce capital and operational expenses for the pretreatment step, and also the first pass and the brine disposal system. For example, if the recovery ratio is increased from 50% to 60%, the flow rate of the feed and brine are reduced by 16.66% and 33.33%, respectively.

High recovery and usage of high-flux membranes result in higher B and NaCl passage to the first pass permeate, which eventually require larger and more expensive post-treatment schemes to reach the B concentration criteria.[14] Given the trends in membrane development and B requirement discussed above, it seems that the problem of increased NaCl passage can be overcome by development in membrane technology, leaving B passage as the main limitation for further price reduction. This is true today, especially for plants that are required to produce product water with a low B concentration in addition to Total Dissolved Solids (TDS) concentration of ~500 mg/L (corresponding to chloride concentration of ~250 mg/L).[15]

To conclude, the prospect of cost-effectively reducing B passage during the SWRO step is: (1) eliminating or significantly decreasing the second pass; (2) reducing the energy consumption of the plant by using high-flux membranes; (3) significantly reducing the size of the pretreatment and reject disposal steps resulting from applying an increased recovery value.

14.2 APPROACH CHALLENGES AND POTENTIAL SOLUTIONS

14.2.1 Scaling Control at High pH SWRO Operation

Achieving overall B rejection >90% in the SWRO step while operating at high flux and high recovery requires a reduction in the boric acid species concentration in the seawater feed. The most straightforward, studied and applied method is to increase the pH value by dosing a strong base to the feed water, commonly NaOH, to convert boric acid to borate. However, operation at high pH can promote irreversible chemical scaling of calcium carbonate at pH values just above seawater's natural pH and of magnesium hydroxide at pH > 9.4.[16] The challenge of B removal in the first pass is thus interlinked with the problem of scaling, caused by chemical precipitation from seawater brine. In the following sections we describe some of the methods suggested to control scaling in the first SWRO pass operated at high pH conditions. We discuss the potential benefits and the technical difficulties related to each method in the context of B removal. Note that although other methods were suggested in the literature for water types other than seawater, only works addressing high pH seawater B removal are covered in this chapter.

14.2.1.1 Antiscalants

The conventional method for minimizing $CaCO_3$ scaling is the dosage of antiscalants. The option of elevating the seawater feed pH, while applying antiscalants, to achieve single pass B removal was considered in several works. A work done with a Doosan hydro membrane type showed that by adjusting the seawater feed to pH9.0, a total B rejection of 96% could be obtained.[17] No permeate flux reduction was observed during 45-day operation, although a laboratory experiment showed that $CaCO_3$ precipitated at this pH value even in the presence of antiscalants. SWRO operation at high pH was tested at full scale in the Larnaca SWRO desalination plant in Cyprus. The pH was elevated from 8.02 to 8.60 and first stage permeate B concentration was consequently reduced from 0.96 to 0.6 mgB/L. An increased dose of antiscalants was added to the feed water and a thorough monitoring campaign showed that the membrane had not scaled; however, operation period lasted only 5 days.[18] Various configurations for seawater B removal were economically assessed as part of a work initiated by the U.S. Bureau of Reclamation. Single-pass operation at pH8.5 was found to be the most cost-effective alternative for arriving at a permeate concentration of <1.0 mgB/L.[19]

According to Hasson et al.,[20] scaling propensity in the first pass is negligible at pH < 9.0 and pH < 8.7 with and without antiscalant addition. These conclusions, however, were based on induction time lab experiments which may not properly represent real SWRO conditions, e.g., areas of stagnant flow and multiple types of granulation surfaces (e.g., feed spacer and small particulate matter). Computer projections, using membrane manufacturers' software, resulted in single pass B removal

(<0.3 mgB/L in this case) for temperature below 16 °C at pH 8.7 and for all ambient temperatures at pH 9.[20]

Although antiscalants have long been proven as an effective scale control measure in many cases, several disadvantages are associated with their use: (1) the price of antiscalant chemicals is high (can amount to 4000 $/ton); (2) antiscalants are disposed to the sea along with the brine, raising environmental concerns[21]; (3) some antiscalants promote biofouling of RO membranes[22]; (4) at high supersaturation, as obtained in seawater at pH > 9.0, antiscalants may prove ineffective. At high pH conditions, at which increased antiscalant dosage is needed, these listed drawbacks become more prominent, thus an alternative, unconventional method of scale control may be required.

14.2.1.2 Nanofiltration (NF) as a Pretreatment Step

Application of NF as a pretreatment step prior to SWRO has been intensively studied in recent years by several research groups worldwide.[23–29] NF membranes cover the range between RO and ultrafiltration (UF) membranes. NF efficiently removes residual bacteria and very fine turbidity via size exclusion.[23] Ions are also partly rejected following a combination of size and charge related mechanisms, not yet fully understood, resulting in significantly higher rejection of multivalent over monovalent ions. NF permeate is therefore superior as SWRO feed compared to the conventionally used UF or dual sand media filtrate resulting in lower salt concentration, lower tendency for particulate fouling, biofouling and also to some extent chemical scaling. Thus, NF pretreatment can potentially reduce energy consumption, increase recovery, prolong membrane life and improve the permeate quality in the SWRO step.

The NF pretreatment concept was first suggested by Hassan et al. (1998)[24] as part of research done in the Saline Water Conversion Corporation (SWCC) in Saudi Arabia. The idea was to allocate a train of NF elements between the sand filtration and the SWRO steps as presented in Figure 14.3. In ongoing research, the NF-SWRO process was studied at the pilot scale and later in a demonstration plant,[25] showing both technical feasibility and economic advantage over the conventional approach.

Llenas et al. (2011) tested B rejection from seawater by NF as part of work dedicated to the NF-SWRO process.[23] They reported close to zero or negative B rejection by five out of the six NF membranes used. The only membrane presenting a significant positive B rejection (~30%), was a rather loose membrane (Dow's NF-270), with the second highest water permeability out of the membranes tested. This result is surprising and should be rechecked, as it differs from the results obtained thus far for polyamide membranes, usually showing a strong link between water and B permeability (see Figure 14.1). Although most NF membranes can reject very little, if any, B from seawater, its selective rejection toward divalent ions could assist in more efficient B removal in the SWRO step. As pointed out by Llenas et al., pH of the seawater, softened by NF, could be elevated with reduced risk of chemical scaling, thus enabling better B removal.

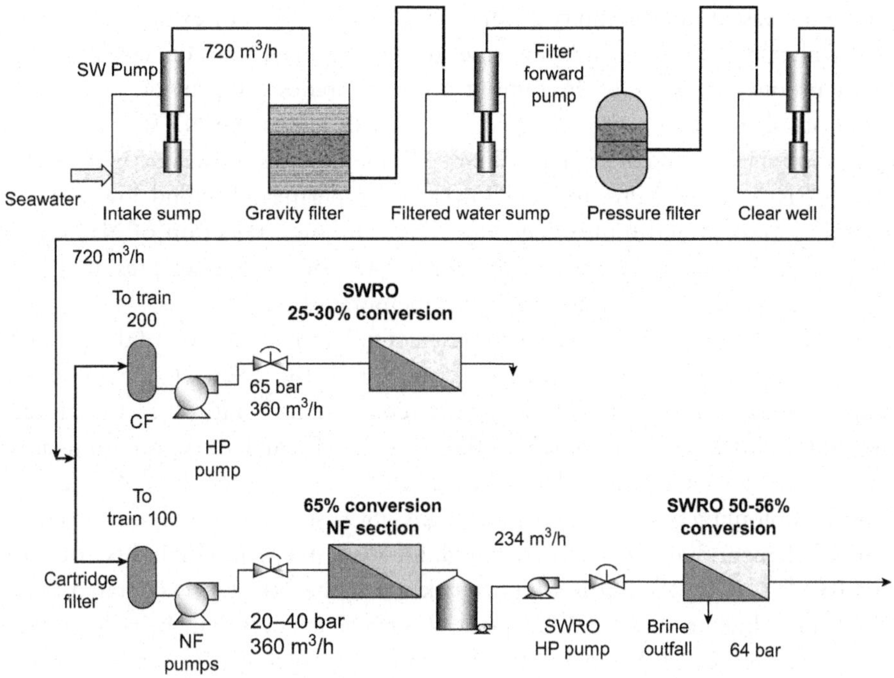

Figure 14.3 Schematic flow diagram of Ummlujj SWRO plant after incorporating NF pretreatment. *(Ref. [29]).*

Seawater NF can be operated at low pressure and high recovery due to the high water permeability and the high passage of Na^+ and Cl^- ions, decreasing the osmotic pressure difference between the concentrate and the permeate. Since water and ions permeability significantly varies between commercial NF membranes, much of the work on this subject focused on finding the best membrane for the process.[23,26,27] The reduction of chemical scaling risk depends on the rejection of specific ions. Most of commercial NF membranes are negatively charged, resulting in an almost complete SO_4^{2-} removal, and decreasing $CaSO_4$ scale tendency in the SWRO step. However, according to the most reliable seawater solubility model (Harvie−Muller−Weare Pitzer based model), gypsum ($CaSO_4{:}2H_2O$) supersaturation occurs only when seawater is concentrated by a factor of ~3. Considering SWRO recovery in the range of 50−60%, $CaSO_4$ scaling conditions are unlikely to develop even when concentration polarization (CP) is taken into account.

$CaCO_3$, on the other hand, is supersaturated even at natural seawater conditions, and is the main chemical fouling agent in the SWRO step. $CaCO_3$ scaling prevention through NF pretreatment depends on the rejection of inorganic carbon species and Ca^{2+} ions by the specific NF membrane used. The studies cited in this chapter reported Ca^{2+} rejection in the range 40−95% while HCO_3^- rejection was in the range 10−85%.[23−29] If high pH operation for high B removal is desired, the required NF rejection must make up for the increase in the solubility product ($[Ca^{2+}][CO_3^{2-}]$). For example, adjusting pH to pH9 will result in a ~10-fold increase in the carbonate ion concentration and the solubility product alike. Only the tightest NF membranes can provide such combined rejection of calcium and carbonate species, especially when the NF step is operated at higher, more practical, recovery ratio. Tight NF membranes, however, are not ideal for the NF-SWRO process, due to their lower water permeability and lower selectivity between monovalent and multivalent ions.[26–28] The high rejection of NaCl increases the transmembrane osmotic pressure difference and imposes higher pressure operation and lower recovery, compared to looser NF membranes.

As reported by the SWCC, after eight years of full-scale operation of the NF-SWRO process, the NF pretreatment effectively maintained the SWRO step performance, without need for cleaning or RO membrane replacement.[29] However, the problem of fouling and scaling has been relocated to the NF step. The high-flux conditions at which the NF step was operated (especially in the lead element) increased the CP phenomenon, promoting biofouling and particulate fouling. $CaSO_4$ supersaturation could develop in the tail NF element due to the relatively high recovery (65%) and high flux compared to the SWRO.[28] $CaCO_3$ scaling is a higher risk when the NF step is operated at natural seawater pH or higher. In the SWCC plant, this problem was solved by reducing the pH of the NF feed to pH6.8, however, this allows more dissolved CO_2 to permeate the NF membrane. In the case of high pH operation in the SWRO pass, the result will be a less effective $CaCO_3$ scaling reduction.

14.2.1.3 Decarbonation of the Seawater Feed

Decarbonation of the seawater feed, i.e., the removal of practically all inorganic carbon concentration (C_T), was suggested by a few authors as a pretreatment approach. C_T in natural seawater amounts to ~2.3 mM, a low concentration compared to the standard seawater concentrations of Mg^{2+} (54.74 mM), SO_4^{2-} (29.79 mM) and Ca^{2+} (10.74 mM), targeted by the NF pretreatment step. Once C_T is effectively removed, $CaCO_3$ is maintained undersaturated and the pH of the SWRO feed can be safely increased up to ~pH9.4, where it is limited only by $Mg(OH)_2$ supersaturation in the brine. At pH > 9 an overall B rejection >90% is achievable even when one uses high-flux SWRO elements and operates at high recovery ratios.[15]

Two distinct approaches for decarbonation of seawater, found in the relevant literature, are discussed: (1) $CaCO_3(s)$ removal by precipitation at high pH; (2) $CO_2(g)$ removal by aeration at low pH. At the natural pH of seawater (~pH8) bicarbonate is dominant among the dissolved inorganic carbon species (~90%). The removal of C_T as solid $CaCO_3$ precipitant requires strong base dosage to convert bicarbonate to carbonate, while the removal of C_T as gas requires dosage of strong acid to convert bicarbonate to dissolved CO_2. The classic method of hard water softening is the addition of $Ca(OH)_2$ (lime), which is a relatively cheap basic chemical. However, lime solubility in water is limited and its dissolution is relatively slow. Consequently, lime dosage is a rather elaborate procedure, requiring large reactors for creating the lime suspension known as "Milk of Lime". When contacted with the treated influent, only part of the lime dissolves, while some remain as a suspension or act as a nucleation site for supersaturated solids. The lime process is thus characterized by a large amount of sludge, high area demand, and increased effluent turbidity, which can be problematic in the operation of large scale SWRO plants.

El-Manharawy and Hafez (2002) suggested NaOH dosage as an alternative to lime for the purpose of seawater softening prior to SWRO. NaOH could be stored and added as a highly concentrated solution (up to 32%), allowing a more simple and controlled process.[30] The main drawback in this approach is the high and fluctuating price of NaOH, ranging between U.S. $300 and 800 per ton of pure NaOH. A dose of $\sim$2.5 eq/m^3 of strong base is required to convert all C_T to carbonate (boric acid species consume $\sim$0.4 eq/m^3 of strong base plus $\sim$2.1 eq/m^3 that is consumed by bicarbonate). Assuming 60% recovery and NaOH price of 550 $/ton, this approach will add $\sim$9 cent/m^3 to the product water cost. Furthermore, $CaCO_3$ precipitation is bound to be accompanied by an increase in turbidity that will have to be completely removed prior to the SWRO stage, increasing the load on the filtration step.

Aguinaldo (2009)[31] suggested that a hybrid process comprising lime dosage followed by low pressure UF could be used as a pretreatment for high pH SWRO operation. The process was studied in a pilot plant running on hard municipal water. Feasibility was shown using an immersed UF membrane which was periodically cleaned by air back-wash. This approach seems appealing due to the relatively low price of lime and the effective removal of fine turbidity by UF. However, the process may be difficult to implement with seawater due pH control difficulty because at pH > 10 $Mg(OH)_2$ starts precipitating, substantially increasing both the turbidity load and lime consumption.

Decarbonation of the seawater feed by air-stripping, as a $CaCO_3$ scaling prevention measure at high pH operation, was suggested by Nir et al. (2012).[16] A new process, aimed at single-pass B removal, while reducing the cost of the remineralization, post-treatment step. A schematic description of the process is shown in Figure 14.4. The process begins with the acidification of the seawater feed to pH4.3–4.4, at which range >95% of the inorganic carbon is in the form of dissolved CO_2. The acidified seawater, highly supersaturated with CO_2(aq), is now subjected to a two-stage packed bed degasifying step (using conventional stripping towers). The first stage is aimed at removing 20–30% of the CO_2 mass by vacuum and a low air-to-water flow ratio (Qg/Ql). The goal is to form an air stream rich in CO_2(g) which will subsequently be reused for $CaCO_3$ (s) dissolution in the PT stage. The rest of the CO_2 mass is removed in the second degassing stage using an air blower and a high air-flow to water-flow ratio. The aim of this step is to reduce the inorganic carbon concentration (C_T) to below 4% of the original value in seawater (overall CO_2 removal in both stages >96%). After most of

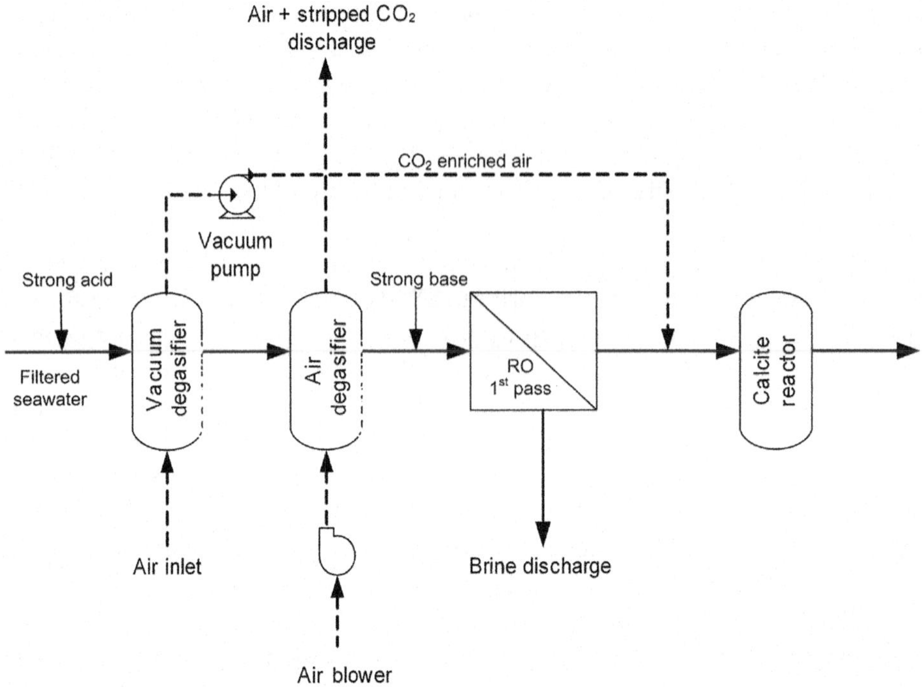

Figure 14.4 Schematic flow diagram of the CO_2 degassing pretreatment process, including CO_2 reuse in the re-mineralization step. *(Ref. [16])*.

the carbonate system concentration had been removed, feed pH is raised to 9.0—9.45 by the addition of a relatively small amount of strong base, prior to entering the RO step. $CaCO_3$ scaling in the RO step is avoided due to the water's low-to-negligible C_T concentration, while $Mg(OH)_2(s)$, another potential scaling agent at this pH range, is intentionally maintained slightly undersaturated. The permeate produced in the RO stage is mixed with the CO_2-enriched air from the vacuum (first) degassing stage, before it is transferred to the calcite dissolution reactor. That is, the CO_2-rich air is used to acidify the permeate stream, enabling calcite dissolution at a reasonable rate.

As opposed to the high amount of base needed for the precipitation process, chemical demands for the stripping process comprise mainly a strong acid (H_2SO_4 or HCl), which is significantly cheaper than equivalent mass of strong base (usually NaOH). Packed bed degassers are simple to operate, and require very little maintenance compared to $CaCO_3$ crystallizers which require sludge handling, regular cleanings, and maintaining optimal nucleation seed concentration in the reactor.

Nir et al. (2012) tested B rejection in a pilot-scale experimental system comprising a single 4″ standard SWRO element (Dow's SW30HRLE), using decarbonated seawater as feed. The concentrate was recycled through the membrane module, while the

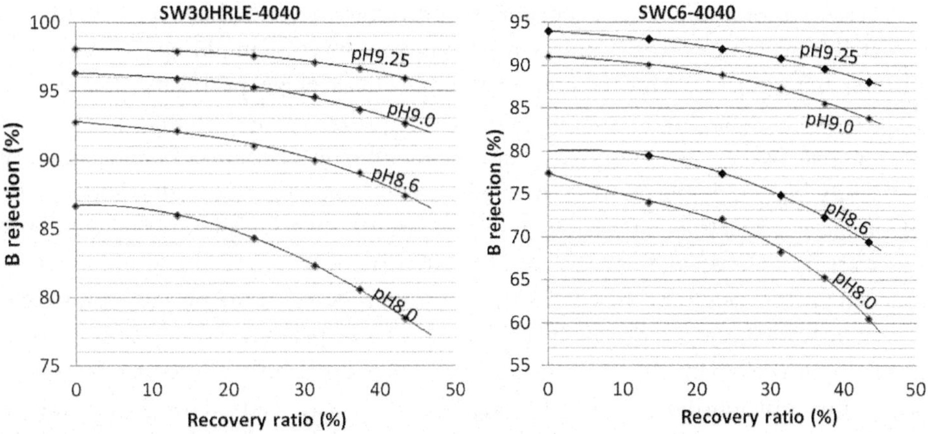

Figure 14.5 Overall B rejection by two types of SWRO elements, plotted versus recovery ratio at four feed pH values. *(Ref. [15]).*

permeate was collected in a separate tank until a 43.3% recovery ratio was reached. Overall rejection >90% was attained at pH 9.0, as presented on the left hand side of Figure 14.5, showing feasibility for achieving B permeate concentration of <0.5 mgB/L in a single pass. The average chloride concentration obtained by this membrane was ~130 mg/L as seen on the left hand side of Figure 14.6, which suits plants requiring low B concentrations well, but are allowed to supply water with chloride concentration in the range 200–300 mg/L. Examples for such plants can be found in Spain, Australia, and the United States (California).

In a following study, the rejections of B (Figure 14.5, right hand side) and Cl^- (Figure 14.6, right hand side) by a high-flux SWRO membrane (Hydranautics' SWC6)

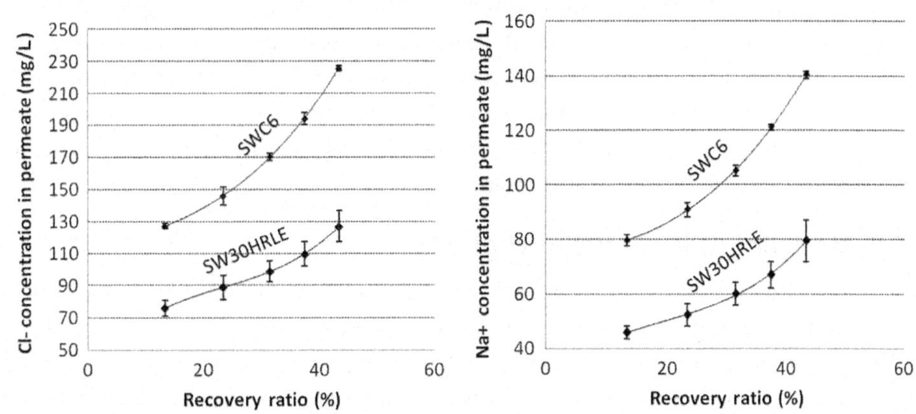

Figure 14.6 Average Cl^- and Na^+ concentrations in SWRO permeate versus recovery for two types of membrane elements (n = 8 for SW30HRLE; n = 3 for SWC6). *(Ref. [15]).*

were measured using the same experimental procedure described above.[15] At pH 9.25, the overall B rejection was 88%, implying that further increase in pH will be required for achieving 0.5 mgB/L permeate concentration. The average permeate flux was 18% higher with the high-flux membrane, compared to the standard membrane, indicating the potential savings in energy or required membrane elements. NaCl passage was significantly higher, demonstrating the well-known tradeoff between flux and rejection.

Computer projections of a large scale seawater desalination plant were used by Nir et al. (2013) to study the energy saving potential of first pass high pH operation. Advanced process design of the SWRO step was considered in the projections, in which different membrane elements are placed in the same pressure vessel. This approach, termed "hybrid element design" or "internally staged design", is used to decrease the unequal flux distribution (causing fouling hazards in the lead element and improper hydraulic conditions in the rear elements) associated with high-flux elements.[32] A second pass for further NaCl removal was also considered for the most strict B and Cl^- requirements (B <0.3 mg/L and Cl^-<50 mg/L). The results suggested that significant energy savings and higher recovery could be achieved at high pH operation, while meeting the B and Cl^- product quality requirements.

It has to be mentioned, though, that the widely used design software provided by membrane manufacturers, can provide reliable projections for standard operation conditions only and has a limited ability to predict B rejection under nonstandard conditions, such as discussed in this chapter. The challenge of modeling B rejection is discussed in the next section.

The main downfall of the degassing pretreatment approach is the considerable amount of acid required for the acid-stripping process (~125 g H_2SO_4 per m^3 of seawater). This goes against the popular trend of reducing chemicals' consumption within desalination, while increasing the strong-acid handling equipment and adding to the chemical cost. The increase in chemical cost could be partly compensated by the reuse of CO_2 in the calcite dissolution step, as previously described. However, some technical difficulties are associated with the operation of a calcite dissolution reactor using an air stream mostly comprising of N_2. Work is currently underway to address the engineering challenges and find the optimal conditions for this brand new concept of CO_2 recycling from seawater to posttreatment.

14.2.2 Modeling B Removal at High pH

The choice of the most cost-effective process and its optimal design requires a reliable B cross-membrane transport simulation procedure, which produces accurate results in a wide range of operational conditions and feed water compositions. In the previous section of this chapter, different process schemes for operating SWRO at high pH for improved B removal were reviewed. These advanced processes alter the chemical composition of the feed, affecting B equilibria and B transport during the SWRO step.

However, most of the existing B transport models were developed specifically for seawater, and cannot account for such effects. Moreover, the dynamic nature of acid–base equilibria, indicated by the change in the pH of the feed along the membrane, is commonly unaccounted for.

RO concentrate during filtration is characterized by complex interrelations between transport phenomena (e.g., permeation, convection, and CP) and chemical processes (e.g., ion interactions, acid–base equilibria, and complexation). Therefore, the accurate prediction of the pH dependent B permeation needed for design is challenging and requires a modeling tool which can describe both transport and chemical equilibrium phenomena. B Transport models, B speciation models, their coupling and application for full-scale high SWRO B transport simulation are discussed in the following sections.

14.2.2.1 Boron Transport Models

Following the pioneer work of Taniguchi et al. (2001),[33] most of the B permeation simulations appearing in the literature follow the Spiegler–Kedem transport approach,[34–37] which is based on irreversible thermodynamics and the CP film-layer model. This model requires at least three empirical coefficients to describe the characteristics of the applied membrane: the permeability of water (L_W) through the membrane, the permeability of B (L_B) and the B reflection coefficient. Conversely, application of the solution-diffusion model for B transport,[38] requires (as a minimum prerequisite) to experimentally determine the two permeability coefficients. Under the assumption that the reflection coefficient is ~1 (generally correct for SWRO membranes, as recognized in[33]) the Spiegler–Kedem model is reduced to the solution-diffusion model. The film-layer model for the CP phenomena requires a value for the mass transfer coefficient, which can be determined experimentally[33,34] or derived from the hydrodynamic conditions using empirical equations developed for spiral wound.[35,36] All the model parameters (permeability and mass transfer coefficients) are affected by change in temperature. Permeability coefficients' temperature dependency was incorporated in B transport models using an exponential empirical correlation,[34] while the mass transfer coefficient is affected through changes in density, viscosity, and diffusivity affected by temperature and salinity.

$B(OH)_3$ RO transport can be accurately described by the Spiegler–Kedem or solution-diffusion models combined with the film model. In the early work of Taniguchi,[33] pH dependence was not treated and the seawater feed was maintained at pH 7. Since practically all boron is in the form of $B(OH)_3$ at pH 7, B transport depended only on the solution-diffusion film model. The experimental B rejections fitted well to this model as shown in Figure 14.7, indicating the transport model applicability for this case. The low pH results of Mane et al.,[36] matched the model predictions well (see Figure 14.8.), using model coefficients obtained by independent measurements, also indicating the reliability of the transport model. However, the model used in this work

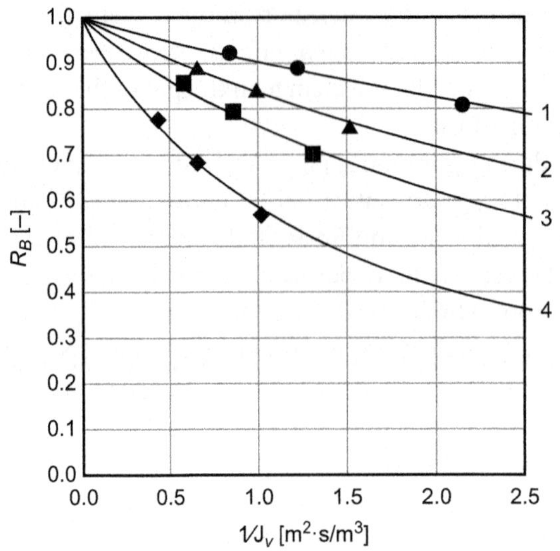

Figure 14.7 Experimental rejections of boric acid (discrete points) by four SWRO membranes fitted to the solution-diffusion film model. *(Ref. [33]).*

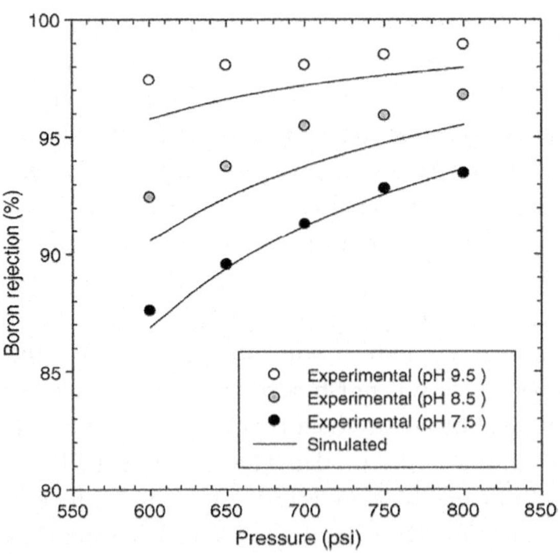

Figure 14.8 Experimental rejections of boric acid (discrete points) at different pH values membranes fitted to the Spiegler–Kedem film model. *(Ref. [36]).*

was less successful in predicting the results obtained at high-pH, suggesting limitations in the model calculations employed for boron speciation.

14.2.2.2 Boron Speciation Models

The dependence of boron rejection on pH, is a result of the high permeability associated with the acidic $B(OH)_3$ species, as compared to the basic species. Determining B speciation is thus essential for modeling B cross-membrane transport under various pH conditions. Given the boric acid dissociation pK′ in seawater and brine conditions, pH highly affects B permeation in the pH range of 7.5–10.5. In water with low ionic strength (I <0.1 M) and low B concentration (<20 mgB/L), the B acid–base equilibria can be fully described by the mass action law, shown in Eqn (14.2), and the mass balance on the most dominant B species (Eqn (14.3)).

$$K_B = \frac{(H^+)\gamma_{B(OH)_4^-}\left[B(OH)_4^-\right]}{\gamma_{B(OH)_3}\left[B(OH)_3\right]} \tag{14.2}$$

$$B_T = \left[B(OH)_3\right] + \left[B(OH)_4^-\right] \tag{14.3}$$

where $(H^+) = 10^{-pH}$; K_B = the thermodynamic acid dissociation constant; γ = activity coefficient; B_T = total B concentration [M].

B_T and (H^+) can be measured analytically while the activity coefficients can be calculated using a simple ion activity model (function of I and T only), which accounts for solution nonideality. K_B can be corrected for temperature effects using the reaction enthalpy of boric acid dissociation. Equations (14.2) and (14.3) can then be solved simultaneously, giving rise to the following expressions:

$$\left[B(OH)_3\right] = \frac{B_T(H^+)}{K_B' + (H^+)} \tag{14.4}$$

$$\left[B(OH)_4^-\right] = \frac{B_T K_B'}{K_B' + (H^+)} \tag{14.5}$$

where K_B' is the apparent mixed acid dissociation constant given by:

$$K_B' = K_B \gamma_{B(OH)_3} / \gamma_{B(OH)_4^-} \tag{14.6}$$

This simple representation (Eqns (14.4)–(14.6)) is not sufficiently accurate when attempting to model concentrated mixed electrolyte solutions, such as seawater. In such solutions, specific, close range interactions, particularly between opposite charged ions, significantly affect the activity of ions (neutral species are also moderately affected). Calculating B speciation in this case generally requires the inclusion of weakly associated ion-pairs and the use of a specific ion interaction model to calculate activity coefficients.

This approach is appropriate for mixed electrolyte solutions of various compositions up to high ionic strength. Incorporation of such models into B SWRO transport simulation codes, discussed later in this chapter, involves some technical difficulties and requires relatively large computing resources.

For seawater, however, there is an alternative method to characterize the borate acid–base system, using empirical equations for K'_B. Seawater media, including concentrated or diluted seawaters, has been extensively studied over the years and K'_B was measured at a wide range of temperatures (T) and salinities (S). Edmond and Gieskes (1970)[39] fitted experimental data obtained by several researchers and derived the following equation:

$$pK'_B = 2291.90/T + 0.01756T - 3.3850 - 0.32051(S/1.80655)^{1/3} \qquad (14.7)$$

Equation (14.12) was originally presented in terms of T(K) and chlorinity (‰Cl), which can be converted to salinity (S) by the term ‰S = 1.80,655‰Cl. The data correlated by this equation was obtained by acid–base titration methods in seawater solutions at a salinity range of 5%–95‰ and temperature range of 280.65–305.65 K. pH during titration was measured using a glass electrode calibrated according to the National Bureau of Standards (NBS) (today called National Institute of Standards and Technology (NIST)) procedure.

Combining Eqns (14.4), (14.5) and (14.7), boric acid and borate concentrations can be calculated at any point along the membrane, or at any recovery ratio, provided that B_T, pH, T and S are known. Hyung and Kim (2006)[34] have used this approach in their mechanistic model development, which was later used for simulating B rejection at a full-scale SWRO plant.[36,37] They introduced two separate permeability coefficients for the two major B species $B(OH)_3$ (boric acid), and $B(OH)_4^-$ (borate anion). It is noteworthy that the borate concentration calculated in this way may include also cation-borate ion-pairs. However, similarly to borate, the permeability of these highly hydrated species is assumed to be negligible (rejection >99%), compared to $B(OH)_3$.

The same approach was adopted by Hung et al. (2009)[38] which used the following correlation term, proposed by Dickson (1990)[40]:

$$\ln(K'_B) = \left(-8966.90 - 2890.53S^{1/2} - 77.942S + 1.728S^{3/2} - 0.0996S^2\right)/T + \left(148.0248 + 137.1942S^{1/2} - 0.2474\right)\ln(T) + \left(0.053105S^{1/2}\right)T \qquad (14.8)$$

The data fitted by Eqn (14.8) was obtained using electromotive force (EMF) measurements of a highly accurate electrochemical cell containing synthetic seawater to which borax ($Na_2B_4O_7 \cdot 10H_2O$) was added. Measurements were conducted in a salinity range of 5–45‰ and in temperature range of 273.15–318.15 K. Since standard seawater has a salinity of 35‰, Eqn (14.8) is valid only for recovery ratios <22% (corresponding to S ~ 45‰), which is not practical for a full-scale SWRO plant. Moreover, even when considering low salinity seawater feed (e.g., estuaries) or low recovery, caution must be exercised in using Eqn (14.8) since K'_B is defined on a different pH scale and is not consistent with the K'_B appearing in Eqns (14.4)–(14.6).

Several pH scales were defined for seawater. The "total pH scale" (pH_{TOT}) used in Dickson's work is defined as follows:

$$pH_{TOT} = -\text{Log}\{m(H^+) + m(HSO_4^-)\} \quad (14.9)$$

where $m(H^+)$ and $m(HSO_4^-)$ are the molal (m/kg-H_2O) concentration of the proton and the bisulfate ion respectively. The molality based pH scales were developed to address the constraints regarding the applicability of the pH_{NBS} (i.e., calibration with dilute NIST pH standards) to high ionic strength solutions. pH_{NBS} does not have a clear thermodynamic meaning and is subjected to uncertainties resulting from the difference in the liquid junction potential between the NBS dilute buffer and the concentrated measured solution. In seawater, the pH error stemming from this phenomenon is estimated to be ~0.08[41] and is expected to increase at higher salinities, e.g., SWRO brines. Conversely, pH_{TOT} is a well-defined thermodynamic quantity, which can be accurately measured using an electrochemical cell without a liquid junction (i.e., the "Harned Cell"), or by using the glass electrode with special seawater buffers, traceable to the "Harned cell". Unfortunately these procedures are currently available only in several marine analytical chemistry laboratories, engaged in oceanic acid–base equilibria research. pH_{TOT} is lower by ~0.1 pH units in seawater, as compared to pH_{NBS}, Thus, introducing the pH_{NBS} value into Dickson's equation (as done in[38]) results in underestimation of the $B(OH)_3$ concentration and therefore in underestimation of the predicted B permeation values. For further discussion in seawater pH scales see.[42]

Empirical acid dissociation constants are mainly available for seawater and NaCl solutions. When the feed composition differs from the relative composition of seawater, a full aqueous speciation model is required for the characterization of acid–base equilibria. Such models are based on a system of equations comprising of mass action laws (e.g., Eqn (14.2)) and mass balance (e.g., Eqn (14.3)). Optionally, charge balance and proton balance is employed. Additionally, an ion activity model is needed for estimating the activity coefficients, included in the mass action equations. At low ionic strength solutions ($I < 0.1$ M) the activity coefficients of all ions can be calculated by the Davies model or the extended Debye–Huckel model with sufficient accuracy. These models, which are simple functions of temperature and ionic strength, only account for long range electrostatic effect.

However, short range interactions become more significant as the concentration of ions increases, while high valence ions ($z > 1$) are affected even in dilute solutions.

The most popular method used to account for the ionic interactions in natural waters is the ion association model.[43] In this approach ions of different charge are assumed to form ion pairs which are treated as separate species. This adds more mass action equations and mass balance terms to the model, resulting in increased computational complexity. Stability constants for ion-pair formation are commonly determined by indirect potentiometric methods thus their calculated value depends on the activity model chosen. Moreover, the number of possible ion-pair formations is very large and it is often ambiguous which should be included in the model (see Table 14.1.) Although the ion-association approach was repeatedly and adequately applied for determining metal speciation at low ionic strength waters, it was less successful, in itself, for modeling concentrated solutions.[43]

A different approach for speciation calculation in concentrated solutions is the specific ion interaction approach. In this sort of treatment, all the ion interactions under consideration are lumped into the activity coefficient model. The most widely used among these are the SIT (Specific ion Interaction Theory) and the Pitzer theory. The SIT was originated by Guggenheim (1922) and brought to its current version, i.e., Eqn (14.9), by Bronsted (1935)[44]:

$$\log \gamma_j = -z_j^2 \frac{0.5\sqrt{I}}{1 + 1.5\sqrt{I}} + \sum_k \varepsilon_{jk} m_k \tag{14.10}$$

where z is the valence of the jth ion; ε the specific interaction coefficient; m the molality of oppositely charged kth ion.

The first, Debye—Huckel-type term in Eqn (14.9) accounts for the long range electrostatic effects, while the second term accounts for the specific interactions between oppositely charged ions. For further information on the SIT model and its implementation in seawater, the reader is referred to Whitfield (1973).[44]

Relative to the SIT approach, the Pitzer ion interaction model (Pitzer 1973)[45] was developed on a more rigorous theoretical basis. It is based on a virial expansion for the excess Gibbs energy for nonideal solution (i.e., the difference in Gibbs energies between

Table 14.1 Acid Dissociation Constants for Boric Acid and Ion-Pair Stability Constants for Borate with the Major Seawater Cations, Taken from Different Thermochemical Databases Embedded in PHREEQC.

	Database:		
Reaction:	**Minteq.V4**	**SIT**	**Pitzer**
$B(OH)_3 + H_2O \leftrightarrow B(OH)_4^- + H^+$	$pK_B = 9.236$	$pK_B = 9.24$	$pK_B = 9.239$
$Mg^{2+} + B(OH)_4^- \leftrightarrow MgB(OH)_4^+$	$pK = 1.54$	$pK = 1.6$	$pK = 1.4$
$Ca^{2+} + B(OH)_4^- \leftrightarrow CaB(OH)_4^+$	$pK = 1.76$	$pK = 1.8$	$pK = 1.65$
$Na^+ + B(OH)_4^- \leftrightarrow NaB(OH)_4^0$	$pK = 0.213$	$pK = 1.226$	Not considered

an ideal and a nonideal solution of the same composition). Differentiation of the virial expansion by the concentration of a specific ion gives the following expression for its activity coefficient (e.g., for an anion):

$$\ln(\gamma_X) = z_X^2 \begin{pmatrix} -A^{\phi}\left[\dfrac{\sqrt{I}}{1+b\sqrt{I}} + \dfrac{2}{b}\ln\left(1+b\sqrt{I}\right)\right] + \\ \sum_c \sum_a m_c m_a B'_{ca} + \sum_c \sum_{c'} m_c m_{c'} \phi'_{cc'} + \sum_a \sum_{a} m_a m_{a'} \phi'_{aa} \end{pmatrix} \tag{14.11}$$

$$+\sum_c m_c(2B_{cX} + ZC_{cX}) + \sum_a m_a\left(2\phi_{Xa} + \sum_c m_c \psi_{Xac}\right)$$

$$+\sum_c \sum_{c'} m_c m_{c'} \psi_{cc'X} + |z_X| \sum_c \sum_a m_c m_a C_{ca}$$

Subscripts X, a, and c stand for the relevant anion, other anions, and cations, respectively.

In Eqn (14.10), the first term (consisting of a Debye–Huckel term and terms for interactions between all other possible ion pairs) accounts for long range electrostatic effects; the second term describes specific interactions between the relevant anion with all the cations; the third term describes the interactions between the relevant anion and all other anions (ϕ) and triplet interactions between the relevant anion and all the possible anion–cation pairs (ψ); the fourth term describes interactions between the relevant anion and all possible cation–cation pairs; the fifth term describe the effect of interactions between other cations and anions on the activity of the relevant anion. The virial coefficients B, B', C, ϕ, ϕ', and ψ appearing in Eqn (14.10) are functions of ionic strength and can be correlated to experimental results, obtained from binary and ternary solutions. For a full definition of the Pitzer equations and their implementation, the reader is referred to Pitzer (1973)[45] and Harvie and Weare[46] (1980).

The combination of the ion interaction and ion association models, as first suggested by Whitfield (1975)[47] is now the common paradigm for chemical equilibrium models of concentrated solutions. Speciation of multicomponent electrolyte solutions requires solving a large set of nonlinear equations and a considerable amount of thermochemical data. Fortunately these are included in various available computer codes, incorporating numerical solvers. The embedded algorithms for formulating and solving multicomponent systems are mathematically equivalent for all these programs, while the major differences are in the database and interface.[48] The most widely used computer programs as indicated by the number of search results in Scopus are PHREEQC (3229 entries in Scopus), MINEQL (1072 entries in Scopus), and MINTEQ (715 entries in Scopus). MINEQL, a commercial software, and MINTEQ, developed by the U.S. Environmental

Protection Agency (USEPA) were developed for low to moderate ionic strength solutions ($I < 0.5$ M). Both programs utilize the USEPA thermodynamic database, which is comprehensive and fully documented. PHREEQC, developed by the United States Geological Survey (USGS) allows a choice between several thermodynamic databases, including USEPA's (minteq.v4.dat).

For high ionic strength solutions, PHREEQC provides two database files, "sit.dat" and "pitzer.dat". "sit.dat", utilizes the SIT activity model, developed by the Organisation for Economic Co-operation and Development (OECD) nuclear agency. It includes a large number of ion-pairs, for which stability constants were determined by correlations based on the SIT approach. "pitzer.dat" is largely based on the Harvie–Muller–Weare 25 °C speciation model,[49] utilizing the Pitzer activity model. It includes a limited number of ion-pairs, which stability constants were determined by correlations based on Pitzer's approach. Consequently, the stability constants of ion-pairs as well as the types of ion-pairs included in the aqueous model are not universal, but rather depend on the activity model applied. The ion-pairs stability constants of borate with the major seawater cations, taken from the three mentioned PHREEQC databases, are shown in Table 14.1. The variations in the pK's of ion association reactions are much higher than the differences in pK_B's, while the $NaB(OH)_3^0$ ion pair is not considered in the Pitzer database. Generally, the number of ion-pairs required in Pitzer-based models is considerably reduced, simplifying the speciation calculation.

The Pitzer database is assumed to be the most reliable choice, for three reasons: (1) it invokes the Pitzer approach which is considered the most accurate activity model; (2) it comes largely from one source and is therefore highly internal-consistent; (3) it was tested and verified against empirical results in a wide range of solutions of various compositions and ionic strength values. Although Pitzer's interaction parameters are only available for a limited number of species as compared to SIT, the database is sufficient for most types of natural waters including seawater. Pitzer's interaction parameters for the borate system were added to the Harvie–Muller–Weare model by Felmy and Weare (1986),[50] who also included two polyborate species (i.e., $B_3O_3(OH)_4^-$ and $B_4O_5(OH)_4^{2-}$), becoming significant at $B_T \sim >22$ mgB/L.[51] More recent determinations of B Pitzer parameters, made by Hershey et al.[52] (1986) and Simonson et al. (1987, 1988),[53,54] were not included in the Pitzer database of PHREEQC. Although it is possible for the user to modify the Pitzer parameters, caution must be exercised, for the parameters are interdependent and their modification may result in inconsistency.[55] The model is reliable in the temperature range of 0–60 °C as claimed by its developers.[55] Temperature effects are taken into account using the reaction enthalpy. Additionally, some of the Pitzer parameters are given as a function of temperature.

As previously mentioned here, the pH of high ionic strength solutions is subject to errors when one uses the NBS calibration procedure. As a result, the measured pH can be significantly different from the theoretical pH calculated by Eqn (14.10). In an attempt to minimize these differences all the ion activities in PHREEQC's Pitzer model, including pH, are scaled according to the Macinnes convention. In the speciation process, all concentrations and activities are first determined using Pitzer's equations. The activity coefficient of Cl^- (γ_{Cl}) is then set to be equal to the mean activity coefficient of a KCl solution at the same ionic strength ($\gamma_{Cl(Mac)}$) as the modeled solution. Each ion (i) is then scaled as follows:

$$\gamma_{i(Mac)} = \gamma_i\left(\gamma_{Cl}/\gamma_{Cl(Mac)}\right)^{z_i} \tag{14.12}$$

This procedure was found to by Harvie et al.[49] to better reproduce the measured pH_{NBS} at high ionic strength solutions. For synthetic seawater, for example, the difference between the calculated and scaled pH and the measured pH_{NBS} was found to be 0.05. Since each activity is scaled by the same normalized factor, thermodynamically measurable quantities such as ion activity products are unaffected.

Chemicals equilibrium programs can be coupled with RO transport models for the purpose of B transport simulations, as suggested by Nir et al. (2013).[56] This approach can be applied on any B containing feed water, e.g., seawater conditioned by an advanced pretreatment process (discussed in Sections 2.1.2 and 2.1.3.), brackish water, geothermal water, oil & gas produced water, etc. The concept was tested on seawater by manually coupling a MATLAB® code implementing the solution-diffusion film model with PHREEQC utilizing the pitzer.dat database. Automation of this coupling process is currently being finalized, aiming at releasing a free, open source computer code for the simulation of B transport. The code implementing the physical models and the coupling procedure was improved and rewritten in Python, which is a free, open sourced (unlike MATLAB®), and fully interpreted (i.e., does not require compiling to run similar to MATLAB®) programming language. Many analytical, numerical and graphical tools are available in Python through designated libraries. Coupling PHREEQC to Python is relatively easy thanks to efforts made by the creators of PHREEQC in USGS to develop free modules used specifically for this purpose.[57] Featuring the numeric engine and thermodynamic database of PHREEQC, these modules (also called COM objects) can be operated from within the python environment using predetermined keywords.

14.2.2.3 Boron Transport Simulation at Full-Scale SWRO

In the two previous sections we discussed B transport models that are valid only for very low recovery ratios. Predicting B permeate concentration at higher, more practical, recovery values based on these transport-speciation models require iteration-based procedures. This usually involves the division of the membrane to either n 1D length segments[33,35] or n·m 2D length and width segments,[36,37] although segments of recovery ratios may also be defined.[57] The initial parameters, i.e., solution composition (e.g., B_T, pH, salinity), operational conditions (e.g., pressure, flow-rate, temperature) and membrane properties (e.g., permeability coefficients, module geometry) are used by the model to calculate the permeate flow and permeate concentrations of B and other solutes for the first segment. This step is followed by mass balances to determine the retentate flow and retentate solutes concentrations, which are used by the model in the next segment. These procedure repeats itself until the desired recovery is achieved or the length of the modeled membrane train is reached.

The change in solution composition (e.g., B_T, pH, salinity) with increasing recovery along the membrane's train requires the calculation of B species concentration at the membrane wall for each numerical segment. As thoroughly discussed in this chapter, pH (as an indicator of acid—base equilibria) is a key parameter affecting B speciation and permeation. However, with the exception of Nir et al. (2012),[57] all full scale B simulation works published thus far[33–38] are based on the assumption that the pH of the brine, from the raw seawater to the brine at the outlet of the SWRO step, is constant. B speciation was therefore calculated considering only the change in pK_B' induced by the increasing salinity. In stark contrast with this assumption, significantly different pH values are measured regularly at the feed and brine in SWRO applications.[58] Acknowledging that the pH value changes along the feed water (brine) path, accurate determination of B species concentration at each numerical point requires recalculation of the pH value at each step throughout the simulation. Nir et al. (2012)[57] used this approach in an attempt to predict B permeate concentrations obtained by pilot-scale SWRO system using decarbonated seawater. As shown in Figure 14.9, the predictions better matched the experimental results compared to a simulation based on the (false) constant pH assumption. The simulation results were obtained using permeability coefficients derived from standard test conditions as reported by the membrane manufacturer.

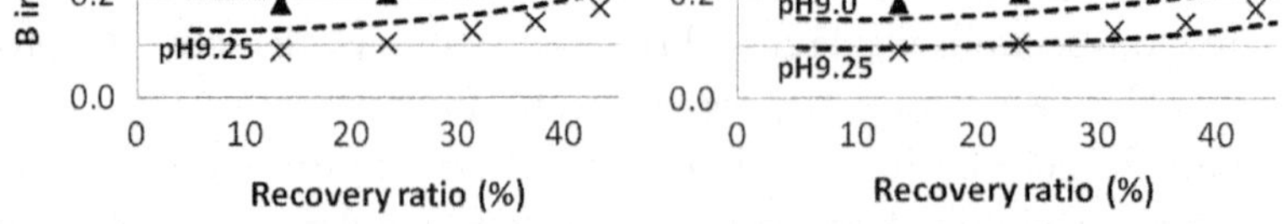

Figure 14.9 Experimental results (discrete points) vs. simulation predictions (curves) of permeate B concentrations as a function of recovery value and initial feed water pH. Predictions based on the coupled approach appear in the left hand graph. Results of the traditional simulation approach are shown on the right. *(Ref. [56]).*

Table 14.2 Major Processes Resulting in Acid–Base Spatial Variance within the Membrane Module During SWRO.

pH Decrease	pH Increase
Acidic species retention: $B(OH)_3$, HCO_3^-	Acidic species permeation: $B(OH)_3$, CO_2
Basic species activity decrease: short and long range interactions	Basic species retention: $B(OH)_4^-$, CO_3^{2-}, HCO_3^-

Any process involving a change in acid–base parameters may potentially induce pH variations. The main processes affecting pH during SWRO filtration are described in Table 14.2. Since the carbonate system is the dominant weak-acid system in seawater, it has a major influence on pH variations and thus on B membrane transport. Published works attempting to theoretically calculate the pH of SWRO brines are scarce and focused on prediction of scaling potential only (with the exception of Nir et al.). Huff (2003)[59] used a PHREEQC evaporation model to calculate saturation indices of calcite and silicate minerals in SWRO brines. Only water permeation was modeled, neglecting other potentially significant effects such as $B(OH)_3$ and CO_2 permeation. No details were provided regarding the database used. Waly et al. (2011)[58] have calculated the pH of the brine using the apparent pK's of the carbonate system, given as empirical expressions of temperature and salinity. The results were compared with computer projections, including a PHREEQC Pitzer based evaporation model and with empirical results. However, the pH was measured on the NBS scale, where the empirical equations used were based on concentration pH scales e.g., the "total" scale and the "free proton" scale. In brine conditions, this incongruence may cause errors approaching 0.2 pH units. Moreover, bicarbonate and carbonate concentrations in the brine were determined by a simple mass balance on these ions, disregarding the fact that these are not conservative parameters.

Calculating the pH along the membrane's train for accurate full-scale B rejection simulation requires a more comprehensive treatment toward acid–base equilibria, which will include all weak-acid systems in significant concentrations. Such an approach is included in our new Python- PHREEQC coupled simulation tool previously mentioned. The simulation algorithm is illustrated in Figure 14.10. In compliance with the film CP model, the feed bulk, CP layer and permeate are treated as distinct regions. CP and membrane permeation are calculated for each species separately according to its diffusion coefficient, permeability and concentration gradient. A mass balance is made on the conservative parameters C_T, B_T, and alkalinity on every cross between regions (i.e., bulk, CP layer, and permeate) and between numerical segments (i.e., recovery ratios). The conservative parameters are used to determine the pH and species distribution in every region in the membrane module using PHREEQC's Pitzer model. This approach could be easily extended to other types of RO feed waters containing boron and additional weak-acid systems affecting the pH.

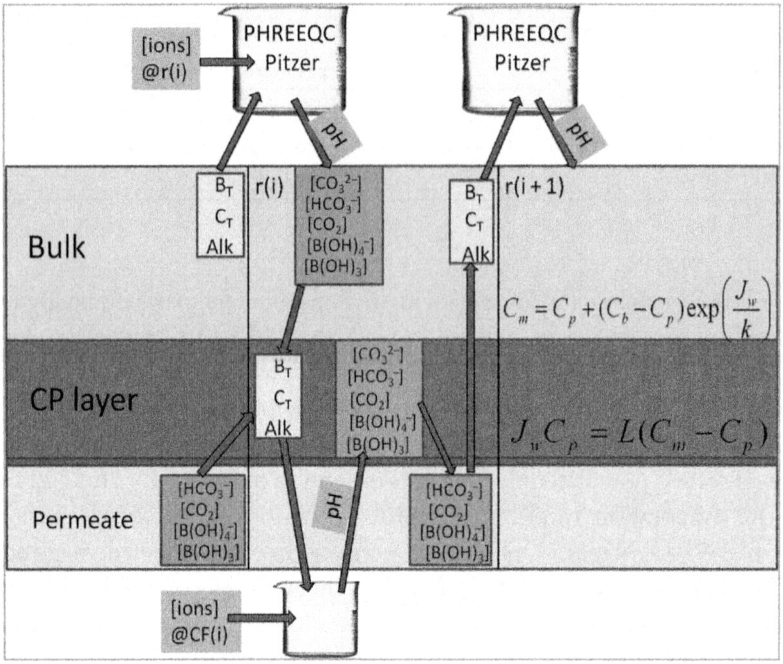

Figure 14.10 Schematic illustration of a new B membrane transport modeling approach, accounting for pH variations along the membrane element train and valid for a wide range of feed water compositions.

REFERENCES

1. Peñate B, García-Rodríguez L. Current trends and future prospects in the design of seawater reverse osmosis desalination technology. *Desalination* January 4, 2012;**284**(0):1–8.
2. Yermiyahu U, Tal A, Ben-Gal A, Bar-Tal A, Tarchitzky J, Lahav O. Environmental science: rethinking desalinated water quality and agriculture. *Science* 2007;**318**(5852):920–1.
3. Hilal N, Kim GJ, Somerfield C. Boron removal from saline water: a comprehensive review. *Desalination* 2011;**273**(1):23–35.
4. Kabay N, Güler E, Bryjak M. Boron in seawater and methods for its separation—a review. *Desalination* 2010;**261**(3):212–7.
5. Chillón Arias MF, Valero i Bru L, Prats Rico D, Varó Galvañ P. Approximate cost of the elimination of boron in desalinated water by reverse osmosis and ion exchange resins. *Desalination* June 15, 2011;**273**(2–3):421–7.
6. Lomax I. Experiences of Dow in the field of seawater reverse osmosis. *Desalination* April 15, 2008;**224**(1–3):111–8.
7. Buonomenna MG. Nano-enhanced reverse osmosis membranes. *Desalination* 2013;**314**:73–88.
8. Elimelech M, Phillip WA. The future of seawater desalination: energy, technology, and the environment. *Science* 2011;**333**(6043):712–7.
9. Tang CY, Zhao Y, Wang R, Hélix-Nielsen C, Fane AG. Desalination by biomimetic aquaporin membranes: review of status and prospects. *Desalination* 2013;**308**:34–40.
10. Cohen-Tanugi D, Grossman JC. Water desalination across nanoporous graphene. *Nano Lett* 2012;**12**(7):3602–8.

11. Bartels C, Franks R, Bates W. *Design advantages for SWRO using advanced membrane technology, conference proceeding published 11/01/2009 by American Water Works Association.*
12. Zhu A, Christofides PD, Cohen Y. Effect of thermodynamic restriction on energy cost optimization of RO membrane water desalination. *Ind Eng Chem Res* 2009;**48**(13):6010–21.
13. Molina VG, Busch M, Sehn P. Cost savings by novel seawater reverse osmosis elements and design concepts. *Desalin Water Treat* 2009;**7**(1–3):160–77.
14. Bartels CR, Rybar S, Andes K, Franks R. *Optimized removal of Boron and other specific contaminants by SWRO membranes. IDA World Congress-Dubai UAE*, vol. 9; 2009. 7–12.
15. Nir O, Herzberg M, Lahav O. A new, energy-efficient approach for boron removal from SWRO plants. *Desalin Water Treat* 2013;**51**(7–9):1651–6.
16. Nir O, Herzberg M, Sweity A, Birnhack L, Lahav O. A novel approach for SWRO desalination plants operation, comprising single pass boron removal and reuse of CO_2 in the post treatment step. *Chem Eng J* 2012 4/1;**187**(0):275–82.
17. Sanchis EF, Lee C, Dela Cruz M, Chul Choi PMY. *High rate Boron rejection in sea water desalination using single-pass RO*; 2008. http://www.doosanhydro.com/.
18. Andrews B, Davé B, López-Serrano P, Tsai S-, Frank R, Wilf M, et al. Effective scale control for seawater RO operating with high feed water pH and temperature. *Desalination* 2008;**220**(1–3): 295–304.
19. Kim J, Hoon H, Wilf M, Jong-Sang P, Brown J. Boron rejection by reverse osmosis membranes: national reconnaissance and mechanism study. In: *Desalination and Water Purification Research and Development Program Report no. 127*; 2009. http://www.usbr.gov/pmts/water/publications/reportpdfs/report127.pdf.
20. Hasson D, Shemer H, Brook I, Zaslavschi I, Semiat R, Bartels C, et al. Scaling propensity of seawater in RO boron removal processes. *J Membr Sci* November 11, 2011;**384**(1–2):198–204.
21. Roberts DA, Johnston EL, Knott NA. Impacts of desalination plant discharges on the marine environment: a critical review of published studies. *Water Res* 2010;**44**(18):5117–28.
22. Sweity A, Oren Y, Ronen Z, Herzberg M. The influence of antiscalants on biofouling of RO membranes in seawater desalination. *Water Res* 2013;**47**(10):3389–98.
23. Llenas L, Martínez-Lladó X, Yaroshchuk A, Rovira M, de Pablo J. Nanofiltration as pretreatment for scale prevention in seawater reverse osmosis desalination. *Desalin Water Treat* 2011;**36**(1–3):310–8.
24. Hassan AM, Al-Sofi MAK, Al-Amoudi AS, Jamaluddin ATM, Farooque AM, Rowaili A, et al. A new approach to membrane and thermal seawater desalination processes using nanofiltration membranes (part 1). *Desalination* 1998;**118**(1–3):35–51.
25. Hassan AM, Farooque AM, Jamaluddin ATM, Al-Amoudi AS, Al-Sofi MA, Al-Rubaian AF, et al. Demonstration plant based on the new NF-SWRO process. *Desalination* 2000;**131**(1–3):157–71.
26. Llenas L, Ribera G, Martínez-Lladó X, Rovira M, de Pablo J. Selection of nanofiltration membranes as pretreatment for scaling prevention in SWRO using real seawater. *Desalin Water Treat* 2013;**51**(4–6): 930–5.
27. Song Y, Xu J, Xu Y, Gao X, Gao C. Performance of UF-NF integrated membrane process for seawater softening. *Desalination* 2011;**276**(1–3):109–16.
28. Song Y, Gao X, Gao C. Evaluation of scaling potential in a pilot-scale NF–SWRO integrated seawater desalination system. *J Membr Sci* September 15, 2013;**443**(0):201–9.
29. Al-Hajouri AA, Al-Amoudi AS, Farooque AM. Long term experience in the operation of nanofiltration pretreatment unit for seawater desalination at SWCC SWRO plant. *Desalin Water Treat* 2013;**51**(7–9):1861–73.
30. El-Manharawy S, Hafez A. Study of seawater alkalization as a promising RO pretreatment method. *Desalination* 2003;**153**(1–3):109–20.
31. Aguinaldo JT. Application of integrated chemical precipitation and ultrafiltration as pre-treatment in seawater desalination. *Desalin Water Treat* 2009;**2**(1–3):113–25.
32. Peñate B, García-Rodríguez L. Reverse osmosis hybrid membrane inter-stage design: a comparative performance assessment. *Desalination* 2011;**281**(1):354–63.
33. Taniguchi M, Kurihara M, Kimura S. Boron reduction performance of reverse osmosis seawater desalination process. *J Membr Sci* March 1, 2001;**183**(2):259–67.

34. Hyung H, Kim J. A mechanistic study on boron rejection by seawater reverse osmosis membranes. *J Membr Sci* December 15, 2006;**286**(1–2):269–78.
35. Sagiv A, Semiat R. Analysis of parameters affecting boron permeation through reverse osmosis membranes. *J Membr Sci* November 1, 2004;**243**(1–2):79–87.
36. Mane PP, Park P, Hyung H, Brown JC, Kim J. Modeling boron rejection in pilot- and full-scale reverse osmosis desalination processes. *J Membr Sci* August 10, 2009;**338**(1–2):119–27.
37. Park P, Lee S, Cho J, Kim J. Full-scale simulation of seawater reverse osmosis desalination processes for boron removal: effect of membrane fouling. *Water Res* 2012;**46**(12):3796–804.
38. Hung PVX, Cho S, Moon S. Prediction of boron transport through seawater reverse osmosis membranes using solution–diffusion model. *Desalination* 2009;**247**(1–3):33–44.
39. Edmond JM, Gieskes JMTM. On the calculation of the degree of saturation of sea water with respect to calcium carbonate under in situ conditions. *Geochim Cosmochim Acta* 1970;**34**(12):1261–91.
40. Dickson AG. Thermodynamics of the dissociation of boric acid in synthetic seawater from 273.15 to 318.15 K. Deep sea research part A. *Oceanogr Res Pap* 1990;**37**(5):755–66.
41. Bagg J. Temperature and salinity dependence of seawater-KCl junction potentials. *Mar Chem* 1993;**41**(4):337–42.
42. Marion GM, Millero FJ, Camões MF, Spitzer P, Feistel R, Chen C-A. pH of seawater. *Mar Chem* September 20, 2011;**126**(1–4):89–96.
43. Millero FJ, Pierrot D. A chemical equilibrium model for natural waters. *Aquat Geochem* 1998;**4**(1):153–99.
44. Whitfield M. A chemical model for the major electrolyte component of seawater based on the Brønsted-Guggenheim hypothesis. *Mar Chem* 1973;**1**(4):251–66.
45. Pitzer KS. Thermodynamics of electrolytes. I. Theoretical basis and general equations. *J Phys Chem* 1973;**77**(2):268–77.
46. Harvie CE, Weare JH. The prediction of mineral solubilities in natural waters: the Na-K-Mg-Ca-Cl-SO4-H2O system from zero to high concentration at 25° C. *Geochim Cosmochim Acta* 1980;**44**(7):981–97.
47. Whitfield M. The extension of chemical models for seawater to include trace components at 25 °C and 1 atm pressure. *Geochim Cosmochim Acta* 1975;**39**(11):1545–57.
48. Schecher WD, McAvoy DC. MINEQL+: a software environment for chemical equilibrium modeling. *Comput Environ Urban Syst* 1992;**16**(1):65–76.
49. Harvie CE, Møller N, Weare JH. The prediction of mineral solubilities in natural waters: the Na-K-Mg-Ca-H-Cl-SO_4-OH-HCO_3-CO_3-CO_2-H_2O system to high ionic strengths at 25 °C. *Geochim Cosmochim Acta* 1984;**48**(4):723–51.
50. Felmy AR, Weare JH. The prediction of borate mineral equilibria in natural waters: application to Searles Lake, California. *Geochim Cosmochim Acta* 1986;**50**(12):2771–83.
51. Tu KL, Nghiem LD, Chivas AR. Boron removal by reverse osmosis membranes in seawater desalination applications. *Sep Purif Technol* October 13, 2010;**75**(2):87–101.
52. Hershey JP, Fernandez M, Milne PJ, Millero FJ. The ionization of boric acid in NaCl, NaCaCl and NaMgCl solutions at 25°C. *Geochim Cosmochim Acta* 1986;**50**(1):143–8.
53. Simonson JM, Roy RN, Roy LN, Johnson DA. The thermodynamics of aqueous borate solutions I. Mixtures of boric acid with sodium or potassium borate and chloride. *J Solut Chem* 1987;**16**(10):791–803.
54. Simonson JM, Roy RN, Mrad D, Lord P, Roy LN, Johnson DA. The thermodynamics of aqueous borate solutions. II. Mixtures of boric acid with calcium or magnesium borate and chloride. *J Solut Chem* 1988;**17**(5):435–46.
55. Plummer LN, Parkhurst DL, Fleming GW, Dunkle SA. A computer program incorporating Pitzer's equations for calculation of geochemical reactions in brines, US Geological Survey. *Water-Resour Invest Rep* 1988:88–4153.
56. Nir O, Lahav O. Coupling mass transport and chemical equilibrium models for improving the prediction of SWRO permeate boron concentrations. *Desalination* 2013;**310**:87–92.

57. Charlton SR, Parkhurst DL. Modules based on the geochemical model PHREEQC for use in scripting and programming languages. *Comput Geosciences* 2011;**37**(10):1653–63.
58. Waly T, Kennedy MD, Witkamp G, Amy G, Schippers JC. Predicting and measurement of pH of seawater reverse osmosis concentrates. *Desalination* March 10, 2011;**280**(1–3):27–32.
59. Huff GF. Use of simulated evaporation to assess the potential for scale formation during reverse osmosis desalination. *Desalination* January 30, 2004;**160**(3):285–92.

CHAPTER 15

Seawater Reverse Osmosis Permeate: Comparative Evaluation of Boron Removal Technologies

Amos Bick[1], **Gideon Oron**[2,3,4]

[1]7 Harey-Jerusalem St. Ganey-Tikva, Israel
[2]J. Blaustein Institutes for Desert Research, Ben-Gurion University of the Negev, Kiryat Sde-Boker, Israel
[3]Faculty of Industrial Engineering and Management, Ben-Gurion University of the Negev, Beer Sheva, Israel
[4]The Environmental Engineering Program, Ben-Gurion University of the Negev, Beer Sheva, Israel

15.1 INTRODUCTION

Reverse osmosis (RO) is a valuable method for removing boron from the drinking water supply and can reduce the boron concentration to less than 0.5 mg/L, which is a drinking water standard value under the World Health Organization (WHO) drinking water quality guidelines.[1,2] However, there are some disadvantages concerning RO treatment of seawater: (1) it is a pH dependent[3,4] (the boric acid is able to diffuse through the membrane in a similar way to that of carbonic acid or water itself). The process concerning alkaline dosage is good from the economical point of view, but it sometimes fails to qualify the boron regulation due to the limit of pH, which should be less than 10 or 11, for the durability of RO membranes; (2) low boron removal from additional passes for permeate desalination; and (3) energy wasted by the process. In order to increase the removal efficiency of boron several technologies can be used: (1) split partial[5,6]; (2) multipass RO system[7]; (3) ion exchange resins (IX)[8,9]; (4) biological treatment[10]; and (5) electrocoagulation.[10]

A RO system is divided into groups of pressure vessels, called stages. In each stage pressure vessels are connected in parallel, with respect to the direction of the feed/concentrate flow. The design methodology is based on the term 'stage' and 'pass' and it is important to understand the difference between a single and a double stage RO and a single and a double pass RO. In a one stage RO system, the feed water enters the RO system as one stream and exits the RO as either concentrate or permeate. Additional stages increase the recovery and in a two stage system the concentrate (or reject) from the first stage then becomes the feed water to the second stage. The permeate that is collected from the first stage is combined with the permeate from the second stage. In order to explain the term 'pass', one has to think of a 'pass' as a stand-alone RO system. In a double pass RO, the permeate from the first pass becomes the feed water to the second pass (or second RO) and producing a much higher quality permeate. A double

Boron Separation Processes
ISBN 978-0-444-63454-2, http://dx.doi.org/10.1016/B978-0-444-63454-2.00015-0

Table 15.1 Boron Removal Technologies Concerning Seawater RO (SWRO) Analyzed in the Study

Design Code	Number of Stages	Number of Passes	Desalination Seawater RO (SWRO)	Boron Removal Process	Additional Treatment References
1st2p	1	2	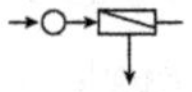	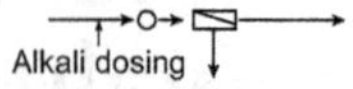	2,6
1st2pIXp	1	2	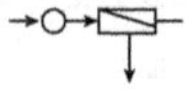	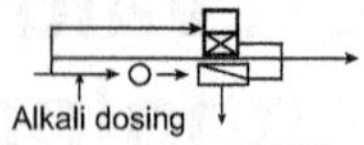	Ion exchange in parallel[2]
1st2pIXs	1	2	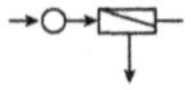	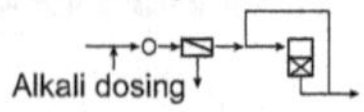	Ion exchange in series[2]
1st4p	1	4	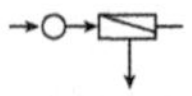	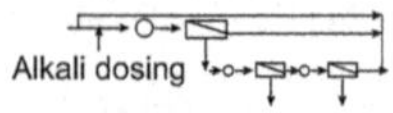	3,7
2st2p	2	2	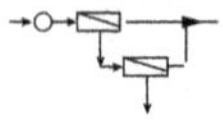	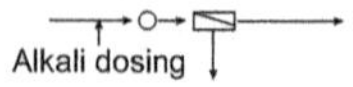	2,6
2st2pIXp	2	2	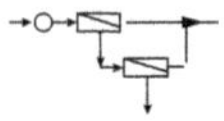	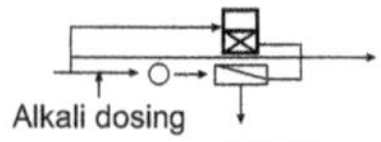	Ion exchange in parallel[2]
2st2pIXs	2	2	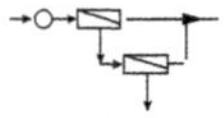	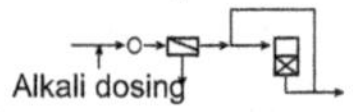	Ion exchange in series[2]
2st4p	2	4	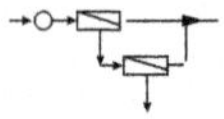	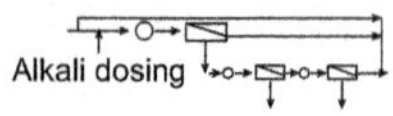	3,7

pass system also allows the opportunity to remove boron and carbon dioxide gas from the permeate by injecting caustic between the first and the second pass. Several technologies for boron removal that are classified according to the number of stages and passes are presented in Table 15.1.

15.2 MATERIALS AND METHOD

15.2.1 Management Modeling

The amount of data and information collected and retained by organizations and businesses is constantly increasing, due to advances in data collection, computerization of transactions, and breakthroughs in storage technology. In order to extract useful information from such large datasets, it is necessary to be able to identify patterns, trends, and relationships in the data and visualize their global structure to facilitate decision making.

There are four elements or characteristics common to all management modeling problems: (1) analysis of a finite and generally small set of discrete and predetermined options or alternatives, it is important that it will include only those attributes that vary significantly among one or more alternatives and for which the decision maker considers this variation to be important[21]; (2) no single alternative exhibits the most preferred available value or performance for all attributes; (3) the attributes will generally not all be measurable in the same units; and (4) a problem can generally be characterized by a "decision matrix." The decision matrix indicates both the set of alternatives and the set of attributes being considered in a given problem, and it summarizes the "raw" data available to the decision maker at the start of the analysis. A decision matrix has a row corresponding to each alternative being considered and a column corresponding to each attribute being considered. A problem with a total of m alternatives characterized by n attributes is described by an $m \times n$ matrix. Decision-making techniques used for the ranking of various options on the basis of more than one attribute are strictly dependent on the attributes setting and thus can be completely different for different settings.[21]

In recent years ranking strategies have been widely applied with different purposes: (1) evaluation of aquatic toxicological tests; (2) analysis of waste disposal sites; (3) ranking chemicals for environmental hazard; (4) comparison among ecosystems; and (5) ranking of contaminated sites.[22] Several evaluation methods that define a ranking parameter generating a total order ranking have been proposed in literature; those more frequently used are desirability functions, utility functions, dominance functions, absolute functions, and concordance analysis. These scalar functions always rank elements in an ordered sequence, even if there is an ambiguity within the attributes.[22]

15.2.2 The Analytical Hierarchical Process

AHP falls within the broader class of methods known as "additive weighting methods," which are best known and most widely used owing to their simple and intuitive logic, their multipurpose functionality, and their incorporation of compensatory trade-offs among attributes.[23] The AHP has been widely reviewed and applied in the literature, and its use is supported by several commercially available user-friendly software packages.[24] It is a powerful and flexible decision-supporting process that helps in setting priorities and making the best decision when both qualitative and quantitative aspects are considered,[25–27] is designed for subjective evaluation of a set of options based on multiple attributes, and is arranged in a hierarchical structure.[28]

There are three important ways in which the AHP method extends the basic additive weighting method and they are its use of pairwise comparisons, the principal eigenvector method, which allows the decision maker to provide judgments that are perfectly "consistent," and hierarchy.[29] Specifying pairwise comparison judgments in AHP requires the decision maker to answer a series of questions of the form: "how much more important is Attribute A than Attribute B? Verbal mediation allows a decision maker to select (from a pre-specified list) a verbal answer for each judgment using the fundamental Saaty's Scale,[21] a 1–9 numerical scale with verbal equivalents (Table 15.2).

The fundamental scale of absolute numbers has been proven in practice and validated by physical and decision problem experiments and has been shown to be a scale that captures individual preferences with respect to quantitative and qualitative attributes just as well or better than other scales.[21] It converts individual preferences into ratio scale weights that can be combined into a linear additive weight for each alternative. The resultant can be used to compare and rank the alternatives and, hence, assist the decision

Table 15.2 Fundamental Saaty's Scale for Pairwise Comparison[21]

Numerical Values	Verbal Term	Explanation
1	Equally important	Two elements have equal importance
3	Moderately more important	Experience or judgment slightly favors one element
5	Strongly more important	Experience or judgment strongly favors one element
7	Very strongly more important	Dominance of one element proved in practice
9	Extremely more important	The highest order dominance of one element over another
2,4,6,8	Important intermediate values	Compromise is needed

maker in making a choice. The AHP can be considered to be both a descriptive and prescriptive model of decision making and uses three basic functions: (1) structuring complexity; (2) measuring on a ratio scale; and (3) synthesizing.

Inconsistencies in assessment responses can occur and are attributed to human errors along the audit process. Perfect cardinal transitivity means that for any three attributes A, B, and C, if A is judged to be x times as important as B, and B is judged to be y times as important as C, then A must be $(x * y)$ times as important as C. The statistics and implementation of a judgment matrix[30] was implemented in this work.

The major advantage of AHP is based on a detailed structured and systematic decomposition of the overall problem into its fundamental components and interdependencies, with a large degree of flexibility.[21] AHP formalizes the use of hierarchy to reduce the number of paired comparisons and to neglect much of the multi-attribute complexity of the problem. In such cases, an attempt is made to "boil down" the large number of detailed requirements and objectives into a smaller number of very general, perhaps even vague attributes. This approach may fail to make use of available information or strong judgmental preferences, which distinguish alternatives only at a finer level of detail.

By structuring a complex problem hierarchically, a decision maker can include detailed attributes where they are useful or essential, while still keeping the number of pairwise comparisons generally manageable. An example hierarchy of attributes and subattributes is illustrated in Figure 15.1. Note that the overall goal of the analysis (e.g., "select the best") is positioned at the top of the hierarchy. At the outermost point of each branch of the hierarchy are the subattributes or so-called "leaf" attributes.

In AHP, the alternatives are compared or scored only with respect to each of the subattributes. Once the problem has been structured hierarchically, the attribute-weighting problem becomes one of finding normalized weights for each set of attributes (and subattributes, etc.) in the hierarchy. Operationally, the use of hierarchy in AHP

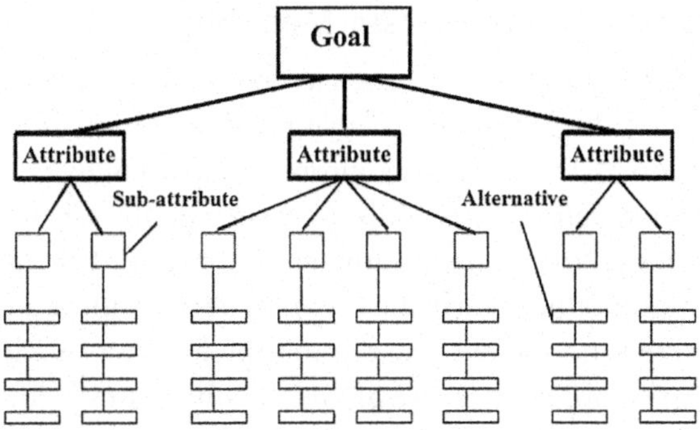

Figure 15.1 Demonstrating the hierarchical structure of the analytical hierarchical process.

generally calls for keeping the number of attributes in a set less than or equal to seven. The limit of seven attributes per set is not a hard constraint of the AHP theory, but has been a limit posed by some software packages that facilitate the use of AHP. Newer versions of some software packages allow up to nine attributes per set. Still, keeping the number of attributes per set small helps limit the total number of pairwise comparisons as the total number of attributes becomes large.

AHP provides a well-tested method, which allows design, investment, and management analysts to include consideration of multiple, conflicting, nonmonetary attributes of alternatives into their decision making. Another important strength related to AHP is the availability of a well-tested, flexible, and user-friendly software package to facilitate its application.

The primary limitations of AHP are of a more theoretical nature, and have been the subject of some debate in the technical literature.[31] Critics of AHP cite a number of limitations including the assumption of independence among the evaluation criteria, and the difficulty in scaling the approach to problems with a large number of comparisons, which would burden the experts. Several authors have pointed out that, to be meaningful, the attribute weighting questions must be answered with respect to the average performance levels of the alternatives. Others have noted the possibility for rank reversal among remaining alternatives after one is deleted from consideration. Defenders of AHP respond that rank reversal is not a flaw because real-world decision making occasionally exhibits this property as well. While acceptance of AHP is not universal, its combination of flexibility and ease of use have contributed to its application in a large variety of practical problems.[24]

Since the construction of the hierarchy structure and the determination of the comparison matrix are strongly dependent on the expert judgment, several experts' opinions are usually required to get objective conclusions. In this work, the novelty of AHP consists in the consultation of three experts, who are working in the desalination industry and were asked to build their own hierarchies and to express their overall importance of the attributes with regard to the defined goal.

15.2.3 Partial Order Theory and the Hasse Diagram Modeling

This study refers to the problem of ranking options (objects such as chemicals, databases, projects, and strategies) when a number of attributes are available for these objects that convey different comparative information.[32] In complex environmental datasets, it is often necessary to analyze multi-indicator systems. Commonly, to these examples is that each object can be characterized by more than one parameter.[33] Objects that are characterized by several quantities often cannot necessarily be ordered, because there are conflicts between their attributes.[34] There is no unique way to rank these objects while taking all attributes into account and one technique is to use discrete mathematics[35]: it is often necessary to compare different objects on the basis of a set of different attributes and partial order theory is a discipline of discrete mathematics and one may consider partial order theory as an example of mathematics without arithmetic.[36]

Partial order ranking is a vectorial approach, which recognizes that not all elements can be directly compared with all other elements because, when many criteria are used, contradictions in the ranking are bound to exist; in fact the higher the number of criteria, the higher is the probability that contradictions in the ranking exist. The vectorial approach not only ranks elements but also identifies contradictions in the criteria used to rank: some "residual order" remains when many criteria are considered and this fact motivates the term "partial order."[22]

Partial order ranking is a simple principle, which a priori include "$\leq$" as the only mathematical relation among the objects (no assumptions about linearity or distribution are made).[36] Object a is considered intrinsically "better" than object b, which we denote $a > b$, if all indicators rate object a greater than or equal to object b with at least one indicator considering object a strictly greater than object b. If object a is neither intrinsically better nor intrinsically worse than object b, then we consider the two objects incomparable, denoted by $a||b$. Conversely, we consider objects a and b comparable when if either $a > b$ or $b > a$. For every pair of objects whose indicator values are not identical, one of the following must be true, $a > b$, $a < b$, or $a||b$.

HDT is a graphical visualization approach for the mathematical concepts of partial order and can be able to perform ranking decisions from the information available without using any aggregation criteria. The number of incomparable elements in the Hasse diagram may obviously constitute a limitation in the attempt to rank the objects according to their attributes. To a certain extent this problem can be remedied through the application of the so-called linear extensions of the partial order ranking. However, the number of linear extensions goes with $N!$ (concerning database with N elements). Therefore, it becomes a computational problem and some approximations are necessary.[36] An example is shown in Table 15.3.

There are five objects (a, b, c, d, e) characterized by two attributes q_1 and q_2. The objective function is to find the optimal object with respect to all properties (maximization

Table 15.3 Illustrative Example for Hasse Diagram

Object	Attribute q_1	Attribute q_2
a	3.0	20
b	4.5	25
c	3.5	30
d	5.0	33
e	4.0	22

of q_1 and q_2 attributes). One way to solve this problem is to arrange the five objects according to their attributes (Figure 15.2) in order to obtain a permutation diagram.[37]

Concerning this type of diagram the observer may be confused if many objects are included and especially if more than two attributes characterize the objects. In that case a corresponding number of sequences may arise because this troublesome procedure leads to $m*(m-1)/2$ pairs of permutation diagrams (m attributes are used).[38]

One technique to solve this problem is the use of HDT that provides a useful tool for visualization (if the number of objects is not too large, i.e., between 50 and 100, just to give an idea at hand). A typical analysis includes favorably up to 50 data points.[39,40] This technique focuses on individual objects and their relation to each other: (1) each object is represented by a circle in the two-dimensional plane; and (2) the order relation of each pair of objects is either shown by a line between them or by a sequence of lines followed upward or (exclusively) downward. An example of the partially ordered set of the five objects (Table 15.3) is visualized in a Hasse diagram (Figure 15.3).

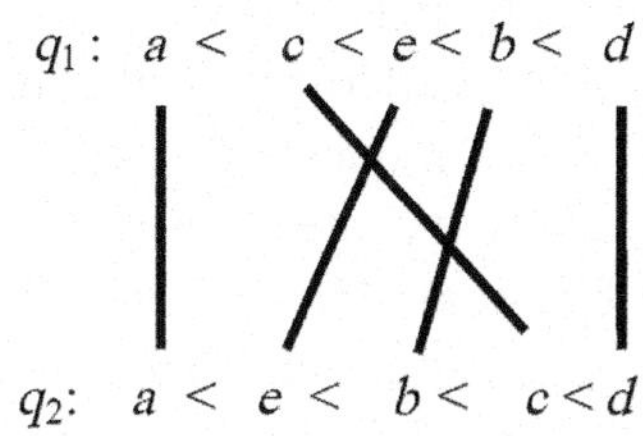

Figure 15.2 Illustrative scheme of two sequences of objects, "permutation diagram" (according to Table 15.3 data matrices).

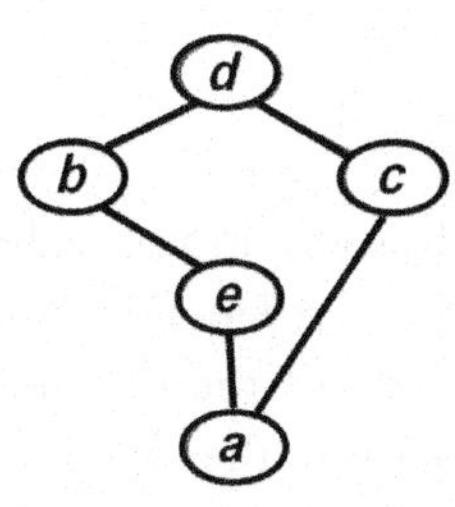

Figure 15.3 The Hasse diagram concerning the illustrative example (Table 15.3). Besides the vertical structure of levels the objects are located as symmetrical as possible.

The circles at the top of the drawing plane (Fig. 15.3) are called maximal elements. If there is only one maximal element, then this is called greatest element. Objects that have no successors are called minimal elements: they are often located near the bottom of the drawing plane. If there is only one minimal element, then this is called least element. Elements that have no predecessors and no successors are simultaneously maximal and minimal elements. These elements are called isolated elements. In Figure 15.3 element d is a predecessor of elements b and c, and element a is a successor of elements e and c. The connecting lines show that $d > b$, $d > e$, $d > a$, $d > c$, $b > e$, $b > a$, $e > a$, $c > a$, $c||b$, and $c||e$. If the ordering is represented by transitivity, unnecessary lines are avoided: for example $d > c$ and $c > a$, then we do not draw a line between objects d and a. If c is incomparable with b, then no line between the two objects is drawn.

This procedure is repeated until all of the tiers are established and lines drawn. A chain is a set of comparable elements, whereas an antichain is a set of incomparable elements. The analysis of chains ("vertical analysis of Hasse diagrams") is most often of more interest, as chains can be seen as an interim result of ranking. Antichains are of less interest, albeit they are the obvious indication that different quantities cannot necessarily be measured on one scale, i.e., they are incommensurable, but even an antichain can have a certain kind of structure.[41]

To construct a Hasse diagram, the following procedure is sufficient: (1) draw lines from each item to those others that it is objectively greater than (where all indicators agree); (2) position the items so that all lines point downward: the greatest items are then at the top of the diagram (this implies that the diagram has a certain number of levels); and (3) remove those lines implied by transitivity. HDT has some relevant advantages: (1) evaluation can be represented as a graph and (2) the underlying mathematics is very simple.

Nevertheless there are some relevant drawbacks: (1) if many elements are to be evaluated, preliminary multivariate statistic techniques, like cluster analysis, are needed to get a readable diagram; (2) it has to be assured that any two elements ordered by ">" can be considered as physical and numerical significantly different, i.e., they should have numerical significant differences of data; and (3) when data are characterized by large errors of measurements the "quantitative" information cannot be used; in these cases, the original variables can be replaced by their rank orders and it obviously results in a loss of information.[22]

15.3 CASE STUDIES

The comparative performance of boron removal technologies by the various alternative methods is very complex, and is highly depending on various site-specific operational and economic attributes. An analysis of the salt and boron removal efficiency in 14 commercial seawater membranes from the main manufacturers illustrates that it is possible to obtain 0.75 mg/L boron after 3 years of operation in a single pass at the expense of consuming more energy.[6]

Boron removal with RO alone is based on the development of high boron rejection membranes that produce an acceptable concentration without requiring additional RO system[42] and is done at pH 9. At this pH, the boron is partially negatively charged and the rejection can be up to 90% with seawater RO, and 75% with brackish water RO elements.[10] The high pH can increase the scaling potential where precipitates of calcium carbonate and magnesium hydroxide are formed on the surface of the membrane. To achieve a limit of below 0.5 mg/L boron in the permeate stream, an additional RO treatment of the first permeate is necessary and the pH adjustment is done in front of the second pass (the configuration is characterized by two pass or partial two pass system: 1st2p, 2st2p).[43]

An alternative for this approach is a posttreatment with four passes or IX (1st4p, 1st2pIXp, 1st2pIXs, 2st4p, 2st2pIXp, and 2st2pIXs). The process with IX treatment has some technical advantages for low boron concentration (0.2–0.3 mg/L)[44] and is preferred for the second or third pass because the specific costs for permeate using IX is dependent on the salt concentration of the feed water.[44]

In order to analyze the different configurations, the design goal must be defined by an objective function: (1) minimization of batch operation; (2) maximization of boron removal; and (3) minimization of investment cost. In order to support adequate selection, the decision maker has to define a utility function (Z_T).

$$Z_T = \begin{pmatrix} \text{Min} \\ \text{Batch} \\ \text{Operation} \end{pmatrix} + \begin{pmatrix} \text{Max} \\ \text{Boron} \\ \text{Removal} \end{pmatrix} + \begin{pmatrix} \text{Min} \\ \text{Investment} \\ \text{Cost} \end{pmatrix} \tag{15.1}$$

It is straightforward that different experts build different hierarchies and create different results: a comparison and arithmetic mean defined on the basis of previous experience and/or information obtained from three skilled experts is introduced in Table 15.4.

Table 15.4 Boron Removal Technologies Scores (AHP) by Three Skilled Experts

Design Code	Min. Batch Operation	Max. Boron Removal	Min. Investment Cost	Remarks
1st2p	9	4	9	
1st2pIXp	3	8	4	Ion exchange in parallel
1st2pIXs	6	9	5	Ion exchange in series
1st4p	9	8	7	
2st2p	9	5	9	
2st2pIXp	3	8	4	Ion exchange in parallel
2st2pIXs	6	9	5	Ion exchange in series
2st4p	9	8	7	

Data is based on integer arithmetic mean.

15.4 AHP AND HASSE DIAGRAM IMPLEMENTATION

From the matrix obtained in Table 15.4, the geometric mean (w_i, approximates the product of the elements in each row concerning a matrix of n rows and n columns) and the normalized geometric mean (p_i) are determined according to Eqns (15.2) and (15.3), respectively and introduced in Table 15.5.

$$w_i = \left(\prod_{j=1}^{n} a_{ij}\right)^{1/n} \quad i = 1, \ldots, n \tag{15.2}$$

$$p_i = \left(\prod_{i=1}^{n} a_{ij}\right)^{1/n} \Big/ \sum_{i=1}^{n}\left(\prod_{j=1}^{n} a_{ij}\right)^{1/n} \quad i = 1, \ldots, n \tag{15.3}$$

where a_{ij} is an assessed value, i is the index for rows (alternative, $i = 1, \ldots, n$), and j is the index for columns (quality attribute, $j = 1, \ldots, n$). Table 15.5 shows that for each decision attribute chosen, the importance of the design alternatives varies. It indicates how the options are prioritized over others with respect to each objective as well as to the overall objective.[45]

Table 15.6 compares maximization of boron removal in regards to each of the options and according to the subjective comparison, the configurations with ion exchange in series (1st2pIXs and 2st2pIXs) are preferred because of low boron concentration. A similar comparison is conducted for each of the attributes. The resulting set of weights for each of the technology (options) with respect to each attributes is presented in Table 15.7.

The output of Hasse diagram using DART software concerning the design configurations is shown in Figure 15.4. According to this figure the options are arranged in three levels: design configurations 1st2pIXs, 2st2p, 1st4p, and 2st4p are maximal (and are of special concern), and are not covered by any other element (obviously the maximals are mutually incomparable). The configurations 1st2p and 1st2pIXp are minimal and, especially, a least element.

Table 15.5 Variability in Importance across Design Options

Attribute	Design Option
Min. batch operation	2st4p ≅ 2st2p ≅ 1st4p ≅ 1st2p >2st2pIXs ≅ 1st2pIXs> 2st2pIXp ≅ 1st2pIXp
Max. boron removal	2st2pIXs ≅ 1st2pIXs> 2st4p ≅ 1st4p ≅ 2st2pIXp ≅ 1st2pIXp>2st2p ≅ 1st2p
Min. investment cost	2st2p ≅ 1st2p>2st4p ≅ 1st4p>2st2pIXs ≅ 1st2pIXs> 2st2pIXp ≅ 1st2pIXp

The notations > and ≅ symbolize the option preceding the sign in, "preferable to" and "equal to" the one after the sign, respectively.

Table 15.6 AHP Pairwise Evaluation of Max. Boron Removal (Numbers Are Based on Saaty's Scale and an Expert Subjective Point of View)

Design	1st2p	1st2p-IXp	1st2p-IXs	1st4p	2st2p	2st2p-IXp	2st2p-IXs	2st4p	Geometric Mean (a)	Normal-ization (b)
1st2p	1	1/5	1/7	1/5	1/3	1/5	1/7	1/5	0.239	0.03
1st2pIXp	5	1	1/2	1	2	1	1/2	1	1.121	0.12
1st2pIXs	7	3	1	2	4	2	1	2	2.144	0.23
1st4p	5	1	1/2	1	2	1	1/2	1	1.121	0.12
2st2p	3	1/2	1/4	1/2	1	1/2	1/4	1/2	0.573	0.06
2st2pIXp	5	1	1/2	1	2	1	1/4	1	1.028	0.12
2st2pIXs	7	2	1	2	4	2	1	2	2.144	0.23
2st4p	5	1	1/2	1	2	1	1/2	1	1.121	0.12
Total									9.495	

[a]For example, the geometric mean of 2st2pIXs is $(7 \cdot 2 \cdot 1 \cdot 2 \cdot 4 \cdot 2 \cdot 1 \cdot 2)^{1/8} = 2.144$.
[b]For example, the normalization of 2st2pIXs is $2.144/9.495 = 0.23$.

Table 15.7 AHP Pairwise Results of Normalized Attribute Weights

Design	Min. Batch Operation	Max. Boron Removal	Min. Investment Cost
1st2p	0.21	0.03	0.26
1st2pIXp	0.02	0.12	0.03
1st2pIXs	0.07	0.23	0.07
1st4p	0.21	0.12	0.14
2st2p	0.21	0.06	0.26
2st2pIXp	0.02	0.12	0.07
2st2pIXs	0.07	0.23	0.03
2st4p	0.21	0.12	0.14

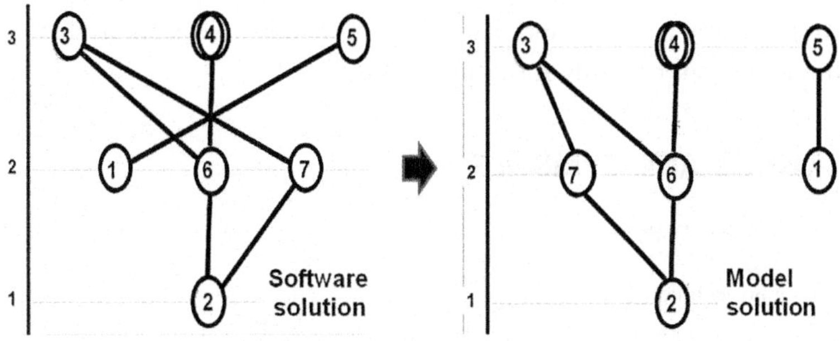

Figure 15.4 Illustrative scheme of Hasse diagram (using DART software) relating to boron removal technologies (Object legend: (1) 1st2p; (2) 1st2pIXp; (3) 1st2pIXs; (4) 1st4p; (5) 2st2p; (6) 2st2pIXp; (7) 2st2pIXs; (8) 2st4p). Configuration 1st4p "equal to" 2st4p.

Concerning Figure 15.4, it is a directed graph having two components: (1) six treatments with "high-control" technology (four passes or ion exchange) and (2) two "low-tech" treatment concerning two passes for boron removal (1st2p and 2st2p). This is a strong indication that for a "simple" management strategy one stage or two stages with two passes should be considered.

15.5 CONCLUSIONS

Selecting an appropriate boron removal technology is often a subjective task, especially in the water sector. Integrating quantitative methods into the evaluation procedure enables decision makers to identify the most appropriate option objectively and efficiently.

This study adopts the AHP method and establishes an evaluation model for production technology and has clearly demonstrated a total ranking based on partial order methodology done without any preassumptions concerning possible relations between the attributes and also introduced using HDT.

This study uses the user-friendly DART software[46], however it does not present any elaborate analysis compared to PyHasse, which is an upgraded version of partial order software.[36] According to the structure of the Hasse diagram two stages of RO are preferred and can produce four permeate streams with different boron concentrations by using permeate split. Treatment in a series of low quality permeate by ion exchange in series (1st2pIXs) can improve the performance of a single stage.

This study demonstrates the possibility and appropriateness of using the AHP method for the selection of optimal boron removal technologies and provides a systematic decision making framework with several characteristics: (1) different technological performances can be evaluated using multiple attributes—both quantitative and qualitative—rather than profitability alone; (2) the use of ratings makes it possible to evaluate the applicability of different options for the end user; (3) the use of AHP method provides an effective way of documenting the managerial process; (4) HDT is a useful tool with an easy visualization of the obtained results; and (5) the proposed approach forms the basis for a continuous process of planning and managing technology selection, so that the priorities of the technologies can be easily modified and updated.

ABBREVIATIONS AND SYMBOLS

AHP Analytical hierarchical process
a_{ij} An assessed value
HDT Hasse diagram technique
i Index for row
IX Ion exchange
j Index for columns
n Matrix size (n rows and n columns)

NF Nanofiltration
p_i Normalized geometric mean
RO Reverse osmosis
WHO World Health Organization
w_i Geometric product of the elements in each row

REFERENCES

1. Xu Y, Jiang JQ. Technologies for boron removal. *Ind Eng Chem Res* 2008;**47**:16–28.
2. Hilal N, Kim GJ, Somerfield C. Boron removal from saline water: a comprehensive review. *Desalination* 2011;**273**:23–35.
3. Tu KL, Nghiem LD, Chivas AR. Boron removal by reverse osmosis membranes in seaweater desalination application. *Sep Purif Technol* 2010;**75**:87–101.
4. Yavuz E, Arar O, Yuksel M, Yüksel U, Kabay N. Removal of boron from geothermal water by RO system-II-effect of pH. *Desalination* 2013;**310**:135–9.
5. Rybar S, Boda R, Bartels C. Split partial second pass design for SWRO plants. *Desalin Water Treat* 2010;**13**:186–94.
6. Dominguez-Tagle C, Romero-Termero VJ, Delgado-Torres AM. Boron removal efficiency in small seawater reverse osmosis systems. *Desalination* 2011;**265**:43–8.
7. Faigon M, Hefer D. Boron rejection in SWRO at high pH conditions versus cascade design. *Desalination* 2011;**273**:421–7.
8. Arias MFC, Brue LVI, Rico DPP, Galvan PV. Approximate cost of the elimination of boron in desalinated water by reverse osmosis and ion exchange resins. *Desalination* 2011;**273**:421–7.
9. İpek IY, Kabay N, Yuksel M. Modeling of fixed bed column studies for removal of boron from geothermal water by selective chelating ion exchange resins. *Desalination* 2013;**310**:151–7.
10. Ezechi EH, Isa MH, Kutty SRBM. Boron in produced water: challenges and improvement: a comprehensive review. *J Appl Sci* 2012;**12**(5):402–15.
11. Caputo AC, Pelagagge PM, Salini P. AHP-based methodology for selecting safety devices of industrial machinery. *Safety Sci* 2013;**53**:202–18.
12. Bruggemann R, Patil GP. Multicriteria prioritization and partial order in environmental sciences. *Environ Ecol Stat* 2010;**17**:383–410.
13. Bick A, Bruggemann R, Oron G. Assessment the intake and the pretreatment design in existing seawater reverse osmosis (SWRO) plants by Hasse diagram technique, (HDT). *Clean–Soil Air, Water* 2011;**39**(11):933–40.
14. Bick A, Oron G. Boron removal from seawater reverse osmosis permeate: a Hasse diagram analysis of current technologies. *Desalination* 2013;**310**:34–8.
15. Carlsen L, Bruggemann R, Sailaukhanuly Y. Application of selected partial order tools to analyze fate and toxicity indicators of environmentally hazardous chemicals. *Ecol Indic* 2013;**29**:191–202.
16. Sailaukhanuly Y, Zhakupbekova A, Amutova F, Carlsen L. On the ranking of chemicals based on their PBT characteristics: comparison of different ranking methodologies using selected POPs as an illustrative example. *Chemosphere* 2013;**90**(1):112–7.
17. Tsakovski T, Simeonov V. Hasse diagram technique as exploratory tool in sediment pollution assessment. *J Chemom* 2011;**25**(5):254–61.
18. Voigt K, Bruggemann R, Scherb H, Cok I, Mazmanci B, Mazmanci MA, et al. Evaluation of organochlorine pesticides in breast milk samples in Turkey applying features of the partial order technique. *Int J Environ Health Res* 2012:1–21.
19. Voyslavov T, Tsakovski S, Simeonov V. Hasse diagram technique as a tool for water quality assessment. *Anal Chim Acta* 2013;**770**:29–35.
20. Bruggemann R, Halfon E, Welzl G, Voigt K, Steinberg C. Applying the concept of partially ordered sets on the ranking of near-shore sediments by a battery of tests. *J Chem Inf Comp Sci* 2001;**41**:918–25.

21. Saaty TL. Decision-making with the AHP: why is the principal eigenvector necessary. *Eur J Oper Res* 2003;**145**(1):85–91.
22. Pavan M, Todeschini R. New indices for analysing partial ranking diagrams. *Anal Chim Acta* 2004;**515**:167–81.
23. Srdjevic B, Srdjevic Z. Synthesis of individual best local priority vectors in AHP-group decision making. *Appl Soft Comput* 2013;**13**(4):2045–56.
24. Subramanian N, Ramanathan R. A review of applications of analytic hierarchy process in operations management. *Int J Prod Econ* 2012;**138**:215–41.
25. Bick A, Oron G. Post-treatment of seawater reverse osmosis plants boron removal technology selection for potable water production and environmental control. *Desalination* 2005;**178**:233–46.
26. Tzfati E, Sein M, Rubinov A, Raveh A, Bick A. Pre-treatment of wastewater: optimal coagulant selection using partial order scaling analysis (POSA). *J Hazard Mater* 2011;**190**:51–9.
27. Zhang R, Zhang X, Yang J, Yuan H. Wetland ecosystem stability evaluation by using analytical hierarchy process (AHP) approach in Yinchuan Plain, China. *Math Comput Model* 2013;**57**(3–4):366–74.
28. Kayastha P, Dhital MR, De Smedt F. Application of the analytical hierarchy process (AHP) for landslide susceptibility mapping: a case study from the Tinau watershed, west Nepal. *Comput Geosci* 2013;**52**: 398–408.
29. Bozoki S, Fulop J, Koczkodaj WW. An LP-based inconsistency monitoring of pairwise comparison matrices. *Math Comput. Model* 2011;**54**(1–2):789–93.
30. Yang X, Yan L, Zeng L. How to handle uncertainties in AHP: the cloud delphi hierarchical analysis. *Inf Sci* 2013;**222**:384–404.
31. Costa CAB, Vansnick JC. A critical analysis of the eigenvalue method used to derive priorities in AHP. *Eur J Oper Res* 2008;**187**:1422–8.
32. Restrepo G, Weckert M, Bruggemann R, Gerstmann S, Frank H. Ranking of refrigerants. *Environ Sci Technol* 2008;**42**:2925–30.
33. Simon U, Bruggemann R, Behrendt H, Shulenberger E, Pudenz S. METEOR: a step-by-step procedure to explore effects of indicator aggregation in multi criteria decision aiding-application to water management in Berlin, Germany. *Acta Hydroch Hydrob* 2006;**34**:126–36.
34. Patil GP, Taillie C. Multiple indicators, partially ordered sets, and linear extensions: multi-criterion ranking and prioritization. *Environ Ecol Stat* 2004;**11**:199–228.
35. Restrepo G, Bruggemann R, Voigt K. Partially ordered sets in the analysis of alkanes fate in rivers. *Croat Chem Acta* 2007;**80**:261–70.
36. Voigt K, Bruggemann R, Kirchner M, Schramm KW. Influence of altitude concerning the contamination of humus soils in the German Alps: a data evaluation approach using PyHasse. *Environ Sci Pollut Res Int* 2010;**17**(2):429–40.
37. Urrutia J. Partial orders and euclidian geometry. In: Rival I, editor. *Algorithms and order.* Dordrecht: Kluwer Academic Publisher; 1989. pp. 387–434.
38. Simon U, Bruggemann R, Pudenz S. Aspects of decision support in water management—example Berlin and Potsdam (Germany) I-spatially differentiated evaluation. *Wat Res* 2004;**38**:1809–16.
39. Pudenz S. ProRank-Software for partial ordering. *Match Commun Math Co* 2005;**54**(3):611–22.
40. Annoni P, Bruggemann R, Saltelli A. Partial order investigation of multiple indicator systems using variance-based sensitivity analysis. *Environ Model Soft* 2011;**26**:950–8.
41. Bruggemann R, Voigt K. Antichains in partial order, example: pollution in a German region by lead, cadmium, zinc and sulfur in the herb layer. *Match* 2012;**67**:731–44.
42. Rahmawati K, Ghaffour N, Amy GL. Boron removal efficiency from red sea water using different SWRO/BWRO membranes. *J Membr Sci* 2012;**423–424**:522–9.
43. Farhat A, Ahmad F, Hilal N, Arafat HA. Boron removal in new generation reverse osmosis (RO) membranes using two-pass RO without pH adjustment. *Desalination* 2013;**310**:50–9.
44. Lipnizki J, Adams B, Okazaki M, Sharpe A. Water treatment: combining reverse osmosis and ion exchange. *Filtr Separat* 2012;**49**(5):30–3.
45. Cay T, Uyan M. Evaluation of reallocation criteria in land consolidation studies using the analytic hierarchy process (AHP). *Land Use Policy* 2013;**30**(1):541–8.
46. Bruggemann R, Patil GP. Partial order and software. *Environ Ecol Stat* 2011;**5**:279–89.

CHAPTER 16

Hybrid Adsorption—Microfiltration Process with Plug Flow of Microparticulate Adsorbent for Boron Removal

Marek Blahušiak[1], Štefan Schlosser[1], Nalan Kabay[2]
[1]Slovak University of Technology, Institute of Chemical and Environmental Engineering, Radlinského, Bratislava, Slovakia
[2]Ege University, Department of Chemical Engineering, Izmir, Turkey

16.1 INTRODUCTION

Ion-exchange (IEX) and adsorption are widely used techniques for the removal of metals and other solutes from aqueous solutions, including waste waters. IEX became an attractive alternative for the removal of boron from drinking water produced by reverse osmosis (RO) process[1–6] or from geothermal waters.[7–10] The world's first desalination plant combining one stage RO and IEX to prepare potable water with boron content below 0.5 mg/dm^3 with the capacity of 50,000 m^3/day was started up in May 2006.[3] Kinetics of adsorption or IEX is in most cases strongly influenced by diffusion resistance in particles of the adsorbent. This resistance can be decreased by using smaller particles, which is, however, limited by the increasing pressure drop in an adsorption column. Application of a microparticulate adsorbent in suspension combined with microfiltration (MF) presents an alternative approach to the recovery of metals and bioproducts as it has been discussed in the literature.[11,12] Intensification of the copper removal from water by the microparticulate ion-exchanger Lewatit S100 with the mean particle size of $d_{32} = 5.25$ μm was reported in published papers[12,13] where a hybrid process combining adsorption and MF was also suggested. The adsorption time needed to achieve 90% of the equilibrium in copper adsorption by the ion-exchange resin Lewatit S100 (strongly acidic cation exchange resin) at the initial Cu concentration of 98 mg/dm^3 was 31 min and 30 s for particles with the mean size of 324 and 5.25 μm, respectively. Boron removal by adsorbent microparticles with a mean diameter of 4.7 μm was reported.[5,14] This was also used for boron removal from aqueous solutions simulating geothermal water by an adsorbent with the mean particle size of 50 μm.[4,15,16] This shows a great potential for adsorption intensification using microparticulate adsorbents which have to be separated from the treated solution by MF, regenerated and recirculated.

Boron Separation Processes
ISBN 978-0-444-63454-2, http://dx.doi.org/10.1016/B978-0-444-63454-2.00016-2

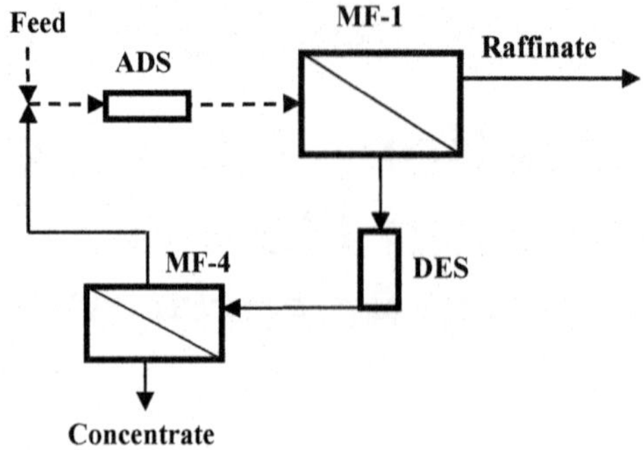

Figure 16.1 Simplified scheme of a hybrid AMF process for the removal of solute from the feed by adsorption into microparticulate adsorbent in the flow through an adsorber with a regeneration (desorption) loop of the adsorbent.[12,13]

In previous works, several approaches have been used in the hybrid adsorption-membrane separation process with a microparticulate adsorbent or ion exchange beads:

1. Flow through an adsorber with plug flow of the microparticulate adsorbent: Feed is mixed with concentrated suspension of the adsorbent and flows through the adsorber. Raffinate is removed by MF and adsorbent suspension enters the regeneration cycle with desorption and removal of the solute concentrate by MF. Regenerated adsorbent is returned back to the adsorption process as shown in Figure 16.1.[5,11–14,17]
2. Flow through a mixed adsorber with an addition of adsorbent: Fresh relatively diluted suspension of the microparticulate adsorbent is fed into the mixed adsorber and at the same time partly saturated suspension is withdrawn for regeneration[4,9,10,15,18] as shown in Figure 16.2. Purified permeate is withdrawn from the mixed adsorber through a submerged[9,10] or in the circulation loop operating[4,15] MF/ultrafiltration (UF) module(s).

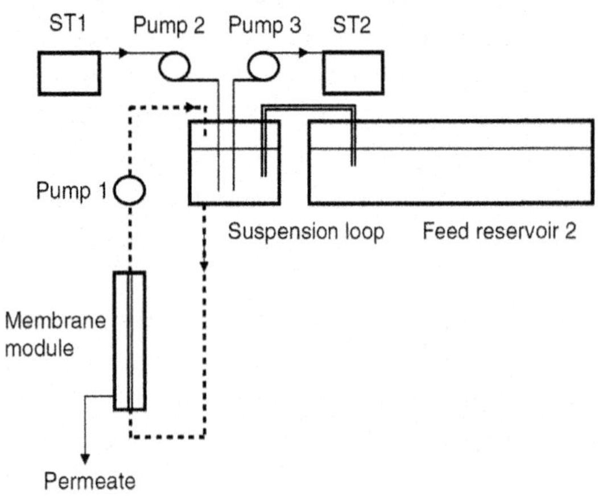

Figure 16.2 Scheme of a hybrid AMF process for the removal of solute from the feed by adsorption in the flow through a mixed adsorber with the microparticulate adsorbent. Suspension of fresh adsorbent is fed from container ST1 to the adsorber and the same volume of the suspension of partly saturated adsorbent is removed to container ST2.[4,15]

3. Feed flows through a well-mixed vessel with a batch of adsorbent and the raffinate is recovered through a MF membrane.[19,20]
4. Intensive adsorption by adsorbent microparticles, which can also be microorganisms, in a layer or cake at the MF membrane[21,22] with collection of permeate as raffinate.
5. Dialysis mode of adsorber operation, where the feed and adsorbent suspension were separated by a membrane, was considered in a paper.[23]

The aim of this chapter is to analyze the application potential of a hybrid process which combines adsorption of a target solute on microparticulate adsorbent in an adsorber with plug flow of the suspension combined with MF. This is used for the removal of boron from RO permeate in desalination to prepare drinking or irrigation water.

16.2 AMF PROCESS

The basic concept of a hybrid adsorption—microfiltration (AMF) process with a microparticulate adsorbent suspension is to achieve a fast adsorption of a solute from a solution. Suspension with the loaded adsorbent is preconcentrated by MF preferably in submerged modules with polymeric capillary membranes and raffinate is recovered as permeate analogously to the MBR systems for water cleaning. The loaded adsorbent is regenerated in a closed loop and returned back to the process. A flow-sheet of the AMF process is shown in Figure 16.3.[5,14] Feed, stream 1, is mixed with concentrated suspension of the regenerated microparticulate adsorbent, stream 22. The suspension of feed and adsorbent mircroparticles flows through retention volume ADS, for saturation of the adsorbent with boron from the feed. Raffinate is recovered by MF as a permeate stream 3. MF1 modules can be with advantage divided into several segments (three segments are shown in Figure 16.3) to decrease the membrane area needed in MF due to the decreased suspension concentration in the first segments and thus the increased permeate flux in them. The stripping agent, usually an acid solution in stream 13, is added to a concentrated suspension in stream 12, to decrease pH and desorb boron from the adsorbent. This is achieved in a relatively short time of less than 1 min in retention volume DES. A suspension of the regenerated adsorbent is further concentrated by MF in the cross-flow tubular module MF4. A concentrate of the adsorbate (boron) is obtained as permeate, stream 16. Most of the solute and excess of acid can be removed from suspension by diafiltration (washing) with a small part of the raffinate, stream 8, in a subsequent step in the cross-flow module MF5 as suggested in papers.[12–14] A base is added, stream 21, to the concentrated regenerated suspension, stream 20, to adjust its pH to a value favorable for adsorption and the suspension stream 22 is returned to the feed. The retention volumes in adsorption, ADS, and desorption, DES, should be constructed so as to achieve plug flow of the suspension and adequate residence time to achieve efficient boron removal and loading the adsorbent close to its saturation value.

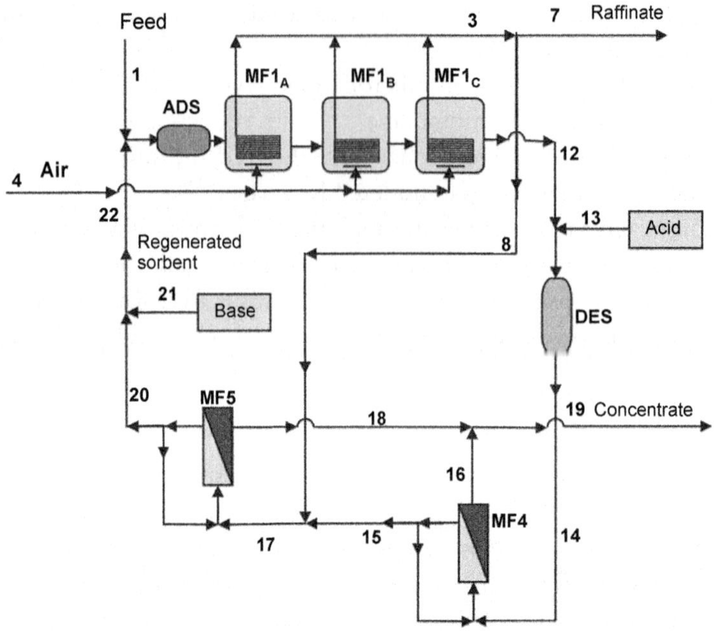

Figure 16.3 Flow-sheet of a hybrid AMF process for the removal of boron from the RO permeate into the microparticulate adsorbent with a regeneration loop of the adsorbent.[14] ADS, retention volume to achieve saturation of the sorbent; DES, retention volume to achieve desorption of boron from the adsorbent; MF1, submerged microfiltration modules (split into three well mixed segments in series); MF4 and MF5, cross-flow tubular MF modules for concentration MF4 and diafiltration MF5 of the adsorbent suspension in the regeneration loop.

Using an adsorbent with a mean particle size below 10 μm enables achievement of better than 90% of the equilibrium adsorbent saturation with a residence time shorter than 1 min, preferably shorter than 30 s. Retention volume for this residence time in an adsorption unit with the capacity of 100 m^3/h is of about 0.8 to 0.4 m^3, respectively. In the desorption loop, with much lower volumetric flow rate, the retention volume needed is only about 20 to 10 L, respectively, what can be the connecting pipe leading to MF4.

Concentration of the adsorbent in the suspension, which can be achieved in hollow fiber or capillary submerged modules, is limited and is of great importance for the economy of the AMF process. Furthermore, low pH tolerance of the MF steps in the regeneration loop is required. Thus, suspension preconcentrated in submerged modules, MF1, can be further concentrated after desorption of the solute to the final concentration of the regenerated suspension in, e.g., a tubular cross-flow module, MF4. These should be preferably equipped with more expensive ceramic membranes with good chemical resistance to low pH needed in desorption of adsorbent, because of the small membrane area required, as will be shown later.

16.3 MF OF ADSORBENT SUSPENSIONS

Data on MF of microparticulate adsorbent suspensions in submerged and cross-flow tubular membrane modules are necessary when evaluating the potential of the hybrid AMF process. For this purpose, grinded commercially available selective ion exchange resin Dowex XUS-43594 (Dow, US) for boron removal, sample three designated as XUS-43594-G3, with the Sauter mean particle diameter $d_{32} = 4.7$ μm (measured in aqueous suspension) was used in experiments.[24,25] Distribution of swollen particle diameters was relatively wide, from 0.4 to 120 μm. The proportion of small particles with the diameter below 0.4 and 0.8 μm was 0.1% and 1%, respectively.

16.3.1 MF in a Submerged Membrane Module

A laboratory MF system stand with a vertical submerged membrane element containing a bundle of MF capillaries with the total surface area of 104 cm^2 with an introduction of aeration was built and was described in more details before.[24] Geometry of the submerged system mimicked the situation in industrial submerged modules. The membrane tested was a proprietary hydrophilic capillary membrane (Zenon, now GE) with the pore size of 0.04 μm, which is designed for submerged modules for water treatment including membrane bioreactors for biological treatment of wastewaters. Specific consumption of air used in the experiments with the hydrophilic membrane was about 0.42 $Nm^3/m^2/h$.

Volumetric flux of the permeate through a membrane, J_v, is related, at least in a limited interval of smaller fluxes, to the transmembrane pressure difference, Δp, by a linear relation:

$$J_v = \Delta p/R = L_p \Delta p \tag{16.1}$$

where R is the overall mass-transfer resistance of the membrane and the adjacent filtration cake and its reciprocal value, $1/R$, is the permeability, L_p.

An example of the time dependence of the transmembrane pressure difference at constant permeate flux through the hydrophilic membrane for a suspension with 7 mass % of dry adsorbent is shown in Figure 16.4. From this figure, it is evident that Δp is more or less constant with time up to the value of the critical permeate flux of about $11.7 \cdot 10^{-6}$ m/s (41.9/$dm^3/m^2/h$); above this value, Δp increases with the filtration time because of the intensive fouling of the membrane surface by microfiltered components. This reflects the continuing accumulation of a filtration cake on the membrane surface. The slope of this dependence increases with the increasing permeate flux.

A very positive feature of the submerged module with GE membranes is a mild flux decline with the increasing suspension concentration, observed in a relatively wide concentration interval, up to about 7 mass %, Figure 16.5(b).

Pressure dependencies of the permeate flux for various suspension concentrations are presented in Figure 16.5(a). Data in these plots show that the critical flux is achieved at

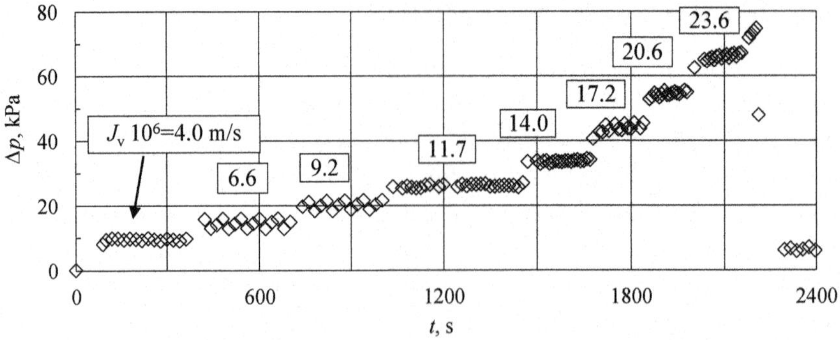

Figure 16.4 Time dependence of the transmembrane pressure difference at given constant permeate fluxes in the microfiltration of suspension with the XUS-43594-G3 concentration of 7 mass % through a GE Zenon membrane.

Δp of about 35 kPa. The value of critical flux at higher concentrations is lower and at the suspension concentration of 9 mass % is about 11.6 m/s ($41.8 \cdot dm^3/m^2/h$). However, working in the subcritical flux region does not guarantee the most efficient conditions at constant Δp for a longer operation time. The concept of higher sustainable flux is suggested aiming at a flux value at which the flux decline over time is operationally acceptable.[26] One way of prolonging the operation time of MF modules at higher flux values is backwashing of the membrane with reversed flow of the permeate. Also, the intensity of aeration is important for the accumulated filtration cake removal from the membrane surface and its intensity and way of operation should be tested and optimized as well.

Results of the MF experiments with permeate backwash performed with the suspension of 10 mass % of XUS-43594-G3 in a model RO permeate at the permeate flux of $8.7 \cdot 10^{-6}$ m/s, which is slightly above the critical permeate flux, are presented in

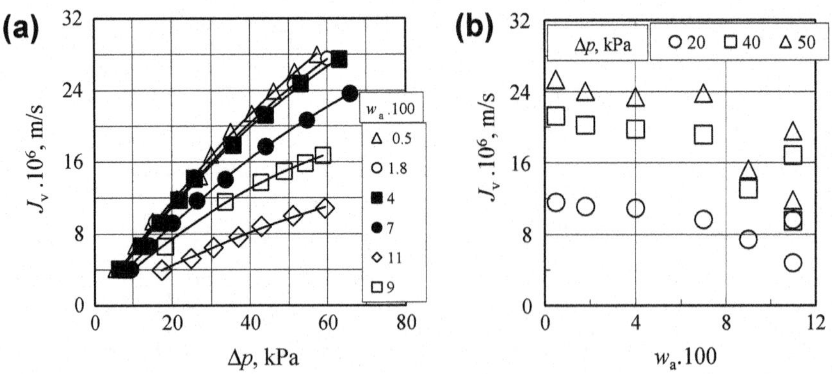

Figure 16.5 Permeate flux through a hydrophilic Zenon membrane in the submerged module SM4 in microfiltration of suspensions of adsorbent in model seawater for various suspension concentrations ($w_a \cdot 10^2$, mass %) vs. the pressure difference (a) and the suspension concentration (mass %) for three pressure differences in kPa (b).

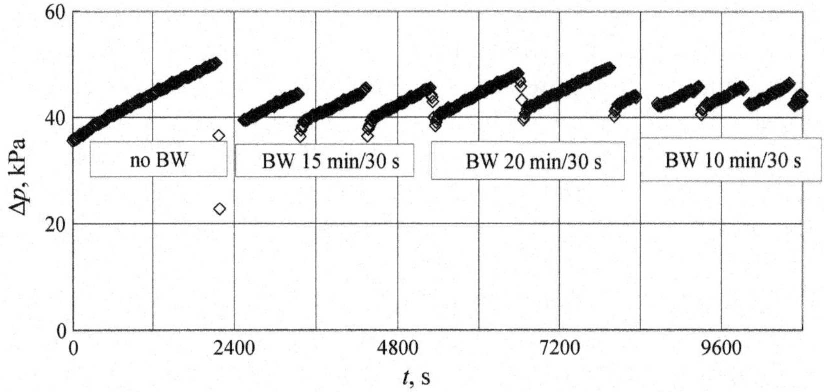

Figure 16.6 Time dependence of the transmembrane pressure in microfiltration of a 10 mass % adsorbent suspension in model RO permeate at permeate flux of $8.7 \cdot 10^{-6}$ m/s through a Zenon membrane. After the first period without backwashing (BW), three cycles with backwash and different suction period were performed at the backwash flux of $9.5 \cdot 10^{-6}$ m/s, $\Delta p_{BW} = 50$ kPa and the back wash duration of 30 s.

Figure 16.6. The time dependence of the transmembrane pressure shows a linear increase (initial part of Figure 16.6) referring to a constant cake growth. However, by the introduction of backwashing with the reversed flow of permeate in reasonable intervals, the membrane performance can be held stable for a longer time. This is shown in the second part of Figure 16.6, where a 30 s backwashing was started at $\Delta p_{BW} = 50$ kPa (maximum recommended by supplier) and flux of $9.75 \cdot 10^{-6}$ m/s after a 15 min period of MF. The value of Δp after the backwash returned almost to the initial value. When the MF time was increased to 20 min, the initial value of Δp started to drift slightly to higher values in repeated cycles after the backwash. However, setting the operation time of the forward flow to 10 min while keeping the backwash period of 30 s resulted in almost constant initial Δp at the beginning of each cycle. A slight drift in this value can be ascribed to irreversible fouling of the membrane as discussed previously.[26] The net permeate flux was equal to $7.8 \cdot 10^{-6}$ m/s compared to $8.7 \cdot 10^{-6}$ m/s of the suction flow rate. These results show the potential of this technique and should be tested for a prolonged time of operation together with the cleaning of the membrane with backwash after approaching the lower limit of effective permeate flux.

As a result, suspensions of the microparticulate Dowex XUS-43594 adsorbent can be effectively concentrated by hydrophobic polypropylene hollow fiber membranes to a relatively high concentration of about 11 mass %.

16.3.2 MF in a Tubular Ceramic Membrane Module

A laboratory cross-flow MF unit with a circulation loop of filtered suspension was tested for MF of more concentrated Dowex XUS-43594 suspensions. Details on the experimental setup are shown in[27] and precise experimental conditions are presented in.[25] An asymmetric tubular ceramic MF membrane with the inner diameter of 7 mm, length of 0.25 m, the mean pore size of the membrane active layer of 0.1 μm and the effective area of 50 cm^2 was used. The active layer was made of α-Al_2O_3 deposited on a durable porous alumina support. The membrane was resistant against mechanical strain and temperatures up to 150 °C. The working range of pH for the membrane used was from 0.5 to 13.5, which enabled cleaning of the membrane with aggressive chemicals.

With the increasing transmembrane pressure difference, Δp, the permeate flux increases significantly at lower suspension concentrations, but at the mass fractions higher than approximately 0.2, the flux is nearly independent of Δp (Figure 16.7(a)). At a relatively low transmembrane pressure difference, below approximately 40 kPa, the critical permeate flux is achieved. The limiting flux, when no increase of the flux with the increasing pressure can be observed, was reached at the pressure of about 100 kPa; however, at higher concentrations, the permeate flux did not increase with the higher pressure even at about 60 kPa. With the increasing concentration of the suspension, the permeate flux decreased, but in the interval of up to about 16 mass %, this decrease was milder (Figure 16.7(b)). At higher adsorbent concentrations, its further accumulation in the cake became dominant and the flux dropped down more steeply. These results show that the resistance of the particle cake is decisive. Permeate flux for concentrated suspension with 20 mass % of the adsorbent is still feasible. Permeability of the membrane for seawater after experiments with the suspensions was lower, indicating membrane fouling. After a single cleaning step, the membrane permeability was partially recovered.

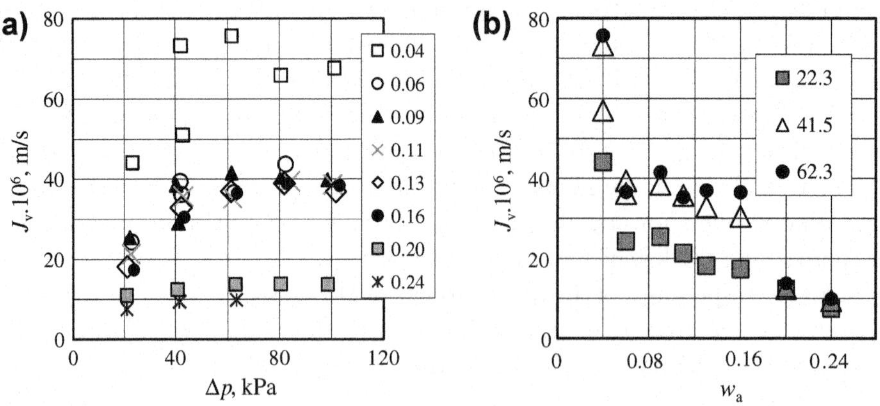

Figure 16.7 Permeate flux through the tubular ceramic membrane vs. the transmembrane pressure difference with the mass fraction of dry adsorbent in the suspension as parameter (a) and permeate flux vs. the mass fraction of dry adsorbent in the suspension with Δp as parameter (b) in MF of the suspension of Dowex XUS-43594 in model seawater. Velocity of the suspension was 2.9 m/s.

In total, suspensions of the microparticulate Dowex XUS-43594 adsorbent can be effectively concentrated by MF through a ceramic membrane to relatively high concentrations, of about 24 mass %, of dry adsorbent.

16.4 SIMULATION OF HYBRID AMF PROCESS

Estimation of the influence of various parameters on AMF process performance was based on mass balance equations describing the AMF process, adsorption equilibrium and mass-transport equations in suspension MF using submerged and tubular membrane modules. It was assumed that predefined saturation and desorption ratios were achieved in a short time. Estimation of the membrane areas was based on experimental dependences of permeate flux as a function of suspension concentration measured in a laboratory tubular ceramic module and in a submerged capillary module. In the formulation of the simulation model presented in paper[14] it was assumed, that:

1. The amount of boron in the adsorbent at pH = 8.2 is a linear function of boron concentration in water at low boron aqueous concentrations:

$$q = X_a q_{eq} = X_a k \rho \tag{16.2}$$

   where X_a is the saturation ratio of the adsorbent in the adsorption approaching to the equilibrium with typical value supposed 0.90 for 90% saturation of adsorbent comparing to the equilibrium saturation, q_{eq}. For Dowex XUS-43594, $k = 5.6$ m^3/kg at concentrations below 5 mg/dm^3. The value of the saturation ratio depends on the adsorbent suspension residence time in the retention volume, ADS, before MF1, which is assumed to be below 30 s for the adsorbent microparticles used.

2. Incomplete desorption by hydrochloric acid with the concentration of 0.2 $kmol/m^3$ (pH = 0.7) in the suspension stream 14 (Figure 16.3) was assumed. Residual concentration of boron in the adsorbent in the desorption process is defined by the relation:

$$q_d = (1 - X_d) \cdot q \tag{16.3}$$

   A typical assumed value of the desorption ratio, X_d, was 0.90 for the 90% desorption of boron from the adsorbent.

3. Permeate flux in the submerged module with a GE membrane and in the cross-flow tubular module with a ceramic membrane was estimated from empirical equations correlating experimental data of the dependence of flux on the relative concentration of Dowex XUS-43594 in suspension. Experimental data of these dependences, presented above, were correlated using a third order interpolating polynomial.

4. Volumetric flow rate of air needed for suspension mixing by aeration in submerged MF1 modules was estimated from the amount of air used in the laboratory MF unit

with submerged module described in paper.[24] In the first approximation, the same air flow rate as that used in a laboratory module with a 29 cm long membrane was assumed for a module with the membrane length of 1.8 m (as in an industrial module), supposing that airflow does not depend on the membrane length. Based on this assumption, the air flow rate per unit of the membrane area in MF1 modules is 0.067 N $m^3/m^2/h$.

16.4.1 Diafiltration

Diafiltration of the regenerated adsorbent by its washing with raffinate, stream 8 decreases the concentration of boron in the regenerated suspension, stream 20, which decreases the flux of the adsorbent needed in the recirculation loop to achieve the requested boron concentration in the raffinate (permeate), as shown in Figure 16.8(a). The adsorbent flow rate in the suspension recirculation loop rapidly decreased with the increasing volumetric flow rate of washing water, stream 8, up to about 0.5 m^3/h. A further increase is not very beneficial. Introduction of the diafiltration of the regenerated suspension dramatically decreased the consumption of chemicals in the regeneration process as shown in Figure 16.9(a) for acid consumption and in Figure 16.9(b) for alkali consumption. This is caused by the decrease of the amount of adsorbent needed and thus by the decreased volumetric flow rate of the suspension in the regeneration loop. Diafiltration also removes an excess of the acid present in the regenerated suspensions, which decrease the alkali consumption even more significantly. On the other hand, the concentration factor of boron in the concentrate, stream 19, rapidly decreases with the increasing flow rate of the diafiltration, stream 8, as shown in Figure 16.9(b). Increased amount of the concentrate increases the costs of its disposal. Higher concentration factor could be of interest if the concentrate utilization is considered.

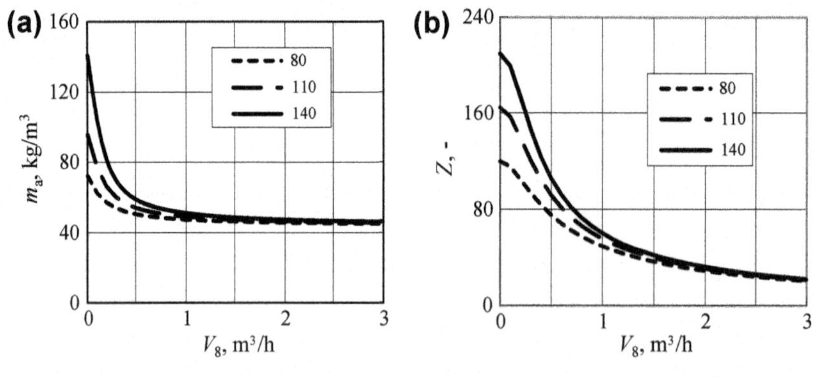

Figure 16.8 Dependence of the dry boron adsorbent flux in the recirculation loop of the adsorbent suspension (a) and boron concentration factor in stream 19 (b) on the volumetric flow rate of diafiltration water for various relative mass concentrations of the adsorbent in the suspension entering the regeneration loop (stream 12), kg/m^3.

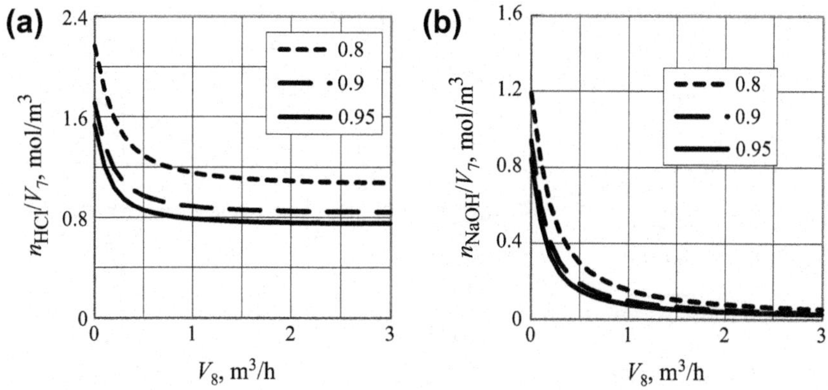

Figure 16.9 Specific consumption of acid (a) and base (b) in the desorption of the adsorbent as a function of the volumetric flow rate of diafiltration water for various adsorption and desorption ratios ($X_a = X_d$).

16.4.2 Suspension Concentration in MF Modules

Increased concentration of the suspension in MF4 and MF5 modules has a positive effect on all process parameters. Due to higher investment costs of units operating with ceramic membranes, optimization of the relative mass concentration of adsorbent in the suspension leaving MF4 and MF5 modules is based on minimum membrane surface area in MF4 and MF5 depending on the relative concentration of the suspension leaving these modules (Figure 16.10(b)). This minimum is close to the relative concentration of about 200 kg/m^3 and is caused by two opposing effects, i.e., the

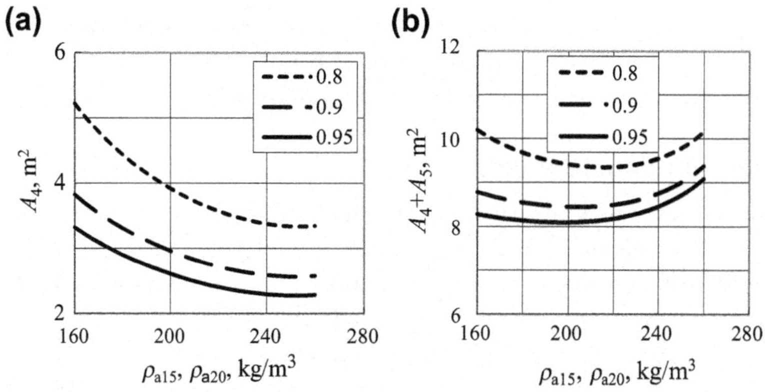

Figure 16.10 Membrane surface area in module MF4 (a) and total membrane surface area in MF4 and MF5 modules (b) vs. the adsorbent concentration in the streams leaving ceramic modules for selected adsorption and desorption ratios ($X_a = X_d$).

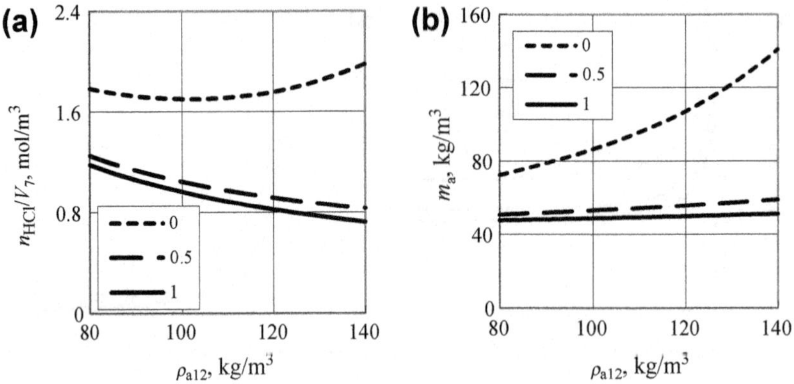

Figure 16.11 Dependence of specific consumption of acid (a) and adsorbent (dry) flux in the recirculation loop of the adsorbent suspension (b) on the concentration of the adsorbent concentration in the streams leaving submerged MF1 module, stream 15, and MF5 module, stream 20, for various flow rates of the diafiltration stream 8 in m^3/h.

reduction of the amount of circulated adsorbent needed when the concentration of the suspension in these modules increases, which is more pronounced for MF4 (Figure 16.10(a)), and by lowering the flow rate of permeate in these modules with increasing suspension concentration.

Increased concentration of the suspension leaving the submerged MF1 module has positive influence on the acid consumption when diafiltration is used, as shown in Figure 16.11(a). On the other hand, the amounts of base and adsorbent needed (Figure 16.11(b)) as well as the membrane surface area in the submerged MF1 module also slightly increase. According to these dependences, the value of 110 kg/m^3 was chosen as the optimal mass concentration of the adsorbent in the stream leaving the MF1 module. The optimized values of basic process parameters along with values of other parameters at these conditions are presented in Table 16.1.

16.4.3 MF Unit Segmentation

The selected values of preliminary optimized AMF process parameters presented in Table 16.1 were provided before.[14] As shown, segmentation of the submerged MF1 module is an effective way of reducing the membrane surface area needed in MF1. In this case, the unit divided into three segments was compared with a single compartment of the MF modules. Ideal mixing in MF tanks with submerged modules was assumed due to intensive aeration. Decrease in the membrane area in segmented MF can be explained by the concentration dependence of flux in the MF modules as shown in Figure 16.5(b). Possibly, the MF1 module can be divided into more segments, which further decreases the membrane surface area.

Table 16.1 Results of Simulation: Optimal Values of the AMF Process Parameters in the Model Unit[14]

Feed	$\dot{V}_1 = 100\ m^3/h$, $\rho_1 = 1.2\ g/m^3$		
Raffinate:	$\rho_7 = 0.4\ g/m^3$		
$\dot{V}_8$, m^3/h	0.5		
$\rho_{a,\ 12}$, kg/m^3	110		
$\rho_{a,\ 15} = \rho_{a,\ 20}$, kg/m^3	200		
Mass Flow Rate of Adsorbent and Chemicals Consumption			
$\dot{m}_a$, kg/h	54.3		
$\dot{n}_{HCl}/\dot{V}_7$, mol/m^3	0.97		
$\dot{n}_{NaOH}/\dot{V}_7$, mol/m^3	0.19		
Air Consumption in MF1 (Three Segments id. Mixing)			
$\dot{V}_4$, Nm^3/h	60		
Surface Area of Membranes and Permeate Flow Rate in MF1 (GE Zenon)			
Three segments (ideal mixing in segments)	$\rho_{a,\ Af} = 3.18\ kg/m^3$	$A_{1A} = 1080\ m^2$	$\dot{V}_{3A} = 83.2\ m^3/h$
	$\rho_{a,\ Bf} = 18.7\ kg/m^3$	$A_{1B} = 193\ m^2$	$\dot{V}_{3B} = 14.2\ m^3/h$
	$\rho_{a,\ Cf} = 110\ kg/m^3$	$A_{1C} = 41\ m^2$	$\dot{V}_{3C} = 2.4\ m^3/h$
	Total: $A_1 = 1315\ m^2$		
One segment (ideal mixing in segment)	$\rho_{a,\ f} = 110\ kg/m^3$ $A_1 = 1695\ m^2$, $\dot{V}_7 = 99.8\ m^3/h$		
Surface Area of Tubular Ceramic Membranes in MF 4 and MF 5			
$A_4 = 3.0\ m^2$; $A_5 = 5.5\ m^2$; total: $A_4 + A_5 = 8.5\ m^2$			
Theoretical Energy Consumption per 1 m^3 of Permeate			
Pumps in MF modules	$N_p/\dot{V}_7 = 0.0112\ kWh/m^3$		
Air blower in MF1 modules (three segments, ideal mixing)	$N_b/\dot{V}_7 = 0.0069\ kWh/m^3$		
Overall power input	$N/\dot{V}_7 - 0.0181\ kWh/m^3$		

16.5 COMPARISON OF AMF PROCESS WITH CLASSICAL IEX IN COLUMNS

Plants with a capacity of 100 m^3/h are considered in this comparison and basic process characteristics for both processes are compared in Table 16.2. An impressively low flow rate of the adsorbent is needed in the adsorbent recirculation loop of the AMF process which in this unit is only 54.3 kg/h. The amount of the adsorbent needed, consumption of chemicals, and theoretical power input per unit volume of the permeate drinking water in this processes are substantially lower as shown in Table 16.2.

This comparison shows high potential for the AMF process. It is clear that the presented data are based on laboratory experiments and should be proved by experiments using larger equipment and prolonged operation. Substantially lower consumption of chemicals and power input make the AMF process a reasonable alternative to existing processes.

Table 16.2 Parameters of the AMF Process and the Process in IEX Columns for a Plant with the Feed Capacity of 100 m^3/h.[5,14] Consumption of Chemicals and Power Input are Provided per 1 m^3 of raffinate

	AMF Process[a]	IEX in Columns[b]
Amount of wet adsorbent in plant, kg	~300	3120
Acid consumption, kg/m^3	0.035	0.06
Alkali consumption, kg/m^3	0.008	0.03
Overall power input, kWh/m^3	0.018	0.06

[a]Results from simulation based on laboratory data and[14] theoretical power input are shown.
[b]Two parallel lines of IEX columns are assumed, resin height of 1.4 m, feed rate of 30 BV/h, 24% bypass, working capacity of the adsorbent of 1.5 kg/m^3. Data on the chemicals consumption and power input were obtained from the industrial unit presented in paper.[3]

16.6 CONCLUSIONS

Suspensions of microparticulate Dowex XUS-43594 adsorbent can be effectively concentrated by MF to relatively high concentrations of dry adsorbent of about 24 mass % for a ceramic membrane in a cross-flow tubular module and of 11 mass % for a hydrophilic capillary membrane in a submerged module. Backwashing has been proved to be an efficient way of achieving prolonged MF operation at the flux higher than its critical values.

Application potential of the AMF process with a microparticulate adsorbent with the mean particle size of about 5 μm was analyzed on basis of laboratory data with a promising outcome. Diafiltration of a regenerated adsorbent suspension with only 0.5% of the raffinate is an effective way of decreasing the chemicals' consumption and adsorbent flow rate in the desorption loop of the adsorbent. The flow rate of the microparticulate adsorbent in the regeneration loop in a model unit for boron removal from the RO permeate of seawater with a capacity of 100 m^3/h is only 54.3 kg/h. The amount of adsorbent needed in this unit is only about 300 kg compared to 3120 kg in the IEX columns. Lower consumption of chemicals in the desorption loop and the overall theoretical power demand were estimated, which shows that the AMF process is a competitive alternative to a classical process with a fixed bed IEX column.

16.7 NOMENCLATURE

A_1, A_4, A_5 membrane surface area in module MF1, MF4 or MF5, m^2
J_v volumetric permeate flux through membrane, m/s
k slope of the linear part of the adsorption isotherm, m^3/kg
L_p permeability, m/s/kPa
$\dot{m}_a$ mass flow of adsorbent in suspension, kg/h
$\dot{n}$ molar flow of chemicals for pH adjustment, $kmol/m^3$
N theoretical overall power input, W

N_c theoretical power input to air blower, W
N_p theoretical power input to MF pumps, W
q adsorbate (boron) concentration in adsorbent, kg/kg
R mass transfer resistance, 1/m s kPa
$\dot{V}_i$ volumetric flow rate of the liquid in stream i (stream numbers are shown in Figure 16.1), m^3/h
$\dot{V}_P$ volumetric flow rate of permeate in a module or its segment, m^3/h
X_a saturation ratio of adsorbent in adsorption, Eqn (16.1), -
X_d desorption ratio of adsorbent in desorption, Eqn (16.2), -
Δp transmembrane pressure, kPa
ρ_a relative concentration of dry adsorbent per unit volume of liquid, kg/m^3
$\rho_{a,i}$ relative concentration of dry adsorbent per unit volume of liquid in stream i (stream numbers are shown in Figure 16.1), kg/m^3
ρ_i boron concentration in aqueous phase in stream i (stream numbers are shown in Figure 16.1), kg/m^3

ACKNOWLDGMENT

Support of the Slovak Grant Agency VEGA project No. 1/1184/11 is acknowledged. Experiments presented were partly supported by the MEDRC project No. 04-AS-004.

REFERENCES

1. Kabay N, Yilmaz I, Bryjak M, Yuksel M. Removal of boron from aqueous solutions by a hybrid ion exchange-membrane process. *Desalination* 2006;**198**:74–81.
2. Glueckstern P, Priel M. Boron removal in brackish water desalination systems. *Desalination* 2007;**205**:178–84.
3. Jacob C. Seawater desalination: boron removal by ion exchange technology. *Desalination* 2007;**205**:47–52.
4. Kabay N, Bryjak M, Schlosser Š Kitis M, Avlonitis S, Matejka Z, et al. Adsorption-membrane filtration hybrid process for boron removal from seawater: an overview. *Desalination* 2008;**223**:38–48.
5. Schlosser Š, Blahušiak M, Kabay N. A new hybrid adsorption-microfiltration process for water treatment with micro-particulate adsorbent, invited lecture at. In: *Proc. Of 12th Aachener Membran Kolloquium, Aachen (DE)*; 2008. pp. 49–56.
6. Kabay N, Guler E, Bryjak M. Boron in seawater and methods for its separation - a review. *Desalination* 2010;**261**:212–7.
7. Kabay N, Yilmaz I, Yamac S, Samatya S, Yuksel M, Yuksel U, et al. Removal and recovery of boron from geothermal wastewater by selective ion exchange resins. *I. Lab Tests* 2004;**60**:163–70.
8. Kabay N, Yilmaz-Ipek I, Soroko I, Makowski M, Kirmizisakal O, Yag S, et al. Removal of boron from Balcova geothermal water by ion exchange-microfiltration hybrid process. *Desalination* 2009;**241**:167–73.
9. Kabay N, Koseoglu P, Yapici D, Yuksel U, Yuksel M. Coupling ion exchange with ultrafiltration for boron removal from geothermal water-investigation of process parameters and recycle tests. *Desalination* 2013;**316**:17–22.
10. Kabay N, Koseoglu P, Yavuz E, Yuksel U, Yuksel M. An innovative integrated system for boron removal from geothermal water using RO process and ion exchange-ultrafiltration hybrid method. *Desalination* 2013;**316**:1–7.
11. Sluys JTM, Bakkenes HW, Creusen RJM, Schneiders LHJM, Hanemaaijer JH. Membrane assisted affinity separations: some techniques and applications. In: Crespo JG, Boedekker KW, editors. *Membrane processes in separation and purification*. Kluwer Acad. Publ; 1994. pp. 395–414.
12. Bakalár T, Schlosser Š, Búgel M. Hybrid adsorption and microfiltration process for removal of metals from aqueous solutions. In: *Proc. of CHISA 2004, full text on CD ROM, Praha (CZ). OSCHI*; 2004. p. 15.

13. Bakalár T, Schlosser Š, Búgel M. Hybrid process for removal of copper from aqueous solutions by adsorption on microparticles combined with microfiltration. In: *Proc. of PERMEA 2003, full text of lecture on CD ROM, Tatranské Matliare (SK). SSCHI*; 2003. p. 15.
14. Blahušiak M, Schlosser Š. Simulation of the adsorption - microfiltration process for Boron removal from RO permeate. *Desalination* 2009;**241**:156–66.
15. Yilmaz I, Kabay N, Bryjak M, Yuksel M, Wolska J, Koltuniewicz A. A submerged membrane-ion-exchange hybrid process for boron removal. *Desalination* 2006;**198**:310–5.
16. Kabay N, Bryjak M, Schlosser Š, Yuksel M, Kitis M, Semiat R, et al. *Study of the adsorption-membrane filtration (Amf) hybrid process for removal of Boron from seawater, final report of the joint project MEDRC 04-AS-004.* Izmir (TR): University of Ege; 2007. p. 191.
17. Blahušiak M, Schlosser Š. Simulation of the adsorption - microfiltration process for Boron removal from RO permeate. In: *Proc. of PERMEA 2007, full text on CD ROM.* Siófok (HU): Pannonia University; 2007.
18. Flores V, Cabassud C. A hybrid membrane process for Cu(II) removal from industrial wastewater - comparison with a conventional process system. *Desalination* 1999;**126**:101–8.
19. Holdich RG, Cumming LW, Perni S. Boron mass transfer during seeded microfiltration. *Chem Eng Res Des* 2006;**84**:60–8.
20. Ipek IY, Holdich R, Kabay N, Bryjak M, Yuksel M. Kinetic behaviour of boron selective resins for boron removal using seeded microfiltration system. *React Funct Polym* 2007;**67**:1628–34.
21. Koltuniewicz A, Bezak K. Engineering of membrane biosorption. *Desalination* 2002;**144**:219–26.
22. Koltuniewicz AB, Witek A, Bezak K. Efficiency of membrane-sorption integrated processes. *J Membr Sci* 2004;**239**:129–41.
23. Shao XM, Hu SX, Govind R. Continuous membrane dialysis using ion-exchange resin suspension for extracting metal-ions. *Ind Eng Chem Res* 1991;**30**:1231–9.
24. Blahušiak M, Onderková B, Schlosser Š, Annus J. Microfiltration of suspensions of micro-particulate boron adsorbent in submerged hollow fibre and capillary modules. *Desalination* 2009;**241**:138–47.
25. Onderková B, Schlosser Š, Blahušiak M, Búgel M. Microfiltration of suspensions of micro-particulate boron adsorbent through ceramic membrane. *Desalination* 2009;**241**:148–55.
26. Guglielmi G, Chiarani D, Judd SJ, Andreottola G. Flux criticality and sustainability in a hollow fibre submerged membrane bioreactor for municipal wastewater treatment. *J Membr Sci* 2007;**289**:241–8.
27. Stopka J, Bugan SG, Broussous L, Schlosser S, Larbot A. Microfiltration of beer yeast suspensions through stamped ceramic membranes. *Sep Purif Technol* 2001;**25**:535–43.

CHAPTER 17

Boron Uptake from Aqueous Solution by Chelating Adsorbents: A Statistical Experimental Design Approach

Esra Bilgin Simsek[1,2], **Bahire Filiz Senkal**[3], **Ulker Beker**[2]
[1]Chemical & Process Engineering Department, Yalova University, Yalova, Turkey
[2]Chemical Engineering Department, Yildiz Technical University, Istanbul, Turkey
[3]Department of Chemistry, Istanbul Technical University, Istanbul, Turkey

17.1 INTRODUCTION

The development of boron-specific chelating resins with large capacity, high selectivity, and high uptake rate has held much interest for removal of borate from geothermal waters and boron containing wastewaters. It was shown that chelating resins containing ligands having three or more hydroxyl groups, located in the cis position, the so-called "*vic—cis* diols," show a high selectivity to boron. They can bind boric acid tightly by forming either neutral boron esters or borate complex anion with a proton as counter ion.[1,2]

Having six hydroxyl functions, sugar derivatives such as sorbitol and mannitol have exceptionally high boron binding abilities. In accordance with this key principle, poly(styrene)-based resins with *N*-methyl D-glucamine functions which emerged in the mid-1960s, have been reported as boron specific sorbents. Bicak et al.[2] prepared terpolymers of glycidyl methacrylate (GMA)—methyl methacrylate (MMA)—divinyl benzene (DVB) and used the terpolymers as a support for boron specific resins possessing *N*-methyl-D-glucamine as the functional group. From sorption and elution tests, it was found that the resin showed good stability in terms of particle disintegration for long-term uses and better performance of regeneration in comparison to common polymeric boron sorbents. Wang et al.[3] developed a chelating polymeric sorbent having *N*-methyl-D-glucamine groups as the functional group and poly GMA-*co*-TRIM as the support. The resin showed low swelling degree and a high capacity of boron sorption. Although the capacity was the same as that of a commercially available *N*-methyl-D-glucamine type polystyrene resin, the rate of sorption was more rapid. It was found that fast sorption kinetic is caused by its low swelling characteristic, permanent macroporous structures, and more hydrophilic characteristics.

Although utilization of these polymeric sorbents is effective in boron adsorption, the classical method of adsorption technique does not adequately describe the process in terms of the effects of independent variables and it does not depict their interactions on the dependent variable. The traditional methods can be time consuming and require a

Boron Separation Processes
ISBN 978-0-444-63454-2, http://dx.doi.org/10.1016/B978-0-444-63454-2.00017-4

number of experiments to determine optimum levels. Application of a statistical method in adsorption phenomena using experimental design techniques could provide mathematical models for designing chemical and physical processes regarding the interactions among the input parameters. The response surface method is one of the statistical methods developing an empirical model of the process and obtaining data more precisely than univariate strategies. This methodology results in a reduced number of assays necessary to optimize the process and improved product yields and closer confirmation of the output response to nominal and target requirements.

In this paper, two different kind of sorbents were synthesized: (1) polymer supported core–shell type iminodipropylene glycol functions (GMA-PVC) and (2) *N*-methyl-D-glucamine modified poly(styrene)-based core-shell type sorbent (VBC-NMG). The objectives of the study were the application of a three factor, three-level Box–Behnken experimental design for maximizing boron removal by polymeric resins and examination of effects of three independent variables (solution pH, temperature and initial concentration). The combined effects of operating parameters on boron adsorption capacity were critically investigated. The adsorption of boron has been modeled using six adsorption isotherm models by nonlinear analysis and thermodynamic studies were conducted. The findings provided new insight into the availability of statistical design for boron removal and better understanding of structural differences among resins effecting boron removal.

17.2 MATERIALS AND METHODS

17.2.1 Preparation of VBC-NMG Sorbent

The poly(vinylbenzyl chloride) (poly(VBC)) beads were prepared by suspension polymerization as described in a previous study.[4] Graft copolymerization of glycidyl methacrylate (GMA) on the beads surface was achieved through benzyl chloride initiation sites of the poly(VBC). The synthesis route was given in Figure 17.1. Poly(VBC) beads (5.0 g) were transferred into a reactor and the following chemicals (monomer GMA, CuBr, bipyridine, and dioxane) were added. After the polymerization reaction, the poly (VBC-g-GMA) beads were modified with *N*-methyl-D-glucamine according to the method described in the literature.[2] The resultant sample was coded as VBC-NMG.

17.2.2 Preparation of GMA-PVC Sorbent

The polymer beads of were prepared according to the procedure described elsewhere.[5] The preparation procedure is presented in Figure 17.2. Graft polymerization of poly(GMA) was achieved through chlorine initiation sites on the cross-linked polyvinyl chloride (PVC) by atom transfer radical polymerization (ATRP) method. The

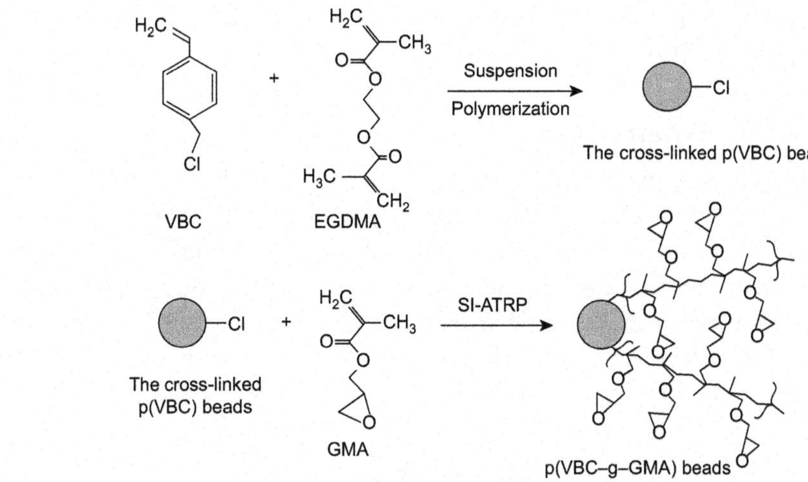

Figure 17.1 The synthesis route of VBC-NMG sorbent.

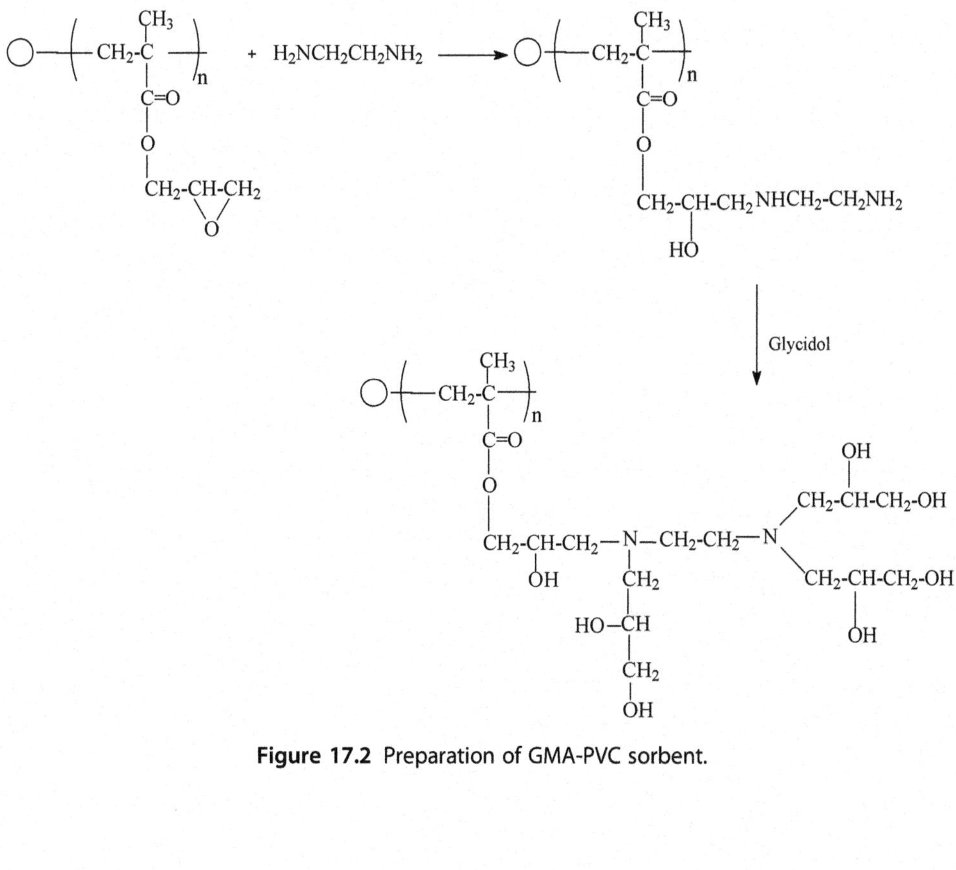

Figure 17.2 Preparation of GMA-PVC sorbent.

poly(GMA) graft copolymer was modified with ethylenediamine. The resultant sample was designated as GMA-PVC.

17.2.3 Adsorption Experiments

Boron adsorption design experiments were conducted with three different boron concentrations (10, 20, 30 mg/L) at different pH's (3.0, 6.5, and 10.0) and temperatures (25, 45, and 65 °C). Adsorption isotherms were obtained by varying the amount of adsorbent at different temperatures. Equilibrium boron concentration was determined by atomic absorption spectrophotometer (Analytik Jena ContrAA 700 TR). The formula used to calculate the amount of boron adsorbed per unit mass is shown in Eqn. (17.1):

$$q_e = \frac{(C_i - C_e) \times V}{m} \tag{17.1}$$

where q_e is the adsorption capacity (mg/g), V is the solution volume (L), m is the adsorbent dosage (g), C_i and C_e are the initial and equilibrium concentration of boron (mg/L), respectively.

17.2.4 Response Surface Methodology

Response surface methodology (RSM) is a collection of statistical and mathematical techniques for optimization of several process parameters by a minimum number of experiments. RSM is useful for analysis of problems in which a response of interest is influenced by several variables and the objective is to optimize this response.[6]

Box–Behnken design (BBD) is a spherical design based on three-level incomplete factorial designs including a central point and the middle points of the edges of the cube surrounded on the sphere.[7] Among the RSM designs (central composite (CCD), Doehlert matrix (DM), and three-level full factorial design) the BBD and DM designs are more efficient than the central composite design.[8] Moreover, the BBD requires less runs than the others allowing calculations of the response function at intermediate levels.

The Box-Behnken model supplies an empirical relationship between the dependent and independent variables. The mathematical relationship between response (Y_i) and factors (x_1, x_2, and x_3) can be approximated by the quadratic equation as follows:

$$Y_i = \beta_0 + \sum_{i=1}^{k} \beta_i x_i + \sum_{i=1}^{k} \beta_{ii} x_i^2 + \sum_{i=1}^{k-1} \sum_{j=2}^{k} \beta_{ij} x_i x_j + \varepsilon \tag{17.2}$$

where, Y_i is the response, β_0 is the constant, β_i is the slope or linear effect of the input factor, β_{ii} is the quadratic effect, β_{ij} is the 2-way linear by linear interaction effect, x_i and

x_j are the independent parameters, ε is the random error, and k is the number of the independent variables.

In the present study, the BBD methodology was used in order to investigate the effect of selected process variables: pH (3.0—10.0), temperature (25—65 °C), and initial boron concentration (10—30 mg/L). The statistical design required 17 experimental runs including five replicates at the center points. The empirical quadratic model was defined by three parameters: pH (x_1), temperature (x_2), and initial concentration (x_3). Boron adsorption capacities of polymeric adsorbents (mg/g) were designated as dependent variables (Y_i). Table 17.1 shows the range and levels of three independent adsorption conditions.

Statistica (Ver. 8.0, StatSoft Inc., USA) software package was used for regression and graphical analyses. The statistical significance of variables was determined using the analysis of variance (ANOVA) and Pareto chart (Student's *t*-test) with 95% confidence limits ($\alpha = 0.05$). The validity of the constructed model was determined by lack of fit, coefficient of determination (R^2), adjusted coefficient of determination (R^2_{adj}) and Fisher's *F*-test. The distributions of residuals and correlation between error terms were examined by Anderson—Darling normality test. This test plays an important role by detecting all departures from normality values. If A^2 is less than the critical value, which is 0.787 at 95% confidence level- and *p*-value is greater or equal to 0.05; the hypothesis of normality could be accepted for that significance level.

17.2.5 Evaluation of Adsorption Isotherms

Adsorption isotherms are the basic source of information about the adsorption process. The description of isotherms can be based on models with two, three or more parameters. There is a growing interest in the derivation of isotherm models due to the limited application of two-parameter models.[9] In the present study, the relationship between adsorbed boron and the concentrations at equilibrium was described by two- and three-parameter isotherm models: Freundlich, Langmuir, Dubinin—Radushkevich, Redlich—Peterson, Sips, and Toth using Statistica (Ver. 8.0, StatSoft Inc., USA) software package.

Table 17.1 Independent Variables and Levels of Each Factor For Box—Behnken

Variables	Symbol	Level		
		Low (−1)	Middle (0)	High (+1)
pH	x_1	3.0	6.5	10.0
Temperature	x_2	25	45	65
Initial boron concentration	x_3	10	20	30

17.2.5.1 Langmuir Model

The Langmuir model is based on the homogeneous adsorption, in which each molecule have equal enthalpies and sorption activation energy.[10] The model is expressed by the following equation:

$$q_e = \frac{Q \cdot b \cdot C_e}{1 + b \cdot C_e} \tag{17.3}$$

where q_e is the adsorbed amount at equilibrium (mg/g), C_e the equilibrium concentration of the adsorbate (mg/L), Q is the Langmuir monolayer sorption capacity (mg/g), and b is the Langmuir equilibrium constant (L/mg) related to the energy of adsorption and affinity of the adsorbent. Webber and Chakkravorti[11] defined a dimensionless constant, R_L, which describes the type of isotherm and represented as:

$$R_L = 1/(1 + b \cdot C_i) \tag{17.4}$$

where C_i is the initial adsorbate concentration (mg/L). The magnitude of R_L determines the adsorption nature to be either unfavorable ($R_L > 1$), linear ($R_L = 1$), favorable ($0 < R < 1$), or irreversible ($R_L = 0$).

17.2.5.2 Freundlich Model

The model assumes multilayer adsorption, with nonuniform distribution of adsorption heat and affinities over the heterogeneous surface. The model is not limited to the formation of a monolayer as with the Langmuir theory. It can be written as:

$$q_e = K_F \cdot C_e^{1/n} \tag{17.5}$$

where K_F and n are indicative isotherm parameters of adsorption capacity and intensity, respectively. $1/n$ is a sign of surface heterogeneity, ranges between 0 and 1, becoming more heterogeneous as it gets closer to zero.

17.2.5.3 Dubinin–Radushkevich (D-R) Isotherm Model

The D-R model is based on the assumption of multilayer character of adsorbent surface and is generally applied to distinguish physical and chemical adsorption mechanism.[12] The isotherm is expressed by:

$$q_e = q_m \cdot \exp\left(-\beta \cdot \varepsilon^2\right) \tag{17.6}$$

$$\varepsilon = R \cdot T \ln\left(1 + 1/C_e\right) \tag{17.7}$$

where R indicates gas constant (8.314 J/mol/K), T and C_e are absolute temperature (K) and equilibrium concentration (mg/L), respectively.

The mean energy of adsorption is the free energy of transfer when 1 mol of solute from solution is transferred to the surface of zeolite and it is calculated by the following equation:

$$E = \frac{1}{(2\beta)^{0.5}} \tag{17.8}$$

where β is the constant related to the sorption energy and can be calculated by using D–R equation.

17.2.5.4 Redlich–Peterson Model

The Redlich–Peterson model is an empirical hybrid isotherm featuring Langmuir and Freundlich isotherms. The theory combines the equations of two models and does not assume ideal monolayer adsorption.[13] The empirical equation is:

$$q_e = \frac{K_{RP} \cdot C_e}{\left(1 + \alpha_{RP} \cdot C_e^{\beta_{RP}}\right)} \tag{17.9}$$

α_{RP} and K_{RP} are the Redlich–Peterson isotherm constants, β_{RP} is the exponent which ranges between 0 and 1. The model fits Freundlich the isotherm model at high concentrations ($\beta_{RP} \approx 0$), while it approaches the ideal Langmuir condition at low concentrations ($\beta_{RP} \approx 1$).

15.2.5.5 Toth Model

The Toth isotherm is the form of the Langmuir equation modified by reducing the errors between experimental and theoretical data of adsorption.[14] The equation is derived from the potential theory and suited to multilayer adsorption similar to BET isotherms, which is a special type of Langmuir isotherm.[15] The Toth equation is given as:

$$q_e = \frac{q_{m_T} \cdot C_e}{\left(\frac{1}{K_T} + C_e^{m_T}\right)^{1/m_T}} \tag{17.10}$$

where q_{m_T} is the Toth maximum adsorption capacity (mg/g), K_T and m_T are Toth equilibrium constant and the Toth model exponent, respectively.

17.2.5.6 Sips Model

The Sips isotherm is a combination of the Langmuir and Freundlich models predicting nonuniform surfaces on adsorption system.[16] The theory has a limitation of increasing concentration of the Freundlich equation.

$$q_e = \frac{q_{m_S} \cdot K_S \cdot C_e^{m_s}}{\left(1 + K_S \cdot C_e^{m_s}\right)} \tag{17.11}$$

where q_{m_S} is the indicator of Sips maximum adsorption capacity (mg/g), K_S and m_s are the Sips equilibrium constant (L/mg) and model exponent, respectively. The model reduces to the Freundlich isotherm at low concentrations, while at high concentrations it assumes a monolayer adsorption capacity characteristic of the Langmuir isotherm.[17]

In order to evaluate the conformity of applied isotherms to experimental data, the coefficient of determination (R^2) (Eqn. (17.12)) and Chi-square values (χ^2) (Eqn. (17.13)) were calculated by using the following equations:

$$R^2 = \frac{\sum \left(q_{cal} - q_{a\ \exp}\right)^2}{\sum \left(q_{cal} - q_{a\ \exp}\right)^2 + \left(q_{cal} - q_{\exp}\right)^2} \tag{17.12}$$

$$\chi^2 = \sum \left\| \frac{\left(q_{cal} - q_{\exp}\right)^2}{q_{cal}} \right\| \tag{17.13}$$

17.2.6 Adsorption Thermodynamics

Thermodynamic parameters such as Gibbs free energy (ΔG°), enthalpy (ΔH°), and entropy (ΔS°) were calculated from experimental isotherms at different temperatures. The values were determined by using the equilibrium adsorption constant (K_C) which is defined as:

$$K_C = \frac{(C_i - C_e) \cdot V}{C_e \cdot m} \tag{17.14}$$

where m (mg) is the amount of adsorbent; V (ml) refers the volume of solution. The changes in thermodynamic parameters (ΔG°, ΔH° and ΔS°) for each temperature were calculated using Eqn. (17.15).

$$\ln K_C = -\frac{\Delta G^o}{RT} = \frac{\Delta S^o}{R} - \frac{\Delta H^o}{RT} \tag{17.15}$$

where R (8.314 J/mol/K) is the gas constant and T (K) is the absolute temperature.

17.3 RESULTS AND DISCUSSION

17.3.1 Characterization of Polymeric Resins

As can be seen in Figure 17.3, an increase in the grafting time from three to 18 h leads to an increase of more than 250% in the grafting percentage of poly(GMA) on the poly(VBC) beads. Epoxy group content of the beads was determined as 7.1 ± 0.2 mmol/g beads by the pyridine–HCl method.[18] The amount of amine content of VBC-NMG was found to be about 6.5 mmol/g beads by using titrimetric method.

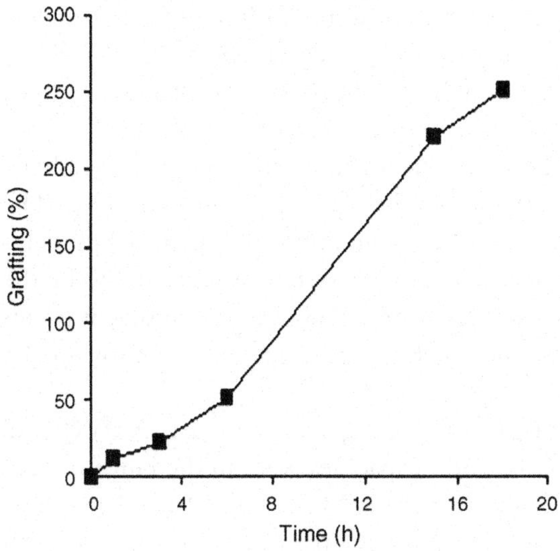

Figure 17.3 Grafting efficiency of VBC-NMG sample.

Thermal dehydrochlorination of commercial PVC was carried out by heating PVC in concentrated NaOH containing isopropanol–water mixture. This method gave a black product. Partial dehydrochlorination of PVC generates new double bonds, and these increase the availability of allylic chlorine atoms in the backbone. This shows that thermal stability of PVC decreases upon an increase in dehydrochlorination. Unstable chlorine atoms in PVC in conjunction with suitable catalyst provide initiation sites for grafting.

Core-shell types of polymers (GMA-PVC) with poly (glycidyl methacrylate) shells were obtained by ATRP method. Graft polymerization of glycidyl methacrylate can be achieved from chlorine atoms onto cross-linked PVC. The grafting percentage of PGMA onto cross-linked PVC reached to 200% for 16 h. Epoxy content of the resin determined by the pyridine–HCl method was found as 7.7 mmol/g sorbent. The reaction of epoxy rings in grafted PGMA with an excess of ethylenediamine gives amine containing resin with 6.25 mmol/g amine functions. Modification of the amino groups with equivalent amount of glycidol, yields iminopropylene glycol functions.

17.3.2 Development of Regression Model and Statistical Analysis

In the first step of experimental design, multiple response functions (linear, quadratic, interactive, and cubic models) on the design matrix were applied to obtain an optimum fitted model to represent boron removal. The lack-of-fit test was designed to determine whether the selected model is adequate to describe the observed data, or whether a more complicated model should be used. Since the *p*-value for lack of fit is greater or equal to 0.05, the model appears to be adequate for the observed data at the 95.0% confidence level. As shown in Table 17.2, for both adsorbents the quadratic model provides the best fit to the experimental data with *p*-values for lack of fit ($p_{GMA\text{-}PVC} = 0.0513$, $p_{VBC\text{-}NMG} = 0.1554$). The R^2 values indicate that the quadratic model explains nearly 99.4–99.8% of the variability in adsorption capacity. The adjusted determination coefficient (R^2_{adj}) statistic also demonstrated the fitted model is more suitable. While the R^2 and R^2_{adj} values of the cubic system were found to be the highest among the others, this model was not recommended due to insufficient points to predict the coefficients. Consequently, the quadratic model was chosen for further analysis.

The observed and predicted boron adsorption capacities of GMA-PVC and VBC-NMG are presented in Table 17.3. Sorption capacities were in the range of 1.29 and 5.8 mg/g for GMA-PVC; 1.0 and 5.66 mg/g for VBC-NMG. When the whole runs were examined together VBC-NMG was found to be the more effective adsorbent when compared to GMA-PVC.

The normal probability plot of residuals is one of the important diagnostic tools detecting the deviations from the assumptions of which errors are distributed homogeneously or not. Figure 17.4 shows plots of normal probability of the residuals indicating almost no serious violation between observed and predicted values. Anderson–Darling normality test indicated that residuals are normally distributed and that the error variance is homogeneous according to the calculated *p* (0.5873 and 0.0875) and A^2 (0.2983 and 0.6564) values.

The interacting factors affecting boron removal were determined by performing an ANOVA (Table 17.4). Results were checked by descriptive statics, such as sum of squares, probability (p), R^2, and R^2_{adj} values. Sum of squares (SS) of each factor determines its importance in the process. When the value of the SS increases the significance of the corresponding factor in the undergoing process also increases. Moreover, the *p*-value is a considerable factor for determining which interaction is significant or not. The effects and interactions of pH, temperature and concentration were considered as potentially significant according to the *p*-values (<0.05). The R^2 values of polynomial regression models were found as 0.9947 and 0.9981 for GMA-PVC and VBC-NMG, respectively, indicating the accuracy and general availability of the model. High values of R^2_{adj} were obtained (0.9880 for GMA-PVC and 0.9958 for VBC-NMG) attributing an excellent correlation between factors and responses.

The magnitude of t values was found significant according to Student's *t*-test. The Pareto chart (Figure 17.5) presents the effects of the independent variables and their interactions on the B adsorption capacity. Linear terms of initial concentration (x_3) ($t = 59.649$), pH (x_1) of solution ($t = 26.100$), and temperature (x_2) ($t = 11.137$) were

Table 17.2 Adequacy of the Tested Models

Lack-of-Fit Tests

	Sum of Squares			Mean Square		F-Value		*p*-Value	
	GMA-PVC	VBC-NMG	df	GMA-PVC	VBC-NMG	GMA-PVC	VBC-NMG	GMA-PVC	VBC-NMG
Linear	0.70172	1.10955	9	0.07797	0.12328	13.594	40.194	0.011515	0.001443
2FI	0.52767	1.05257	6	0.08795	0.17543	15.333	57.194	0.009847	0.000784
Quadratic	0.11153	0.02797	3	0.03718	0.00932	6.482	3.040	0.051368	0.155423
Cubic	0.000	0.000	0						
Pure error	0.02294	0.01227	4	0.005736	0.003067				

Model Summary Statistics

	R^2		R^2_{Adj}	
	GMA-PVC	VBC-NMG	GMA-PVC	VBC-NMG
Linear	0.97186	0.94947	0.96537	0.93781
2FI	0.97862	0.95204	0.96579	0.92326
Quadratic	0.99478	0.99819	0.98806	0.99586
Cubic	0.99911	0.99945	0.99857	0.99779

Table 17.3 Observed and Predicted Sorption Capacities

	Independent Variables						Boron Sorption Capacity (mg/g)			
	pH, x_1		Temperature, x_2		Concentration, x_3		GMA-PVC		VBC-NMG	
Run	Coded	Actual	Coded	Actual	Coded	Actual	Observed	Predicted	Observed	Predicted
1	−1	3.0	−1	25.0	0	20.0	2.507	2.463	2.435	2.373
2	1	10.0	−1	25.0	0	20.0	3.590	3.604	4.310	4.365
3	−1	3.0	1	65.0	0	20.0	2.817	2.803	2.193	2.137
4	1	10.0	1	65.0	0	20.0	4.413	4.457	3.937	4.000
5	−1	3.0	0	45.0	−1	10.0	1.296	1.212	1.041	1.105
6	1	10.0	0	45.0	−1	10.0	2.506	2.363	2.919	2.865
7	−1	3.0	0	45.0	1	30.0	4.017	4.160	3.480	3.534
8	1	10.0	0	45.0	1	30.0	5.719	5.804	5.691	5.627
9	0	6.5	−1	25.0	−1	10.0	1.824	1.952	2.632	2.630
10	0	6.5	1	65.0	−1	10.0	2.233	2.331	2.496	2.487
11	0	6.5	−1	25.0	1	30.0	5.028	4.929	5.375	5.383
12	0	6.5	1	65.0	1	30.0	5.872	5.743	4.922	4.924
13	0	6.5	0	45.0	0	20.0	3.353	3.403	3.366	3.347
14	0	6.5	0	45.0	0	20.0	3.459	3.403	3.294	3.347
15	0	6.5	0	45.0	0	20.0	3.305	3.403	3.429	3.347
16	0	6.5	0	45.0	0	20.0	3.411	3.403	3.349	3.347
17	0	6.5	0	45.0	0	20.0	3.490	3.403	3.298	3.347

found to be more effective than their quadratic variances and their interactions for GMA-PVC (Figure 17.5(a)). The Pareto chart of VBC-NMG suggests that linear terms of initial concentration (x_3) had a major and positive effect (t = 66.274) on adsorption capacity. While positive coefficients of temperature (x_2) were observed for GMA-PVC; the term of VBC-NMG sample was found as negative (x_2, t = −7.675) indicating temperature has an unfavorable or antagonistic effect on the removal capacity of VBC-NMG adsorbent. Moreover, for VBC-NMG the term of $x_1.x_2$ was insignificant in prediction of the adsorption efficiency (p-value of $x_1.x_2$ was found 0.3045).

Based on the regression coefficients the second-order polynomial models of GMA-PVC and VBC-NMG, after neglecting the insignificant terms, were established as:

$$\begin{aligned} q_{GMA-PVC} = {} & 0.482376 + 0.272788\,x_1 - 0.017393\,x_1^2 - 0.039738\,x_2 + 0.000354\,x_2^2 \\ & + 0.034682\,x_3 + 0.001943\,x_3^2 + 0.001837\,x_1x_2 + 0.003517\,x_1x_3 \\ & + 0.000544\,x_2x_3 \end{aligned} \tag{17.16}$$

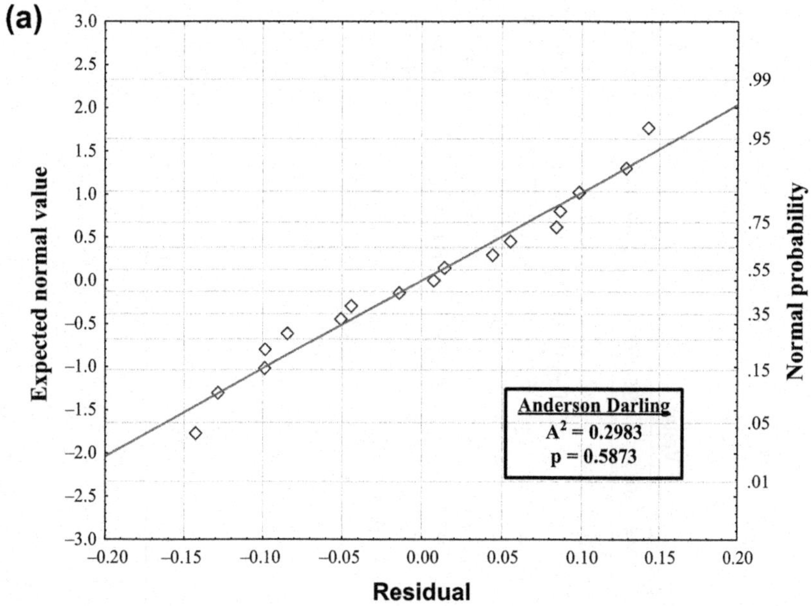

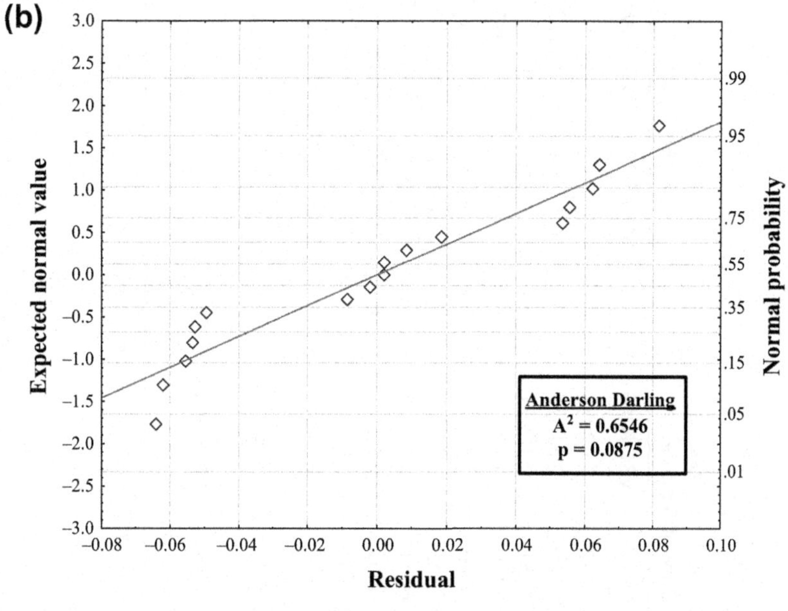

Figure 17.4 Normal probability plots of residuals. (a) GMA-PVC and (b) VBC-NMG.

Table 17.4 ANOVA of Boron Sorption

Factors	Sum of Squares		df	Mean Square		F-Value		*p*-Value	
	GMA-PVC	VBC-NMG		GMA-PVC	VBC-NMG	GMA-PVC	VBC-NMG	GMA-PVC	VBC-NMG
x_1	3.90732	7.42833	1	3.90732	7.42833	681.228	2421.825	0.000013	0.000001
x_1^2	0.19113	0.51832	1	0.19113	0.51832	33.324	168.984	0.004471	0.000202
x_2	0.71151	0.18071	1	0.71151	0.18071	124.050	58.916	0.000370	0.001549
x_2^2	0.08429	0.20853	1	0.08429	0.20853	14.696	67.985	0.018561	0.001180
x_3	20.40780	13.47209	1	20.40780	13.47209	3558.030	4392.246	0.000000	0.000000
x_3^2	0.15889	0.34549	1	0.15889	0.34549	27.702	112.640	0.006241	0.000446
x_1x_2	0.06611	0.00425	1	0.06611	0.00425	11.526	1.384	0.027402	0.304576
x_1x_3	0.06062	0.02774	1	0.06062	0.02774	10.569	9.043	0.031347	0.039658
x_2x_3	0.04731	0.02499	1	0.04731	0.02499	8.248	8.148	0.045377	0.046186
Lack of fit	0.11153	0.02797	3	0.03718	0.00932	6.482	3.040	0.051368	0.155423
Pure error	0.02294	0.01227	4	0.00574	0.00307				
Total SS	25.75129	22.20295	16						

GMA-PVC: $R^2 = 0.99478$; $R^2_{Adj} = 0.98806$; VBC-NMG: $R^2 = 0.99819$; $R^2_{Adj} = 0.99586$.

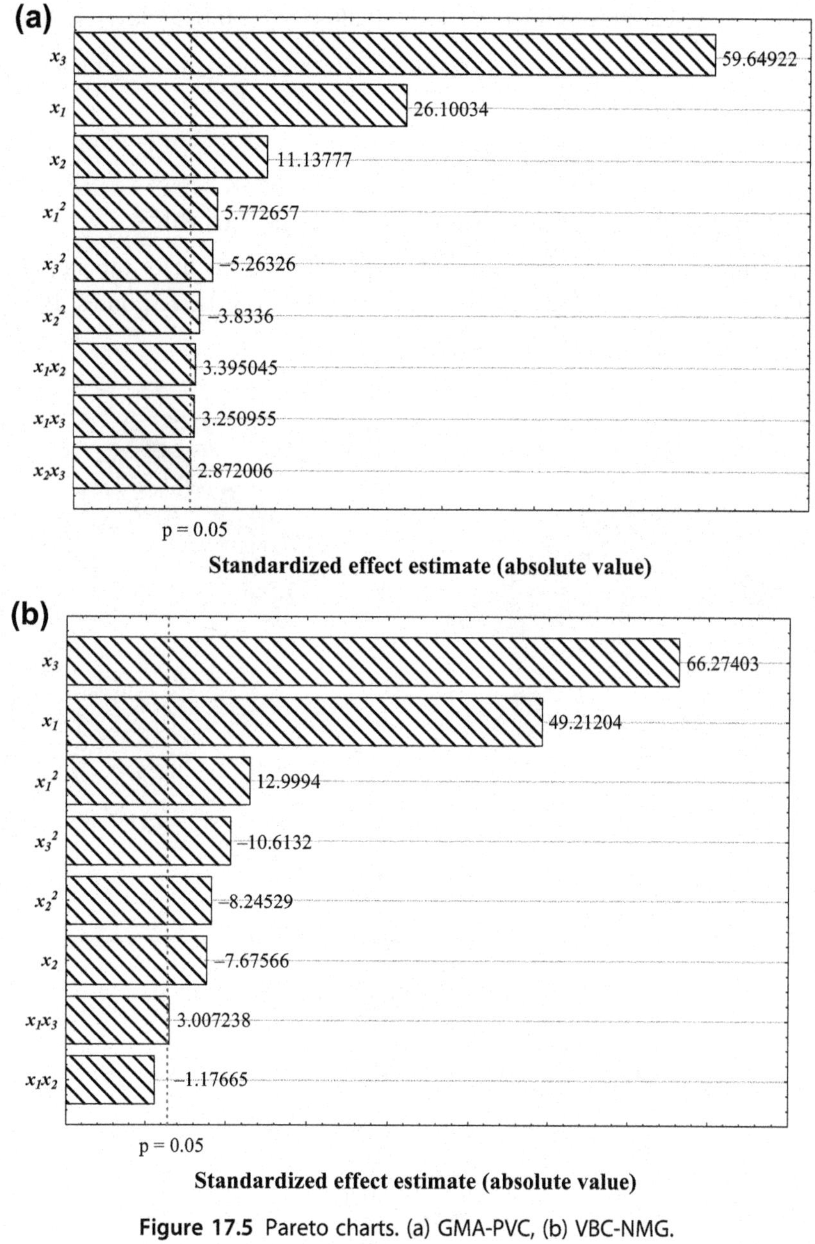

Figure 17.5 Pareto charts. (a) GMA-PVC, (b) VBC-NMG.

$$q_{VBC-NMG} = 0.316271 + 0.600069\,x_1 - 0.028641\,x_1^2 - 0.049682\,x_2 + 0.000556\,x_2^2 + 0.017508\,x_3 + 0.002865\,x_3^2 + 0.002379\,x_1x_3 + 0.000395\,x_2x_3 \quad (17.17)$$

17.3.3 Main and Interaction Effects of Process Variables

The three dimensional (3D) response surface plots could facilitate the straightforward examination of the effects of the experimental variables on the responses. By analyzing

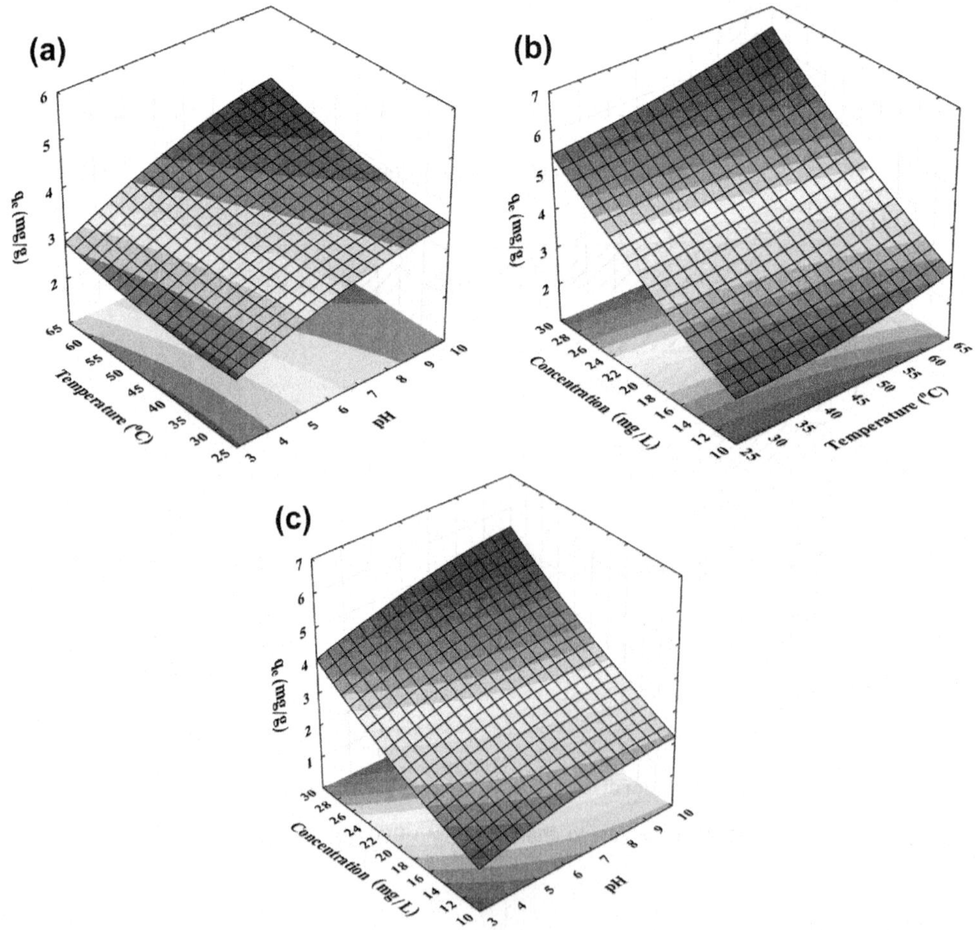

Figure 17.6 Response surfaces for combined effect of (a) temperature and pH at constant initial concentration of 20 mg/L; (b) initial concentration and temperature at constant pH of 10.0; and (c) pH and initial concentration at constant temperature of 25 °C; on adsorption capacity of GMA-PVC.

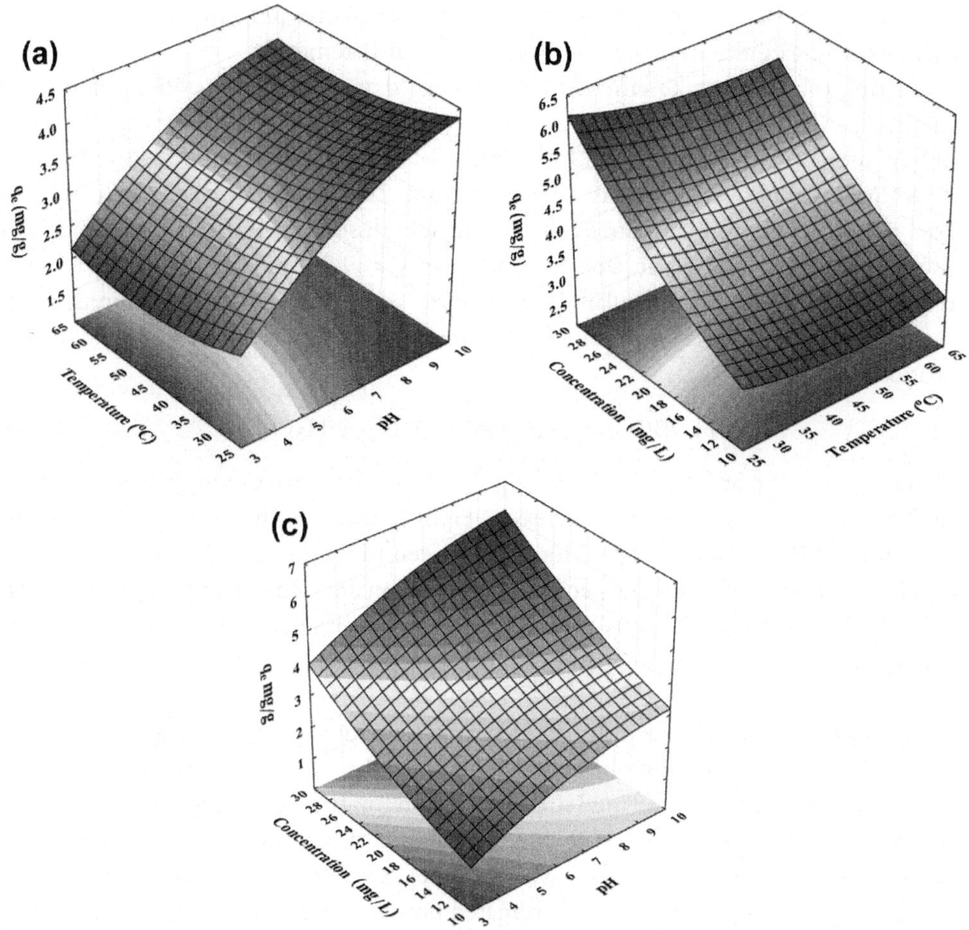

Figure 17.7 Response surfaces for combined effect of (a) temperature and pH at constant initial concentration of 20 mg/L; (b) initial concentration and temperature at constant pH of 10.0; and (c) pH and initial concentration at constant temperature of 25 °C; on adsorption capacity of VBC-NMG.

the plots, the best response range can be calculated. The surface plots were formed based on the model equations as a function of two factors by holding the other factor at certain level (Figures 17.6 and 17.7).

17.3.3.1 Effect of Solution pH

The influent pH is an important parameter in affecting the performance of the adsorption process. The relation between the solution pH and the removal capacity of GMA-PVC and VBC-NMG are given in Figure 17.6(a) and (c) and Figure 17.7(a) and (c), respectively. Batch experiments were conducted in the pH range of 3.0–10.0 in order to investigate the influence of solution pH on B removal.

Maximum boron uptake efficiency was observed at pH 10.0 for both samples. The capacity of GMA-PVC was found as 2.50 mg/g (65.3% removal) at pH 3 (Run #1) while it increased to 3.59 mg/g (90.1% removal) at pH 10 (Run #2). Similarly for VBC-NMG sample, the capacity increased from 2.43 to 4.31 mg/g when pH was raised to 10.0. This could be attributed to the acid dissociation of the boric acid. The dissociation constant of boric acid is equal to $pK_a = 9.24$ and below this pH boron predominantly exists as molecular species (boric acid).[19] In aqueous solutions, boric acid acts as a very weak and monobasic acid which dissociates by the following equation:

$$H_3(BO_3) + 2H_2O \Leftrightarrow B(OH)_4^- + H_3O^+ \quad (17.18)$$

In the pH range of 3.0 and below the pK_a values, boron ions exist predominantly in the form of $B(OH)_3$. When pH increases, H_3BO_3 reacts with water resulting in the production of $B(OH)_4^-$ and H_3O^+ forms. Between pH 9 and 10, the mole fraction of $B(OH)_3$ decreases and $B(OH)_4^-$ species become dominant. Particularly at pH 11, boric acid exists completely as the $B(OH)_4^-$ form which has higher affinity to the hydroxyl groups than H_3BO_3.[20,21] Similar pH effects on the removal of boron from water were also reported by other researches.[6,20,22]

The adsorption mechanism is considered to be strong complexation reactions between $B(OH)_4^-$ and diol functions incorporated into the polymeric structures. The stability of the borate complex formed strongly depends on the type of diol used. According to Power and Woods,[23] a strong complex is formed when the diol used involves the hydroxyl groups oriented in such a way that they accurately match the structural parameters required by tetrahedrally coordinated boron. For example, a stable complex is formed by the reaction of boron with compounds possessing *cis*-diol system such as D-mannitol, D-sorbitol, and D-ribose.[24] In this case, stable borate complexes, *cis*-diol monoborate ester or bis-diol borate complex is formed, which are shown in Figure 17.8.

The functional group has a tertiary amine end and a polyol end, which is shown in Figure 17.9. The role of the tertiary amine in the functional group is to neutralize the proton brought by the formation of tetraborate complex. Amine protonation, shown in Eqn. (17.19), is critical to prevent the decrease of pH by proton released from the dissociation of borate esters.

$$- CH_2 - N(CH_3) - CH_2 + H^+ \Leftrightarrow - CH_2 - N^+K(CH_3) - CH \quad (17.19)$$

As far as hydroxyl groups are concerned, there are five hydroxyl groups in *N*-methyl-D-glucamine. This allows the formation of a strong complex with boron and improves the possibility of complexation by offering several sites for boron. On the other hand,

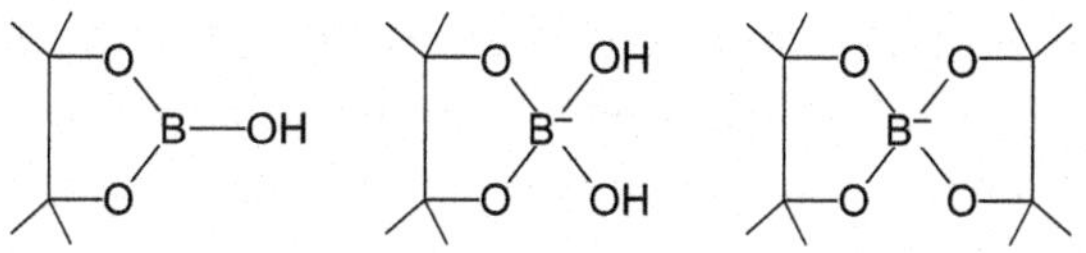

Figure 17.8 Schematic drawings of the neutral *cis*-diol monoborate ester (left), the monoborate complex (middle) and the bis(diol) borate complex (right).

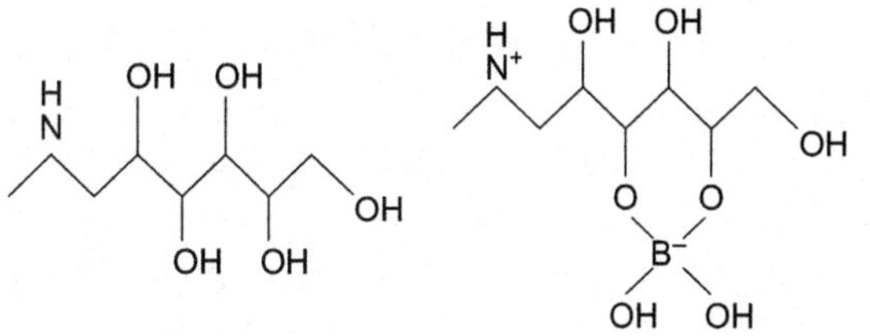

Figure 17.9 Structural formula of *N*-methyl-D-glucamine (left) and monoborate complex (right).

NMG groups might capture boron through a covalent attachment and formation of an internal coordination complex might occur.

17.3.3.2 Effect of Temperature

Variations of adsorption capacities of samples with the temperature at different pH and concentration were depicted in Figures 17.6(a,b) and 17.7(a,b). The equilibrium B adsorption capacity of GMA-PVC increased gradually from 2.507 mg/g at 25 °C (Run #1) to 2.817 mg/g at 65 °C (Run #3). The results indicate the endothermic behavior of the adsorption process of GMA-PVC. The enhancement in adsorption efficiency with increasing temperature might be attributed to the breaking of bonds on the adsorbent surface and increased the diffusion rate of adsorbate ions across the external boundary layer and into the internal pores of adsorbent.[25] However, the adsorption affinity of VBC-NMG decreased from 5.375 mg/g (Run #11) to 3.895 mg/g (Run #12) by virtue of exothermic nature of adsorption process. The decrease in adsorption efficiency with temperature could be related to the excess energy supply that promote desorption.

17.3.3.3 Effect of Initial Boron Concentration

Batch experiments were carried out at different initial B concentrations ranging from 10 to 30 mg/L in order to examine the effect of concentration on adsorption. The combined effects of initial B concentration with temperature and pH are visualized in Figures 17.6(b,c) and 17.7(b,c). Initial B concentration had a marked favorable effect on the amount of B adsorbed onto both adsorbents. The adsorption capacities of GMA-PVC were found to be 1.296 (Run #5) and 4.017 mg/g (Run #7) for initial B concentration of 10 and 30 mg/L, respectively. On the other hand, for VBC-NMG sample, the capacity was observed as 1.041 mg/g at 10 mg/L where it raised to the 3.480 mg/g at 30 mg/L. At higher concentrations, sites which are energetically less favorable get involved while Aksu and Gönen[26] underlined the fact that higher initial concentrations provide faster transport with increasing diffusion or mass transfer coefficient.

17.3.4 Validation and Confirmation of the Regression Model

Optimization studies were carried out by a multiple response method namely "desirability function" to optimize different combinations of process parameters such as pH, temperature and initial concentration. According to the optimum points of the process parameters, a series of three-point calibration studies were carried out. Table 17.5 presents the validation results of applied model for both samples. The predicted uptake capacity of GMA-PVC and VBC-NMG were found to be 5.95 and 7.35 mg/g, while experimental capacities were 5.87 and 7.21 mg/g, respectively. The confirmatory experiments demonstrated that the generated model has sufficient accuracy to estimate the adsorbed amount of boron and it could be further employed to determine the method significance for B adsorption onto GMA-PVC and VBC-NMG samples.

17.3.5 Isotherm Analysis

The parameters obtained from the experimental data fitted to the two- and three-parameter models are summarized in Tables 17.6 and 17.7. The plots of isotherm models are represented in Figures 17.10 and 17.11. Considering the coefficients of correlation ($\geq$0.990) and Chi-square values ($\leq$0.09) the most appropriate isotherm was found to be the D-R model for both samples. Moreover, among the three-parameter models the Sips equation describes the equilibrium data with satisfaction. The obtained results for both adsorbents showed that the best-fitted adsorption isotherm models were in the order of: D-R > Sips > Freundlich > Redlich-Peterson > Toth > Langmuir. Maximum adsorption capacities determined by using the Langmuir model were calculated higher than those of D-R, Freundlich, Toth, Sips, Redlich-Peterson. The Langmuir theory assumes homogeneous surface and single sorption layer; however, boron adsorption onto VBC-NMG and GMA-PVC samples was occurred on heterogeneous surface according to the D-R and Sips models.

Table 17.5 Model Confirmation

Sample	pH	Temperature (°C)	Concentration (mg/L)	Capacity (mg/g)	
				Predicted	Experimental
GMA-PVC	9.33	55.92	29.61	5.95	5.87 $\pm$ 0.14
VBC-NMG	10.0	52.55	30.0	7.35	7.21 $\pm$ 0.16

Table 17.6 Isotherm Constants for B Adsorption onto GMA-PVC (pH 10.0, C_i: 20 mg/L)

Isotherm Model		298 K	318 K	338 K
Langmuir	Q:	4.489	4.638	5.740
	b:	3.961	15.299	80.580
	R_L:	0.012	0.003	0.0006
	R^2:	0.783	0.648	0.488
	χ^2:	0.302	0.211	1.044
Freundlich	K_F:	3.207	3.843	4.732
	n:	6.378	9.702	8.020
	R^2:	0.975	0.973	0.979
	χ^2:	0.136	0.028	0.176
Dubinin−Raduskevich	Q:	1.0 E−04	9.4 E−05	1.2 E−04
	β:	1.1 E−09	6.7 E−10	6.7 E−10
	R^2:	0.993	0.997	0.987
	χ^2:	3.6 E−06	6.0 E−07	4.0 E−06
Toth	q_{mT}:	2.0 E+11	4.0 E+07	7.3 E+10
	K_T:	5.412	8.691	7.048
	m_T:	0.006	0.006	0.005
	R^2:	0.963	0.952	0.937
	χ^2:	0.141	0.030	0.182
Redlich−Peterson	K_{RP}:	6.9 E+09	2.6 E+08	6.7 E+09
	α_{RP}:	2.1 E+09	6.7 E+07	1.4 E+09
	β_{RP}:	0.843	0.896	0.875
	R^2:	0.965	0.953	0.939
	χ^2:	0.135	0.029	0.176
Sips	q_{ms}:	2.7 E+05	1.4 E+10	1.4 E+3
	K_S:	0.001	0.0002	0.991
	m_S:	0.156	0.103	0.004
	R^2:	0.986	0.984	0.986
	χ^2:	0.136	0.028	0.064

The model exponents of Sips (m_S) for samples were found closer to zero than unity (ranges between 0.004 and 0.135). This supports the fact that B adsorption on modified zeolites was fitted more to Freundlich theory than to Langmuir.

17.3.6 Thermodynamic Parameters

The calculated values of ΔG°, ΔH° and ΔS° are shown in Table 17.8. The Gibbs free energy values were found as negative for both adsorbents indicating the spontaneous nature of the adsorption process. For GMA-PVC sample, ΔG° values were found more negative with increasing temperature, indicating a greater driving force to the adsorption process and hence reactions take place more spontaneous. The positive enthalpy value of GMA-PVC (ΔH° = 93.33 kJ/mol) supported the endothermic

Table 17.7 Isotherm Constants for B Adsorption onto VBC-NMG (pH 10.0, C_i: 20 mg/L)

Isotherm Model		298 K	318 K	338 K
Langmuir	Q:	5.253	3.000	2.274
	b:	3.286	1.958	3.697
	R_L:	0.014	0.024	0.013
	R^2:	0.745	0.860	0.895
	χ^2:	0.351	0.164	0.028
Freundlich	K_F:	3.634	1.936	1.762
	n:	6.780	6.091	10.346
	R^2:	0.963	0.963	0.961
	χ^2:	0.056	0.029	0.016
Dubinin–Radushkevich	Q:	1.2 E−04	7.3 E−05	4.3 E−05
	β:	1.2 E−09	1.1 E−09	6.2 E−10
	R^2:	0.997	0.994	0.974
	χ^2:	9.3 E−07	4.0 E−07	1.7 E−07
Toth	q_{mT}:	3.643	3.095	2.731
	K_T:	5.694	4.633	6.055
	m_T:	0.012	0.070	0.344
	R^2:	0.952	0.964	0.956
	χ^2:	0.057	0.030	0.010
Redlich–Peterson	K_{RP}:	1.3 E+08	7.5 E+05	20.48
	α_{RP}:	3.5 E+07	3.9 E+05	10.61
	β_{RP}:	0.852	0.836	0.936
	R^2:	0.958	0.967	0.965
	χ^2:	0.057	0.029	0.010
Sips	q_{ms}:	3.750	8.098	2.597
	K_S:	0.001	0.312	0.006
	m_S:	0.147	0.235	0.190
	R^2:	0.973	0.975	0.971
	χ^2:	0.056	0.028	0.016

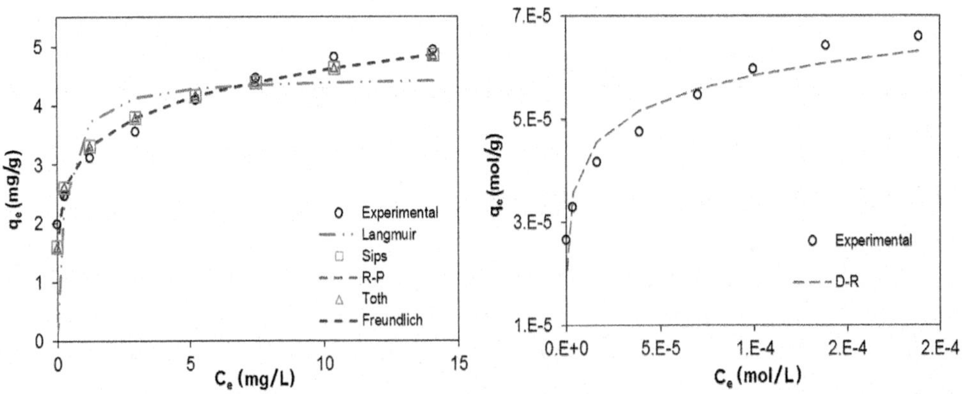

Figure 17.10 Nonlinear isotherms of GMA-PVC sample (pH 10.0, C_i: 20 mg/L, 25 °C).

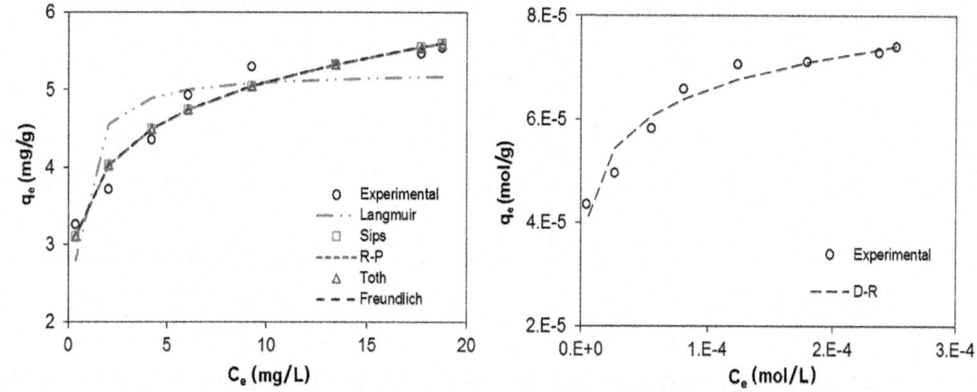

Figure 17.11 Nonlinear isotherms of VBC-NMG sample (pH 10.0, C_i: 20 mg/L, 25 °C).

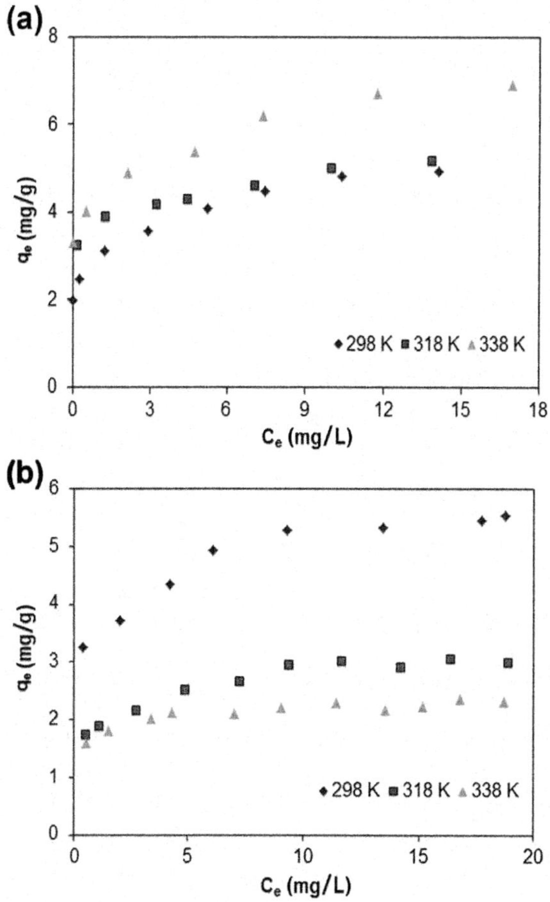

Figure 17.12 Effect of temperature on the adsorption of B on (a) GMA-PVC and (b) VBC-NMG at pH 10.0 with 20 mg/L initial B concentration.

Table 17.8 Thermodynamic Data for Boron Adsorption

Sample	T (K)	E (kJ/mol)	ΔG° (kJ/mol)	ΔH° (kJ/mol)	ΔS° (kJ/mol)	R^2
GMA-PVC	298	21.22	−19.320	93.33	0.378	0.996
	318	27.22	−26.881			
	338	27.19	−34.441			
VBC-NMG	298	20.39	−22.315	−72.27	−0.167	0.980
	318	20.50	−18.962			
	338	28.29	−15.610			

nature while boron adsorption onto VBC-NMG sample follows an exothermic path (ΔH° = −71.27 kJ/mol) (Figure 17.12). The negative value of ΔS° confirm a greater order of reaction during adsorption and forming stable complexes between borate ions and surface functional groups. The positive ΔS° value reflects the increase in the randomness on the solid—liquid interface.

17.4 CONCLUSIONS

In this study, iminopropylene glycol-modified polymeric sorbent (GMA-PVC) and NMG-modified sorbent (VBC-NMG) were synthesized to prepare alternative adsorbents for removal of boron from wastewater. The epoxy rings in grafted PGMA reacted with an excess of ethylenediamine to give an amine containing sorbent (GMA-PVC). VBC-NMG sorbent was prepared by the reaction between PGMA grafted sorbent and *N*-methyl-D-glucamine. RSM was utilized to investigate the role of process parameters on boron adsorption capacity for both sorbents. The solution pH (3.0—10.0), temperature (25—65 °C), and initial B concentration (10—30 mg/L) were selected as independent variables. The lack-of-fit test indicated that the quadratic model was highly significant, providing the best fit to the experimental data with *p*-values for lack of fit. The Anderson—Darling test showed that residuals were normally distributed and the error variance was homogeneous. ANOVA results indicated that interactions of pH, temperature and concentration were highly significant according to the *p*-values. A high degree of precision and a good deal of the reliability of the proposed model were indicated by confirmation experiments. Process pH and initial concentration have an adverse effect on the response for both resins. Under optimum conditions, VBC-NMG has better boron removal potential (7.35 mg/g, 96.3%) than GMA-PVC (5.95 mg/g, 92.4%). For both adsorbents, D-R isotherm model fitted the equilibrium data better than Langmuir isotherm. The positive enthalpy value of GMA-PVC supported the endothermic nature while boron adsorption onto VBC-NMG sample follows an exothermic path. The resulting polymeric resins have been demonstrated to be an efficient specific sorbent for removal of boron in parts per million (ppm) levels.

Abbreviations and Symbols

A^2	Anderson–Darling normality test value	
ANOVA	Analysis of variance	
ATRP	Atom transfer radical polymerization	
b	Langmuir equilibrium constant	L/mg
BBD	Box–Behnken design	
C_i	Initial concentration	mg/L
C_e	Equilibrium concentration	mg/L
CCD	Central composite	
DM	Doehlert matrix	
D-R	Dubinin–Radushkevich model	
df	Degree of freedom	
E	Mean energy of adsorption	kJ/mol
EDTA	Ethylenediaminetetraacetic acid	
GMA	Glycidyl methacrylate	
GMA-PVC	Glycidyl-methacrylate-polyvinyl chloride	
K_T	Toth equilibrium constant	
K_S	Sips equilibrium constant	L/mg
K_C	Equilibrium adsorption constant	
K_F	Freundlich adsorption capacity	mg/g
K_{RP}	Redlich–Peterson isotherm constant	
k	Number of the independent variables	
m	Adsorbent dosage	g
m_s	Sips model exponent	
m_T	Toth model exponent	
n	Freundlich equilibrium constant	
NMG	*N*-Methyl-D-glucamine	
PGMA	Poly glycidyl methacrylate	
PVC	Polyvinyl chloride	
Q	Langmuir monolayer sorption capacity	mg/g
q_e	Adsorption capacity	mg/g
q_{mS}	Sips maximum adsorption capacity	mg/g
q_{mT}	Toth maximum adsorption capacity	mg/g
R_L	Langmuir constant	
R	Gas constant (8.314)	J/mol/K
RSM	Response surface methodology	
R^2	Coefficient of determination	
R^2_{adj}	Adjusted coefficient of determination	
SS	Sum of squares	
T	Absolute temperature	K
WHO	World Health Organization	
V	Solution volume	L
VBC	Vinyl-benzyl-chloride	
VBC-NMG	*N*-Methyl-D-glucamine modified form of VBC	
x_i, x_j	Independent parameter	

(Continued)

Abbreviations and Symbols—cont'd

Y_i	Response term	
ΔG°	Gibbs free energy	kJ/mol
ΔH°	Enthalpy	kJ/mol
ΔS°	Entropy	kJ/mol
α_{RP}	Redlich–Peterson isotherm constant	
β_0	Quadratic equation constant	
β_i	Slope or linear effect of the input factor,	
β_{ii}	Quadratic effect	
β_{ij}	Two-way linear by linear interaction effect	
β_{RP}	Redlich–Peterson model exponent	
β	Constant related to the sorption energy	
ε	Random error	
χ^2	Chi-square	

REFERENCES

1. Kabay N, Yilmaz I, Yamac S, Samatya S, Yuksel M, Yuksel U, et al. Removal and recovery of boron from geothermal wastewater by selective ion exchange resins. I. Laboratory tests. *React Funct Polym* 2004;**60**:163–70.
2. Bicak N, Bulutcu N, Senkal BF, Gazi M. Modification of crosslinked glycidyl methacrylate-based polymers for boron-specific column extraction. *React Funct Polym* 2001;**47**:175–84.
3. Wang L, Qi T, Gao Z, Zhang Y, Chu J. Synthesis of *N*-methylglucamine modified macroporous poly(GMA- co-TRIM) and its performance as a boron sorbent. *React Funct Polym* 2007;**67**: 202–9.
4. Cheng Z, Zhu X, Shi ZL, Neoh KG, Kang ET. Polymer microspheres with permanent antibacterial surface from surface-initiated atom transfer radical polymerization. *Ind Eng Chem Res* 2005;**44**: 7098–104.
5. Yavuz E, Gürsel Y, Senkal BF. Modification of poly(glycidyl methacrylate) grafted onto crosslinked PVC with iminopropylene glycol group and use for removing boron from water. *Desalination* 2013;**310**:145–50.
6. Montgomery DC. *Design and analysis of experiments*. 7th ed. New York: Wiley; 2009.
7. Box GEP, Behnken DW. Some new three-level designs for the study of quantitative variables. *Technometrics* 1960;**2**:455–75.
8. Ferreira SLC, Bruns RE, Ferreira HS, Matos GD, David JM, Brando GC, et al. Box–Behnken design: an alternative for the optimization of analytical methods. *Anal Chim Acta* 2007;**597**:179–86.
9. Ruthven DM. *Principles of adsorption and adsorption processes*. New York: Wiley; 1984.
10. Langmuir I. The adsorption of gases on plane surfaces of glass, mica and platinum. *J Am Chem Soc* 1918;**40**:1361–403.
11. Webber TW, Chakkravorti RK. Pore and solid diffusion models for fixed-bed adsorbers. *AIChE J* 1974;**20**:228–38.
12. Toth J. *Adsorption theory modeling and analysis*. New York: Marcel Dekker; 2002.
13. Redlich O, Peterson DL. A useful adsorption isotherm. *J Phys Chem* 1959;**63**:1024–6.
14. Toth J. Calculation of the BET-compatible surface area from any type I isotherms measured above the critical temperature. *J Colloid Interface Sci* 2000;**225**:378–83.
15. Anirudhan TS, Radhakrishnan PG. Kinetic and equilibrium modelling of Cadmium(II) ions sorption onto polymerized tamarind fruit shell. *Desalination* 2009;**249**:1298–307.
16. Yang RT. *Adsorbents: fundamentals and applications*. New Jersey: John Wiley & Sons; 2003.

17. Pérez-Marín AB, Meseguer Zapata V, Ortuno JF, Aguilar M, Sáez J, Llorens M. Removal of cadmium from aqueous solutions by adsorption onto orange waste. *J Hazard Mater* 2007;**B139**:122–31.
18. Basset J, Denney RC, Jeffery GH, Mendham J. *Vogel's textbook of quantitative inorganic analysis*. 4th ed. London: Longman Group Ltd.; 1978. pp. 319–334.
19. Greenwood NN. *The chemistry of boron*. New York: Pergamon Press; 1975.
20. Cengeloglu Y, Aslan G, Tor A, Kocak I, Dursun N. Removal of boron from water by using reverse osmosis. *Sep Purif Technol* 2008;**64**:141–6.
21. Hilal N, Kim GJ, Somerfield C. Boron removal from saline water: a comprehensive review. *Desalination* 2011;**273**:23–35.
22. Kavak D. Removal of boron from aqueous solutions by batch adsorption on calcined alunite using experimental design. *J Hazard Mater* 2009;**163**:308–14.
23. Power PP, Woods WG. The chemistry of boron and its speciation in plants. *Plant Soil* 1997;**193**:1–13.
24. Geffen N, Semiat R, Eisen MS, Balazs Y, Katz I, Dosoretz CG. Boron removal from water by complexation to polyol compounds. *J Membr Sci* 2006;**286**:45–51.
25. Polowczyk I, Ulatowska J, Koźlecki T, Bastrzyk A, Sawiński W. Studies on removal of boron from aqueous solution by fly ash agglomerates. *Desalination* 2013;**310**:93–101.
26. Aksu Z, Gönen F. Biosorption of phenol by immobilized activated sludge in a continuous packed bed: prediction of breakthrough curves. *Process Biochem* 2004;**39**:599–613.

INDEX

Note: Page numbers followed by "f" indicate figures; "t", tables; "b", boxes.

F

G

H

I

T

U

V

W

CPSIA information can be obtained at www.ICGtesting.com
Printed in the USA
BVOW07*0837091015

421238BV00005B/29/P